Mathematics

Beyond
the Numbers

George T. Gilbert
Texas Christian University

Rhonda L. Hatcher
Texas Christian University

John Wiley & Sons, Inc.
New York • Chichester • Weinheim • Brisbane • Singapore • Toronto

Dedicated to Nolan and Alex

Executive Editor   Ruth Baruth
Senior Development Editor   Nancy Perry
Senior Marketing Manager   Peter Ryttel
Senior Production Manager   Lucille Buonocore
Senior Production Editor   Robin Factor
Senior Designer   Laura Boucher
Illustration Editor   Sigmund Malinowski
Senior Photo Editor   Hilary Newman
Production Assistant   Eugene McDonald
Chapter Opening Art   Norm Christianson
Cover Art   Norm Christianson
Electronic Illustrations   Lineworks, Inc.

This book was set in Janson by York Graphic Services and printed and bound by Von Hoffmann Press. The cover was printed by Phoenix Color.

This book is printed on acid-free paper. ∞

Library of Congress Cataloging-in-Publication Data

Gilbert, George T., 1957–
  Mathematics beyond the numbers / George T. Gilbert, Rhonda L. Hatcher.
    p.   cm.
  Includes index.
    ISBN 0-471-13934-3 (cloth : alk. paper)
    1. Mathematics.   I. Hatcher, Rhonda L.   II. Title.
QA39.2.G513   1999   99-32789
510--dc21   CIP

About the Authors

George Gilbert attended Washington University in Saint Louis, where he earned a B.A. in mathematics in 1979. He went on to earn a Ph.D. in mathematics from Harvard University in 1984, specializing in number theory. He has held teaching positions at the University of Texas in Austin and Saint Olaf College, and he is now a faculty member at Texas Christian University. Professor Gilbert is active in the area of problem solving, having served on the William Lowell Putnam questions committee, the AIME contest committee, and as the editor of the Problems Section of *Mathematics Magazine*.

Rhonda Hatcher earned a B.A. in mathematics from the University of Colorado at Boulder in 1980 and a Ph.D. in mathematics from Harvard University in 1987. Her research specialty is number theory. She has taught at Saint Olaf College and is currently on the faculty of Texas Christian University. Professor Hatcher won the 1994 Deans' Teaching Award, the 1997 Honors Professor of the Year Award, and the Division of Natural Sciences 1998 Distinguished Teaching Award at TCU. In 1998, the Mathematical Association of America awarded her the Deborah and Franklin Tepper Haimo Award for Distinguished College or University Teaching of Mathematics.

Preface

Economist Oskar Morgenstern, one of the founders of the theory of games, once wrote, "An experience of human development, is that where mathematics has once entered in a truly significant way, it has never been driven out again." It is our hope that the students who use this text will come to appreciate this statement and gain a sense of the essential role that mathematics plays in the world.

In selecting a text a few years ago for the liberal arts mathematics course we teach at Texas Christian University, we were searching for one that included a wide selection of interesting mathematical topics and real-world applications and that engaged students in doing mathematics through well-chosen examples and exercises. Unable to find a text that satisfied all of our needs, we set out to write one ourselves. When deciding what to include, we looked for topics that were either very applied so that students could appreciate their practicality or that simply displayed the beauty and fun of mathematics. Our approach is to teach mathematical problem-solving skills through examples so that students can carry out what they have learned. By carefully constructing examples and including a more extensive and varied collection of exercises than are found in many liberal arts mathematics texts, we believe we have succeeded.

Mathematics Beyond the Numbers is written in a conversational style that is inviting to students. A great deal of historical background has been included so that students can see the development of mathematics over time. Some of the topics included might be particularly applicable to a student's field of study. For instance, voting methods and apportionment should be of interest to a political science major, whereas tilings and polyhedra might appeal strongly to an art major. The wide use of real data and applications also help to make the topics come to life for the students. Through numerous examples and exercises, students will be actively doing mathematics, gaining a deeper understanding in the process. As implied by the title of this book, our goal is to lead students to see mathematics "beyond the numbers," as a rich, applicable, and ever-changing subject.

Mathematics Beyond the Numbers is intended for a one-semester or two-semester mathematics course for students majoring in the liberal arts or other fields that do not have a specific mathematics requirement. It could also be used in a mathematics course for elementary or secondary education majors.

Prerequisites

The prerequisite for the text is a basic high school algebra background. Only Chapter 3 on the mathematics of money includes frequent use of algebra, and the required algebra background is reviewed in the first section of this chapter.

A Flexible Organization

The chapters are independent of one another and can be presented in any order. We have included more material than can be covered in a one-semester course, and perhaps even a two-semester course, so that instructors can select the

chapters they like and omit those they do not wish to cover. The level of difficulty of the course can be adjusted by the selection of topics and by the exercises
assigned within a chapter. The chapters themselves need not be covered entirely.
There are several natural stopping places within each chapter, and in some cases
certain topics or even entire sections can be omitted.

- **Chapter 1: Voting Methods**
  This chapter introduces several voting methods: plurality, plurality with a runoff,
  Borda's method, head-to-head comparisons, and approval voting. The strengths
  and weaknesses of each of the methods are considered along with the possible
  effects of strategic voting. The chapter concludes by considering the elusive
  search for an ideal voting method and Kenneth Arrow's Impossibility Theorem.
  This chapter provides students with an understanding of the complexities and
  paradoxes of voting.

- **Chapter 2: Apportionment: Sharing What Cannot Be Divided Arbitrarily**
  This chapter considers the Hamilton, Lowndes, Jefferson, Webster, and
  Hill–Huntington apportionment methods. The methods are presented against
  the backdrop of the history of the apportionment of the U.S. House of Representatives. The chapter concludes with a look at some desirable properties of
  an ideal apportionment method and the work of Balinski and Young, showing
  that no apportionment method fulfills all of the desired properties. In this chapter, students see an area of mathematics that has developed over time and the
  roles many of our famous political leaders played in its development.

- **Chapter 3: The Mathematics of Money**
  This chapter focuses on the issues of personal finance. It considers both simple and compound interest, systematic savings plans, and amortized loans. The
  chapter begins with a review of the algebra needed throughout the chapter. In
  this chapter, students gain an understanding of financial issues that will affect
  them throughout their lives.

- **Chapter 4: Probability**
  Many ideas in elementary probability are covered in this chapter, including odds,
  the addition rule, conditional probability, the multiplication rule, counting techniques, expected value, and the applications of probability to genetics. There is
  a great deal of flexibility in the coverage of this chapter. Section 4.2 on odds is
  optional, and Section 4.4 on conditional probability and the multiplication rule
  can be omitted unless Section 4.8 on genetics is covered. Furthermore, the
  counting techniques covered in Sections 4.5 and 4.6 appear again only in some
  examples and exercises in Section 4.7 on expected value. This chapter provides
  students with an appreciation and understanding of probability and the role that
  chance plays in our lives.

- **Chapter 5: Statistics**
  This chapter presents many of the fundamental ideas of statistics. It begins with
  the presentation of data and computations of the mean, median, mode, range,
  and standard deviation of both ungrouped and grouped data. It considers the
  normal distribution, margins of error, and confidence intervals. The chapter
  concludes with a look at the misuse of statistics. This chapter provides students
  with an understanding of the statistics that they will encounter daily in their
  jobs or through the news media.

- **Chapter 6: Paths and Networks**
  Starting with the Königsberg bridge problem, the basic notions of graph theory are introduced. The chapter considers Eulerian paths and circuits and eulerization. The Traveling Salesman problem is introduced along with the Nearest Neighbor Algorithm and the Greedy Algorithm for finding approximate solutions. The chapter also introduces the minimal network problem and the solution provided by Prim's Algorithm. This chapter exposes students to a fun yet important area of applied mathematics.

- **Chapter 7: Tilings and Polyhedra**
  This chapter considers the mathematics behind tiling patterns and polyhedral structures. It presents regular, semiregular, and monohedral tilings along with techniques for constructing Escher-like tilings. Regular and semiregular polyhedra, Euler's formula, and the elegant geometry of the buckminsterfullerene molecule are also introduced. In this chapter, students will come to appreciate the geometry lying behind beautiful structures and patterns.

- **Chapter 8: Number Theory**
  This chapter begins with an introduction to divisibility and primes and then focuses on modular arithmetic and its applications, including divisibility tests, check digits, tournament scheduling, and cryptography. Several different choices can be made regarding the presentation of material in this chapter. Only Sections 8.1 and 8.2 on divisibility and primes and modular arithmetic are required. From there, the instructor can select from among any of the remaining sections, except that Section 8.3 is a prerequisite for Section 8.4 and Section 8.7 on advanced encryption methods depends on Section 8.6, which introduces cryptology. In this chapter, students see how modular arithmetic, a seemingly odd form of arithmetic, has a wide range of practical applications.

- **Chapter 9: Game Theory with an Introduction to Linear Programming**
  This chapter provides a broad introduction to game theory. It begins with alternate move games and game trees and then considers simultaneous move games, both with and without communication, that can be presented in matrix form. An introduction to linear programming in the context of zero-sum games is included. Section 9.1 on alternate move games is optional and Section 9.6, which covers simultaneous move games with communication, depends only on Section 9.2 on simultaneous move games and dominance. In this chapter, students gain insight into the deductive analysis involved in strategic situations arising, for instance, in business and politics. They also gain experience in mathematical modeling.

Detailed suggestions for omitting optional topics within specific sections and different instructional tracks that can be taken in each chapter are included in the Instructor's Manual.

Learning Through Solving Problems

In order to truly appreciate and understand mathematics and its importance, students need to actively *do* mathematics. For this reason, the emphasis of this text is on solving problems. Each chapter begins with an overview so that students will appreciate the importance and applicability of the topic and then moves on to teaching students to do the mathematics themselves. Ideas are presented through examples that illustrate the main concepts and help guide the students in solving problems. To make the text more helpful to students, important definitions, formulas, theorems, and techniques are placed in highlighted boxes. In addition, key terms are printed in boldface for easy identification, and all key terms are listed at the end of each chapter with page references to the places in the text where the terms are defined.

Real-World Applications and Data

Each chapter includes real-world applications of the mathematics, such as the application of modular arithmetic to check digit methods and cryptology and the role that probability plays in analyzing drug-testing results. To bring the mathematics to life, real data are used extensively throughout the text. For instance, the voting results of the colorful presidential election of 1912 are analyzed in Chapter 1 on voting methods. In Chapter 5 on statistics, students are asked to apply statistical methods to a variety of interesting real data—for example, to finding the mean length of service of the U.S. Supreme Court justices or to estimating the percentage of women bearing children who fall in a given age range. The use of real data and applications not only makes the text more interesting to students, but it also reinforces how vital a role mathematics plays in many areas of our lives.

Exercises

Mathematics Beyond the Numbers includes a large selection of exercises at the end of each section, allowing for a wide range of student abilities. It includes more exercises than texts with similar coverage, and great care was taken to ensure that the exercises cover the important mathematical points and details. There are enough exercises to allow instructors to assign only even or only odd exercises if so desired. Each chapter includes challenging exercises requiring problem-solving skills; these are highlighted in red. A set of chapter review exercises provides an overview of the ideas presented in each chapter and could serve as a practice test. Short answers to most odd-numbered exercises and to all chapter review exercises are provided in the back of the book.

Writing Exercises and Projects

To build the important skill of communicating mathematical ideas, each chapter includes several writing exercises in which students are asked to explain something or give verbal responses to questions. Also included at the end of each chapter are several projects that afford students the opportunity to apply the mathematics they have learned to situations that stretch beyond the boundaries of shorter exercises. The projects range from difficult exercises with a strong problem-solving emphasis to experimental approaches to mathematics. The projects can be assigned to individual students or as group projects.

The Use of Technology

Although each topic presented in the text can be taught in a course in which students have access only to handheld scientific calculators, for instructors who wish to teach a more technology-based course, the opportunities are wide-ranging.

Technology Tips boxes appear throughout the text to help guide the instructor and the students to the possibilities for using technology. In Chapter 1 on voting methods, a spreadsheet can be used to tally votes easily. The calculations involved in performing the apportionments in Chapter 2 can be simplified by using either a spreadsheet or a calculator that manages lists. In Chapter 5, the statistical calculations can be performed using special features on a calculator or using computer software. The tilings studied in Chapter 7 are most easily constructed with a computer drawing program or with specialized software, and many mathematical software packages perform the modular arithmetic used in Chapter 8. A graphing calculator is helpful in the linear programming section of Chapter 9. Most of the applications mentioned here are included in a set of Java applets written specifically to accompany *Mathematics Beyond the Numbers*. A Web site icon in the margin indicates that there is a Java applet directly relevant to that portion of the text.

Branching Out and A Closer Look

Students will have a richer learning experience by reading the special feature boxes—*Branching Out* and *A Closer Look*. As the name implies, *Branching Out* boxes broaden the scope of the material in the main text. In some cases they expand on the mathematics, as in the discussion of the Steiner points in Chapter 6 on paths and networks. In others, an important application is presented, such as the issue of Roth IRAs versus traditional IRAs in Chapter 3 on the mathematics of money. Finally, some are simply for further historical background or fun. In contrast, *A Closer Look* boxes focus on mathematical details. For instance, in the section introducing divisibility and primes in Chapter 8, Euclid's proof that there exist infinitely many prime numbers is presented.

Supplements

The supplementary material for students and instructors has been driven by the same beliefs as the textbook, providing a well-integrated learning system. It provides instructors with tools to create an active learning environment and encourages students to take an active role in the course.

- The **Web site http://www.wiley.com/college/gilbert** hosts supplementary on-line resources for both instructors and students. One of the highlights of the site is a set of Java applets developed by John Orr of the University of Nebraska, Lincoln that are designed to simplify computations or to demonstrate important mathematical ideas. A description of each applet is included in the Instructor's Manual. The site also contains an annotated collection of Internet addresses (URLs) providing up-to-date mathematical information, real-world data, and software links related to topics in the text. This site provides an interactive learning environment with a variety of tools and simulations that will further enhance the real data in the book and bring the mathematics to life.

- **Wiley Web Tests** form a customizable and expandable on-line testing system that allows for administering and grading tests in a variety of modes. The software is platform independent and resides on the local server. Developed by John Orr and the University of Nebraska, Lincoln.

- **WebCT** is a powerful web-based program that allows professors to set up and administer an on-line course, including delivering course content, and providing communication tools such as e-mail, chat-rooms, and bulletin boards. WebCT enables online quizzing and course administration, with features such as student tracking and grading. Additional material keyed to the text will be provided to help you integrate this software into your course.

- The **Student Solutions Manual** contains solutions to each odd-numbered exercise in the text and all chapter review exercises.

- The **Instructor's Manual** contains teaching tips, Java applet descriptions, solutions to selected projects, and overhead transparency masters to assist instructors in organizing their courses.

- The **Instructor's Solutions Manual** contains solutions to all exercises in the text and the chapter review exercises.

- The **Test Bank** contains a wide range of test items and their solutions arranged according to section.

Acknowledgments

There are many people we would like to thank for their help in developing and reviewing this book. We are especially grateful to Peter Renz, who as a developmental editor provided invaluable guidance and deserves credit for many of the ideas used in writing the text. We would also like to thank the following faculty members of Texas Christian University along with their students for using the early drafts of this book and providing helpful suggestions: Kathy Coleman, Roy Combrink, Ze-Li Dou, Peng Fan, Bob Ferguson, Ken Richardson, Tom Schwartz, and Sue Staples. Appreciation is also due to the following reviewers of the text:

James B. Barksdale, Jr.
Western Kentucky University

Mark B. Beintema
Southern Illinois University

Joan Bookbinder
Johnson & Wales University

Steven Butcher
University of Central Arkansas

Larry N. Campbell
Southwest Missouri State University

Melissa Cass
State University of New York–New Paltz

Timothy P. Chappell
Penn Valley Community College

Richard DeCesare
Southern Connecticut State University

Greg S. Dietrich
Florida Community College–Jacksonville

Guesna Dohrman
Tallahassee Community College

John Emert
Ball State University

Sandra Fillebrown
St. Joseph's University

Ruth E. Hanna
Louisiana Technical University

Carol Partenheimer
Rock Valley College

Timothy J. Hodges
University of Cincinnati

Dennis Pence
Western Michigan University

Susan Knights
University of Wyoming

Morgan A. Phillips
Valencia Community College

Kathy Lavelle
Westchester Community College

David Quarles
St. Petersburg Junior College

Deborah Levine
Nassau Community College

Robyn E. Serven
University of Central Arkansas

C. Michael Lohr
Virginia Commonwealth University

Emilio Toro
University of Tampa

Daniel P. Munton
Santa Rosa Junior College

Sheela Whelan
Westchester Community College

We greatly appreciate Carol Brown, Nolan Gilbert, Mony Harden, Sharma Hatcher, Todd Holmberg, and Sean Ingram who helped us track down illustrative materials, and Janice Taylor and Shirley Doran who cheerfully provided administrative help. We want to thank Texas Christian University and our department chair, Robert Doran, for their support of our work, including sabbatical leaves.

Finally, we would like to thank the wonderful staff at Wiley, especially our executive editor, Ruth Baruth; our production editor, Robin Factor; our illustration editor, Sigmund Malinowski; our photo editor, Hilary Newman; and our development editor, Nancy Perry. Their work in the development and editing of the text and in acquiring photographs and drawing illustrations was of the highest quality. They were all essential in the production of this text, and we could not have wished for a better team.

A Note to Students

Our primary advice is simply to read the text. We expect that you will be pleasantly surprised by how helpful and instructive it is. This book was written with you in mind, and we have included many features to help your studies:

- Important definitions, formulas, and algorithms are in highlighted boxes.

- Key terms are printed in boldface, and at the end of each chapter you will find a key term list giving the page on which the term is first defined.

- We have included a large number of examples that illustrate the concepts and problem-solving techniques. We recommend not only reading the examples but also working through them yourself. When in the learning phase, you may make mistakes, but as author James Joyce said, "Mistakes are the portals of discovery," and by correcting your mistakes you will be learning mathematics through the discovery process.

- *Mathematics Beyond the Numbers* is filled with applications of mathematics to real-life situations so that you will see the mathematics in action and how it applies to your life.

- *A Closer Look* boxes provide some of the mathematical details underlying what you have learned and will give you a deeper understanding of the material.

- The *Branching Out* boxes present some fascinating mathematical applications and stories. They are the icing on the cake, and your learning experience will be greatly enriched by reading them.

- *Technology Tips* will inform you of the opportunities to use technology in solving problems.

- The Web site for the book at http://www.wiley.com/college/gilbert includes a set of Java applets designed to simplify computations or to demonstrate mathematical ideas. It also includes links to Internet sites providing more information on topics introduced in the text.

- After completing a chapter, you will find it helpful to do all the chapter review exercises. These exercises serve as a test of your understanding, and all of the answers are provided in the back of the text.

Mathematics Beyond the Numbers is written to you, the student. We hope it helps you experience some of the joy we find in learning and doing mathematics. We also hope that you are left with a sense of the incredible beauty of mathematics and the vital role it plays throughout our lives.

George T. Gilbert
Rhonda L. Hatcher

Contents

Chapter 5   *Statistics*   270

Chapter 6   *Paths and Networks*   362

Mathematics Beyond the Numbers

Voting methods lie at the intersection of political science, economics, and mathematics. Whether a group is electing a leader or choosing a course of action, their decision may depend on the voting method that is used as well as on the issues and the preferences of the voters. In this chapter we focus our attention on voting to reach a single decision—for example, electing an officer. We do not consider multiple, linked decisions.

With the two-party system dominating elections in the United States, voters often face the simplest situation: choose between two candidates, and the one receiving the most votes wins. The

Voters at the polls.

procedure of counting ballots treats voters equally and choices equally. Barring a tie, the outcome of an election with two alternatives is always clear and unequivocal. This is the larger purpose of democratic government—to provide a means of reaching public decisions that unites the people behind a reasonable course of action.

However, many elections require making a decision from among more than two choices. For example, the 1968 presidential election featured a strong challenge from third-party candidate George Wallace. In this election, the Democratic nominee, Hubert Humphrey, received 42.7% of the vote, whereas the Republican nominee and winner of the election, Richard Nixon, received 43.4%. If George Wallace, who received 13.5% of the vote, had not run, the election might have yielded a different winner. Presidential primaries and local elections often require voters to choose among a long list of candidates. If there are only two alternatives to choose between, we know how to proceed: majority rules. The procedures for arriving at a group decision when the group has at least three options to choose from are not so simple and raise interesting questions. In this chapter we look at some of these procedures.

Hundreds of people wait in line to vote in Nelson Mandela's home village in the first democratic election in South Africa.

CHAPTER I

VOTING METHODS

1.1   PLURALITY AND RUNOFF METHODS

The simplest, and perhaps most common, method of deciding among three or more alternatives is the plurality method, described as follows.

The Plurality Method

With the **plurality method**, each voter selects one candidate or choice on the ballot. The winner is the candidate or choice with the most votes.

The winner in an election decided by the plurality method is said to have won by a plurality or, in Great Britain, a relative majority. Note that in an election with three or more candidates, a candidate need not have a majority to win by a plurality. The plurality method is illustrated in the following example.

Example 1   *Plurality Method*

A committee must vote for a main dish to be served at a banquet. The choices are chicken, pork, fish, or beef. The committee decides to use the plurality method, and their ballots read as follows: chicken, pork, beef, chicken, fish, beef, fish, beef, chicken, chicken, beef, pork, beef.

(a) Which main dish did they choose?

(b) Just before counting the vote, the committee realizes that pork is not one of the options offered by the caterer. If those who voted for pork replace their ballots, which main dishes could be the group's selection?

Solution:

(a) The person assigned to count the votes would probably sort the ballots into four piles and find that the tally of the original ballots is beef 5, chicken 4, pork 2, fish 2, so beef is chosen as the main dish.

(b) There are only 2 votes for pork, so fish could get at most 4 votes, not catching up to beef. On the other hand, 1 additional vote for beef or 2 additional votes for chicken would result in that dish being selected. Thus, beef or chicken could be selected.

If one of the people voting for pork or fish in Example 1 voted for chicken instead, what would have been the outcome of the election? The result would have been a tie between beef and chicken with 5 votes each. With a large number of voters, ties are uncommon. However, when the number of voters is small, ties can easily occur under all of the voting methods we will study.

Example 2   *Minimum Votes Needed to Win*

Suppose that 60 votes are cast in an election among four candidates: Hughes, Dunbar, Torres, and Ishmael. After the first 40 votes are counted, the tallies are as follows:

Hughes	8
Dunbar	17
Torres	12
Ishmael	3

What is the minimal number of the remaining votes Dunbar can receive and be assured of a win?

Solution: There are 20 votes left. To win the election, Dunbar must have the most votes after the last 20 votes are cast. Torres is in second place behind Dunbar after the first 40 votes are cast. In the worst-case scenario for Dunbar, Torres would get all the votes that did not go to Dunbar. Torres needs 5 votes to catch up with Dunbar, and then 8 of the remaining 15 votes to beat Dunbar. Thus, Torres would need a total of $5 + 8 = 13$ votes. We can conclude that if Dunbar gets 8 of the last 20 votes, then neither Torres, Hughes, nor Ishmael can catch up, and Dunbar would be assured of a win.

Alternatively, we can solve this problem algebraically. Let x be the number of votes Dunbar needs to ensure at least a tie for first place. If Torres gets all the votes that do not go to Dunbar and the race ends with a tie between Torres and Dunbar, we must have

$$17 + x = 12 + (20 - x).$$

Solving this equation for x, we get

$$17 + x = 32 - x,$$
$$x + x = 32 - 17,$$
$$2x = 15,$$
$$x = \frac{15}{2} = 7.5.$$

We conclude that 7 more votes are not enough to assure a win, but that 8 votes will do the job, ensuring that Dunbar has more votes than any other candidate. ∎

If the plurality method is used in an election involving more than two candidates and one of the candidates receives more than 50% of the votes, then that candidate clearly wins the election and it can be reasonably argued that this winner is the popular choice. However, it is often the case in plurality elections that no candidate receives more than 50% of the votes. The 1992 Philippine presidential election shows what can happen. Out of a field of 7 candidates, Fidel Ramos won a 23.6% plurality. He was elected president but lacked a public mandate. Given the importance of popular support, the question arises whether plurality is the best voting method to use when an election has more than two candidates.

BRANCHING OUT 1.1

THE ELECTORAL COLLEGE

Americans elect their presidents, not by popular vote, but instead through the Electoral College. Under this system, each state has a number of electoral votes equal to the number of representatives from the state plus two for the state's two senators. In addition, the District of Columbia is entitled to three electoral votes. The votes are cast by individuals called electors. If a candidate wins a *majority,* not merely a plurality, of the electoral votes, then the candidate is declared the winner. If no candidate wins, then the election is handed over to the House of Representatives, who vote for one of the top three contenders.

The method of electing the president through the Electoral College was adopted at the Constitutional Convention of 1787 after much debate. The plan left the method by which a state's electors would be chosen to the discretion of the state legislatures. Initially, states used a variety of methods for selecting electors, but gradually the states moved toward a standard system of choosing the presidential electors by a statewide, winner-take-all popular vote. Under this system, whichever candidate wins a plurality of the popular vote in the statewide election is entitled to *all* of the state's electoral votes. However, because electors merely pledge to support a particular candidate, they occasionally break the pledge and vote for another candidate. As recently as 1988, Margaret

Leach, a Michael Dukakis elector from West Virginia, voted for Dukakis' running mate, Lloyd Bentsen, for president rather than Dukakis.

Because the electoral votes of a state normally go to the one candidate who wins a plurality of the popular vote in the state, no matter how close the tally, the electoral vote totals often exaggerate the winning margin. For instance, in the 1980 election, Ronald Reagan beat President Jimmy Carter by only 50.7% to 41.0% in the national popular vote, but in the electoral vote the split was 489 to 49, with Reagan taking 92.6% of the electoral votes. The nature of this "landslide election" is shown on the map, where we see that Carter captured only six states and the District of Columbia.

Third-party candidates can have significant support across the country, but will only win electoral votes in states in which they win a plurality of the popular vote. For instance in the 1992 election, even though Ross Perot captured 19% of the popular vote, he did not win any electoral votes because he did not have a plurality of the votes in any state.

It is possible for a candidate who has not even won a plurality of the national popular vote to win the election in the Electoral College. This has happened three times in the history of the country, with the elections of John Quincy Adams in 1824, Ruther-

A central problem with the plurality method is illustrated by the following example. Suppose that an election between two very different candidates, Smith and O'Neil, would result in Smith comfortably defeating O'Neil 60% to 40%. Now suppose that two candidates very similar to Smith, whom we will call Smyth and Smythe, are also in the election and that there is not much to distinguish Smith, Smyth, and Smythe. In this case, 60% of the voters would prefer either Smith, Smyth, or Smythe to O'Neil. Thus, in an election with all four candidates, O'Neil would probably win with 40% of the vote, whereas Smith, Smyth, and Smythe individually would run far behind with roughly 20% each. With many candidates, the plurality system makes it fairly likely that an extreme candidate with a small but intense base of support will defeat a hoard of candidates with

ford B. Hayes in 1876, and Benjamin Harrison in 1888. (For more on the controversial election of Hayes, read Branching Out 2.1.) John Quincy Adams was elected president even though he failed to win a plurality of the Electoral College votes. He had only 84 electoral votes to the 99 for Andrew Jackson. But this election included two other strong candidates: Henry Clay, who captured 37 electoral votes, and William H. Crawford, with 41 votes. Because no candidate had a majority of the 261 electoral votes, the House of Representatives decided the election and gave the victory to John Quincy Adams.

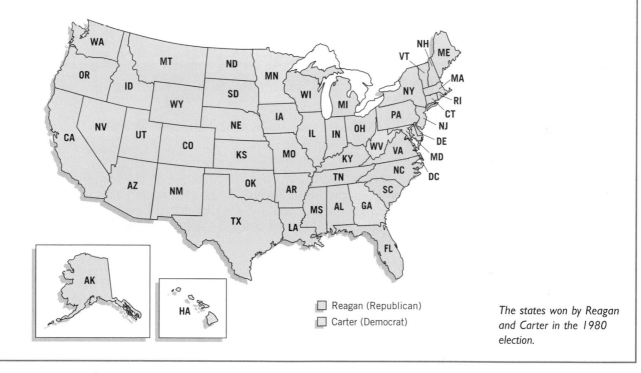

☐ Reagan (Republican)
☐ Carter (Democrat)

The states won by Reagan and Carter in the 1980 election.

broad support and similar views. The propensity of the Democratic Party to select a liberal candidate and of the Republican Party to select a conservative candidate through their presidential primaries and caucuses, rather than more moderate candidates, supports this observation. Among the more memorable examples of this are the Republican Party's nomination of Barry Goldwater in 1964 and the Democratic Party's nomination of George McGovern in 1972.

A second problem with the plurality method is the dilemma that a supporter of a "weak" candidate faces when there are more than two candidates running. Should the voter cast a ballot for his or her first choice, a "wasted" vote, or switch to one of the serious contenders in an effort to affect the outcome of the election? The incentive to switch is an incentive to vote strategically. When a per-

1996 Democratic Party Convention.

son votes in a way that does not reflect his or her true preferences in an attempt
to improve the outcome of the election from that person's point of view, we call
this **strategic voting** and say that the voter is voting strategically.

Runoff Elections

Because of shortcomings in the plurality method, many elections, particularly local elections in which races often have several candidates, have a provision for a
runoff election as described next.

> ### *Runoff Election*
>
> In a **runoff election**, a plurality vote is taken first. If one of the candidates
> has more than 50% of the votes, then that candidate wins. If none of the
> candidates receives a majority of the votes, then a second plurality election
> is held with a designated number of the top candidates. The process is re-
> peated until one candidate has more than 50% of the votes.

In most cases, the top two candidates from the first vote run in the runoff election. Having a provision for a runoff helps in elections in which an extreme candidate wins the plurality vote with fewer than 50% of the votes. A runoff election also lessens the incentive to switch one's vote from a preferred candidate
to a more electable one, because there will be a second chance to vote unless
one candidate receives a majority of all votes, in which case such a decision is
irrelevant.

George Bush, Ross Perot, and Bill Clinton debating during the 1992 presidential election campaign.

Example 3 *1992 Presidential Election*

In the 1992 U.S. presidential election, Bill Clinton received 44,908,254 votes, George Bush received 39,102,343, and Ross Perot received 19,741,065, with some other votes going to other minor candidates. (Recall, however, that U.S. presidents are actually elected through the Electoral College.) Suppose there had been a runoff election between Clinton and Bush. Assuming the same people voted in the original election and the runoff and ignoring the voters supporting minor candidates, answer the following:

(a) In a runoff election between Clinton and Bush, what percentage of Perot supporters would need to vote for Clinton in order for Clinton to receive a majority of the popular vote?

(b) In a runoff election between Clinton and Bush, what percentage of Perot supporters would need to vote for Bush in order for Bush to receive a majority of the popular vote?

(c) Polls showed that Perot took support from Clinton and Bush in roughly equal numbers. How likely does it seem that George Bush could have won the popular vote without Ross Perot in the election?

Solution:
(a) The total number of votes cast for each candidate was as follows:

Clinton	44,908,254
Bush	39,102,343
Perot	19,741,065
Total	103,751,662

Because the total is an even number, a majority is 1 more than half of the total or

$$\frac{1}{2}(103{,}751{,}662) + 1 = 51{,}875{,}831 + 1 = 51{,}875{,}832 \text{ votes.}$$

We assume that all of Clinton's and Bush's supporters in the original election will continue to support them in the runoff. Because Clinton received 44,908,254 votes in the original election, we see that he would need

$$51,875,832 - 44,908,254 = 6,967,578$$

of the Perot supporters' votes to have a majority. This is

$$\frac{6,967,578}{19,741,065} \approx 0.3529 = 35.29\%$$

of the Perot supporters.

(b) We could do a calculation similar to the one we did in part (a), but it is easier to observe that if Clinton needs more than 35.29% of the Perot supporters to receive a majority of the popular vote in the runoff, then Bush would need more than $100\% - 35.29\% = 64.71\%$ to win in the runoff.

(c) In part (b), we saw that more than 64.71% of Perot supporters must switch to Bush in order for Bush to win in the runoff election. Because only about 50% of Perot's supporters would have switched to Bush if Perot had dropped out of the election, we can conclude that it does not seem likely that Bush would have won the popular vote had Perot dropped out.

BRANCHING OUT 1.2

CHOOSING THE SITE OF THE 2000 SUMMER OLYMPICS

The Olympics have always been political, and politics plays a role in the selection of sites for the Olympic games. This is illustrated by the vote for the site of the 2000 Summer Olympics. The selection was made in 1993 by the 89 members of the International Olympic Committee. The cities in contention were Beijing, Berlin, Istanbul, Manchester (England), and Sydney. China pressed hard for Beijing and organized a strong campaign to lobby the Olympic Committee. The appropriateness of choosing Beijing as an Olympic site was questioned by the United States Congress and the European Parliament, both of whom adopted resolutions calling for Beijing's candidacy to be rejected on human rights grounds. Nevertheless,

Preference Rankings

In most voting situations, each voter has an order of preference of the candidates. Such an ordering is called a **preference ranking**. The voter may have to think hard to arrive at this preference ranking, and it may change quickly and easily with time and circumstance. In some voting methods, such as Borda's method which we will see in the next section, the voters actually write their preference rankings on their ballots. In plurality elections, this is not the case. However, if we know the preference rankings of the voters, we can analyze how an election would come out under various voting methods.

It is possible for a voter to be indifferent between two or more alternatives—for instance, if the voter has no knowledge whatsoever of several of the candidates. We assume that whenever indifference among choices exists, the voter will arbitrarily, perhaps randomly, order the choices. Thus, we assume each voter has a preference ranking that orders all candidates from most preferred to least preferred.

when the committee began casting votes, Beijing was viewed as the favorite for selection.

The voting process calls for a plurality election, with a runoff between all of the candidates except the one in last place. This voting method is known as the *Hare method*. The process continues until a city has a majority of the votes. In the first round of voting the tallies were as follows:

First Round	
Beijing	32
Sydney	30
Manchester	11
Berlin	9
Istanbul	7

Thus, Istanbul was eliminated from the competition. The second round of votes yielded the following:

Second Round	
Beijing	37
Sydney	30
Manchester	13
Berlin	9

We see that, most likely, 5 Istanbul votes went to Beijing and 2 to Manchester. With Berlin eliminated, the

third-round votes were as follows:

Third Round	
Beijing	40
Sydney	37
Manchester	11

Note that one voter did not cast a ballot. Interestingly, Manchester lost 2 votes, although one of the lost votes might have been the voter who did not cast a ballot. Sydney gained ground on Beijing, but Beijing was still in the lead. In the final vote between Beijing and Sydney, Beijing fell behind in the voting for the first time, and the result was as follows:

Fourth and Final Round	
Sydney	45
Beijing	43

Again, one voter did not cast a ballot. Based on the results of the first round of voting, it appears that Beijing supporters would have been better off with a simple plurality election. However, it is safe to say that the final result better reflected the desires of the entire Olympic committee.

Source: International Olympic Committee.[1]

We record the preference rankings of all the voters in an election by using a table such as Table 1.1.

Table 1.1	Example of Preference Rankings			
	Number of Voters			
	6	*3*	*1*	*4*
Apples	1	2	2	3
Oranges	3	1	3	2
Pears	2	3	1	1

The three "candidates" in this election are apples, oranges, and pears, and they are listed in the first column on the left. In the next column we see a possible ranking, where apples are ranked first, pears second, and oranges third, as indicated by the numbers recorded in the rows corresponding to each of the fruits. The boldface number at the top of this column indicates the number of voters who have this particular preference ranking. In this case, there were six such voters. The remaining columns include other possible preference rankings and the number of voters with each ranking. We see by adding across the first row that there are a total of 14 voters. Notice that, in this example, not every possible ranking is included. For instance, no voters had a preference ranking of apples first, oranges second, and pears third, or a ranking of oranges first, pears second, and apples third.

Throughout this chapter we make some assumptions about preference rankings. Our first assumption is that if a voter has ranked one candidate higher than another, then if the voter must choose between those two candidates, the voter would choose the higher-ranked one. We also assume that the order of the preferences is not changed by the elimination of one or more candidates—for example, in the case of a runoff election.

Let's now look at an example of how the results of an election decided by a plurality with a runoff can be worked out if we know the preference rankings of the voters.

Example 4   *Election Results from Preference Rankings*

Eleven members of a department are voting on which day to schedule their weekly meetings. Their choices are Tuesday, Wednesday, Thursday, and Friday, and their preference rankings are listed in Table 1.2.

Table 1.2	Preference Rankings for Meeting Day						
	Number of Voters						
	1	*3*	*1*	*1*	*1*	*2*	*2*
Tuesday	1	1	4	3	2	3	4
Wednesday	2	3	1	1	4	2	2
Thursday	4	4	2	4	1	1	3
Friday	3	2	3	2	3	4	1

(a) In a plurality election, which day would be selected?

(b) In a plurality election with a runoff between the top two finishers, which day would be selected?

(c) In a plurality election with a runoff between the top two finishers, could the two voters who ranked Friday first achieve a preferable outcome by voting strategically if the others voted as indicated in the table?

Solution:

(a) To determine the winner of a plurality election, we need only count the first-place rankings. The preference rankings with Tuesday ranked first are shaded in pink in Table 1.3.

Table 1.3	**Those Voting for Tuesday in a Plurality Election**						
	Number of Voters						
	1	*3*	*1*	*1*	*1*	*2*	*2*
Tuesday	1	1	4	3	2	3	4
Wednesday	2	3	1	1	4	2	2
Thursday	4	4	2	4	1	1	3
Friday	3	2	3	2	3	4	1

We see that a total of 4 voters ranked Tuesday first and would therefore vote for Tuesday in a plurality election. Similarly, Wednesday would get 2 votes, Thursday 3 votes, and Friday 2 votes. Therefore, Tuesday would win.

(b) The runoff election choices would be Tuesday and Thursday, with each voter choosing whichever of these two days he or she ranks higher. In Table 1.4, the preference rankings in which Tuesday is ranked above Thursday are shaded pink. These 5 voters would vote for Tuesday in the runoff. The preference rankings in which Thursday is ranked above Tuesday are shaded blue, and these 6 voters would vote for Thursday. Therefore, Thursday would win with 6 votes, with Tuesday receiving 5 votes in the runoff.

Table 1.4	**Runoff Election Between Tuesday and Thursday**						
	Number of Voters						
	1	*3*	*1*	*1*	*1*	*2*	*2*
Tuesday	1	1	4	3	2	3	4
Wednesday	2	3	1	1	4	2	2
Thursday	4	4	2	4	1	1	3
Friday	3	2	3	2	3	4	1

(c) Recall that to vote strategically means to vote in a way that does not reflect one's true preferences, attempting to obtain an outcome preferable to that obtained when voting according to one's true preferences. These two voters prefer both Friday and Wednesday to Thursday, so they prefer that either of these two days wins instead of Thursday. There is no point in voting strategically in the runoff. However, if they vote for Wednesday instead of voting for Friday on the first ballot, then the results of the first ballot would be Tuesday 4, Wednesday 4, Thursday 3, and Friday 0. Therefore, the runoff would be between Tuesday and Wednesday. In the runoff, if all of the voters vote as expected from their preference rankings, then Tuesday would get 5 votes, and Wednesday would get 6 votes and win the election. Thus, we see that our two strategic voters can achieve the preferable outcome of meeting on Wednesday.

■

Sometimes the breakdown of voters according to their preference rankings is given in terms of a percentage of voters rather than a number of voters. We look at an example of this kind next.

Example 5   *Preference Rankings Broken Down by Percentage*

Suppose that a university is considering whether to expand, decrease, or maintain the current level of career counseling service. Student fees are affected by the level of service, with more service resulting in higher fees. The preference rankings of the student body regarding the level of service are as follows:

	Percentage of Voters			
	37	*23*	*21*	*19*
Expand	1	3	2	3
Decrease	3	1	3	2
Maintain	2	2	1	1

(a) Which option would the students choose using the plurality method?

(b) Which option would they choose using the plurality method followed by a runoff between the first- and second-place finishers?

Solution:

(a) Counting only the first-place votes, we see that in a plurality election 37% would vote to expand service, 23% would vote to decrease service, and 21% + 19% = 40% would vote to maintain service at its current level. Therefore, maintaining the level of service would win.

(b) The runoff would be between maintaining service at its current level and expanding service. Because all those with decreasing service as a first choice had maintaining the level of service as a second choice, the results of the runoff election would be 37% for expanding services and 40% + 23% = 63% for maintaining service at its current level. Once again, we see that maintaining the level of service would win.

■

We saw that when there are more than two candidates, the plurality method can result in an outcome that is less desirable to a majority of the voters than some other particular outcome. We will see later that the same problem can occur with the plurality method with a runoff. In addition, both methods are susceptible to manipulation by strategic voting. Neither of these methods allows voters to express their relative preferences among *all* the candidates. In the next three sections we look at voting methods that allow voters to do this.

EXERCISES FOR SECTION 1.1

1. A group of friends is deciding what type of food to go out for on Sunday night. The votes are Mexican, Italian, Chinese, Mexican, Italian, French, Chinese, Chinese, Indian. Which type of food will they eat if the decision is by plurality?

2. A committee decides to elect a chair by a plurality election. The three candidates are Gorman, Hwang, and Page. The ballots read Hwang, Hwang, Page, Gorman, Page, Page, Hwang, Gorman, Hwang, Page, and Hwang. Who wins the election?

3. Suppose there are 100 votes cast in an election between three candidates: Flores, Payne, and Bronowski. The election is to be decided by plurality. After the first 70 votes are counted, the tallies are as follows:

Flores	31
Payne	23
Bronowski	16

   (a) What is the minimal number of remaining votes Flores needs to be assured of a win?

   (b) What is the minimal number of remaining votes Payne needs to be assured of a win?

   (c) What is the minimal number of remaining votes Bronowski needs to be assured of a win?

4. Suppose there are 80 votes cast in an election between three candidates—Donahue, Garza, and Weis—to be decided by plurality. After the first 55 votes are counted, the tallies are as follows:

Donahue	24
Garza	18
Weis	13

   (a) What is the minimal number of remaining votes Donahue needs to be assured of a win?

   (b) What is the minimal number of remaining votes Garza needs to be assured of a win?

   (c) What is the minimal number of remaining votes Weis needs to be assured of a win?

5. If 302 votes are cast, what is the smallest number of votes a winning candidate can have in a five-candidate race that is to be decided by plurality?

6. If 203 votes are cast, what is the smallest number of votes a winning candidate can have in a four-candidate race that is to be decided by plurality?

7. The 1824 presidential election included four competitive candidates, all from the De-

mocratic-Republican Party. Andrew Jackson received 151,271 votes, John Quincy Adams received 113,122 votes, Henry Clay received 47,531 votes, William H. Crawford received 40,856 votes, and 13,053 voters voted for other candidates. Because Jackson failed to get a majority in the Electoral College, the House of Representatives decided the election by choosing John Quincy Adams.

(a) If those voting for Crawford and miscellaneous candidates had voted for one of the top three, who could have won a plurality of the votes?

(b) What percentage of these voters would Jackson need to ensure a plurality?

8. Suppose that three candidates—Rosen, Brown, and Wheatley—are running in an election that will be decided by the plurality method with a runoff between the top two finishers if none of the candidates receives a majority of the votes. The results of the first ballot are given next.

Rosen	2346
Brown	5784
Wheatley	6230

In a runoff election between Wheatley and Brown, what percentage of Rosen supporters would need to vote for Brown in order for Brown to win the election?

9. On June 10, 1997, the Conservative members of the British Parliament held an election to find a successor to John Major as leader of the Conservative Party. The election was to be decided by the plurality method with a runoff if none of the candidates received a majority of the votes. The results of the first ballot are given next.

Kenneth Clarke	49
William Hague	41
Michael Howard	23
Peter Lilley	24
John Redwood	27

On June 17, 1997, a runoff election between Clarke, Hague, and Redwood was held, and the results were as follows:

Kenneth Clarke	64
William Hague	62
John Redwood	38

For the final runoff on June 19, 1997, John Redwood threw his support to Kenneth Clarke. What percentage of Redwood supporters would have needed to vote for Clarke in order for Clarke to win the final runoff? (As it turns out, Hague won the runoff by a margin of 22 votes.)

10. The procedure for enacting legislation in the U.S. House of Representatives and Senate is to consider amendments to a bill one at a time, and then to vote on passage of the entire bill at the end. In 1955, the Senate considered a highway appropriations bill introduced by Albert Gore of Tennessee, father of the future vice president. One amendment was to remove the so-called Davis–Bacon clause that mandated fair pay standards for workers on federal construction projects. The Senate could be divided into three groups, none forming a majority. One group, primarily Republicans, favored no bill of any sort, but favored keeping the Davis–Bacon clause in the bill. A

second group, mainly Southern Democrats, favored the bill if and only if the Davis–Bacon clause were removed. The third group, mainly other Democrats, favored the bill in either form, but preferred inclusion of the Davis–Bacon clause.

(a) What would have resulted from a vote on the amendment and then on the bill?

(b) Could one of the three groups obtain a result preferable to the outcome in (a) by voting strategically on one of the votes?

(What actually happened is that the Democratic leadership of the Senate, which included future President Lyndon Johnson, was able to remove the clause without a formal vote, and thus secured passage of the highway bill.)

11. Nine members of a committee must decide what kind of flooring to install in a community room. The preference rankings of the committee members are listed here.

	Number of Voters				
	3	*1*	*2*	*1*	*2*
Carpet	1	1	3	2	3
Ceramic tile	2	3	1	3	2
Wood	3	2	2	1	1

(a) Which type of flooring would be chosen using the plurality method?

(b) Which type of flooring would be chosen using the plurality method followed by a runoff between the first- and second-place finishers?

12. Eleven girls on a basketball team want to choose the type of food to be served at a team party. Each of the girls has an order of preference for the different types of food. The preference rankings of the girls are listed in the following table.

	Number of Voters					
	3	*1*	*1*	*1*	*3*	*2*
Pizza	1	2	3	4	2	3
Hamburgers	2	1	1	3	3	2
Hot dogs	3	3	2	1	4	4
Chicken	4	4	4	2	1	1

(a) Which type of food would win using the plurality method?

(b) Which type of food would win using the plurality method followed by a runoff between the top two finishers?

13. A poll of members of a community asked people about their opinions on a curfew time for young people under the age of 18. Their preference rankings broke down into the following percentages.

	Percentage of Voters							
	9	*2*	*6*	*20*	*4*	*17*	*10*	*32*
10 P.M.	1	2	3	4	3	4	4	4
12 midnight	2	1	1	1	2	2	3	3
1 A.M.	3	3	2	2	1	1	1	2
No curfew	4	4	4	3	4	3	2	1

(a) Which option would win using the plurality method?

(b) Which option would win using the plurality method followed by a runoff between the top two finishers?

14. Suppose three candidates—Whitney, Pronzini, and O'Kane—are running for mayor, and the preference rankings of the voters, broken down into percentages, are given in the following table.

	Percentage of Voters					
	21	*14*	*20*	*19*	*16*	*10*
Whitney	1	1	2	3	2	3
Pronzini	2	3	1	1	3	2
O'Kane	3	2	3	2	1	1

(a) In a plurality election, who would win?

(b) In a plurality election with a runoff between the top two finishers, who would win?

15. A college sorority must decide how many hours of community service should be required of its members each year. The preference rankings of the members are listed here.

	Number of Voters							
	3	*1*	*1*	*1*	*2*	*1*	*2*	*8*
0 hours	1	2	3	3	5	3	5	5
5 hours	2	1	1	2	3	2	4	4
10 hours	3	3	2	1	1	1	3	3
20 hours	4	4	4	4	2	4	1	2
40 hours	5	5	5	5	4	5	2	1

(a) Which choice would be selected in a plurality election?

(b) Which choice would be selected in a plurality election with a runoff between the top two finishers?

(c) In a plurality election, could the three members who ranked 0 hours first achieve a preferable outcome by voting strategically if the other members voted as shown in the table?

(d) In a plurality election with a runoff between the top two finishers, could the three members who ranked 0 hours first achieve a preferable outcome by voting strategically if the other members voted as shown in the table?

16. Three candidates are running for president of a service organization. The preference rankings of the voters are as follows:

	Number of Voters					
	13	*5*	*10*	*7*	*10*	*6*
Rosenthal	1	1	2	3	2	3
Salinas	2	3	1	1	3	2
Milner	3	2	3	2	1	1

(a) Who would win in a plurality election?

(b) Who would win in a plurality election with a runoff between the top two finishers?

(c) In a plurality election, could the six voters who ranked Milner first and Salinas second achieve a preferable outcome by voting strategically if the others voted as shown in the table?

(d) In a plurality election with a runoff between the top two finishers, could the six voters who ranked Milner first and Salinas second achieve a preferable outcome by voting strategically if the others voted as shown in the table?

Exercises 17–26 demonstrate how critical the method or order of voting may be to the outcome. In each of these questions we consider the situation where 12 people at a picnic want to select an activity for the afternoon. Their preference rankings of the possibilities are listed here.

	Number of Voters							
	1	*1*	*1*	*1*	*3*	*2*	*2*	*1*
Football	1	1	2	2	3	2	3	4
Soccer	2	4	1	4	2	4	2	3
Softball	3	2	3	1	1	3	4	2
Volleyball	4	3	4	3	4	1	1	1

17. (a) Which activity would win a plurality election?

(b) In a plurality election, could the two whose first choice was football have achieved a preferable outcome by voting strategically if the other ten voted as shown in the table?

18. If the group did not have volleyball equipment, which of the remaining choices would win a plurality of the votes?

19. If the group did not have enough gloves for softball, which of the other three choices would win a plurality of the votes?

20. (a) In a plurality election with a runoff between the top two finishers, which activity would be selected?

(b) In a plurality election with a runoff between the top two finishers, could the two voters who ranked volleyball first, soccer second, football third, and softball last achieve a better outcome by voting strategically if the other ten voted as shown in the table?

21. If the vote for activity is sequential, where the picnickers decide between football and soccer, then between the winner of that vote and softball, then between the winner of that vote and volleyball, which activity will they choose (if it is not dark by the time they decide)?

22. If the vote for activity is sequential, where the picnickers decide between softball and volleyball, then between the winner of that vote and soccer, then between the winner of that vote and football, which activity will they choose?

23. Is it possible to determine a sequence of votes, such as those described in Exercises 19 and 20, that yields soccer as the winner?

24. Suppose the vote is tournament style, with football matched against soccer and softball matched against volleyball, with a vote between winners determining the activity. Which activity wins?

25. Suppose the vote is tournament style, first with football matched against volleyball and soccer matched against softball, followed by a vote between the two winners. Which activity wins?

26. For each of the four activities, is it possible to arrange a tournament as in Exercises 24 and 25 so that activity wins?

1.2   BORDA'S METHOD: A SCORING SYSTEM

Elections determined by the plurality method or the plurality method with a runoff do not take into account the voters' relative preferences for *all* the candidates. A voter's top choice among the alternatives is revealed in a plurality election, but the balloting reveals nothing about the voter's last choice. In analyzing plurality methods with or without runoff, we used the assumption that we knew the preference rankings of the voters to determine which candidate particular voters would support in a given round of voting. However, in plurality elections the voters are not asked to give their preference rankings. In this and the following section, we assume that voters are asked to list their full set of preferences on a ballot, and we look at methods that use all this information.

One method of voting that requires voters to list their preference ranking for all choices is a scoring system called Borda's method, named for the Frenchman Jean-Charles de Borda (1733–1799). The method works as follows:

Borda's Method

With **Borda's method**, voters rank the entire list of candidates or choices in order of preference from their first choice to their last choice. After the votes have been cast, they are tallied as follows: On a particular ballot, the lowest-ranked candidate is given 1 point, the second lowest is given 2 points, and so on, to the top candidate who receives points equal to the number of candidates. The number of points given each candidate is summed across all ballots. We call this total the **Borda count** for the candidate. The winner is the candidate with the highest Borda count.

This method is illustrated in the following example.

Example 1   *Borda's Method*

Suppose a five-member committee needs to select a chair from among three candidates named Coleman, Horowitz, and Taylor. They decide to use Borda's method. The preference rankings of the five committee members are recorded in Table 1.5.

Table 1.5 **Preference Rankings for Committee Chair**				
	Number of Voters			
	1	*2*	*1*	*1*
Coleman	1	2	2	3
Horowitz	2	1	3	2
Taylor	3	3	1	1

Who will be the winner using Borda's method?

Solution: The Borda count for Coleman is given by

(no. of 1st-place votes)3 + (no. of 2nd-place votes)2 + (no. of 3rd-place votes)1
$$= 1 \cdot 3 + 3 \cdot 2 + 1 \cdot 1 = 3 + 6 + 1 = 10.$$

Similarly, the Borda count for Horowitz is

$$2 \cdot 3 + 2 \cdot 2 + 1 \cdot 1 = 6 + 4 + 1 = 11$$

and the Borda count for Taylor is

$$2 \cdot 3 + 0 \cdot 2 + 3 \cdot 1 = 6 + 0 + 3 = 9.$$

We see that Horowitz is the winner of the election. ■

> ## Technology Tip
>
> A simple spreadsheet can do all the arithmetic required by Borda's method given the voters' preference rankings.

There are other equivalent ways to determine the Borda winner. One way is to give 0 points to the person last in the voter's ranking, 1 to the next-to-last person, and so on, up to the top-ranked person who receives points equal to one less than the number of candidates. The Borda method can also be carried out by giving 1 point to the top-ranked candidate, 2 to the candidate ranked second, and so forth, the Borda winner being the candidate with the lowest point total.

Notice that in Example 1, if a plurality election had been held, the result would have been a tie between Horowitz and Taylor. If a runoff between Horowitz and Taylor were held, we would expect their supporters to vote for them again, and we would expect the Coleman supporter to vote for the candidate that he or she ranked second. This would result in a runoff win for Horowitz, and we see that a plurality with a runoff and Borda's method give the same result. However, these two methods do not always result in the same winner. In fact, as we see in the next example, the Borda's method winner may not even make it to the runoff election.

Example 2 *Comparison of Voting Methods*

Suppose that, in a survey, people were asked to rank the ice cream flavors chocolate, vanilla, and strawberry in order from their first to last choice, with the results given in Table 1.6.

Table 1.6	**Ice Cream Preference Rankings**					
	Percentage of Voters					
	33	*3*	*10*	*20*	*7*	*27*
Chocolate	1	1	2	3	2	3
Vanilla	2	3	1	1	3	2
Strawberry	3	2	3	2	1	1

(a) Which flavor would win using Borda's method?

(b) Which flavor would win using the plurality method?

(c) Which flavor would win using the plurality method with a runoff between the first- and second-place finishers in the plurality election?

Solution:

(a) The voting data are given in the form of percentages, but we can simply treat the percentages as we would voters and apply Borda's method as usual. The Borda count for chocolate is

(pct. of 1st-place votes)3 + (pct. of 2nd-place votes)2 + (pct. of 3rd-place votes)1
$= (33 + 3) \cdot 3 + (10 + 7) \cdot 2 + (20 + 27) \cdot 1 = 36 \cdot 3 + 17 \cdot 2 + 47 \cdot 1 = 189.$

Similarly, the Borda count for vanilla is

$$30 \cdot 3 + 60 \cdot 2 + 10 \cdot 1 = 220$$

and the Borda count for strawberry is

$$34 \cdot 3 + 23 \cdot 2 + 43 \cdot 1 = 191.$$

We see that the winner using Borda's method is vanilla.

(b) In a plurality election, the people vote for the flavor they rank as a first choice, so the results of the election would be

chocolate	36%
vanilla	30%
strawberry	34%

The winning flavor in a plurality election is chocolate. We see that vanilla, the winner using Borda's method, comes in last under the plurality method.

(c) Because chocolate and strawberry were the first- and second-place winners in the plurality election, they will be the two available choices in the runoff election. Notice that vanilla, the Borda's method winner, did not even make it into the runoff. In the runoff, the people who ranked vanilla first will now vote for their second-choice flavor. Tabulating these numbers, we arrive at the following runoff election result:

chocolate	36% + 10% = 46%
strawberry	34% + 20% = 54%

and we see that the winner in this case is strawberry.

Just as with the plurality and plurality with runoff methods, with Borda's method voters may be able to get a preferable result by voting strategically. This phenomenon is illustrated in the next example in the context of Borda's method.

Example 3   *Borda's Method and Strategic Voting*

Ten members of the families of a couple planning a wedding want to decide what kind of music will be played at the wedding reception. They have narrowed down the choices to country, jazz, polka, or rock and roll. For family harmony, they decide to vote using Borda's method, and the results are in Table 1.7.

Table 1.7	Wedding Music Preference Rankings							
	Number of Voters							
	1	*1*	*1*	*3*	*1*	*1*	*1*	*1*
Country	1	1	1	2	2	3	4	4
Jazz	2	3	3	4	4	2	2	3
Polka	3	2	4	1	3	4	3	2
Rock and roll	4	4	2	3	1	1	1	1

(a) Which musical style will win using Borda's method?

(b) Can the individual who ranked the music in the order rock and roll first, country second, polka third, and jazz fourth vote strategically in such a way that the Borda winner would be rock and roll if the others voted as shown in the table?

(c) The two who ranked country music last are probably quite disappointed, particularly if they are the bride and groom. Could they have achieved a preferable outcome by voting strategically if the other eight voted as shown in the table?

Solution:

(a) Country music received 3 first-place votes, 4 second-place votes, 1 third-place vote, and 2 fourth-place votes, so its Borda count is

$$3 \cdot 4 + 4 \cdot 3 + 1 \cdot 2 + 2 \cdot 1 = 28.$$

We compute the other Borda counts similarly, and the final tally is country 28, jazz 19, polka 26, and rock and roll 27. Therefore, country music is the Borda winner.

(b) Suppose the individual who ranked rock and roll first, country second, polka third, and jazz last instead ranked rock and roll first, jazz second, polka third, and country last. To see how this would affect the outcome of the election, we could completely recompute all of the Borda counts. However, it is easier to start with the original counts from part (a) and find the new Borda counts by adding or subtracting the appropriate amount from the original counts. For instance, by putting country last rather than second, the strategic voter would reduce its Borda count by 2, so country music's Borda count would be $28 - 2 = 26$. Similarly, by moving jazz from fourth place to second place, the strategic voter would increase jazz's count by 2, resulting in a count of $19 + 2 = 21$. Because polka and rock and roll remained in the same places in the strategic voter's rankings, their Borda counts of 26 and 27, respectively, would remain the same. Therefore, we see that the final tally would be country 26, jazz 21, polka 26, and rock and roll 27, resulting in a win for rock and roll.

(c) Suppose both of the voters ranking country music last had ranked polka first, rock and roll second, jazz third, and country last. We could recompute all of the Borda counts to see what happens, but we again compute the new Borda counts by adjusting the original counts accordingly. Because neither of the strategic voters would have changed their ranking of country last, country's original Borda count of 28 would be unchanged. Because both voters moved rock and roll from first to second place, each of them would reduce rock and roll's counts by 1, resulting in a new Borda count of $27 - 2 \cdot 1 = 25$. By ranking polka first, one of the strategic voters moved it up one place, whereas the other moved it up two places. Therefore, polka's new Borda count would be $26 + 1 + 2 = 29$. Jazz's ranking was moved down one by one of the strategic voters and was unchanged by the other, so its new Borda count would be $19 - 1 = 18$. Therefore, the new Borda count would be country 28, jazz 18, polka 29, and rock and roll 25, resulting in a win for polka. Note that although both of the strategic voters would prefer rock and roll to polka, they cannot lower country's count or raise rock and roll's, so they must settle for the smaller improvement from country to polka. ∎

In using Borda's method, there are so many computations that you may want to check your results. Of course, one way to check is to redo the calculation, but there is also an easy partial check. Suppose the election has c candidates and v voters. Then each voter casts c numerical votes totaling

$$1 + 2 + \cdots + c = \frac{c(c + 1)}{2}.$$

Therefore, the sum of all the Borda counts will total $vc(c + 1)/2$.

A CLOSER LOOK 1.1

SUMMING THE FIRST n INTEGERS

To arrive at the formula for the sum of all the Borda counts, we used the fact that the sum of the first n integers is $n(n + 1)/2$. Let us see why this formula is true by looking at the sum of the first four integers. First we write this sum as $1 + 2 + 3 + 4$ and then in the reverse order as $4 + 3 + 2 + 1$. Then we add these together and see by the arithmetic and the corresponding picture that $2 \cdot (1 + 2 + 3 + 4) = 4 \cdot 5$, or $1 + 2 + 3 + 4 = \frac{4 \cdot 5}{2}$.

$$
\begin{array}{r}
1 + 2 + 3 + 4 \\
4 + 3 + 2 + 1 \\
\hline
5 + 5 + 5 + 5 = 4 \cdot 5
\end{array}
$$

The same argument shows that

$$1 + 2 + \cdots + n = \frac{n(n + 1)}{2}.$$

> In a Borda's method election with c candidates and v voters, the sum of all the Borda counts will be
> $$\frac{vc(c+1)}{2}.$$

If the election results are given in the form of percentages, we set the number of voters, v, in the above formula equal to 100.

Sometimes we would like to use Borda's method, but the number of candidates or choices the voters must rank is rather overwhelming. In this case, a modified version of Borda's method may be used in which the voters rank only their top few choices. The voting method for the Heisman Memorial Trophy Award is of this type. On the Heisman ballot, shown in Figure 1.1, voters are asked to rank only their top three choices from among all college football players in the United States. The Borda count for each candidate is computed by giving 3 points for each first-place vote, 2 points for each second-place vote, and 1 point for each third-place vote. Any candidate not listed on a ballot will get 0 points from that ballot. The winner is declared to be the candidate with the highest Borda count. In sports polls where this form of voting is commonly used, the voters may know a lot about the top teams or players and be able to rank them, but may not know enough to rank all eligible candidates, so lumping all but the top candidates together with 0 points simplifies the process for the voters.

Borda's method is just one voting method that takes into account the entire preference rankings of voters. Some methods that are similar to Borda's method assign a wider spread of points to the different rankings. For instance, in voting for the National Basketball Association Most Valuable Player award, 116 mem-

The Heisman Memorial Trophy Award
Football's Greatest Individual Award
DOWNTOWN ATHLETIC CLUB
62nd Annual Award

1996 OFFICIAL BALLOT-PLEASE PRINT **Ballot No.** 0130

I hereby designate ________________ (NAME) ________________ (COLLEGE)

as My First Choice to receive the HEISMAN MEMORIAL TROPHY awarded to the Outstanding College Football Player of the United States for 1996. To the best of my knowledge, he conforms to the rules governing the vote.

My Second Choice is: ________________ (NAME) ________________ (COLLEGE)

My Third Choice is: ________________ (NAME) ________________ (COLLEGE)

Ballots are Void unless Signed and the First, Second and Third Choices are indicated.

RULES GOVERNING VOTE: In order that there will be no misunderstanding regarding the eligibility of a candidate, the recipient of the award MUST be a bona fide student of a recognized college or university including the United States Service Academies. Football players on service and professional teams and those who had already completed MORE than four years of college varsity football prior to the 1996 season are not eligible.

Name: ________________ Signature: ________________

(Paper, Radio or TV Station) ________________

City ________________ State ________________ Zip ________________

All Ballots are to be returned for compilation to DELOITTE & TOUCHE LLP, Certified Public Accountants, P.O. Box 3318, New York, N.Y. 10277-0464. The HEISMAN Poll will close at 5:00 PM on Thursday, December 12, 1996, and Ballots received after that time will not be included in the tabulation.

Figure 1.1 • Ballot for the Heisman Memorial Trophy Award.

BRANCHING OUT 1.3

DETERMINING WHO IS NUMBER ONE IN COLLEGE FOOTBALL

In 1936, Alan Gould, the sports editor of the Associated Press (AP), came up with the idea of the AP college football poll to satisfy his customers—the subscriber newspapers of the Associated Press. The newspapers were looking for a way to arouse interest in, and controversy about, college football. Gould's idea of a national poll fit the bill.

The voters in the AP poll are newspaper, radio, and television sports reporters throughout the country. In 1997, there were 70 voters. The rankings are determined by a Borda-type scheme. Each voter ranks his or her choice for the top 25 football teams in Division I-A in order. A first-place vote is worth 25 points, a second-place vote is worth 24 points, and so on, down to a 25th-place vote, which is worth 1 point. There are 112 teams in Division I-A football, and each voter can choose the top 25 from among any of these 112 teams. Teams not included in a voter's ranking get no points from that voter. The final ranking of the teams is determined by the teams' point total across all of the voters. The results are published in newspapers across the country. The AP Top 25 published on November 3, 1997 is shown in the table.

bers of the media list their first through fifth choices for the award. Each first-place vote receives 10 points, each second-place vote 7 points, each third-place vote 5 points, each fourth-place vote 3 points, and each fifth-place vote 1 point.

The advantage of Borda's method over plurality methods is that voters are able to express their opinions about candidates other than just their first choice

AP Top 25 in College Football, November 3, 1997 (First-place votes are in parentheses.)

	Record	Points
1. Nebraska (46)	8-0	1719
2. Penn State (16)	7-0	1643
3. Florida State (5)	8-0	1627
4. Michigan (1)	8-0	1561
5. North Carolina (2)	8-0	1481
6. Washington	7-1	1362
7. Ohio State	8-1	1324
8. Tennessee	6-1	1312
9. Georgia	7-1	1138
10. UCLA	7-2	1089
11. Kansas State	7-1	1013
12. Iowa	6-2	904
13. Florida	6-2	902
14. LSU	6-2	837
15. Arizona State	6-2	811
16. Washington State	7-1	775
17. Auburn	7-2	515
18. Toledo	8-0	449
19. Mississippi State	6-2	347
20. Virginia Tech	6-2	304
21. Texas A&M	6-2	299
22. Syracuse	6-3	286
23. Purdue	6-2	255
24. Southern Mississippi	6-2	253
25. Oklahoma State	6-2	125

Others receiving votes: West Virginia 99, Missouri 93, Colorado State 49, Wisconsin 42, Ohio University 35, Michigan State 32, Virginia 29, Louisiana Tech 11, Georgia Tech 10, Brigham Young 7, Marshall 4, New Mexico 3, Air Force 2, Mississippi 2, Southern California 1.

In 1950, United Press International started a second college football poll, in which coaches, rather than media people, were the voters. A similar coaches poll in use today is the *USA Today*/ESPN college football poll. It uses a voting method identical to the AP poll. The voters in the AP and *USA Today*/ESPN polls wield a great deal of power over the fortunes of college football teams. Unlike most collegiate sports, Division I-A football does not have a playoff to determine a national champion, and instead relies on a formula based on the AP and *USA Today*/ESPN poll rankings, three computer rankings, schedule strength, and win-loss records to determine the top teams. Beginning with the 1998 season, the teams ranked number 1 and 2 according to this formula at the end of the regular season have automatically played each other in a bowl game. Because millions of dollars ride on the possibility of playing in this national championship bowl game, the AP and *USA Today*/ESPN polls are quite important. This leads to the ethical question of whether or not the football rankings should be affected by the votes of media people and coaches who may have a vested interest in particular teams. Of particular concern is the temptation to manipulate the results by use of strategic voting. Because a Borda-type method is used in the polls, a single voter has a great deal of power to affect the outcome. For instance, if a voter's favored team is in a close contest for first place with another team, the voter can damage the other team's prospects significantly by ranking them very low on the ballot or not even ranking them all. It is unlikely that the controversy surrounding the national championship of college football will be resolved by anything other than the institution of a playoff that includes all teams that might reasonably be considered in the running for the title.

candidate. Because of this, a candidate rated highly, but not necessarily first, by most voters will often win an election decided by Borda's method. However, this advantage is somewhat offset by the fact that Borda's method is more susceptible to manipulation by strategic voters than are the plurality and plurality with runoff methods.

*E*XERCISES FOR **S**ECTION **1.2**

1. A 14-member board used Borda's method to elect a chair. The four candidates were Cardona, Pitts-Jones, De Plata, and Vincent. The preference rankings on the 14 ballots are listed here.

	Number of Voters							
	1	*4*	*1*	*2*	*2*	*1*	*1*	*2*
Cardona	1	1	1	3	4	2	2	3
Pitts-Jones	2	3	3	1	1	3	4	2
De Plata	3	2	4	2	2	1	1	4
Vincent	4	4	2	4	3	4	3	1

Who was the winner using Borda's method?

2. The returning members of a basketball team decide to select a team captain for the next season using Borda's method. The three candidates are Thomas, Walker, and Goodman. The preference rankings on the nine ballots are listed here.

	Number of Voters					
	3	*1*	*1*	*1*	*2*	*1*
Thomas	1	1	2	3	2	3
Walker	2	3	1	1	3	2
Goodman	3	2	3	2	1	1

Who is the winner using Borda's method?

3. An eight-member committee must decide on a school mascot. The choices have been narrowed down to the following: Bears, Falcons, Trojans, Mustangs, and Horned Frogs. The preference rankings of the committee members are listed here.

	Number of Voters							
	1	*1*	*1*	*1*	*1*	*1*	*1*	*1*
Bears	1	2	3	2	4	3	5	5
Falcons	3	1	1	4	3	2	4	4
Trojans	5	4	5	1	1	4	2	3
Mustangs	4	5	4	5	5	1	3	2
Horned Frogs	2	3	2	3	2	5	1	1

Which mascot is the winner using Borda's method?

4. A publishing company plans to open a new office in one of the following six cities: Chicago, New York, Philadelphia, Seattle, Los Angeles, or Denver. An 11-member executive board must decide on the city. The preference rankings of each board member are given next.

						Number of Voters					
	1	*1*	*1*	*1*	*1*	*1*	*1*	*1*	*1*	*1*	*1*
Chicago	4	5	5	3	3	3	5	2	2	3	4
New York	1	1	1	6	5	6	4	4	4	2	3
Philadelphia	3	4	3	1	2	2	2	5	5	5	6
Seattle	6	3	4	4	4	5	3	3	6	6	2
Los Angeles	5	6	2	2	1	1	1	6	3	4	5
Denver	2	2	6	5	6	4	6	1	1	1	1

Which city is the winner if Borda's method is used?

5. A small company decided to use Borda's method to determine on what day of the week to have a company picnic. The three choices were Monday, Wednesday, and Thursday. Employees ranked their choices as shown here.

		Number of Voters				
	8	*10*	*8*	*7*	*4*	*8*
Monday	1	1	2	3	2	3
Wednesday	2	3	1	1	3	2
Thursday	3	2	3	2	1	1

Which day won using Borda's method?

6. Sixty-four people were registered to attend a conference. The person in charge of planning the conference decided to use Borda's method to let the participants vote on a group recreational activity. The four choices were ballet, opera, baseball game, and horse racing. The preference rankings of the people are given here.

			Number of Voters					
	7	*14*	*5*	*6*	*7*	*5*	*10*	*10*
Ballet	1	1	2	2	3	4	3	4
Opera	2	2	1	1	4	3	4	3
Baseball game	3	4	3	4	1	1	2	2
Horse racing	4	3	4	3	2	2	1	1

Which activity won using Borda's method?

7. Three candidates—Kahn, Papini, and Tilson—ran for a seat on city council. Polls indicate that the preference rankings of the voters broke down into the following percentages.

		Percentage of Voters				
	26	*13*	*5*	*22*	*18*	*16*
Kahn	1	1	2	3	2	3
Papini	2	3	1	1	3	2
Tilson	3	2	3	2	1	1

Who would have been the winner of the election if Borda's method had been used?

8. Candidates Mijares, Zapata, and Schwartz are running for president of a professional society. The election is to be decided using Borda's method, and the preference rankings of the voters break down into the following percentages.

	Percentage of Voters					
	17	14	22	15	17	15
Mijares	1	1	2	3	2	3
Zapata	2	3	1	1	3	2
Schwartz	3	2	3	2	1	1

Who is the winner using Borda's method?

9. The preference rankings of a class of 22 college students for their favorite national news anchor are as follows:

	Number of Voters					
	3	3	6	2	5	3
Tom Brokaw	1	1	2	3	2	3
Peter Jennings	2	3	1	1	3	2
Dan Rather	3	2	3	2	1	1

(a) Which anchor has the top Borda count?
(b) Which anchor wins a plurality of the vote?
(c) Which anchor wins using the plurality method with a runoff between the top two finishers?

10. An engineering team needed to select a team leader from among the candidates Tate, Kummer, Dobbins, and Coscia. Dobbins was not well liked and was ranked last by everyone but himself. Of course, he ranked himself first. The preference rankings of the members were as follows:

	Number of Voters						
	7	3	8	4	1	4	3
Tate	1	1	2	3	2	2	3
Kummer	2	3	1	1	4	3	2
Dobbins	4	4	4	4	1	4	4
Coscia	3	2	3	2	3	1	1

(a) Which candidate would have won using Borda's method?
(b) Which candidate would have won using the plurality method?
(c) Which candidate would have won using the plurality method with a runoff between the first- and second-place finishers?

11. A 25-member division needs to decide which sport to play at their picnic. The preference rankings of the division members follow:

Number of Voters

	4	3	5	3	6	4
Soccer	1	1	2	3	2	3
Baseball	2	3	1	1	3	2
Football	3	2	3	2	1	1

(a) Which sport wins using the plurality method?

(b) Which sport wins using the plurality method with a runoff between the first- and second-place finishers?

(c) Which sport wins using Borda's method?

(d) If Borda's method is used, can the four voters who ranked soccer first, baseball second, and football third obtain a preferable result by voting strategically if the others vote as shown in the table?

12. A soccer team wants to select a color for their uniforms. The preference rankings of the team members are listed in the following table.

Number of Voters

	4	3	5	2	3	1
Blue	1	1	2	3	2	3
Green	2	3	1	1	3	2
Red	3	2	3	2	1	1

(a) Which color wins using the plurality method?

(b) Which color wins using the plurality method with a runoff between the top two finishers?

(c) Which color wins using Borda's method?

(d) Could the members whose first choice was green obtain a preferable result in an election decided by Borda's method by voting strategically if the other members voted as shown in the table?

13. An English department with ten members is selecting a new chair. Six members of the department are eligible and willing to run for the position. The preference rankings of the department members are given next.

Number of Voters

	1	1	2	1	1	1	1	2
LeBreton	1	2	6	6	4	5	5	3
Frye	5	1	1	1	5	4	6	6
Reeves	2	6	4	2	1	6	2	4
Keen-Sims	4	5	3	5	6	1	1	5
Cullors	6	3	2	3	2	3	3	2
Allen	3	4	5	4	3	2	4	1

(a) Who has the top Borda count?

(b) If Borda's method is used, could the members who ranked Allen first obtain a preferable result by voting strategically if the other members voted as shown in the table?

(c) If Borda's method is used, could the member who ranked Keen-Sims first and Reeves second obtain a preferable result by voting strategically if the other members voted as shown in the table?

14. A PTA board voted to select how to spend $20,000 they had earned in a fund-raiser. The preference rankings of the voting members appear next.

	Number of Voters								
	4	*3*	*1*	*2*	*2*	*1*	*1*	*6*	*2*
Computers	1	1	2	2	3	3	3	2	3
Library books	2	3	1	1	1	2	4	3	2
Playground equipment	4	4	3	4	4	1	1	4	4
Science lab	3	2	4	3	2	4	2	1	1

(a) Which choice has the top Borda count?

(b) If Borda's method is used, could the member who ranked playground equipment first and science lab second obtain a preferable result by voting strategically if the other members voted as shown in the table?

(c) If Borda's method is used, could the two members who ranked library books first, science lab second, computers third, and playground equipment last obtain a preferable result by voting strategically if the other members voted as shown in the table?

Exercises 15 and 16 consider two of the several alternate definitions of Borda's method.

15. Show that when, for each preference ballot, the lowest-ranked candidate is given 0 points, the second lowest is given 1 point, and so on, to the top candidate who receives points one less than the number of candidates, and the results are summed across all ballots, the candidate with the most points is always the winner of the Borda count according to the definition in the text.

16. Show that when, for each preference ballot, the top candidate is given 1 point, the second highest candidate 2 points, and so forth, and the results are summed across all ballots, the candidate with the fewest points is always the winner of the Borda count according to the definition in the text.

1.3 HEAD-TO-HEAD COMPARISONS

If we know the preference rankings of the voters, as we must to apply Borda's method, we can also see who would be the winner were any two of the candidates pitted against each other in a plurality election in the absence of the other candidates. We call such a plurality election between any two of the candidates a **head-to-head comparison**. To see how this is done, we refer back to an earlier example.

Example 1 *Head-to-Head Comparisons*

In the election between Coleman, Horowitz, and Taylor that we looked at in Example 1 of the previous section, the preference rankings were listed as in Table 1.8.

Table 1.8	Preference Rankings for Committee Chair			
	Number of Voters			
	1	*2*	*1*	*1*
Coleman	1	2	2	3
Horowitz	2	1	3	2
Taylor	3	3	1	1

(a) Who would be the winner in a head-to-head comparison between Coleman and Horowitz?

(b) Who would be the winner in a head-to-head comparison between Coleman and Taylor?

(c) Who would be the winner in a head-to-head comparison between Horowitz and Taylor?

Solution:

(a) In Table 1.9 those ballots on which Coleman is ranked above Horowitz are shaded pink. These two voters would vote for Coleman in a head-to-head comparison with Horowitz. The ballots on which Horowitz is ranked above Coleman are shaded blue. These three voters would vote for Horowitz in a head-to-head comparison between Coleman and Horowitz. Therefore, Horowitz would win with 3 votes to Coleman's 2 votes in a head-to-head comparison.

Table 1.9	Head-to-Head Comparison Between Coleman and Horowitz			
	Number of Voters			
	1	*2*	*1*	*1*
Coleman	1	2	2	3
Horowitz	2	1	3	2
Taylor	3	3	1	1

(b) Again, by comparing the rankings of Coleman and Taylor on each of the ballots, we see that Coleman would win in a head-to-head comparison with Taylor by receiving 3 votes, whereas Taylor would receive 2 votes.

(c) From the rankings, we see that Horowitz would receive 3 votes in a head-to-head comparison with Taylor, whereas Taylor would receive the remaining 2 votes, so Horowitz would win.

■

In Example 1 we saw that Horowitz beat both of the other candidates in head-to-head comparisons. Horowitz is an example of a Condorcet winner, named after the Frenchman Marie Jean Antoine Nicolas Caritat, the Marquis de Condorcet (1743–1794), and defined as follows.

Condorcet Winner

A candidate who is the winner of a head-to-head comparison with every other candidate is called a **Condorcet winner**. If a candidate beats or ties every other candidate in head-to-head comparisons, we call that candidate a **weak Condorcet winner**. A given election may or may not have a Condorcet winner or a weak Condorcet winner.

Many have agreed with Condorcet that a Condorcet winner deserves to be elected. Charles Dodgson, better known to us as Lewis Carroll, the author of *Alice's Adventures in Wonderland*, was a lecturer in mathematics at Christ Church College, Oxford. He supported Condorcet's view as did the Welsh economist Duncan Black, whose 1963 book, *The Theory of Committees and Elections*, was one of the pioneering works on voting methods.

Technology Tip

A spreadsheet or computer program can be set up to make all of the head-to-head comparisons required to determine a Condorcet winner, given the voters' preference rankings.

Although many view the election of a Condorcet winner as desirable and some view it as mandatory, a problem remains. There may be no Condorcet winner!

Example 2   *An Election with No Condorcet Winner*

This example originated with Condorcet in 1785. The preference rankings of three voters deciding between candidates A, B, and C are listed in Table 1.10.

Table 1.10	Preference Rankings Without a Condorcet Winner		
	Number of Voters		
	1	*1*	*1*
A	1	3	2
B	2	1	3
C	3	2	1

Show that there is no Condorcet winner.

Solution: In a head-to-head match-up between candidates A and B, the rankings show that A beats B by a score of 2 to 1. Similarly, C beats A by a score of 2 to 1, and B wins over C by a score of 2 to 1. We see that no candidate beats every other candidate in head-to-head match-ups, and therefore there is no Condorcet winner.

■

Marie Jean Antoine Nicolas Caritat, the Marquis de Condorcet (1743–1794).

In Examples 1 and 2, we had the preference rankings of all the voters and hence we could find the Condorcet winner or determine that there was none. Preference rankings cannot be completely worked out from the results of most elections, but combining information about the nature of the candidates with information gathered from exit polls may allow us to draw reasonable conclusions about the voters' preferences and hence about whether or not there is a Condorcet winner. Hypothesizing preference rankings from limited data is an example of *mathematical modeling*, in which real-life problems are analyzed by deducing or hypothesizing a mathematical structure. This is an important problem-solving strategy. The validity of a model can be measured by how well it agrees with real-life observations. We have more confidence in the predictions based on a model if they are not significantly affected by small changes in the model's most debatable assumptions. As an example of this type of analysis, we consider the 1990 election for governor of Louisiana.

Example 3   *1990 Louisiana Gubernatorial Election*

The three main candidates in the 1990 election for governor of Louisiana were David Duke, formerly a Grand Wizard in the Ku Klux Klan; Edwin Edwards, ex-governor, twice indicted though not convicted for fraud; and incumbent Buddy Roemer, whose switch from the Democratic to the Republican Party and efforts to reform state government alienated many voters. The election was conducted using the plurality method followed by a runoff election between the top two finishers. On the initial ballot, Edwards won 34% of the vote, Duke 32%, and Roemer 27%, with the remaining 7% going to other candidates. After receiving a grudging endorsement from Roemer for the runoff, Edwards prevailed over Duke 60% to 40%.

To guess at a reasonable distribution of preference rankings or second choices, we begin by scaling the initial vote to 100%. Because the initial percentages for the three candidates total 93%, we multiply each candidate's share of the vote by

Duke, Edwards, and Roemer (left to right).

100/93. After rounding, this gives us Edwards 37%, Duke 34%, and Roemer 29%.

Because Edwards only received 37% of the (adjusted) initial vote and he received 60% of the vote in the runoff, it is reasonable to suppose that 60% − 37% = 23% of the voters originally supported Roemer but switched to Edwards in the runoff. These voters would rank the candidates in the order Roemer, Edwards, and Duke from first to last choice.

Similarly, because Duke received 34% in the initial vote and 40% in the runoff, we would suppose that 6% of the voters supported Roemer originally and switched to Duke in the runoff. These voters would rank the candidates in the order Roemer first, Duke second, and Edwards last.

We can only guess the second and third choices for Edwards and Duke supporters. In light of the split of the Roemer supporters and of the candidates' political baggage, one approximation of the breakdown of preference rankings is given in Table 1.11.

Table 1.11	Possible Preference Rankings for the 1990 Louisiana Governor's Election					
	Percentage of Voters					
	9	*28*	*11*	*23*	*23*	*6*
Edwards	1	1	2	3	2	3
Duke	2	3	1	1	3	2
Roemer	3	2	3	2	1	1

Based on this table, who is the Condorcet winner?

Solution: Using our figures, in a head-to-head comparison of Roemer with Edwards, Edwards would have his 37% plus the 11% who ranked him second behind Duke, giving him 37% + 11% = 48% of the vote. Roemer would get the remaining 52% of the vote, and thus Roemer would have beaten Edwards 52% to 48%. Similarly, Roemer would defeat Duke 57% to 43%. Thus, Roemer is the Condorcet winner.

■

Note that the conclusions in this example depended on the suppositions we made about the breakdown of voter rankings. We assumed about 67.6% of Duke's supporters ranked Roemer second. If instead we had assumed fewer than 62% of Duke's supporters had Roemer as a second choice, then Edwards would have been the winner in a head-to-head comparison with Roemer.

The situation in Example 3 shows that a Condorcet winner may not even make the runoff in a plurality with runoff election. It is also true that a Condorcet winner may not win an election under the Borda method. However, in arguing in favor of the Condorcet winner, one could regard the roles of all other candidates as spoilers, because none would beat the Condorcet winner in a head-to-head match-up. In horse racing, middle- to long-distance races in track meets, and bicycle races, it is common for a "rabbit" or pacer with no chance of winning to be entered in order to increase the chances of the contender on the "team." In many ways, an election or choice among several alternatives may be viewed in a similar light, even if the effect is unintentional.

How common is an election with no Condorcet winner? With three candidates and an odd number of voters (to avoid ties) who order the candidates at random, the probability that there is a Condorcet winner decreases as the number of voters increases, but is always above 91%. Similarly, the probability of a Condorcet winner when an odd number of voters randomly order four candidates exceeds 82%. For five and six candidates, the percentages exceed 74% and 68%, respectively. In real-life situations there are often some strong candidates and some weaker candidates, so the likelihood of there being a Condorcet winner is significantly higher.

In a race with c candidates, there are $c(c-1)/2$ head-to-head comparisons. Computing all of them could take a bit of time. A more efficient way to look for a Condorcet winner is to begin with two strong candidates, perhaps the first- and second-place finishers by the plurality method or Borda's method. The winner of this contest would be compared with a third candidate, the winner of this with a fourth, and so on. Once all candidates except one have been defeated in this way, the remaining candidate is faced off against candidates not yet compared with until this candidate loses or defeats all candidates. Proceeding in this way only requires from $c-1$ to $2c-3$ head-to-head comparisons, depending on the particular outcomes.

Single-Peaked Preference Rankings

Before deciding to use the Condorcet method, it would be nice to be assured that a Condorcet winner exists. There is a condition that guarantees a Condorcet winner when the number of voters is odd or a weak Condorcet winner when the number of voters is even. We now examine this condition in the case of an election with four candidates, A, B, C, and D. Suppose that the candidates can be ordered in a line, for instance in the order C, D, A, B, in such a way that in ranking the candidates every voter makes a first choice and then moves only outward along the line from this first choice in ranking the remaining candidates. Three examples of rankings of this sort are shown in Figure 1.2.

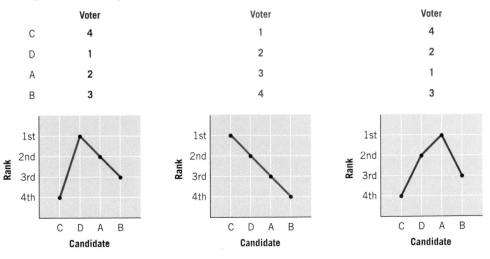

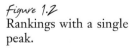
Figure 1.2
Rankings with a single peak.

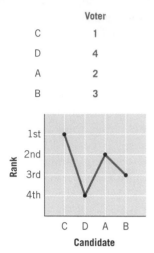

	Voter
C	1
D	4
A	2
B	3

figure 1.3
A ranking with two peaks.

Graphically, we see that these kinds of preference rankings have single peaks. For instance, the preferences of the blue voter have a single peak at candidate D, those of the green voter have a single peak at candidate C, and the preferences of the red voter have a single peak at Candidate A. In any election, if *all* voters' preference rankings have a single peak with respect to some *fixed* ordering of the candidates, we say that preference rankings are **single-peaked** with respect to that order. We cannot call the system of preferences single-peaked if even one voter gives a ranking that is not single-peaked with respect to the fixed order of the candidates, such as the ranking graphed in Figure 1.3, with peaks at Candidates C and A with respect to the order C, D, A, B.

Economist Duncan Black showed that any election in which the preference rankings are single-peaked has a weak Condorcet winner. The only time there will not be a Condorcet winner when preference rankings are single-peaked is if two or more candidates tie head-to-head against each other and defeat the rest of the candidates. Observe that the number of voters must be even in this case.

Preferences are often single-peaked in situations involving money, such as deciding the amount to commit to some project or cause, or determining a salary. This makes sense, because a voter probably has an ideal amount of money in mind and moves outward from that ideal in ranking other alternatives. Preferences may also be single-peaked or nearly single-peaked in an election with a single dominant issue. In some situations, such as ranking a favorite fruit from among bananas, apples, oranges, and grapes, you would not expect the preferences to be single-peaked.

The following example illustrates a situation in which the single-peaked pattern occurs, and, as expected, a Condorcet winner exists.

Example 4   *Single-Peaked Preference Rankings and a Condorcet Winner*

Suppose a nine-member board is deciding the salary of a company's chief executive officer and has the options of $200,000, $250,000, $300,000, $350,000, and $400,000. The rankings of the members' choices of salaries are listed in Table 1.12.

Table 1.12	**Preference Rankings for CEO's Salary**			
	Number of Voters			
	4	*2*	*2*	*1*
$200,000	1	4	5	5
$250,000	2	3	4	4
$300,000	3	1	1	3
$350,000	4	2	2	1
$400,000	5	5	3	2

Find the Condorcet winner.

Solution: Five of the nine board members favor $300,000 over $200,000 or $250,000, and eight of nine favor $300,000 over $350,000 or $400,000. Therefore, $300,000 is the Condorcet winner. Notice that the preferences are single-peaked with respect to the natural order $200,000, $250,000, $300,000, $350,000, $400,000.

■

An election procedure from which a Condorcet winner, if there is one, emerges victorious requires a back-up voting procedure to use in the case that no Condorcet winner exists. Such a process is called a **Condorcet completion process**. In 1876, Charles Dodgson suggested choosing the Condorcet winner if one exists and in the case of no Condorcet winner, he suggested several alternatives for deciding the election. In 1963, Welsh economist Duncan Black, in his study of voting, advocated using Borda's method when there is no Condorcet winner. Although feasible, this leaves the task of employing two completely independent methods to determine the result of the election. In the late nineteenth century, E. J. Nanson proposed using the Borda counts to eliminate all candidates with below average Borda counts. Voters' rankings of the remaining candidates are then used to compute new Borda counts, those with below average counts are eliminated, and the process continues until a single candidate is left. Exercise 20 asks you to prove that if there is a Condorcet winner, that candidate will be the winner under Nanson's method. Nanson's method has several variations. One is to eliminate only the lowest count before recomputing Borda counts. This has the advantage of being less subject to manipulation by a group of voters, but may require significantly more Borda count computations. This disadvantage is relatively minor when results are tallied by a computer.

A Condorcet winner has a good claim to being the most preferred choice in a multicandidate election and is largely unaffected by strategic voting or additional candidates without realistic prospects for winning. The drawback is that Condorcet winners do not always exist, so that any election procedure that elects a Condorcet winner, if one exists, must be augmented by a method to decide elections without one.

EXERCISES FOR SECTION 1.3

1. Three candidates—Bauer, Sanders, and Donevska—are running for president of a service organization with 12 members. The preference rankings of the members are listed here.

	Number of Voters					
	2	1	4	2	2	1
Bauer	1	1	2	3	2	3
Sanders	2	3	1	1	3	2
Donevska	3	2	3	2	1	1

(a) Who is the winner in a head-to-head comparison between Bauer and Sanders?
(b) Who is the winner in a head-to-head comparison between Bauer and Donevska?
(c) Who is the winner in a head-to-head comparison between Donevska and Sanders?
(d) Which candidate, if any, is the Condorcet winner?

2. Eight friends are trying to decide which movie to rent at the video store. They have narrowed the choices down to a comedy, western, or action movie. The preference rankings of the friends are listed here.

	Number of Voters				
	2	*1*	*2*	*1*	*2*
Comedy	1	2	3	2	3
Western	3	1	1	3	2
Action movie	2	3	2	1	1

   (a) Which is the winner in a head-to-head comparison between comedy and western?
   (b) Which is the winner in a head-to-head comparison between comedy and action movie?
   (c) Which is the winner in a head-to-head comparison between western and action movie?
   (d) Which type, if any, is the Condorcet winner?

3. The preference rankings of a class of college students for their favorite "ethnic" food are as follows:

	Number of Voters											
	1	*1*	*2*	*1*	*1*	*1*	*1*	*2*	*1*	*2*	*3*	*3*	*2*
Cajun	1	1	1	3	4	2	3	2	3	4	3	4	5
Chinese	2	3	4	1	1	4	4	4	2	2	4	3	3
Indian	3	5	5	5	5	5	5	5	5	5	5	5	4
Italian	4	4	3	2	2	1	1	3	4	3	2	2	2
Mexican	5	2	2	4	3	3	2	1	1	1	1	1	1

   Which type of food, if any, is the Condorcet winner?

4. The members of a tennis team are asked to select the most valuable player on the team from among four candidates named Hall, Perez, Moore, and Chihara. The preference rankings of the members are as follows:

	Number of Voters						
	1	*1*	*2*	*1*	*1*	*2*	*1*
Hall	1	1	3	3	3	3	4
Perez	3	3	1	2	4	2	3
Moore	2	4	2	1	1	4	2
Chihara	4	2	4	4	2	1	1

   Which player, if any, is the Condorcet winner?

5. The managers of a catering service meet to decide on their first ever employee-of-the-month. Three employees are suggested: Julia, Paul, and Wolfgang. The managers' preference rankings appear here.

	Number of Voters			
	1	*1*	*3*	*2*
Julia	1	1	3	2
Paul	2	3	1	3
Wolfgang	3	2	2	1

   Which employee, if any, is the Condorcet winner?

6. Twelve members of a search committee in charge of hiring the new president for a group of banks have narrowed their list to three finalists: DeZern, Parker, and Schultz. After much discussion they decided to rank the finalists and offer the position to the Condorcet winner, if one exists. The rankings are given in the table.

	Number of Voters					
	2	*2*	*3*	*1*	*3*	*1*
DeZern	1	1	2	3	2	3
Parker	2	3	1	1	3	2
Schultz	3	2	3	2	1	1

Which candidate, if any, is the Condorcet winner?

7. A small-town newspaper asked its readers to rate their favorites in various categories. In the restaurant category, the choices were the Hungry Boar, the Hungry Dog, and the Hungry Pig. The percentage of those responding with each possible ranking is given in the table.

	Percentage of Voters					
	13	*21*	*17*	*12*	*24*	*13*
Hungry Boar	1	1	2	3	2	3
Hungry Dog	2	3	1	1	3	2
Hungry Pig	3	2	3	2	1	1

Which restaurant, if any, is the Condorcet winner?

8. An AM radio station in financial trouble is seeking a partner for a merger. Three potential partners have made offers. The AM station's stockholders must choose whether to merge with an FM radio station, a television station, or a newspaper. The percentage of stockholders with each possible preference ranking is given in the table.

	Percentage of Voters					
	16	*17*	*20*	*9*	*11*	*27*
FM radio station	1	1	2	3	2	3
Television station	2	3	1	1	3	2
Newspaper	3	2	3	2	1	1

Which option, if any, is the Condorcet winner?

9. The parents of children in an elementary school were asked to vote on having mandatory school uniforms, optional school uniforms, or no school uniforms. The preference rankings of the parents are as follows:

	Number of Voters					
	36	*5*	*16*	*19*	*1*	*42*
Mandatory uniforms	1	1	2	3	2	3
Optional uniforms	2	3	1	1	3	2
No uniforms	3	2	3	2	1	1

(a) Which option has the top Borda count?
(b) Which option would win a plurality of the vote?

(c) Which option would win using the plurality method with a runoff between the first- and second-place finishers?

(d) Which option, if any, is the Condorcet winner?

10. In order to promote a wider viewpoint, a board of directors of a company has decided to increase its size by adding from one to five members. The preference rankings regarding the number of members to add to the 11-member board are as follows:

	Number of Voters					
	2	1	2	1	1	4
One	1	3	5	5	5	5
Two	2	1	1	3	3	4
Three	3	2	2	1	2	3
Four	4	4	3	2	1	2
Five	5	5	4	4	4	1

(a) Which number has the top Borda count?

(b) Which number would win a plurality of the vote?

(c) Which number would win using the plurality method with a runoff between the first- and second-place finishers?

(d) Which number, if any, is the Condorcet winner?

11. A family of six is planning to buy a dog. They have narrowed the choice of breed down to four choices, and their preference rankings are as follows:

	Number of Voters					
	1	1	1	1	1	1
Beagle	1	1	1	3	3	4
Fox terrier	3	4	3	1	2	3
Bulldog	2	2	4	4	1	1
Boxer	4	3	2	2	4	2

(a) Which breed has the top Borda count?

(b) Which breed would win a plurality of the vote?

(c) Which breed would win using the plurality method with a runoff between the top two finishers?

(d) Which breed, if any, is the Condorcet winner?

12. Are the preference rankings in Exercise 10 single-peaked with respect to the ordering of candidates one, two, three, four, five? Would you expect them to be? Explain why or why not.

13. Are the preference rankings in Exercise 11 single-peaked with respect to the ordering of candidates beagle, fox terrier, bulldog, boxer?

14. Are the preference rankings in Exercise 11 single-peaked with respect to any ordering of the candidates? Justify your answer.

15. In January 1925, the U.S. Senate considered development of the Muscle Shoals area of the Tennessee River in Alabama. In successive votes the Senate voted for (1) private development over public development, 48-37; (2) recommitting the bill for further study over private development, 46-33; (3) public development over recommitment, 40-39; (4) private development over public development, 46-33. At this point the motion to recommit was again introduced. Because of White House lobbying, several senators switched votes and private development won over recommitment, 43-38. The bill finally passed in this version. Of the 96 members of the Senate, 75 voted on enough of these amendments and in a consistent manner on the first four votes so that we may surmise a preference ranking among the three alternatives. The results are as follows:

	Number of Voters					
	9	*18*	*4*	*26*	*16*	*2*
Private development	1	1	2	3	2	3
Public development	2	3	1	1	3	2
Recommitment	3	2	3	2	1	1

Source: Congressional Record.[2]

Assume that only these 75 senators voted.

(a) Which alternative would win a plurality vote?

(b) Which alternative would win a runoff election between the top two finishers?

(c) Which alternative is the Borda winner?

(d) Which option, if any, is the Condorcet winner?

(The outcome in (d) was originally observed by John L. Neufeld, William J. Hausman, and Ronald B. Rapoport, under slightly different assumptions. See the Suggested Readings at the end of this chapter.)

16. The 1860 election, precursor to the Civil War, saw four candidates win states. Split along regional lines, the contest was primarily between Republican Abraham Lincoln and Democrat Stephen A. Douglas in the North and Southern Democrat John C. Breckinridge and Constitutional Unionist John Bell in the South. Although no candidate won 40% of the popular vote, in 81% of the counties some candidate won a majority of the vote. Furthermore, Abraham Lincoln would have won the electoral vote even if all opposition votes had gone to a single opponent. Combining information about the candidates' platforms with the state-by-state breakdown of the vote, we estimate one reasonable, if hypothetical, distribution of the 24 possible preference rankings by percentage as shown below. The percentage of voters ranking a given candidate first is known; it is simply the percentage of votes the candidate received in the election.

	Percentage of Voters											
	8.0	*23.9*	*1.7*	*0.3*	*5.6*	*0.3*	*4.4*	*10.3*	*2.6*	*1.8*	*8.0*	*2.4*
Lincoln	1	1	1	1	1	1	2	2	3	4	3	4
Douglas	2	2	3	4	3	4	1	1	1	1	1	1
Breckinridge	3	4	2	2	4	3	3	4	2	2	4	3
Bell	4	3	4	3	2	2	4	3	4	3	2	2

	Percentage of Voters											
	0.2	*0.7*	*0.2*	*3.4*	*2.7*	*10.9*	*0.5*	*0.8*	*0.7*	*3.7*	*1.5*	*5.4*
Lincoln	2	2	3	4	3	4	2	2	3	4	3	4
Douglas	3	4	2	2	4	3	3	4	2	2	4	3
Breckinridge	1	1	1	1	1	1	4	3	4	3	2	2
Bell	4	3	4	3	2	2	1	1	1	1	1	1

Bell, Douglas, Breckinridge, and Lincoln, 1860.

THE NATIONAL GAME. THREE "OUTS" AND ONE "RUN".
ABRAHAM WINNING THE BALL.

   (a) Under our assumptions, who would have received the most votes in a runoff between Lincoln and Douglas?

   (b) Who would be the Borda winner?

   (c) Who, if anyone, is the Condorcet winner?

17. Which of the following choices would you expect to have single-peaked preferences among those voting?

   (a) A review committee's merit ratings of an employee, where the choices are poor, good, excellent, and superior

   (b) A restaurant chain's decision on how many restaurants to open in a region, where the choices are 1, 2, or 3 restaurants

   (c) A group of people's preference rankings of the four vegetables corn, peas, green beans, and broccoli

18. Which of the following choices would you expect to have single-peaked preferences among those voting?

   (a) A school board decision on student–teacher ratios, where the choices are 20, 25, 30, or 35 students per teacher

   (b) A condominium committee's decision on what kind of trees to plant, where the choices are elm, maple, or pecan

   (c) A group of student's preference rankings about the ideal day of the week to have an exam

19. If a candidate wins under plurality and is a Condorcet winner, must the candidate win a plurality election with runoff between the top two finishers, assuming there are no ties for the runoff positions? Give an example where this does not happen or explain why it always does.

20. In the voting method known as Nanson's method, the Borda counts of all candidates are computed. Next, those candidates with below average Borda counts (meaning a Borda count less than $v(c + 1)/2$, where v is the number of voters and c is the number of candidates) are eliminated. After the first round, the original preference rankings are used to construct a new set of preference rankings for the remaining candidates only. These preference rankings are then used to compute new Borda counts, those with below average counts are eliminated, and the process continues until a single candidate is left. Show that Nanson's method will select the Condorcet winner if there is one by relating the different head-to-head vote counts to the Borda count.

21. Suppose that preferences are single-peaked with respect to a given, known ordering of candidates along a line. Explain how to determine the Condorcet winner knowing only the first choices of the voters.

22. Construct an example of an election with three candidates that has a Condorcet winner with fewer first-place votes and fewer second-place votes than the Borda winner.

1.4   APPROVAL VOTING

In the late 1970s, several political scientists and others, writing independently, introduced and analyzed another voting method called approval voting, described as follows.

> ### Approval Voting Method
>
> With the **approval voting method**, voters indicate their approval or disapproval of each of the candidates. A ballot in an approval vote lists all the candidates' names, and voters check off all the candidates of whom they approve. The winner is declared to be the candidate with the highest approval count.

Voters can approve of from none to all of the candidates on a ballot, but to affect who is elected they should approve of at least one candidate and not approve of at least one candidate. Studies of approval voting indicate that voters tend to vote for at most half of all candidates. Approval voting is now used in the election of the Secretary General of the United Nations and the election of officers of some academic and professional societies.

The following example demonstrates how approval voting works.

Example 1   *Approval Voting Method*

An 11-member committee has 5 members—Hart, Alvarez, Wolsey, Hanson, and Park—running for chair. The committee decides to use the approval voting method in the election. The results are listed in Table 1.13, where a ✓ indicates approval of the candidate.

Table 1.13	Approval Ballots in Committee Chair Election						
	Number of Voters						
	1	*1*	*3*	*1*	*2*	*1*	*2*
Hart			✓			✓	✓
Alvarez	✓		✓	✓		✓	✓
Wolsey					✓		✓
Hanson				✓			
Park		✓			✓	✓	✓

Which candidate is the winner of the election?

Solution: The approval vote count is Hart 6, Alvarez 8, Wolsey 4, Hanson 1, and Park 6. Therefore, Alvarez is the winner.

∎

We sometimes want to consider examples that compare the outcomes of elections using different voting methods. In order to incorporate approval voting in a table of preference rankings, we put checks next to the rankings of any candidates approved by the voters, as in the following example.

Example 2   *Comparison of Voting Methods*

A city council needs to select a month in which to hold its first annual fair. The preference rankings of the council members along with the months of which they approve are shown in Table 1.14.

Table 1.14	**Preference Rankings of Month for City Fair**						
	Number of Voters						
	3	*1*	*1*	*2*	*1*	*2*	*1*
June	1 ✓	1 ✓	2 ✓	3	3	2	2 ✓
July	2	3	1 ✓	1 ✓	1 ✓	3	3
August	3	2 ✓	3	2	2 ✓	1 ✓	1 ✓

(a) Which month would be selected if the approval method were used?

(b) Which month would win a plurality election?

(c) Which month is the Borda winner?

Solution:

(a) In an approval election, June would receive 6 votes, July 4 votes, and August 5 votes, so June would win.

(b) The vote tallies in a plurality election would be 4 votes for June, 4 votes for July, and 3 votes for August, so a plurality election would result in a tie between June and July.

(c) June's Borda count is given by

$$4 \cdot 3 + 4 \cdot 2 + 3 \cdot 1 = 12 + 8 + 3 = 23.$$

Similarly, July's Borda count is 22 and August's Borda count is 21, so June is the Borda winner.

∎

Approval voting has entered the picture recently and has many supporters. It is particularly suitable when voters perceive two classes of candidates: those worthy of support and those not. Approval voting is less subject to manipulation than is Borda's method because each individual voter can only change the total vote count of a candidate by one point, whereas with Borda's method, the difference between first and last place can be several points. On the other hand, by grouping all candidates into one of two classes, those approved of and those not approved of, a voter may like one candidate much better than another and yet give them the exact same level of support.

In the 1980 presidential election, Ronald Reagan and Jimmy Carter faced independent candidate John Anderson. At one point in the campaign, Anderson's support was nearly 15%. However, by election day his support faded to 6.6%, as voters deserted him to choose between the two front-runners. In the election, Ronald Reagan won with 50.7% of the popular vote to Jimmy Carter's 41.0%. Thus, had the election been decided by popular vote, Ronald Reagan would have won with a majority in a plurality election. Whether he would have won under other voting methods is less clear. A *Time* magazine poll shortly before the election found that percentages finding each candidate "acceptable," presumably roughly equivalent to an affirmative approval vote, were Reagan 61%, Carter 57%, and Anderson 49%, much closer than the results of the popular vote. Exercises 13 and 14 at the end of this section ask you to build models consistent with these figures that will predict what the outcome of the election might have been had Condorcet's method or Borda's method been used.

In the following example, we look at another political situation to see how all the voting methods we have looked at play out.

Example 3 *1912 Presidential Election*

In 1912, ex-President Theodore Roosevelt challenged incumbent President William Howard Taft for the Republican Party presidential nomination and lost. Roosevelt then ran for president under the Progressive Party or Bull Moose Party banner. The third major candidate in the 1912 presidential election was Democrat Woodrow Wilson. Also running was Socialist Eugene V. Debs, who made his best showing in his five attempts at the presidency. The results of the popular vote are given in Table 1.15.

Table 1.15	Popular Vote in 1912 Presidential Election			
Wilson	*Roosevelt*	*Taft*	*Debs*	*Other*
6,293,152	4,119,207	3,486,333	900,369	241,902

We can roughly place the four candidates on a liberal–conservative spectrum as shown in Figure 1.4.

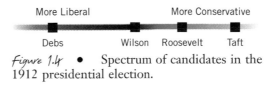

Figure 1.4 • Spectrum of candidates in the 1912 presidential election.

BRANCHING OUT 1.3

THE VOTING METHOD USED BY THE JURY OF THE VAN CLIBURN INTERNATIONAL PIANO COMPETITION

Every four years, 35 of the most talented young pianists in the world are invited to Fort Worth, Texas to compete in the Van Cliburn International Piano Competition. The competition includes three rounds: preliminary, semifinal, and final. After the preliminary round, the field of competitors is narrowed from 35 to 12. After the semifinal round, 6 finalists are selected. After the final round, the first-place winner is awarded a gold medal, the second-place finisher a silver medal, and the third-place finisher a bronze medal.

A jury of 12 members votes to determine who advances in each round and, in the final round, determines the medalists. In both the preliminary and semifinal rounds, the goal of the vote is different from the ones we have discussed in the text in that the jury must select a group of winners, rather than a single winner. There are many ways to do this, and the voting method set forth in the Van Cliburn Competition handbook for the jury can be described as follows.

After the conclusion of the preliminary round, each jury member votes to pass 12 competitors into the semifinals. Any competitor who gets a passing vote from all members of the jury goes on to the next round. The remaining competitors are grouped by the number of votes received. Starting with the group with the highest number of votes and proceeding through groups receiving fewer and fewer votes, all members of a group advance to the semifi-

1997 Van Cliburn gold medalist Jon Nakamatsu performing at the competition.

nal round as long as the members of the group received at least a 51% passing vote and there are enough semifinalist slots left to be filled to accommodate the entire group. If either criterion fails to hold, the process stops.

If any of the 12 slots have not been filled, then a second vote is taken. If the last round was stopped because there were not enough slots to accommodate all the members of a group receiving at least 51% of the vote, the candidates for the next votes are only those from that group. If the process of the previous

We might assume that voters would be likely to vote for those candidates nearest to them on this spectrum. However, there were other pertinent factors. Taft was the incumbent. Former President Roosevelt, with a personality many people loved or hated, ran as an independent and entered under his specially formed Bull Moose Party. His positions on conservation and integration both attracted and alienated voters. Domestic issues were often separate from international policies such as isolationism; many voters focused on one or the other.

For simplicity, we consider only the votes for the top four candidates. Let us also assume that Debs supporters rank the candidates from first choice to last in this order: Debs, Wilson, Roosevelt, Taft. We further assume that the voters who did not vote for Debs would rank him last. Suppose also that 75% of Wilson sup-

From right to left: first- through third-place winners in the 1997 Van Cliburn International Piano Competition, Jon Nakamatsu, Yakov Kasman, and Aviram Reichert.

To determine the medalists, the jury first determines the gold medalist using the plurality method with runoffs between all competitors receiving votes except the one receiving the least votes until one competitor has at least 51% of the vote. This method is known as the *Hare method*. The jury then uses the same method to name the silver medalist and then the bronze medalist.

Of course, ties can occur in the voting methods used by the Cliburn jury. In the event of a tie, the jury members are asked to hold a discussion after which a second vote is taken. If the second vote again results in a tie, the jury chair has the option of declaring a tie or calling for a third vote in which the jury chair's ballot will be counted. For the gold medalist a tie is not allowed.

According to the Cliburn jury handbook, "Some competitions and voting procedures are criticized for selecting as prize winners those pianists who least offend the majority of the jury members, whereas those with originality who impress and offend an equal number of jury members are eliminated." To address this problem, the voting procedures for the preliminary and semifinal rounds have an additional rule. Each jury member is asked to identify his or her top three choices. The names of any competitors who receive three or more such top votes, but do not pass on to the next round, will be brought before the jury for discussion and a possible re-vote.

Not surprisingly, the Cliburn jury voting procedure can take a while. In the 1997 competition, audience members had to wait until after midnight to hear the preliminary round results.

round stopped because a group with less than a 51% passing vote was reached, all remaining contestants who received at least 1 vote are still eligible for selection. The members of the jury now vote for as many candidates as there are slots available. The votes are tabulated as in the first round of voting and additional competitors are passed on to the semifinals. This process is repeated until all 12 slots are filled.

The method for voting to determinine the 6 finalists from among the 12 semifinalists is the same as the process used in determining the semifinalists, except that the goal is to pass 6 competitors out of an initial 12.

porters prefer Roosevelt to Taft, that 80% of Roosevelt supporters prefer Taft to Wilson, and that 85% of Taft supporters prefer Roosevelt to Wilson. Finally, we assume that all the voters approve of their first choices, 60% of the voters would also approve of their second choices, and no voter would approve of his or her third choice. Without accurate polling information, these are only plausible guesses. They are simplified, but still give us a qualitative idea of what *could* have happened. We want to answer the following questions about the 1912 presidential election:

(a) Who would have won a plurality of the popular vote had Roosevelt not run?
(b) Who would have won a plurality of the popular vote had Roosevelt won the Republican nomination, leaving Taft out of the election?

Booker T. Washington
dining with Theodore
Roosevelt at the White
House.

(c) Who would have won a majority of votes in a runoff between Wilson and Roosevelt?

(d) Who would have won by Borda's method?

(e) Would there be a Condorcet winner under the above assumptions?

(f) Who would have won an approval vote?

Solution: To answer all these questions, it is best to begin by determining the preference rankings of the voters. Based on our assumptions, the only possible preference rankings from first to last place are shown in Table 1.16.

Table 1.16	Possible Preference Rankings for the 1912 Presidential Election						
Wilson	1	1	2	3	2	3	2
Roosevelt	2	3	1	1	3	2	3
Taft	3	2	3	2	1	1	4
Debs	4	4	4	4	4	4	1

The 900,369 Debs supporters' preference rankings, based on our assumptions, were Debs first, Wilson second, Roosevelt third, and Taft last. Similarly, the number of voters ranking Wilson first, Roosevelt second, Taft third, and Debs last is

$$0.75 \cdot (6{,}293{,}152) = 4{,}719{,}864,$$

whereas the number of voters ranking Wilson first, Taft second, Roosevelt third, and Debs last is

$$6{,}293{,}152 - 4{,}719{,}864 = 1{,}573{,}288.$$

We compute estimates for the remaining four orderings similarly, and obtain the estimates shown in Table 1.17.

Table 1.17	Hypothetical Preference Rankings for the 1912 Presidential Election						
	Number of Voters						
	4,719,864	*1,573,288*	*823,841*	*3,295,366*	*522,950*	*2,963,383*	*900,369*
Wilson	1	1	2	3	2	3	2
Roosevelt	2	3	1	1	3	2	3
Taft	3	2	3	2	1	1	4
Debs	4	4	4	4	4	4	1

Note that some of the entries in Table 1.17 have been rounded to give whole numbers. We can now answer our questions.

(a) Under our assumptions, had Roosevelt not run, Debs would have received 900,369 votes; Wilson would pick up 823,841 votes, for a total of 7,116,993 votes; and Taft would pick up 3,295,366 votes, for 6,781,699 votes. Therefore, Wilson would have won a plurality election.

(b) Dropping Taft from the election would yield 522,950 additional votes for Wilson, for a total of 6,816,102 votes, and 2,963,383 for Roosevelt, for a total of

7,082,590 votes. Debs would have received 900,369 votes. In this case, Roosevelt would have won a plurality election.

(c) Had Wilson and Roosevelt faced each other in a runoff, we could begin with the totals from part (b) and add Debs' votes to Wilson's total, giving Wilson a total of 7,716,471 votes, and making him the runoff winner.

(d) Wilson's Borda count would be

$$6,293,152 \cdot 4 + 2,247,160 \cdot 3 + 6,258,749 \cdot 2 = 44,431,586.$$

Similarly, Roosevelt's Borda count would be 45,519,783, Taft's Borda count would be 40,539,073, and Debs' Borda count would be 17,500,168. We see that Roosevelt is the Borda winner.

(e) Part (c) showed that Wilson would beat Roosevelt head-to-head. Clearly, Wilson would defeat Debs, who would be routed with only 900,369 votes. Finally, Wilson would defeat Taft, 8,017,362 votes to 6,781,699 votes. Thus, Wilson would be the Condorcet winner.

(f) Giving each candidate his votes plus 60% of his second-place votes, we find the approval vote totals to be Debs 900,369, Wilson 7,641,448, Roosevelt 8,729,155, and Taft 6,407,525. The approval vote winner would be Roosevelt.

∎

Given the simplified and speculative nature of our assumptions, only by trying a variety of equally reasonable assumptions can we get a good idea of the likely outcome of the 1912 presidential election under different methods of election. The more often such outcomes agree, the more likely they are to indicate what would have happened. In the exercises, we consider other sets of assumptions.

Approval voting has the advantages of being easy for voters to understand, being only moderately subject to strategic voting, and possibly leaving more voters satisfied with the outcome of an election. However, the approval voting method does not allow voters to specifically rank their preferences. Therefore, a candidate who is acceptable to, but not strongly liked by, many voters may win over a candidate who is exceptionally desirable to slightly fewer voters.

EXERCISES FOR SECTION 1.4

1. A service club decided to use an approval vote in the election for president of the club. The five candidates for the position were McClain, Snyder, Freeman, Sanders, and Yang, and the results are given in the table.

	Number of Voters									
	1	*1*	*1*	*2*	*2*	*1*	*1*	*1*	*1*	*1*
McClain	✓			✓			✓			✓
Snyder		✓			✓	✓	✓	✓		✓
Freeman			✓		✓		✓	✓	✓	✓
Sanders							✓		✓	✓
Yang				✓		✓				✓

Who is the club's new president?

2. A professional society wants to hold its next annual convention in New Orleans, Las Vegas, San Francisco, or New York. The Executive Council, consisting of 12 members, decides to make the decision by approval ballot, and the results are listed in the table.

	Number of Voters						
	1	2	4	1	2	1	1
New Orleans	✓			✓	✓		✓
Las Vegas		✓		✓			✓
San Francisco					✓	✓	✓
New York				✓		✓	

Which city wins the vote?

3. The Institute of Electrical and Electronics Engineers (IEEE) uses the approval method to elect its president. The results of the 1988 election are shown in the table, where the candidates' names have been replaced by the letters A, B, C, and D.

	Number of Voters														
	10,738	6561	7626	8521	3578	659	6679	1425	1824	608	148	5605	143	89	523
A	✓				✓	✓	✓				✓	✓	✓		✓
B		✓			✓			✓	✓		✓	✓		✓	✓
C			✓			✓		✓		✓	✓		✓	✓	✓
D				✓			✓		✓	✓		✓	✓	✓	✓

Source: Institute of Electrical and Electronics Engineers.[3]

   (a) Find the winner of the election.
   (b) What percentage of the engineers voted in a way that could not help determine the winner regardless of how the others voted?

4. The members of a marching band vote to choose what time they want to hold an evening rehearsal. The choices are 6 P.M., 7 P.M., or 8 P.M. The results of an approval election are recorded here.

	Number of Voters					
	28	33	40	19	9	24
6 P.M.	✓			✓	✓	
7 P.M.		✓		✓		✓
8 P.M.			✓		✓	✓

Which time is the winner?

5. Three candidates are running for president of a union. A poll of the union membership indicates that if an approval election were held, the results would be as recorded here.

	Percentage of Voters					
	17	21	22	17	14	9
De Castro	✓			✓	✓	
Telger		✓		✓		✓
Segura			✓		✓	✓

Which candidate would win an approval election?

6. A toothpaste company decides to let the public determine a new flavor for children's toothpaste. The breakdown of the votes in an approval election are as follows:

	9	*15*	*10*	*8*	*9*	*4*	*6*	*5*	*3*	*1*	*7*	*4*	*9*	*2*	*8*
Grape	✓				✓	✓	✓				✓	✓	✓		✓
Strawberry		✓				✓		✓	✓		✓	✓		✓	✓
Cherry			✓			✓		✓		✓	✓		✓	✓	✓
Orange				✓			✓		✓	✓		✓	✓	✓	✓

Percentage of Voters

Which flavor is chosen?

7. The preference rankings of a class of 16 college students for the minimum voting age and the ages of which they approve are shown in the table.

	1	*1*	*1*	*1*	*1*	*3*	*2*	*2*	*3*	*1*
Sixteen	4	6	5	5	6	6	6	6	6	6
Seventeen	2 ✓	2 ✓	3	3	3 ✓	5	5	5	5	5
Eighteen	1 ✓	1 ✓	1 ✓	1 ✓	1 ✓	1 ✓	1 ✓	1 ✓	4	4
Nineteen	3 ✓	3	2	2 ✓	2 ✓	2	2 ✓	2 ✓	3	3
Twenty	6	4	4	4	4	3	3	3 ✓	2	2 ✓
Twenty-one	5	5	6	6	5	4	4	4	1 ✓	1 ✓

Number of Voters

(a) Which age would win a plurality vote?

(b) Which age would win a plurality vote with runoff between the top two finishers?

(c) Which age would win the Borda count?

(d) Which age, if any, is the Condorcet winner?

(e) Which age would win an approval vote?

8. The preference rankings of a class of 17 college students for the minimum legal drinking age and the ages of which they approve are shown in the table.

	1	*1*	*2*	*2*	*1*	*1*	*1*	*6*	*1*	*1*
Sixteen	1 ✓	3	6	6	6	6	6	6	6	6
Seventeen	2	2	5	5	5	5	5	5	5	5
Eighteen	3	1 ✓	1 ✓	1 ✓	1 ✓	4	4	4	4	4
Nineteen	4	4	2	2 ✓	2 ✓	3	3 ✓	3	3	3 ✓
Twenty	5	5	3	3 ✓	3 ✓	1 ✓	1 ✓	2	2 ✓	2 ✓
Twenty-one	6	6	4	4	4 ✓	2 ✓	2 ✓	1 ✓	1 ✓	1 ✓

Number of Voters

(a) Which age would win a plurality vote?

(b) Which age would win a plurality vote with runoff between the first- and second-place finishers?

(c) Which age would win the Borda count?

(d) Which age, if any, is the Condorcet winner?

(e) Which age would win an approval vote?

9. The partners of a law firm must decide which one of five junior members to promote to partner. The preference rankings of the partners and the candidates of which they

approve are included in the table.

	Number of Voters							
	2	1	1	1	1	3	1	1
Douglas	4	4	5	4	5	4	5	5
Holmes	1 ✓	1 ✓	1 ✓	2	2 ✓	2	2 ✓	3
Cochrane	5	5	4	3	4 ✓	5	4	4
Mason	3	3	2 ✓	1 ✓	1 ✓	3	3 ✓	2 ✓
Bailey	2	2 ✓	3 ✓	5	3 ✓	1 ✓	1 ✓	1 ✓

(a) Which member would win a plurality vote?

(b) Which member would win a plurality vote with runoff between the first- and second-place finishers?

(c) Which member would win the Borda count?

(d) Which member, if any, is the Condorcet winner?

(e) Which member would win an approval vote?

(f) Could the voter whose first choice was Holmes and second choice was Mason have obtained a preferable result in an approval vote by voting strategically if the others voted as shown in the table?

10. Due to prison overcrowding, a parole board must release one of four minor offenders before he has served his full sentence. The board votes to determine which prisoner will be released. The preference rankings of the parole board members and the candidates of which they approve are included in the table.

	Number of Voters						
	1	1	1	1	1	1	1
Ford	1 ✓	2 ✓	4	3	2	3	3
Robbins	3	1 ✓	3 ✓	4	3	2	2 ✓
Cage	2 ✓	3	1 ✓	1 ✓	4	4	4
Bacon	4	4	2 ✓	2 ✓	1 ✓	1 ✓	1 ✓

(a) Which prisoner would win a plurality vote?

(b) Which prisoner would win a plurality vote with a runoff between the top two finishers?

(c) Which prisoner would win the Borda count?

(d) Which prisoner, if any, is the Condorcet winner?

(e) Which prisoner would win an approval vote?

(f) Could the voter whose first choice was Robbins have obtained a preferable result in an approval vote by voting strategically if the others voted as shown in the table?

11. A condominium association must vote to determine what color to paint the wood trim of a building. The three choices are white, green, and ivory. The preference rankings of the association members and the colors of which they approve are shown in the table.

	Number of Voters								
	4	3	2	1	2	2	4	2	1
White	1 ✓	1 ✓	1 ✓	2 ✓	3	3	2	3	3
Green	2	3	3	1 ✓	1 ✓	1 ✓	3	2	2 ✓
Ivory	3	2	2 ✓	3	2	2 ✓	1 ✓	1 ✓	1 ✓

(a) Which color would win a plurality vote?

(b) Which color would win a plurality vote with runoff between the top two finishers?

(c) Which color would win the Borda count?

(d) Which color, if any, is the Condorcet winner?

(e) Which color would win an approval vote?

(f) Could the five voters with first choice white and second choice ivory have obtained a preferable result in an approval vote by voting strategically if the others voted as shown in the table?

12. A PTA board must determine what kind of fund-raiser it will hold. The choices have been narrowed down to holding a carnival, selling magazines, selling wrapping paper, or holding an auction. The preference rankings of the 11 board members and the choices of which they approve are shown here.

	Number of Voters							
	1	2	1	1	1	1	3	1
Holding a carnival	1 ✓	1 ✓	3	4	3	4	2 ✓	2
Selling magazines	4	4	1 ✓	1 ✓	2 ✓	2 ✓	3	3
Selling wrapping paper	3	3	2 ✓	2 ✓	1 ✓	1 ✓	4	4
Holding an auction	2	2 ✓	4	3	4	3 ✓	1 ✓	1 ✓

(a) Which fund-raiser would win a plurality vote?

(b) Which fund-raiser would win a plurality vote with runoff between the top two choices?

(c) Which fund-raiser would win the Borda count?

(d) Which fund-raiser, if any, is the Condorcet winner?

(e) Which fund-raiser would win an approval vote?

(f) Could the three board members whose first choice was holding a carnival have obtained a preferable result in an approval vote by voting strategically if the other members voted as shown in the table?

13. Counting only votes cast for the top three candidates in the 1980 presidential election, Ronald Reagan received 51.6% of these votes, Jimmy Carter received 41.7%, and John Anderson received 6.7%. Assume that these percentages represent the first choices of the voters (even though it is clear that some Anderson supporters voted for either Reagan or Carter in order to have more potential effect on the election). Assume that their approval counts were Reagan 61%, Carter 57%, and Anderson 49%, and that every voter approved of either one or two candidates.

(a) Explain why Reagan must be the Condorcet winner.

(b) What percentage of people had Ronald Reagan as their second choice *and* approved of him?

(c) What percentage of people had Jimmy Carter as their second choice *and* approved of him?

(d) What percentage of people had John Anderson as their second choice *and* approved of him?

(e) Estimate the percentage, rounded to the nearest whole number percentage, of second-place votes each candidate would get from among *all* the voters. To do this use your answers to parts (b) through (d) and assume that the percentage of second-place votes a candidate would get among all the voters is equal to the percentage of second-place votes that candidate got from among those voters who approved of their second choice. (Be sure your percentages add up to 100%.)

(f) Estimate the results of an election by Borda's method.

14. John Anderson's support in the 1980 presidential election was as high as 15% in the preelection polls, before fading to 6.6% in the actual election. We will ignore all candidates other than Anderson, Ronald Reagan, and Jimmy Carter. Assume that the drop-off in the election represented strategic voters who still ranked Anderson first (rather than any disillusionment with Anderson). Assume that the remaining first-place ranks were proportional to the actual vote total, so that Reagan was ranked first by 47% and Carter by 38%. Assume that their approval counts were Reagan 61%, Carter 57%, and Anderson 49%, and that every voter approved of either one or two candidates.

(a) What percentage of people had Ronald Reagan as their second choice *and* approved of him?

(b) What percentage of people had Jimmy Carter as their second choice *and* approved of him?

(c) What percentage of people had John Anderson as their second choice *and* approved of him?

(d) Estimate the percentage, rounded to the nearest whole number percentage, of second-place votes each candidate would get from among *all* of the voters. To do this use your answers to parts (a) through (c) and assume that the percentage of second-place votes a candidate would get among all the voters is equal to the percentage of second-place votes that candidate got from among those voters who approved of their second choice. (Be sure your percentages add up to 100%.)

(e) Estimate the results of an election by Borda's method.

(f) What fraction of those voters ranking Anderson first would need to rank Reagan second in order for Reagan to defeat Carter in a head-to-head comparison?

(g) What fraction of those voters ranking Carter first would need to rank Reagan second in order for Reagan to defeat Anderson in a head-to-head comparison?

(h) In light of your answers to parts (d), (f), and (g), does it seem likely that Reagan would be a Condorcet winner?

In Exercises 15–17, we consider the 1912 presidential election as we did in Example 3. The popular vote count for the top four candidates is shown here.

Wilson	*Roosevelt*	*Taft*	*Debs*
6,293,152	4,119,207	3,486,333	900,369

Again, we just look at these top four candidates. For brevity we use the notation DWRT to denote a preference ranking of Debs first, Wilson second, Roosevelt third, and Taft last, and denote the other possible preference rankings similarly. For the given situation answer the following.

(a) Who would have won a plurality of the popular vote had Roosevelt not run?

(b) Who would have won a plurality of the popular vote had Roosevelt won the Republican nomination, leaving Taft out of the election?

(c) Who would have won the majority of votes in a runoff between Wilson and Roosevelt?

(d) Who would have won by Borda's method?

(e) Which candidate, if any, is a Condorcet winner under the assumptions?

(f) Who would have won an approval vote?

15. Assume that all Debs voters ranked the candidates DWRT. Also assume that 10% of Wilson voters ranked the candidates WDRT, 20% ranked them WTRD, and the rest of the Wilson voters ranked them WRTD. Assume that 75% of Roosevelt voters ranked them RTWD and the rest RWTD, and that 80% of Taft voters ranked them TRWD and the rest TWRD. Finally, assume that all voters approve of their first choices, 25% of the voters approve of their second choices, and no voter approves of his or her third or fourth choices.

16. Assume that all Debs voters ranked the candidates DWRT and that the voters who did not vote for Debs would rank him last. Assume that 45% of Wilson supporters ranked the candidates WRTD, that 70% of Roosevelt supporters ranked the candidates RTWD, and that 70% of Taft supporters had the ranking TRWD. Finally, assume that all voters approve of their first choices, 50% of the voters approve of their second choices, and that no voter approves of his or her third or fourth choices.

17. Assume that all Debs voters ranked the candidates DWRT. Assume that 20% of Wilson voters ranked the candidates WDRT and the rest of the Wilson voters ranked them WRTD, that 85% of Roosevelt voters ranked them RTWD and the rest RWTD, and that 90% of Taft voters ranked them TRWD and the rest TWRD. Finally, suppose that 75% of voters in each group would support their second choice in an approval vote and that none would support their third or fourth choices.

In Exercises 18–21, we look at state elections. The Progressive Party nominated candidates for many offices in 1912. We consider possible outcomes under different voting methods of several governors' races that had moderately close races between the Democratic, Republican, and Progressive Party candidates in 1912. We will ignore other candidates in these races. The votes are given either as raw numbers or as percentages. Assume that 60% of Democratic voters ranked the Progressive candidate second, that 70% of Republican voters ranked the Progressive candidate second, that 65% of Progressive candidates ranked the Republican candidate second, and that 40% of those with a given ranking would vote to approve of only their first and second choice, whereas the remaining 60% would vote to approve of only their first choice in an approval vote. Determine

(a) the winner of a runoff election,

(b) the Borda winner,

(c) the Condorcet winner, if there is one,

(d) the winner of an approval vote.

18. Illinois: E. Dunne (Democrat) 443,120, C. Deneen (Republican) 318,469, F. Funk (Progressive) 303,401

19. Connecticut: S. Baldwin (Democrat) 78,264, J. Studley (Republican) 67,531, H. Smith (Progressive) 31,020

20. Utah: J. Tolten (Democrat) 35.3%, W. Spry (Republican) 41.6%, N. Morris (Progressive) 23.1%

21. Michigan: W. Ferris (Democrat) 37.5%, A. Musselman (Republican) 32.9%, L. Watkins (Progressive) 29.6%

In Exercises 22 and 23, we consider the 1970 New York Senate election, which was decided by the plurality method among three candidates. James Buckley (Conservative) with 39% of the vote defeated Richard Ottinger (Democrat) with 37% of the vote and Charles Goodell (Republican and Liberal Parties) with 24% of the vote. (The Liberal Party consisted mostly of more liberal Republicans, so it would land between the Democrats and the Conservatives on the liberal–conservative spectrum.) Assume that preferences are single-peaked, with Goodell in the middle.

22. Suppose that among Goodell supporters, 60% have Ottinger as a second choice and 40% have Buckley as a second choice. Suppose also that 40% of voters with any particular preference ranking approve of their second choice, that all approve of their first choice, and that none approve of their third choice.

    (a) Who would win a runoff between Buckley and Ottinger?

    (b) Who is the Condorcet winner?

    (c) Who wins an approval vote?

23. Suppose that Goodell supporters split 50%-50% for their second choice and that, for each ranking, half of the voters with this ranking approve of their second choice, all approve of their first choice, and none approve of their third choice.

    (a) Who would win a runoff between Buckley and Ottinger?

    (b) Who is the Condorcet winner?

    (c) Who wins an approval vote?

24. The 1980 New York Senate election, decided by plurality, saw Republican Alphonse D'Amato win with 45% of the vote, trailed by Democrat Elizabeth Holtzman with 44%, and Jacob Javits, running under the Liberal Party banner after losing the Republican nomination, with 11%. Javits was in the middle of the liberal–conservative spectrum. However, concerns about his age and health probably prevented preferences from being single-peaked. An ABC News poll estimated the results of a potential approval vote to be Holtzman 60%, D'Amato 56%, and Javits 49%. Assume that voters always approved of their first choice and never approved of their third choice.

    (a) For each candidate, what percentage of the voters placed that candidate second *and* approved of him or her?

    (b) What is the smallest possible percentage of voters who departed from rankings based on the liberal–conservative spectrum?

    (c) Estimate the percentage of second-place votes each candidate would get from among *all* the voters. To do this use your answer to part (a) and assume that the percentage of second-place votes a candidate would get among all the voters is equal to the percentage of second-place votes the candidate got from among those voters who approved of their second choice. (Be sure your percentages add up to 100%.)

    (d) Estimate the results of an election by Borda's method.

25. Show that you, as a single voter, make the most head-to-head distinctions between pairs of candidates (approving of one candidate and disapproving of the other) when you approve of as close to half of all candidates as possible.

THE SEARCH FOR AN IDEAL VOTING SYSTEM

Each voting method discussed in this chapter can yield outcomes that seem less than ideal. For example, if the plurality method is used with a slate of more than two candidates, then the winner may well be a candidate that the majority of the voters find highly undesirable. We have also seen that a Condorcet winner may not win an election under Borda's method and may not even make the runoff in a plurality election with a runoff. On the other hand, there may not be a Condorcet winner. It seems reasonable to ask for a better voting system, one that always satisfies certain desirable properties. In this section, we follow the ideas of Nobel Laureate Kenneth Arrow, who, beginning in the 1940s, looked at the general properties of ways of ordering choices among public policies. His theory applies equally well to elections or to any issues of group choice. The results of his

Kenneth Arrow
(1921–   ).

work depend only on the general properties that are assumed and are independent of the actual voting scheme used to tally the results.

Arrow set out a list of four properties an ideal voting system should satisfy. His basic assumptions are that individual voters have a consistent order of preference for the choices and that ties in the rankings are permitted.

The first property listed by Arrow is called **universal domain,** and it stipulates that any ordering of the candidates is allowed. In other words, there are no restrictions placed on the rankings of the candidates a voter may choose.

The second property, called **Pareto optimality,** requires that if all the voters prefer candidate A to candidate B, then the group choice should not prefer B to A. This is a minimal condition for the choice to reflect the will of the constituents rather than be imposed by some outside entity.

Along the same lines, a third property is that no one individual voter's preferences totally determine the group choice. This condition is referred to as **nondictatorship.**

The fourth and final property in Arrow's system, called **independence from irrelevant alternatives**, requires that if a group of voters chooses candidate A over candidate B, then the addition or subtraction of other choices or candidates should not change the group choice to B. Of the four properties, this last one is by far the most debatable. In its favor, a choice between A and B should not depend on what other choices happen to be available—other choices that may be arbitrary or deliberately chosen to achieve a particular outcome. On the other hand, it is only by comparison with other possibilities that voters' perceptions of differences between candidates can be brought to light.

Arrow's acclaimed theorem is that no voting method based on rankings can satisfy these four properties. More precisely, it can be stated as follows.

Arrow's Impossibility Theorem

There is no voting method based on rankings that satisfies the properties of universal domain, Pareto optimality, nondictatorship, and independence from irrelevant alternatives.

So if we consider the four properties to be necessary for an ideal voting method, then no ideal voting method exists.

Arrow and others have made changes in the required properties and have also relaxed the condition that ballots be rankings. These four properties are only some of many possibilities. Another possible axiom is **anonymity,** which requires that each voter's ballot receives the same treatment or weight. Thus, interchanging the ballots of two or more voters would not change the outcome of the election. This egalitarian assumption seems a most reasonable one for a democracy. Another, **neutrality,** requires that each candidate receive equal treatment. Thus, interchanging all voters' rankings of two candidates would also exchange the election outcome for the two candidates. For example, favoritism for a particular spot on the ballot is not permitted. All the voting methods we have discussed are both anonymous and neutral.

We have also seen in the text and exercises that voters can sometimes obtain better results by strategic voting—in other words, by casting ballots that do not reflect their true rankings. In the early 1970s, Allan Gibbard and Mark Satterthwaite independently proved that in any voting system, other than a dictatorship, that is based on rankings, there will be instances where one or more voters may obtain a more desirable outcome through strategic voting. One way of proving their theorem is to show that a system in which no one can gain through strategic voting must satisfy all the properties of Arrow's theorem—conditions that cannot simultaneously hold. Determining a voter's best decisions, allowing for strategic voting, falls into the study of game theory, the subject of another chapter of this book.

Example 1   *Approval Voting and Pareto Optimality*

Explain why approval voting satisfies the Pareto optimality property.

Solution: If all voters prefer A to B, then any voter approving of B must also approve of A. Thus, A will get at least as many approval votes as B. (Note that A and B can actually tie in the approval vote. Some authors call this condition *weak Pareto optimality* and the case where A must actually be preferred to B is then called *strong Pareto optimality*.)

Example 2   *Plurality Method and Independence from Irrelevant Alternatives*

Give an example that shows how the plurality method can violate the independence from irrelevant alternatives property.

Solution: Suppose A defeats B in a plurality election by 4 to 3. Now suppose a new candidate C enters the picture and that two of A's supporters rank C first, whereas the other voters all rank C last. The new plurality vote would be B 3, A 2, and C 2, so the group of voters now favors B over A, and B wins the election.

The rankings of the seven voters are shown in Table 1.18. Note that it is consistent with the original 4 to 3 tally for A over B.

Table 1.18	**Preference Rankings for Plurality with C Added**		
	Number of Voters		
	2	*2*	*3*
A	2	1	2
B	3	2	1
C	1	3	3

Although the solution to Example 2 is short and easy to understand, it is deceptively simple because it may take quite a bit of effort to come up with such an example. Solving these kinds of problems usually requires a bit of trial-and-error, using what you learn from any unsuccessful tries. It often pays to look for examples with small numbers. In the case of searching for a solution to Example 2, we started with A barely defeating B so that it would not take much to reverse the result. In fact, if A defeats B by a single vote, A need only lose two votes to a third candidate C in order for B to get more votes than A. If A initially defeated B by 2 to 1, then C would end up winning the election. If A initially defeated B by 3 to 2, then B and C end up tied for the win. Therefore, we moved on to try the example of A defeating B by 4 to 3, which worked.

After learning of impossibility theorems such as those of Arrow and of Gibbard and Satterthwaite, we might be tempted to abandon the search for new and improved voting systems. However, a great deal of current research attempts to find voting methods that satisfy slightly weaker conditions or that satisfy a set of conditions a large proportion of the time.

In practice, the voting methods we have studied work well most of the time. In many elections, almost any voting method one might care to use will result in the same winner. Unfortunately, the plurality method, which is the one most commonly used in the United States, seems to result in controversial outcomes more often than other methods.

EXERCISES FOR SECTION 1.5

1. Explain why the plurality method satisfies the Pareto optimality property.

2. Explain why Borda's method satisfies the Pareto optimality property.

3. Give an example that shows how Borda's method can violate the independence from irrelevant alternatives property.

4. Give an example that shows how the plurality method with runoff between the top two finishers can violate the independence from irrelevant alternatives property.

5. Give an example that shows how the approval method can violate the independence from irrelevant alternatives property.

6. Can an anonymous system (with more than one voter) be a dictatorship? If so, give an example; if not, explain why not.

7. When either the U.S. House of Representatives or the Senate considers a bill, amendments are voted up or down one at a time and after consideration of all amendments, the final bill, amended or not, is voted on.

    (a) Explain why this method is anonymous.

    (b) Give an example to show that this method is not neutral. (Here, neutrality would mean that the final result does not depend on the order of consideration.)

    (c) If a particular bill must pass both House and Senate, show that when one takes into account all representatives and all senators, the process is not anonymous.

8. In approving presidential nominations, the Senate votes yes or no on a particular nomination, the decision being by majority. If the nominee is approved, the process ends. Otherwise, it moves on to a new nomination. In some sense, this is similar to an approval ballot if one assumes that a senator would vote for any candidate of whom he or she approves.

    (a) Is this method anonymous?

    (b) Is this method neutral? (Here, we assume that the order in which nominees are presented, until one is confirmed, is fixed.)

9. Suppose that a deliberative body (with more than one member) only votes on certain monetary matters for which the preferences are single-peaked and that the winning choice is the Condorcet winner. Clearly the property of universal domain is violated.

    (a) Explain why the system satisfies the Pareto optimality property.

    (b) Explain why the system satisfies the independence from irrelevant alternatives property.

Exercises 10–12 examine another desirable property of a voting method. A voting method or system is called *monotone* if a candidate never does worse when one or more voters raise the candidate in their rankings while all other candidates maintain the same positions relative to one another.

10. Explain why the plurality method is monotone.

11. Explain why Borda's method is monotone.

12. Give an example to show that the plurality method with a runoff between the top two finishers is not monotone.

WRITING EXERCISES

1. Suppose an election involves three candidates: one liberal, one moderate, and one conservative. Polls indicate that the liberal has 41% support, the moderate 20%, and the conservative 39%. Explain what you would expect to happen in an election decided by each of the voting methods we studied. Which method do you feel would yield the fairest result in this situation? Support your answer with an explanation.

2. A plurality election followed by a runoff can be conducted by having the voters fill out a preference ballot. However, runoff elections typically are held by having the voters fill out separate ballots for each stage of the process, and there may be several days or weeks between the separate ballots. Discuss the pros and cons of these two options.

3. By asking voters to support all the candidates acceptable to them, approval voting attempts to shift the voter's view from thinking of the glass as half empty to thinking of it positively as half full. Supporters of the approval voting method claim it will reduce negative campaigning, increase voter turnout, and increase the viability of minority candidates. In your view, how likely are these claims to be true?

4. Of the voting methods of plurality, plurality with a runoff between the top two candidates, Borda's method, or the approval method, which do you feel is the best method overall for an election with more than two candidates? Which method do you feel is the weakest? Support your answers with full explanations.

5. All voting methods based on preference rankings violate one of the four properties of Arrow's Impossibility Theorem. Which property would you drop as a requirement of a voting system and why?

6. On the front page of the August 16, 1995 *Los Angeles Times*, writer K. C. Cole wrote "Mathematicians do not agree on the best system. But they have no problem pointing their fingers at the worst: the plurality systems used in most U.S. elections." On page 316 of his book, *The Next American Nation*, author Michael Lind writes, "Our archaic first-past-the-post plurality electoral system preserves a two-party monopoly rejected by a growing number of alienated American voters, forces many Americans to waste

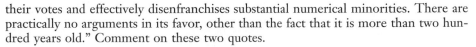

their votes and effectively disenfranchises substantial numerical minorities. There are practically no arguments in its favor, other than the fact that it is more than two hundred years old." Comment on these two quotes.

7. Suppose you asked a group of people to give their preference rankings on the ideal time to begin work, where the choices are 6, 7, 8, 9, or 10 A.M. Would you expect the set of preference rankings to be single-peaked? Support your answer with an explanation.

8. The procedure for determining the first-, second-, and third-place winners in the Van Cliburn International Piano Competition is described in Branching Out 1.4. Do you think the procedure is a good one for its purposes or do you think another procedure might be better? Support your answer with an explanation.

PROJECTS

1. When there are three candidates in an election, there is usually a Condorcet winner. With more candidates, the likelihood of a Condorcet winner decreases, but even with five candidates there is more likely than not a Condorcet winner.
   (a) Think of five issues with three "candidates" for which elections could be held in which you might *not* expect a Condorcet winner. Ask 15 people to fill out preference ballots for each of the five elections to see how often you get a Condorcet winner.
   (b) Repeat the process in part (a) for issues with four candidates.
   (c) Repeat the process in part (a) for issues with five candidates.
   (d) Are the results as you expected? Based on your limited results, what effect did increasing the number of candidates seem to have on the likelihood of a Condorcet winner?

2. Different voting methods sometimes yield the same winner and sometimes yield different winners. To study how they differ in practice, think of 10 issues with four "candidates" for which elections could be held. Ask 15 people to fill out preference ballots and indicate how they would vote on an approval ballot for each of the 10 elections. For each election, find
   (a) the winner of a plurality vote,
   (b) the winner of a plurality vote with runoff between the top two finishers,
   (c) the Borda winner,
   (d) the Condorcet winner, if there is one,
   (e) the winner of an approval vote.
   Discuss the results of the different voting methods for your elections.

3. In 1861, Thomas Hare introduced a voting method now known as the Hare method or the single transferable vote method. This method is simply a plurality with runoff method in which in each stage the candidate (or candidates, if a tie) with the fewest votes is eliminated, and a runoff is held between the remaining candidates.
   (a) Construct examples of elections showing that the Hare method does not satisfy independence from irrelevant alternatives, is not monotone, and may not select the Condorcet winner, if there is one. (For the definition of monotone, see Exercises 10–12 in Section 1.5.)
   (b) Find some examples of elections in which the Hare method is actually used.

4. The 1968 election was the last in which a third-party candidate actually carried a state. Running against Republican Richard Nixon and Democrat Hubert Humphrey, George Wallace, running under the American Independent party label after seeking the Democratic nomination, won five southern states. After dropping the one-third of one percent who voted for some other candidate, the actual vote is given in the first col-

umn of the table. A Gallup poll estimated the breakdown by political party as shown in the final three columns of the table.

	Percentage of All Votes	*Percentage of Republican Votes*	*Percentage of Democratic Votes*	*Percentage of Independent Votes*
Nixon	43.5	86	12	44
Humphrey	42.9	9	74	31
Wallace	13.6	5	14	25

Source: Gallup Opinion Index.[4]

Although Nixon won handily in the Electoral College, the popular vote was razor close.

(a) To analyze the election results, we first find the percentage of voters who are Republican, Democrat, and Independent. Letting R, D, and I denote the percentage of Republican, Democratic, and Independent voters, respectively, we can estimate these percentages by solving the set of equations

$$0.86R + 0.12D + 0.44I \approx 43.5,$$
$$0.09R + 0.74D + 0.31I \approx 42.9,$$
$$0.05R + 0.14D + 0.25I \approx 13.6,$$
$$R + D + I = 100.$$

Explain why R, D, and I should satisfy these four equations.

(b) Show that an approximate solution to the set of equations in part (a) is given by $R \approx 32.6\%$, $D \approx 44.4\%$, and $I \approx 23.0\%$.

(c) Suppose that within each of the three classes of voters, the second choices are proportional to the percentages supporting the other two candidates. For instance, of the Republicans voting for Nixon, $9/(9 + 5)$ would have Humphrey as their second choice. Create a table showing the percentage of voters having each possible preference ranking. Estimate who would have won the popular vote in a runoff between Nixon and Humphrey, and estimate who would have won the popular vote in an election between Nixon and Wallace.

5. Create a spreadsheet or computer program that will calculate the results of an election for one or more of the methods we have studied.

6. In 1956, the House of Representatives considered allocating funds to the states for school construction. New York Democrat Adam Clayton Powell introduced an amendment to withhold funds from any state not complying with Supreme Court decisions, which was particularly aimed at southern states maintaining segregated schools. The Powell amendment passed; the amended bill was defeated. The votes of those representatives voting on both the amendment and the funding bill broke down as follows, with the first vote on the amendment (Y for yes, N for no).

	YY	YN	NY	NN
Republicans	52	96	23	22
Southern Democrats	0	0	3	101
"Northern" Democrats	77	0	39	3

Source: Congressional Record.[5]

There are six possible rankings of the alternatives of the original allocation, the amended bill, and no bill.

(a) Which rankings are possible for those who voted yes on both including the amendment and the amended bill?

(b) Which rankings are possible for those who voted yes on including the amendment and no on the amended bill?

(c) Which rankings are possible for those who voted no on including the amendment and yes on the amended bill?

(d) Which rankings are possible for those who voted no on both including the amendment and the amended bill?

It seems reasonable to assume that no representative would vote for the bill only with the Powell amendment. Further assume, perhaps less reasonably, that when preferences between the original bill and no bill remain unclear, Democrats favor the bill and Republicans oppose it.

(e) Which of the three alternatives wins a plurality?

(f) Which choice would win a runoff election between the top two alternatives?

(g) Which option is the Borda winner?

(h) Is there a Condorcet winner?

(i) Could any group with a given ranking have achieved a preferable result by voting strategically if the others voted as shown in the table?

It was suggested by President Truman and others in the Democratic leadership that the Republicans who voted first for the Powell amendment and then against the entire bill were voting strategically on the amendment in order to defeat the bill and also to embarrass the Democrats. The facts are ambiguous and political scientists do not agree whether there was significant strategic voting. Assume that Republicans voting strategically in this way knew the Powell amendment could not pass on the final vote, and keep the assumptions that precede parts (e)–(i).

(j) What rankings are possible for such a Republican who voted strategically?

(k) For which of these rankings would a strategic vote make sense?

(l) What is the least number of strategic voters that would have had to vote according to their true preferences for the original bill to pass?

7. The Democrats in the 1976 U.S. House of Representatives elected their majority leader by an elimination ballot, where the candidate receiving the least votes in any round of balloting was eliminated, and voting continued until one candidate received a majority of the votes. The results were as follows:

	Ballot 1	*Ballot 2*	*Ballot 3*
James Wright, Jr.	77	95	148
Philip Burton	106	107	147
Richard Bolling	81	93	—
John J. McFall	31	—	—

Source: Samuel Merrill, *Making Multicandidate Elections More Democratic.*[6]

Construct or explain why it is impossible to construct orderings of the candidates, consistent with the data, so that the Condorcet winner

(a) is Wright,

(b) is Burton,

(c) is Bolling,

(d) is McFall,

(e) does not exist.

Many political observers ordered the candidates from liberal to conservative as Burton, Bolling, McFall, Wright.

(f) What is the largest number of voters whose preferences could be single-peaked with respect to this ordering of the candidates?

(g) Can you determine the Condorcet winner assuming single-peaked preferences to the greatest possible extent?

(h) Can you determine the Borda winner assuming single-peaked preferences to the greatest possible extent?

KEY TERMS

anonymity, 59
approval voting method, 45
Arrow's Impossibility Theorem, 59
Borda count, 20
Borda's method, 20
Condorcet completion process, 39
Condorcet winner, 34

head-to-head comparison, 32
independence from irrelevant
   alternatives, 59
neutrality, 59
nondictatorship, 59
Pareto optimality, 59
plurality method, 4

preference ranking, 11
runoff election, 8
single-peaked preference
   rankings, 38
strategic voting, 8
universal domain, 59
weak Condorcet winner, 34

CHAPTER 1 REVIEW EXERCISES

1. If 454 votes are cast, what is the smallest number of votes a winning candidate can have in a six-candidate race that is to be decided by plurality?

2. Suppose there are 140 votes cast in an election among five candidates—Stein, O'Rourke, Cohen, Holt, and Massey—to be decided by plurality. After the first 100 votes are counted, the tallies are as follows:

Stein	12
O'Rourke	23
Cohen	17
Holt	29
Massey	19

(a) What is the minimal number of remaining votes Holt needs to be assured of a win?

(b) What is the minimal number of remaining votes Cohen needs to be assured of a win?

3. The preference rankings of a class of 24 college students for their favorite fruit are as follows:

	Number of Voters											
	2	*2*	*1*	*3*	*2*	*1*	*3*	*2*	*3*	*1*	*2*	*1*	*1*
Apples	1	1	2	3	4	3	4	2	3	4	3	4	2
Bananas	2	3	1	1	1	1	1	4	2	2	4	3	3
Grapes	3	2	3	2	2	4	3	1	1	1	1	1	4
Oranges	4	4	4	4	3	2	2	3	4	3	2	2	1

(a) Which fruit would win a plurality election?

(b) Which fruit would win a plurality election with a runoff between the top two?

(c) Which fruit has the top Borda count?

(d) Which fruit, if any, would be the Condorcet winner?

4. Suppose that a small town must decide whether to build tennis courts, a basketball court, or a baseball field. The residents of the town are polled and their preference rankings are as follows:

Percentage of Voters						
	12	20	11	24	10	23
Tennis courts	1	1	2	3	2	3
Basketball court	2	3	1	1	3	2
Baseball field	3	2	3	2	1	1

(a) Which choice would win a plurality election?

(b) Which choice would win a plurality election with a runoff between the top two?

(c) Which choice has the top Borda count?

(d) Which choice, if any, would be the Condorcet winner?

5. A campus programming committee must decide what kind of act to book for its next engagement. The choices are a comedian, a jazz trio, a pianist, a rock band, and a classical guitarist. The committee members decide to make the decision through an approval election and the resulting ballots are as follows:

Number of Voters								
	1	3	2	1	1	1	2	1	1
Comedian	✓		✓				✓	✓	
Jazz trio			✓	✓	✓		✓		✓
Pianist				✓		✓			✓
Rock band					✓		✓	✓	✓
Classical guitarist		✓				✓		✓	✓

Which act wins the vote?

6. The members of a community theater organization must vote to decide which play they would like to put on. The preference rankings of the members are as follows:

Number of Voters								
	2	1	4	1	1	3	1	2
The Fantasticks	1	1	2	3	4	2	3	4
Romeo and Juliet	3	4	1	1	1	4	2	3
Our Town	4	3	3	4	2	1	1	2
Death of a Salesman	2	2	4	2	3	3	4	1

(a) Which play would win a plurality vote?

(b) Which play would win a plurality vote with a runoff between the top two finishers?

(c) Which play would win under Borda's method?

(d) Which play, if any, is the Condorcet winner?

(e) Could the three voters who ranked *Our Town* first and *The Fantasticks* second achieve a preferable outcome in an election decided by Borda's method by voting strategically if the others voted as shown in the table?

(f) Could the three voters who ranked *Our Town* first and *The Fantasticks* second achieve a preferable outcome in an election decided by the plurality method with a runoff between the top two by voting strategically if the others voted as shown in the table?

7. The eight members of a board of directors of an art museum must select a new museum head. The preference rankings of the board members and the candidates of whom they approve are included in the table.

	Number of Voters					
	1	*1*	*1*	*2*	*1*	*2*
Perella	1 ✓	1 ✓	2 ✓	3	3	3
Mintz	3	3	1 ✓	1 ✓	2	2 ✓
Zukoff	2	2 ✓	3	2	1 ✓	1 ✓

   (a) Which candidate would win a plurality election?
   (b) Which candidate would win a plurality election with a runoff between the top two finishers?
   (c) Which candidate would win the Borda count?
   (d) Which candidate, if any, is the Condorcet winner?
   (e) Which candidate would win an approval vote?
   (f) Could the two voters who ranked Mintz first and Zukoff second achieve a preferable outcome in an election decided by Borda's method by voting strategically if the others voted as shown in the table?
   (g) Could the two voters who ranked Zukoff first and Mintz second and approved of two candidates achieve a preferable outcome in an approval election by voting strategically if the others voted as shown in the table?

8. Explain why the plurality method with a runoff between the top two finishers satisfies the Pareto optimality property.

9. Construct an example of preference rankings for an election with four candidates—A, B, C, and D—so that, in two-person races, A would defeat B, B would defeat C, C would defeat D, and D would defeat A.

SUGGESTED READINGS

Brams, Steven J. "Comparison Voting." In *Political and Related Models*, vol. 2, edited by Steven J. Brams, William F. Lucas, and Philip D. Straffin, Jr. New York: Springer-Verlag, 1983, 32–65. Negative voting, leading naturally to approval voting.

Brams, Steven J. *Game Theory and Politics*. New York: The Free Press, 1975. Strategic analysis, some mathematical, some not, mainly in the context of elections. Voting power. Numerous examples.

Brams, Steven J., and Peter C. Fishburn. *Approval Voting*. Boston: Birkhäuser, 1983. Discussion and analysis supporting approval voting. Final chapter analyzes the 1980 presidential election using linear algebra.

Fishburn, Peter C., and Steven J. Brams. "Paradoxes of Preferential Voting." *Mathematics Magazine* 56:4 (September 1983), 207–214. Examples, engagingly written.

Lind, Michael. *The Next American Nation: The New Nationalism and the Fourth American Revolution*. New York: The Free Press, 1995, 311–319. An argument in support of proportional representation, especially for the U.S. Senate.

Neufeld, John L., William J. Hausman, and Ronald B. Rapoport. "A Paradox of Voting: Cyclical Majorities and the Case of Muscle Shoals." *Political Research Quarterly* 47:2 (June 1994), 423–438. Detailed examination of the setting leading up to Exercise 15 of Section 1.3.

Niemi, Richard G., and William H. Riker. "The Choice of Voting Systems." *Scientific American* 234:6 (June 1976), 21–27. Voting methods, single-peaked preferences, and strategic voting.

Riker, William H. *The Art of Political Manipulation.* New Haven, Conn.: Yale University Press, 1986. Fascinating tales of how defining the issue and setting the agenda lead to favorable results in subsequent votes.

Riker, William H. *Liberalism Against Populism.* Prospect Heights, Ill.: Waveland Press, Inc., 1982. Voting methods and properties of voting systems, intermixed with lots of politics and examples.

Romer, Thomas, and Howard Rosenthal. "Voting Models and Empirical Evidence." *American Scientist* 72:5 (September–October 1984), 465–473. Geometric viewpoint of preferences, agenda control, and data on Congress.

Saari, Donald G. "Are Individual Rights Possible?" *Mathematics Magazine* 70:2 (April 1997) 83–92. Begins with a theorem of Sen pertaining to the title, which is then connected to Arrow's Impossibility Theorem.

Steen, Lynn A. "From Counting Votes to Making Votes Count: The Mathematics of Elections." Martin Gardner's column in *Scientific American* 243:4 (October 1980), 16–26B. Preferences in single-issue elections, voting methods, and voting power.

Straffin, Philip D., Jr. "The Power of Voting Blocs: An Example." *Mathematics Magazine* 50:1 (January 1977), 22–24. A nice example analyzing whether a group of delegates should organize as a bloc.

Taylor, Alan D. *Mathematics and Politics: Strategy, Voting, Power and Proof.* New York: Springer-Verlag, 1995. Includes chapters on measurement of political power and on properties of voting procedures that include plurality and Borda methods.

ISBN 0-471-13935-

*E*ven as a child, you probably faced problems of achieving a fair division, as in the case of dividing seven sea shells among three children. Governments are constantly faced with larger issues of fair allocation or apportionment. An apportionment is a division and allotment of resources among various parties. A state's representation in the U.S. House is apportioned according to its population. The teachers in a school district are assigned to schools based in large part on the number of students in the school. The common feature in all of these examples is that, unlike dividing land or slicing cake, the items to be apportioned may not be divided arbitrarily. A state may have 9 representatives or 10 representatives, but cannot have 9.37 representatives. In this chapter we investigate some of the methods used for such apportionment problems. The methods we discuss have been discovered and rediscovered by various individuals and groups over hundreds of years. We will follow the common practice in the United States of naming them after their major proponent in U.S. political history. All the methods we include have been debated in the U.S. Congress over the past 200 years.

The floor of the U.S. House of Representatives.

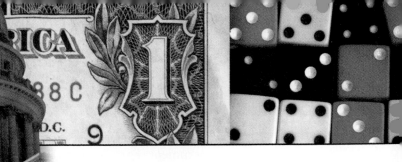

CHAPTER 2

APPORTIONMENT:
Sharing What Cannot Be Divided Arbitrarily

Although we will look at many apportionment problems that do not involve government representation, it is easiest to use the same terminology as we would use in the context of government. Therefore, we call the items to be apportioned **seats**. The total number of seats is called the **house size**. The various parties to whom we will assign the items are called **states**. A state's **population** is some measurement of its size on which the assignment of seats is based. The sum of all the states' populations is called the **total population**.

In the case of the U.S. House of Representatives, the house size is now 435. These 435 seats are apportioned to the 50 states based on their populations. According to the 1990 census, the total U.S. population was estimated to be 248,709,873. To divide the representatives evenly, it should make sense that one representative should go to each

$$\frac{248{,}709{,}873}{435} \approx 571{,}746.8345$$

people. The number 571,746.8345 is called the **natural divisor** for the apportionment. It is the number of people in the United States per seat in the House of Representatives. In the more general context, we have the following definition:

$$\text{natural divisor} = \frac{\text{total population}}{\text{house size}}.$$

Now suppose we want to choose the number of representatives to assign to Pennsylvania, which had an estimated population of 11,881,643 in 1990. Because there are approximately 571,746.8345 people per seat in the House, Pennsylvania should be entitled to

$$\frac{11{,}881{,}643}{571{,}746.8345} \approx 20.7813$$

seats. This number is called its **natural quota**. In the general context, we define a state's natural quota as follows:

$$\text{state's natural quota} = \frac{\text{state's population}}{\text{natural divisor}}.$$

Because Pennsylvania's natural quota is 20.7813, the most natural number of representatives to allot to Pennsylvania is 21, which happens to be what it received.

Throughout this chapter, we round divisors and quotas to four decimal places before recording them. This will provide a degree of accuracy that will suffice for our purposes.

In any apportionment problem, the sum of the *unrounded* natural quotas is equal to the house size. Apportioning by rounding each state's natural quota to the nearest whole number seems logical, and sometimes it works, as we see in the following example.

Example 1 *An Idea for Apportioning*

Suppose that Canada, Mexico, and the United States, parties to the North American Free Trade Agreement (NAFTA), decide to convene 100 business leaders for a conference to make recommendations for future treaties. The conference organizers decide that the number of leaders from each country should be proportional to its population. According to U.S. Bureau of the Census estimates for 1996, Canada's population was about 30 million, Mexico's population was about 95 million, and the population of the United States was about 265 million. Find the natural quota for each country, and apportion the 100 leaders by rounding each country's natural quota to the nearest whole number.

Solution: The total population of the three countries is 30 + 95 + 265 = 390 million people. The natural divisor for this apportionment is 390/100 = 3.9. Canada's natural quota is $30/3.9 \approx 7.6923$ leaders. Similarly, Mexico's natural quota is $95/3.9 \approx 24.3590$ and the natural quota for the United States is $265/3.9 \approx 67.9487$.

Table 2.1	Quotas for NAFTA Countries		
Country	Population (in millions)	Natural Quota	Rounded Natural Quota
Canada	30	7.6923	8
Mexico	95	24.3590	24
United States	265	67.9487	68
Total	390		100

Source: U.S. Bureau of the Census.[1]

If we round Canada's natural quota to 8, Mexico's to 24, and the United States' to 68, as shown in Table 2.1, we get a total of 100 business leaders. So this scheme gives us the right number of leaders. ∎

The ease of our solution in Example 1 may deceive you into thinking that apportionment is an easy problem. However, it was just fortunate that things worked out in this example. In the following example, we see how the method of rounding the natural quotas to the nearest whole number can fail.

Example 2 *Why Simple Rounding Fails for Apportioning*

Suppose that, instead of 100 leaders, the conference planners in Example 1 want a much larger conference with 800 business leaders. Find the natural quota for each country, and try to apportion the 800 leaders by rounding each country's natural quota to the nearest whole number. What happens in this case?

Solution: The natural divisor for the new apportionment is 390/800 = 0.4875.

Canada's natural quota is now $30/0.4875 \approx 61.5385$ leaders. We next compute the natural quotas for Mexico and the United States and record them in Table 2.2.

Table 2.2 **Quotas for NAFTA Countries**			
Country	**Population (in millions)**	**Natural Quota**	**Rounded Natural Quota**
Canada	30	61.5385	62
Mexico	95	194.8718	195
United States	265	543.5897	544
Total	390		801

If we round this time, all three natural quotas are rounded up and we end up allocating too many representatives: $62 + 195 + 544 = 801$. This won't do! ∎

Example 2 shows us that the problem of apportionment is more complex than it might first appear. The sum of the unrounded natural quotas is equal to the house size. However, when we round each country's natural quota to the nearest whole number, the sum of these rounded quotas may differ from the house size. In this section, we will see that one approach to apportionment is to follow some rule for rounding quotas that ensures that the sum of the rounded quotas is equal to the house size.

The problem of apportioning the U.S. House of Representatives is as old as the Constitution, which states:

Representatives and direct Taxes shall be apportioned among the several States which may be included within this Union, according to their respective Numbers, which shall be determined by adding to the whole Number of free Persons, including those bound to Service for a Term of Years, and

The signing of the Constitution.

excluding Indians not taxed, three fifths of all other Persons. The actual
Enumeration shall be made within three Years after the first Meeting of the
Congress of the United States, and within every subsequent Term of ten
Years, in such Manner as they shall by Law direct. The Number of Representatives shall not exceed one for every thirty Thousand, but each State
shall have at Least one Representative; and until such enumeration shall be
made, the State of New Hampshire shall be entitled to chuse three, Massachusetts eight, Rhode-Island and Providence Plantations one, Connecticut
five, New-York six, New Jersey four, Pennsylvania eight, Delaware one,
Maryland six, Virginia ten, North Carolina five, South Carolina five, and
Georgia three.

Thus, every 10 years with each census, there is to be a new apportionment
of representatives. The Constitution does not specify the size of the House or a
precise meaning of "according to their respective Numbers." Such a vague description allows for any method that is politically acceptable, as long as it is sufficiently impartial and mathematically sound to pass judicial scrutiny. As we will
see in the case of the U.S. House of Representatives, it is often the politics that
dominates the mathematics.

Hamilton's Method

In 1791, Alexander Hamilton, then Secretary of the Treasury, supported the
method of apportionment that we will refer to as **Hamilton's method**. It is sometimes called the *method of largest fractional parts*. The **fractional part** of a number is the decimal part. For instance, the fractional part of Canada's natural quota
of 61.5385 in Example 2 is 0.5385. The basic steps of Hamilton's method are
listed next.

Hamilton's Method

- Calculate the natural quota for each state.
- Round each state's natural quota *down* to the nearest whole number that
  does not exceed the quota, and allocate each state this number of seats initially.
- If there are seats left to be assigned, assign them to the states whose natural quotas have the largest fractional parts until all the seats have been
  allocated.

Alexander Hamilton
(1755–1804).

The next two examples illustrate Hamilton's method.

Example 3 *Hamilton's Method*

A city has 21 fire trucks that are to be assigned to one of five fire stations. The
new fire chief decides to allocate the trucks to stations in proportion to the number of fires reported in each station's district over the last three months, given in
Table 2.3. Apportion the fire trucks using Hamilton's method.

Table 2.3	Fires by District
Station	**Number of Fires**
1	154
2	83
3	129
4	217
5	103

Solution: The total population in this case is the sum of the numbers of fires, so it is given by

$$154 + 83 + 129 + 217 + 103 = 686.$$

The natural divisor for the 21 fire trucks is $686/21 \approx 32.6667$. The natural quota for Station 1 is $154/32.6667 \approx 4.7143$. The natural quotas for the remaining stations are computed in a similar way, and all of them are listed in Table 2.4. The initial allocation is found by rounding each natural quota down to the nearest whole number.

Table 2.4	Apportionment of Fire Stations			
Station	**Number of Fires**	**Natural Quota**	**Initial Allocation**	**Final Allocation**
1	154	4.7143	4	5
2	83	2.5408	2	2
3	129	3.9490	3	4
4	217	6.6429	6	7
5	103	3.1531	3	3
Total	686		18	21

In the initial allocation, only 18 fire trucks are assigned, so we have 3 more to allocate. The first extra truck should go to Station 3 because the fractional part of its natural quota, 0.9490, is the largest of the five. The second extra truck should go to Station 1 because it has the second largest fractional part, and the third truck should go to Station 4 because its fractional part is the next largest. The final allocation of trucks is shown in Table 2.4.   ∎

Notice that the natural quotas of all the states that gained seats in Example 3 were rounded up "naturally." In other words, all the quotas that were increased had fractional part above 0.5. However, Station 2 also had a fractional part exceeding 0.5, yet did not gain an extra seat in the final allocation. In the next example, we will see that sometimes states with fractional parts less than 0.5 gain a seat.

Technology Tip

Quotas and apportionments are best computed using a spreadsheet, but even using a calculator with list capabilities makes computations much easier.

Example 4   *Hamilton's Method*

In 1790, the time of Hamilton, Delaware was composed of three counties, as it is today. The populations of these counties based on the 1790 census are listed in Table 2.5. Use Hamilton's method to apportion a hypothetical state legislature of 76 seats among these counties.

Table 2.5	Population of Delaware in 1790
County	**Population**
Kent	18,920
New Castle	19,686
Sussex	20,488

Source: U.S. Census Office.[2]

Solution: The total population of the three counties is

$$18,920 + 19,686 + 20,488 = 59,094.$$

The natural divisor for 76 seats is $59,094/76 \approx 777.5526$. Dividing each county's population by this divisor, we obtain the natural quotas and initial allocation given in Table 2.6.

Table 2.6	Apportionment for Delaware by Hamilton's Method			
County	**Population**	**Natural Quota**	**Initial Allocation**	**Final Allocation**
Kent	18,920	24.3328	24	24
New Castle	19,686	25.3179	25	25
Sussex	20,488	26.3493	26	27
Total	59,094		75	76

In the initial allocation, we are only one seat short of the 76 we need. Notice that all three counties have natural quotas with fractional parts about equal to 1/3, and, if we rounded each to the nearest whole number, all these quotas would round down. However, because we need to allocate 76 seats, one of the quotas will need to be rounded up. Because the natural quota of Sussex County has the largest fractional part, 0.3493, it will be assigned the extra seat. The resulting final allocation is shown in Table 2.6. However, notice that Sussex County received 27 seats even though its natural quota of 26.3493 would round naturally to 26.

■

The proposed bill to use Hamilton's method for apportioning the U.S. House was passed in both the House and Senate in 1791. However, President George Washington vetoed it, the first-ever presidential veto. A method proposed by Secretary of State Thomas Jefferson was adopted instead, and is among those we examine in the next section. Reinvented by Ohio Congressman Samuel Vinton, Hamilton's method was adopted as the official method of apportionment from 1852 until 1900. However, the censuses of 1880 and 1900 revealed some of the problems with Hamilton's method, and it has not been used to apportion the House since. We will look at some of these problems in Section 2.4.

Lowndes' Method

In 1822, Congress discussed, but never adopted, a method similar to Hamilton's method proposed by South Carolina Congressman William Lowndes (1782–1822) shortly before he died at sea. This method is referred to as **Lowndes' method**. With Lowndes' method, the initial allocation of seats is made just as with Hamilton's method by giving each state a number of seats equal to its natural quota rounded down to the nearest whole number. Any remaining seats are then assigned to the states whose initial allocations of seats are the largest percentage short of their natural quotas. To compute this percentage, we would divide the fractional part of the natural quota by the natural quota. However, a slightly simpler calculation that leads to the same apportionment is to compute the **relative fractional part**, which is the fractional part of the natural quota divided by its integral part. For instance, if a state has a natural quota of 14.3734, then the relative fractional part of this quota is given by

$$\frac{0.3734}{14} \approx 0.02667.$$

(Because it is common for 0 to be the first digit of the relative fractional part, we will round the relative fractional parts to 5 decimal places rather than our usual 4 decimal places.) With Lowndes' method, any seats remaining after the initial allocation are assigned to those states with the largest relative fractional parts. Because we divide by a smaller number in computing the relative fractional part of the quota for smaller states, we see that it is possible for a small state whose fractional part is less than that of a large state to have a greater relative fractional part. Thus, in comparing it with Hamilton's method, we see that Lowndes' method favors smaller states. The basic steps of Lowndes' method are summarized next.

Lowndes' Method

- Calculate the natural quota for each state.

- Round each state's natural quota *down* to the largest whole number that does not exceed the quota, and allocate each state this number of seats initially.

- If there are seats left to be assigned, assign them to the states with the largest *relative* fractional parts until all the seats have been allocated. (The relative fractional part is given by the fractional part of the natural quota divided by its integral part.)

One special case arises in Lowndes' method. If a state's initial allocation is 0, then the relative fractional part is undefined because we must divide by 0; essentially it is an infinite relative fraction. Such a state will have the highest priority in receiving an additional seat. If there are more such states than unallocated seats, we allocate the extra seats to those states with the largest populations.

Lowndes' method is sometimes called the *method of largest relative fractions*. We now use Lowndes' method to apportion the same state legislature we apportioned using Hamilton's method in the previous example.

Example 5 *Lowndes' Method*

Recall from Example 4 that the 1790 populations of the three counties of Delaware were Kent County, 18,920; New Castle County, 19,686; and Sussex County, 20,488. Apportion a hypothetical state legislature of 76 seats among these counties using Lowndes' method.

Solution: After computing the natural quotas and initial allocation as we did using Hamilton's method, we see that we have one more seat to assign. We now compute the relative fractional parts for each natural quota. For Kent County, the relative fractional part is $0.3328/24 \approx 0.01387$. We compute the relative fractional parts of New Castle and Sussex similarly, as listed in Table 2.7.

Table 2.7	**Apportionment for Delaware by Lowndes' Method**				
County	*Population*	*Natural Quota*	*Initial Allocation*	*Relative Fractional Part*	*Final Allocation*
Kent	18,920	24.3328	24	0.01387	25
New Castle	19,686	25.3179	25	0.01272	25
Sussex	20,488	26.3493	26	0.01343	26
Total	59,094		75		76

Note that, as we expected, the advantage shifts toward smaller counties. Because Kent County has the largest relative fractional part, it gets the single remaining seat. If a second seat were available, it would go to Sussex County because it has the second largest relative fractional part. The final apportionment using Lowndes' method is listed in Table 2.7. Notice that it is not the same apportionment we obtained in Example 4 using Hamilton's method, where Sussex County got the remaining seat, though sometimes the methods will yield the same apportionment.

■

Sometimes an apportionment must fulfill a special rule that each state must receive at least one seat, as is the case in the U.S. House of Representatives. If the number of states whose natural quotas are less than one is at most the number of seats left to be allocated, we can simply round up the quotas of those states whose natural quotas are less than one first, and then go on to apply Hamilton's or Lowndes' method in the normal manner to assign any remaining seats. We will not consider the somewhat unusual case in which the number of states whose natural quotas are less than one is more than the number of seats left to be allocated, because the different options for resolving the situation would take us too far afield.

This section was called Quota Methods because the two methods we examined satisfy the Quota Property, which can be stated as follows.

> ### *The Quota Property*
>
> An apportionment is said to satisfy the **Quota Property** if each state's allocation is less than one from its natural quota. In other words, the final allocation for each state is equal to its natural quota, rounded either up or down to a whole number. An apportionment method is said to satisfy the Quota Property if *every* apportionment produced by the method satisfies the Quota Property.

Although satisfying the Quota Property is a desirable property for an apportionment method, we will see later that it is not satisfied by all methods. We will look at this problem and other drawbacks of various apportionment methods in Section 2.4.

For simplicity, in this section we have completely avoided any mention of the cases where a tie occurs. We could, of course, build in a tie-breaking procedure such as giving the extra representative to either the bigger or the smaller state, leaving only the case of equal populations. However, because ties are exceedingly rare, we will ignore the possibility of ties throughout our discussion of apportionment methods.

Exercises for Section 2.1

1. Suppose that the five Scandinavian countries want to form an economic council with 45 members. Apportion the council based on the 1994 estimates of their populations given here, using

   (a) Hamilton's method,   (b) Lowndes' method.

Country	Population (in thousands)
Denmark	5,188
Finland	5,069
Iceland	264
Norway	4,315
Sweden	8,778

   Source: U.S. Dept. of Commerce.[3]

2. The first three-way merger in the utility industry occurred in 1995 when IES Industries, Inc., Interstate Power Company, and WPL Holdings, Inc. combined to form the Interstate Energy Corporation, with customers in Iowa, Illinois, Minnesota, and Wisconsin. Apportion 15 directors based on the percentage of stock in the new company held by each company's shareholders, using

   (a) Hamilton's method,   (b) Lowndes' method.

   (c) Do either of these methods produce the actual apportionment of WPL Holdings, Inc. 6, IES Industries, Inc. 6, and Interstate Power Company 3?

Company	Percentage of Shares
WPL Holdings, Inc.	43.9
IES Industries, Inc.	40.9
Interstate Power Company	15.2

   Source: WPL Holdings, Inc.[4]

3. Four software companies decide to merge. They want to apportion the new 26-member board of directors in proportion to the net worth of each company.

  (a) Use Hamilton's method to determine how many members will come from each company.

  (b) Use Lowndes' method to determine how many members will come from each company.

  (c) When switching from Hamilton's method to Lowndes' method, does the shift in seats go from larger companies to smaller companies, or vice versa?

Company	Net Worth (in thousands)
Alpha Software	8310
Beta Technology	1073
Gamma Computing	6757
Delta Development	4541

4. Texas Christian University has an 18-member faculty council. Apportion the representatives to this council based on the number of faculty in each college of the university using

  (a) Hamilton's method, **(b)** Lowndes' method.

College	Number of Faculty Members
Arts & Sciences	186
Fine Arts & Communications	85
Business	39
Education	29
Nursing	25

5. Suppose that a town must apportion its polling places among six districts. Apportion 50 polling places to the districts based on the number of eligible voters in each district using

  (a) Hamilton's method, **(b)** Lowndes' method.

District	1	2	3	4	5	6
Number of Eligible Voters	7478	9003	5397	8825	3562	5936

6. The populations of the seven counties in Vermont in 1790, based on the 1790 census, are listed in the table. Apportion the 150-member House of Representatives among these counties using

  (a) Hamilton's method, **(b)** Lowndes' method.

  (c) When switching from Hamilton's method to Lowndes' method, does the shift in seats go from larger counties to smaller counties, or vice versa?

County	Addison	Bennington	Chittendon	Orange	Rutland	Windham	Windsor
Population	6,449	12,254	7,301	10,529	15,565	17,693	15,748

Source: U.S. Census Office.[5]

7. The populations of the five counties in New Hampshire in 1790, based on the 1790 census, are listed in the table. Apportion a hypothetical state legislature of 28 seats among these counties using

  (a) Hamilton's method, **(b)** Lowndes' method.

(c) What special feature of the fractional parts of the natural quotas would allow you to apportion according to Lowndes' method without actually computing the relative fractional parts explicitly?

County	Cheshire	Grafton	Hillsborough	Rockingham	Strafford
Population	28,772	13,472	32,871	43,169	23,601

Source: U.S. Census Office.[6]

8. Election to Austria's National Council is by parties on a proportional basis, with the number of seats each party wins determined by the percentage of votes the party received. Apportion the National Council of 183 seats among the five major political parties using the percentage of the vote each received in the 1995 election results shown in the table as the parties' populations. Use

(a) Hamilton's method,   **(b)** Lowndes' method.

(c) Does either apportionment agree with the actual apportionment of Social-Democratic 71, People's Party 52, Freedom Thinkers 41, Liberal Forum 10, and Green 9?

Party	*Percentage of Votes*
Social-Democratic	38.1
People's Party	28.3
Freedom Thinkers	21.9
Liberal Forum	5.5
Green	4.8

Source: Wilfried P. C. G. Derksen Web site and the Austrian Parliament Web site.[7]

9. Election to Israel's Knesset is by parties on a proportional basis, with the number of seats each party wins determined by the percentage of votes the party received. Apportion the Knesset's 120 seats among the various major political parties using the percentage of the votes each received in the 1992 election results shown in the table as the parties' populations. Use

(a) Hamilton's method,   **(b)** Lowndes' method.

(c) Does either apportionment agree with the actual apportionment of Labor 44, Likud 32, Energy 12, Zionist 8, Sephardic Jews 6, National Religious 6, United Torah 4, Democratic Front 3, Fatherland 3, and Arab Democratic 2?

Party	*Percentage of Votes*
Labor	34.8
Likud	24.9
Energy	9.2
Zionist	5.7
Sephardic Jews	5.1
National Religious	5.0
United Torah	3.4
Democratic Front (communist)	2.5
Fatherland	2.3
Arab Democratic	1.6

Source: Wilfried P. C. G. Derksen Web site.[8]

10. Apportion the 105 seats in the 1792 U.S. House of Representatives based on the 1790 census figures that follow, using

(a) Hamilton's method, (b) Lowndes' method.

State	Population	State	Population
Connecticut	237,946	New Jersey	184,139
Delaware	59,094	New York	340,120
Georgia	82,538	North Carolina	393,751
Kentucky	73,677	Pennsylvania	434,373
Maine	96,540	Rhode Island	68,825
Maryland	319,728	South Carolina	249,073
Massachusetts	378,787	Vermont	85,539
New Hampshire	141,885	Virginia	747,610

Source: U.S. Census Office.[9]

11. Show algebraically that the natural quota can also be computed by taking the state's fraction of the total population multiplied by the number of seats to be apportioned.

12. (a) Give an example with two states where the apportionments obtained by Hamilton's method and by Lowndes' method are the same.

(b) Give an example with two states where the apportionments obtained by Hamilton's method and by Lowndes' method are different.

13. Show that Lowndes' method favors small states more than Hamilton's method in the sense that if state A is smaller than state B and Hamilton's method rounds A's quota up and B's quota down, then Lowndes' method will not simultaneously round B's quota up and A's quota down.

14. Show that Hamilton's method favors larger states more than Lowndes' method in the sense that if state A is smaller than state B and Lowndes' method rounds A's quota down and B's quota up, then Hamilton's method will not simultaneously round A's quota up and B's quota down.

2.2   EARLY DIVISOR METHODS

After Alexander Hamilton's method of apportionment was vetoed by President Washington in 1791, the rival plan of Thomas Jefferson, then Secretary of State, was enacted. The apportionment method proposed by Jefferson, now known in the United States as Jefferson's method, is an example of a **divisor method**. The basic idea behind divisor methods is that the natural divisor is replaced by a different divisor, and the resulting quotas are then rounded according to some rounding rule. In this section we look at two different divisor methods: Jefferson's method and Webster's method.

Jefferson's Method

Recall that the natural divisor is the total population divided by the house size. Each state's natural quota is then given by dividing its population by the natural divisor. In both Hamilton's and Lowndes' methods, we arrive at an apportionment by rounding some of the natural quotas up. Jefferson's method calls for rounding all the quotas down. However, we cannot do this with the natural quotas, for then we would not allocate all the seats except in the rare situation that all the natural quotas are whole numbers. In order for a procedure that calls for rounding the quotas down to work, the quotas will somehow have to be larger in the first place. Because the natural quotas are calculated by dividing each state's population by the natural divisor, we can get larger quotas by dividing by a new divisor that is smaller than the natural divisor. We call a number that replaces the natural divisor a **divisor**. The new quota that we get by dividing a state's population by a divisor different from the natural divisor will be called a **modified quota**.

The basic idea of **Jefferson's method** is stated next.

Jefferson's Method

Find a divisor such that when all the resulting modified quotas are rounded down to the nearest whole number to give the apportionment, the correct number of seats is allotted.

When the number of seats a state is allocated is divided by its population, the ratio represents the representation per individual. Those ratios greater than the representation per individual in the country as a whole may be viewed as overrepresentation. Jefferson's method minimizes the largest overrepresentation.

The difficulty in using Jefferson's method is in finding a divisor that will work. There will generally be many divisors that will work, and they will all give the same apportionment. If you are using a calculator that can handle lists or a computer with a spreadsheet program, then it is fairly easy to find a suitable divisor through trial-and-error. However, if you have only a basic calculator available, then you will want to be fairly careful when looking for a good divisor so that you can limit the amount of calculation. In the next three examples, we will look at a systematic way to go about finding a suitable divisor with Jefferson's method. When doing all the calculations on a basic calculator, this systematic approach is essential for keeping the amount of work manageable. However, even if you have a calculator that handles lists or a computer with spreadsheet software, the systematic approach will give you a better understanding of what is going on. This is why we do our example systematically, and urge you to study the method used.

Thomas Jefferson
(1743–1826).

Example 1   *Jefferson's Method*

In 1790, there were five counties in Maine. The populations of these counties based on the 1790 census are listed in Table 2.8. Use Jefferson's method to apportion representatives among the five counties for a hypothetical state legislature with 37 seats.

Table 2.8	Population of Maine in 1790
County	*Population*
Cumberland	25,450
Hancock	9,549
Lincoln	29,962
Washington	2,758
York	28,821

Source: U.S. Census Office.[10]

Solution: To apply Jefferson's method, first we find the natural divisor and the natural quotas. The natural divisor is given by

$$D = \frac{\text{total population}}{\text{house size}} = \frac{96{,}540}{37} \approx 2609.1892.$$

We use the letter D to indicate the various divisors we will use. The natural quotas are computed in the usual way by dividing the state size by the natural divisor, and the results are listed in Table 2.9. Rounding all the natural quotas down to the nearest whole number as called for in Jefferson's method gives us the initial allocation in Table 2.9.

Table 2.9	**Initial Allocation for Maine**		
County	*Population*	*Natural Quota* $D = 2609.1892$	*Initial Allocation*
Cumberland	25,450	9.7540	9
Hancock	9,549	3.6598	3
Lincoln	29,962	11.4833	11
Washington	2,758	1.0570	1
York	28,821	11.0460	11
Total	96,540		35

We see that the initial allocation is 2 seats short. We expect to be short with Jefferson's method because we are rounding all quotas down.

Now we need to find a divisor to replace the natural divisor so that when we round down the resulting modified quotas we get an allocation of 37 seats instead of 35 seats. Because the modified quotas have to be larger than the natural quotas, we need to find a divisor that is smaller than the natural divisor. Judging from the fact that Cumberland County has the largest fractional part, we might expect it to get one of the remaining 2 seats. What size would the divisor have to be in order for Cumberland County's modified quota to be 10? If D is the divisor, then we would need

$$\text{Cumberland's modified quota} = \frac{\text{Cumberland's population}}{D} = 10.$$

Solving for D, we have

$$D = \frac{\text{Cumberland's population}}{10} = \frac{25{,}450}{10} = 2545.$$

We will call the number D the threshold divisor for Cumberland to be allocated 10 seats. It is the divisor that gives Cumberland a modified quota of exactly 10 seats. If a smaller divisor than D is used, then Cumberland will get 10 or more seats. If a larger divisor is used, then Cumberland will get 9 or fewer seats.

The two other counties that should be strong contenders for one of the remaining seats are Hancock and Lincoln counties because their quotas have the second and third largest fractional parts from among the five counties. As we will see, it is not clear which of them will gain a seat, so we will compute the threshold divisor for each of them.

The threshold divisor for Hancock County to receive an allocation of 4 seats, one more than its initial 3 seats, is given by

$$D = \frac{\text{Hancock's population}}{4} = \frac{9549}{4} = 2387.25.$$

Similarly, the threshold divisor for Lincoln County to receive an allocation of 12 seats, one more than its initial 11 seats, is

$$D = \frac{\text{Lincoln's population}}{12} = \frac{29{,}962}{12} \approx 2496.8333.$$

In decreasing order, the three threshold divisors we have found are 2545 (Cumberland), 2496.8333 (Lincoln), and 2387.25 (Hancock). As our divisor decreases past each of these numbers, the associated county picks up a seat. Because we are rounding threshold divisors to four decimal places, they may be slightly smaller or larger than the actual threshold divisors. Therefore, to ensure that a state gains a seat with Jefferson's method, we should choose a divisor slightly smaller than the rounded threshold divisor. Because we want to allocate two more seats, we should try a divisor smaller than 2496.8333 (Lincoln), but larger than 2387.25 (Hancock). Let's try 2496.

Using the divisor 2496, we see that the modified quota of Cumberland County is

$$\frac{25{,}450}{2496} \approx 10.1963.$$

Similarly, we find the modified quotas of each of the other counties. Under Jefferson's method we round down the quotas to get the final allocation in Table 2.10. Because 37 seats are allocated, as required, we are done.

Table 2.10	**Apportionment for Maine**				
County	*Population*	*Natural Quota D = 2609.1892*	*Initial Allocation*	*Modified Quota D = 2496*	*Final Allocation*
Cumberland	25,450	9.7540	9	10.1963	10
Hancock	9,549	3.6598	3	3.8257	3
Lincoln	29,962	11.4833	11	12.0040	12
Washington	2,758	1.0570	1	1.1050	1
York	28,821	11.0460	11	11.5469	11
Total	96,540		35		37

We have chosen to illustrate just one of several similar strategies for selecting a suitable divisor. In this example any divisor between 2387.25, the threshold divisor for 4 seats for Hancock County, and 2496.8333, the threshold divisor for 12 seats for Lincoln County, would yield exactly the same apportionment. A divisor above this range would result in a house size smaller than 37, whereas a divisor below this range would result in a house size greater than 37. Figure 2.1 shows the house size for all divisors between 2000 and 3000.

Figure 2.1
How the house size
changes in relation
to the divisor. (The
colors represent the
county gaining the
last seat.)

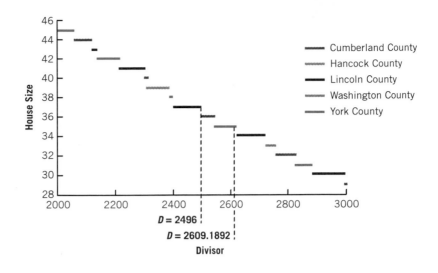

Starting from our natural divisor of 2609.1892, we see in Figure 2.1 that as we decrease the divisor, first Cumberland and then Lincoln gain seats. You might notice that York, not Hancock, would receive the next seat, even though its natural quota was almost as far as possible from the next higher whole number. Looking at the entire range of divisors, observe that the larger counties—Cumberland, Lincoln, and York—stand to gain or lose many more seats than the smaller counties of Hancock and Washington as the divisor changes. The arithmetic behind this is that changing the divisor gives modified quotas that differ from the natural quotas by the same percentage. Thus, the quotas of the larger counties will change much more rapidly than those of the smaller counties. Because Jefferson's method always leads to a divisor smaller than the natural divisor, and hence larger modified quotas, it tends to be an advantageous apportionment method for larger states and a disadvantageous method for smaller states. This effect did not go unnoticed by our country's first political leaders.

In the search for a suitable divisor for Jefferson's method, sometimes the first divisor we arrive at does not quite do the trick and we have to look further. We illustrate this in the next two examples. Before looking at these examples, we first give a more precise definition of threshold divisor.

For any divisor method, we call the divisor at which the number of seats assigned to a particular state increases from $n - 1$ to n the **threshold divisor for n seats**. For Jefferson's method, the threshold divisor is given by

$$\boxed{\text{threshold divisor for } n \text{ seats under Jefferson's method} = \frac{\text{population of the state}}{n}.}$$

If we use a divisor greater than a particular state's threshold divisor for n seats, then the state will be assigned fewer than n seats. If we use a divisor less than the state's threshold divisor, the state will be assigned n or more seats.

We now look at two more applications of Jefferson's method.

Example 2 *Jefferson's Method*

An HMO operates six clinics with a total staff of 14 doctors. The doctors each work five days a week, and each doctor can work at different clinics on different days, but not on the same day. Thus, the HMO has $14 \cdot 5 = 70$ doctor-days available to apportion among the clinics. Use Jefferson's method to allocate the 70 doctor-days to the clinics based on the average weekly patient load given in Table 2.11.

Table 2.11	HMO Patient Loads
Clinic	*Weekly Patient Load*
Central	346
East	299
West	158
South	198
North	213
Lakeside	243

Solution: In this example, the total population is 1457, the sum of the weekly patient loads, and the house size is 70. Therefore, the natural divisor is $1457/70 \approx 20.8143$. We compute the natural quotas by dividing the weekly patient load of each clinic by the natural divisor, and then round the natural quotas down to arrive at the initial allocation in Table 2.12.

Table 2.12	Initial Allocation of Doctors		
Clinic	*Weekly Patient Load*	*Natural Quota D = 20.8143*	*Initial Allocation*
Central	346	16.6232	16
East	299	14.3651	14
West	158	7.5909	7
South	198	9.5127	9
North	213	10.2333	10
Lakeside	243	11.6747	11
Total	1457		67

We see that in the initial allocation, only 67 of the 70 doctor-days have been assigned. Because we have 3 more seats to allocate and we cannot be sure which clinics will eventually get the seats, we compute the threshold divisors for the four clinics with the largest fractional parts. Ranking them in order from the largest threshold divisor to the smallest, we have

$$\text{Central's threshold divisor} = \frac{346}{17} \approx 20.3529,$$

$$\text{Lakeside's threshold divisor} = \frac{243}{12} = 20.25,$$

$$\text{South's threshold divisor} = \frac{198}{10} = 19.8,$$

West's threshold divisor $= \dfrac{158}{8} = 19.75.$

Because 19.8 is the exact threshold divisor for South, we do not have to worry about potential rounding problems and use it as our divisor, knowing this will ensure that Lakeside, Central, and South each get an extra doctor-day, and that West will not. Table 2.13 shows the modified quotas and the resulting allocation when the divisor $D = 19.8$ is used.

Table 2.13	Attempted Allocation of Doctors				
Clinic	*Weekly Patient Load*	*Natural Quota D = 20.8143*	*Initial Allocation*	*Modified Quota D = 19.8*	*Allocation*
Central	346	16.6232	16	17.4747	17
East	299	14.3651	14	15.1010	15
West	158	7.5909	7	7.9798	7
South	198	9.5127	9	10.0000	10
North	213	10.2333	10	10.7576	10
Lakeside	243	11.6747	11	12.2727	12
Total	1457		67		71

In this new allocation Central, South, and Lakeside clinics all gained a doctor-day, but unfortunately so did the East clinic. Looking at the natural quotas, we see that East's natural quota had the second to smallest fractional part, but when we compute East's threshold divisor we find

$$\text{East's threshold divisor} = \frac{299}{15} \approx 19.9333.$$

This threshold divisor is third in size, after Central's and Lakeside's. We now see that we need to pick a divisor that is at least as small as the threshold divisors of Central, Lakeside, and East and larger than those of West and South. We do not use 19.9333 as the divisor because it is the threshold divisor for the East clinic rounded to four decimal places, rather than the exact threshold divisor. To be sure that East is allocated an extra seat, we round down a little to 19.9, being careful to stay above South's threshold divisor of 19.8. Using 19.9 as our new divisor, we see in Table 2.14 that the new modified quotas give the allocation of 70 seats as required.

Table 2.14	Apportionment of Doctors				
Clinic	*Weekly Patient Load*	*Natural Quota D = 20.8143*	*Initial Allocation*	*Modified Quota D = 19.9*	*Final Allocation*
Central	346	16.6232	16	17.3869	17
East	299	14.3651	14	15.0251	15
West	158	7.5909	7	7.9397	7
South	198	9.5127	9	9.9497	9
North	213	10.2333	10	10.7035	10
Lakeside	243	11.6747	11	12.2111	12
Total	1457		67		70

The problem encountered in the last example might spur you to compute all threshold divisors. With experience you will be able to anticipate which states are the likeliest to gain a seat and find a solution with just a few computations. When looking for the states most likely to gain seats, check those with large fractional parts or with very large populations. If you are not sure you have picked the right states, you might want to check at least one more state than you have seats to fill. Note, however, that even if you compute the threshold divisors for all states, you will sometimes encounter problems, as we see in the next example.

Example 3   *Jefferson's Method and Violation of the Quota Property*

A consumer information company wants to poll owners of U.S.-made passenger cars. The company wishes to survey 2500 owners so that the number of owners surveyed who own a car made by a particular company is roughly proportional to the number of cars that company produces. Table 2.15 gives the 1993 breakdown for overall production of U.S.-made cars.

Table 2.15	Car Sales by Company
Company	**Number of Passenger Cars**
Chrysler	494,573
Ford	1,489,699
GM	2,542,455
Honda	403,775
Mazda	219,076
Nissan	292,182
Toyota	356,069
Other	183,152

Source: 1995 Information Please® Almanac.[11]

Use Jefferson's method to determine how many owners of each type of car should be surveyed.

Solution: The total number of passenger cars in the list is 5,980,981, resulting in a natural divisor of 5,980,981/2500 = 2392.3924. We compute the natural quotas and initial allocation in Table 2.16.

Table 2.16	Initial Allocation of Cars for Poll		
Company	**Number of Passenger Cars**	**Natural Quota D = 2392.3924**	**Initial Allocation**
Chrysler	494,573	206.7274	206
Ford	1,489,699	622.6817	622
GM	2,542,455	1062.7249	1062
Honda	403,775	168.7746	168
Mazda	219,076	91.5719	91
Nissan	292,182	122.1296	122
Toyota	356,069	148.8339	148
Other	183,152	76.5560	76
Total	5,980,981		2495

This time we are 5 seats short of the 2500 we need to fill. Observe that Mazda, Nissan, and Other have the smallest fractional parts *and* the smallest populations, so there is no way they can gain a seat. We find the threshold divisors for the other five companies. Ranking them in order from the largest threshold divisor to the smallest, we have

$$\text{GM's threshold divisor} = \frac{2{,}542{,}455}{1063} \approx 2391.7733,$$

$$\text{Ford's threshold divisor} = \frac{1{,}489{,}699}{623} \approx 2391.1701,$$

$$\text{Toyota's threshold divisor} = \frac{356{,}069}{149} \approx 2389.7248,$$

$$\text{Chrysler's threshold divisor} = \frac{494{,}573}{207} \approx 2389.2415,$$

$$\text{Honda's threshold divisor} = \frac{403{,}775}{169} \approx 2389.2012.$$

We need to assign five more seats, so we try 2389.2 as our divisor because it falls just below Honda's threshold divisor and we expect Mazda, Nissan, and Other to have significantly smaller threshold divisors. Using 2389.2 as our divisor and rounding down the modified quotas to allocate the seats, we get the results in Table 2.17.

Table 2.17	Attempted Allocation of Cars for Poll				
Company	*Number of Passenger Cars*	*Natural Quota D = 2392.3924*	*Initial Allocation*	*Modified Quota D = 2389.2*	*Allocation*
Chrysler	494,573	206.7274	206	207.0036	207
Ford	1,489,699	622.6817	622	623.5137	623
GM	2,542,455	1062.7249	1062	1064.1449	1064
Honda	403,775	168.7746	168	169.0001	169
Mazda	219,076	91.5719	91	91.6943	91
Nissan	292,182	122.1296	122	122.2928	122
Toyota	356,069	148.8339	148	149.0327	149
Other	183,152	76.5560	76	76.6583	76
Total	5,980,981		2495		2501

Those companies that we expected to gain a seat did so and Mazda, Nissan, and Other did not, but GM gained two seats, so we are now one seat over the 2500 seats we want to allocate. We know that the threshold divisor for GM to get 1063 seats is about 2391.7733, but what is the threshold divisor for GM to get 1064 seats? We compute this and see

$$\text{GM's threshold divisor for 1064 seats} = \frac{2{,}542{,}455}{1064} \approx 2389.5254.$$

This threshold divisor is larger than the threshold divisors for Chrysler and Honda to gain one more seat, explaining why GM got two extra seats in our last allocation. To reduce the number of seats to 2500, we need to pick a divisor that is smaller than Chrysler's threshold divisor and larger than Honda's threshold divisor. We choose the divisor 2389.24; the resulting modified quotas and final allocation are given in Table 2.18.

Table 2.18	**Apportionment of Cars for Poll**				
Company	*Number of Passenger Cars*	*Natural Quota D = 2392.3924*	*Initial Allocation*	*Modified Quota D = 2389.24*	*Final Allocation*
Chrysler	494,573	206.7274	206	207.0001	207
Ford	1,489,699	622.6817	622	623.5033	623
GM	2,542,455	1062.7249	1062	1064.1271	1064
Honda	403,775	168.7746	168	168.9973	168
Mazda	219,076	91.5719	91	91.6928	91
Nissan	292,182	122.1296	122	122.2908	122
Toyota	356,069	148.8339	148	149.0302	149
Other	183,152	76.5560	76	76.6570	76
Total	5,980,981		2495		2500

It works! With practice, you should usually be able to anticipate which states will gain or lose seats, but even if you guess wrong, the only cost is a bit more computation. Notice that GM's allocation of 1064 automobiles violates the Quota Property, which would restrict its allocation to 1062 or 1063.

∎

The systematic approach for finding a good divisor that we used in the last three examples becomes easier after a little practice.

All divisor methods begin the same way, with the computation of modified quotas based on the choice of a divisor. Where they differ is in how the modified quotas are rounded. Using Jefferson's method, all quotas are rounded down. For any divisor method, *smaller divisors result in larger modified quotas and larger divisors result in smaller modified quotas.*

Jefferson's method remained in effect until 1842. During this time the size of the House increased, so that few states faced the loss of seats. Generally, the House size increased until the physical limitations of the House floor halted the increase in 1932 at 435, which remains its size. In our three examples of Jefferson's method, we have seen that if two states with roughly equal fractional parts in their natural quotas are competing for a remaining seat, the state with the larger population tends to receive it. As we stated earlier, Jefferson's method tends to benefit the larger states over the smaller states. Many opponents of Jefferson's method observed that Virginia, Jefferson's home state, was the largest state of the United States until the 1810 census, when New York claimed the top spot. In 1822, New York's natural quota for the established House size of 181 was 32.5031. Yet Jefferson's method allocated New York 34 seats, illustrating again, as we saw in Example 3, that Jefferson's method does not always satisfy the Quota Property. On the other hand, Jefferson's method will never give a state fewer seats than its natural quota rounded down. The bias of Jefferson's method against the smaller states instigated competing proposals in the early 1800s. In 1822, Lowndes introduced his method; it is quite biased toward the smaller states, and it gathered little support.

Webster's Method

The debate over apportionment in 1832 was especially contentious. On February 15, 1832, no fewer than ten different sizes for the House were discussed. In 1832, Representative and former President John Quincy Adams of Massachusetts proposed a divisor method based on rounding up all quotas. This method favored smaller states for the same reason that Jefferson's method favored larger states. Although Massachusetts was of moderate size, the rest of New England, which had similar political interests, consisted of smaller states. Because the House had already passed its apportionment bill by the time Adams invented his method in 1832, his only recourse was to pass his suggestion on to the Senate, where Daniel Webster, also of Massachusetts, was chair of the committee studying the matter. Webster came up with a different divisor method, perhaps the most natural one. His proposed method, known as Webster's method, is a divisor method similar to those of Jefferson and Adams. However, with Webster's method, instead of rounding the natural quotas and modified quotas consistently downward or consistently upward, the quotas are rounded in the natural way so that fractional parts 0.5 or above round up and fractional parts below 0.5 round down. **Webster's method** is stated next.

> ## *Webster's Method*
>
> Find a divisor such that when all the resulting modified quotas are rounded to the nearest whole number in the natural way (up if the fractional part is greater than or equal to 0.5 and down if the fractional part is less than 0.5) to give the apportionment, the correct number of seats is allotted.

In the discussion of Jefferson's method, we defined the representation per individual for a state to be the ratio of the number of seats that state is allocated

Daniel Webster
(1782–1852).

divided by its population. Webster's method minimizes the greatest difference in representation per individual among the states.

As with Jefferson's method, the key to Webster's method is finding a suitable divisor. With Webster's method, the natural divisor may prove too large, too small, or just right. On the other hand, a suitable divisor for Webster's method is usually closer to the natural divisor than is one for Jefferson's method. We look at Webster's method in the next two examples.

Example 4   *Webster's Method*

The populations of the eight counties in Connecticut based on the 1830 census are given in Table 2.19. Use Webster's method to apportion representatives of the counties to a hypothetical state legislature with 75 seats.

Table 2.19	Population of Connecticut in 1830
County	*Population*
Fairfield	47,010
Hartford	51,131
Litchfield	42,858
Middlesex	24,844
New Haven	43,847
New London	42,201
Tolland	18,702
Windham	27,082

Source: U.S. Census Office.[12]

Solution: The total population of the eight counties is 297,675 people and the natural divisor is 297,675/75 = 3969 (exactly). We next compute the natural quotas for the counties as shown in Table 2.20.

Table 2.20	Initial Allocation for Connecticut		
County	*Population*	*Natural Quota* *D = 3969*	*Initial* *Allocation*
Fairfield	47,010	11.8443	12
Hartford	51,131	12.8826	13
Litchfield	42,858	10.7982	11
Middlesex	24,844	6.2595	6
New Haven	43,847	11.0474	11
New London	42,201	10.6327	11
Tolland	18,702	4.7120	5
Windham	27,082	6.8234	7
Total	297,675		76

Our initial allocation assigns one seat too many. In order to reduce the number of seats assigned, we need to find a larger divisor that will result in slightly smaller quotas. It seems probable that, as we raise the divisor, New London County will be the first to lose a seat because its natural quota of 10.6327 is fairly close to the cutoff point of 10.5. Also, because New London County has a fairly large population, its quota is more strongly affected by a change in the divisor than are the quotas of smaller counties. Remember that New London County's modified quota need only go below 10.5 to have the number of seats allotted it fall from 11 to 10. What divisor D gives New London County a modified divisor of exactly 10.5? If we set

$$\text{New London's modified quota} = \frac{\text{New London's population}}{D} = 10.5,$$

then solving for D yields

$$D = \frac{\text{New London's population}}{10.5} = \frac{42,201}{10.5} \approx 4019.14.$$

This number is the threshold divisor for 11 seats under Webster's method. Because we want New London's modified quota to fall below 10.5, we should choose a divisor just slightly larger than 4019.14. We use the divisor 4020, which results in the final allocation in Table 2.21.

Table 2.21	**Apportionment for Connecticut**				
County	*Population*	*Natural Quota* $D = 3969$	*Initial Allocation*	*Modified Quota* $D = 4020$	*Final Allocation*
Fairfield	47,010	11.8443	12	11.6940	12
Hartford	51,131	12.8826	13	12.7192	13
Litchfield	42,858	10.7982	11	10.6612	11
Middlesex	24,844	6.2595	6	6.1801	6
New Haven	43,847	11.0474	11	10.9072	11
New London	42,201	10.6327	11	10.4978	10
Tolland	18,702	4.7120	5	4.6522	5
Windham	27,082	6.8234	7	6.7368	7
Total	297,675		76		75

In this example, we note that the apportionment we arrive at is the same as the one we would get using Hamilton's method. This is not always the case, as we will see in Example 5.

Before moving on to this example, let's work out the formula for a threshold divisor for Webster's method. Recall that the divisor at which the number of seats assigned to a particular state increases from $n - 1$ to n is called a threshold divisor for n seats. For Webster's method, the threshold divisor is given by

$$\text{threshold divisor for } n \text{ seats under Webster's method} = \frac{\text{population of the state}}{n - 0.5}.$$

Remember that if we use a divisor greater than a particular state's threshold divisor for n seats, then the state will be assigned fewer than n seats. If we use a divisor less than the state's threshold divisor, the state will be assigned n or more seats.

Example 5   *Webster's Method and Violation of the Quota Property*

Statistical data are often presented in terms of percentages. Table 2.22 lists the estimated number of livestock on farms in the United States in 1994.

Table 2.22	Livestock on Farms in 1994
Type of Livestock	*Numer of Animals (in thousands)*
Cattle	101,749
Dairy cows	9,638
Sheep	9,079
Swine	55,630
Chickens	3,397,818
Turkeys	291,750

Source: U.S. Department of Agriculture.[13]

Suppose we want to present these data in terms of percentages of the total livestock. For example, we would want to say what percentage of the total population of livestock is chickens. If these percentages are carried out to one decimal place and we want the percentages to total exactly 100%, then we need to apportion the 1000 values of 0.1 that add up to 100% to the six different kinds of livestock. Use Webster's method to carry out this apportionment, and present the livestock data in the form of percentages carried out to one decimal place.

Solution: The total livestock population is 3,865,664 thousand. We need to apportion the 1000 seats corresponding to the values of 0.1%. The natural divisor for 1000 seats is given by 3,865,664/1000 = 3865.664. After computing the natural quotas, we arrive at the initial allocation in Table 2.23.

Table 2.23	Initial Allocation of Percentages		
Type of Livestock	*Number of Animals (in thousands)*	*Natural Quota D = 3865.664*	*Initial Allocation*
Cattle	101,749	26.3212	26
Dairy cows	9,638	2.4932	2
Sheep	9,079	2.3486	2
Swine	55,630	14.3908	14
Chickens	3,397,818	878.9740	879
Turkeys	291,750	75.4722	75
Total	3,865,664		998

This time the initial allocation is 2 seats short of our desired allocation. In order to increase the number of seats assigned, we need to find a smaller divisor. Because the fractional parts of the natural quotas for dairy cows and turkeys are very close to 0.5, they are strong candidates to get one of the remaining seats. However, because the chicken population is quite large, it is also a candidate for an extra spot even though its natural quota is really not close to the 879.5 quota needed to gain an extra seat. To find the threshold divisors for dairy cows receiving 3 seats under Webster's method, we use the threshold formula for Webster's method with $n = 3$ and get

$$\text{dairy cows' threshold divisor} = \frac{9638}{2.5} = 3855.2.$$

Similarly, the threshold divisors for the chickens and turkeys are given by

$$\text{chickens' threshold divisor} = \frac{3{,}397{,}818}{879.5} \approx 3863.3519,$$

$$\text{turkeys' threshold divisor} = \frac{291{,}750}{75.5} \approx 3864.2384.$$

In decreasing order, the three threshold divisors we found are 3864.2384 (turkeys), 3863.3519 (chickens), and 3855.2 (dairy cows). We see that the turkeys and chickens will get the extra seats, and we should choose a divisor just below the chickens' threshold divisor. If we use 3863 as our divisor, we get an allocation that works, as shown in Table 2.24. Notice that although the chickens had a natural quota of 878.9740, they received 880 seats in the final allocation. Because this allocation differs from the natural quota by more than one, we see that Webster's method does not satisfy the Quota Property.

Table 2.24	**Apportionment of Percentages of Livestock**					
Type of Livestock	*Number of Animals (in thousands)*	*Natural Quota D = 3865.664*	*Initial Allocation*	*Modified Quota D = 3863*	*Final Allocation*	*Percent*
Cattle	101,749	26.3212	26	26.3394	26	2.6
Dairy cows	9,638	2.4932	2	2.4950	2	0.2
Sheep	9,079	2.3486	2	2.3502	2	0.2
Swine	55,630	14.3908	14	14.4007	14	1.4
Chickens	3,397,818	878.9740	879	879.5801	880	88.0
Turkeys	291,750	75.4722	75	75.5242	76	7.6
Total	3,865,664		998		1000	100.0

Because each allocated seat corresponds to 0.1 percent, we can convert the final allocation to the percentages shown in Table 2.24. ∎

Although defeated in 1832, Webster's method was adopted in 1842. The year 1842 was remarkable because it was the only apportionment year in which the size of the House decreased below that of the previous apportionment. Although

BRANCHING OUT 2.1

SHOULD THERE HAVE BEEN A PRESIDENT TILDEN?

In 1872, an illegal apportionment of the House of Representatives was negotiated that did not agree with the mandated Hamilton's method. With the house size at 292, two seats that would have gone to Illinois and New York under Hamilton's method instead went to New Hampshire and Florida. This might not seem particularly significant, but in the year 1876, this discrepancy resulted in a different outcome in the presidential election. In this election, the Democratic nominee Samuel J. Tilden won the popular vote with 4,300,590 votes, whereas his Republican opponent Rutherford B. Hayes received 4,036,298 votes. In the Electoral College, Hayes won by a margin of only one, 185 to 184. (See Branching Out 1.1 for a discussion of the Electoral College.) New York went to Tilden, and Illinois, New Hampshire, and Florida went to Hayes. If Hamilton's method had been used in the congressional apportionment, New York would have had one more vote, taking one of the Hayes votes away from New Hampshire or Florida. Therefore, if the 1872 apportionment of the House of Representatives had been carried out according to the law, Tilden would have won the Electoral College vote and, therefore, the presidency.

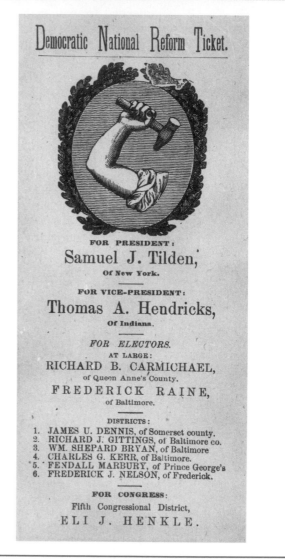

The Democratic National Reform Ticket for the 1876 election year in which Samuel J. Tilden lost to Rutherford B. Hayes in the Electoral College vote.

Congress reverted to Hamilton's method from 1852 to 1900, the size of the House was always chosen so that Hamilton's and Webster's methods agreed. As Example 5 shows, Webster's method does not satisfy the Quota Property. In fact, the Quota Property is not satisfied by any divisor method. This is a consequence of the fact that changing the divisor has a greater effect on the modified quotas of larger states than of smaller states. However, violations of the Quota Property are much rarer with apportionments produced by Webster's method than by Jefferson's method. Unlike Jefferson's method, Webster's method appears to show no favoritism toward either larger or smaller states. With Jefferson's method, the new divisor is always smaller than the natural divisor, so the modified quotas are moving up, which gives the advantage to larger states. With Webster's method, the new divisor is equally likely to be smaller or larger than the natural divisor. If the new divisor is larger than the natural divisor, then the advantage will be to the smaller states. If it is smaller, then the larger states will have an advantage.

The years from 1852 to 1900 periodically revealed problems with Hamilton's method. By 1901, enough members of Congress had become convinced of the inequities in Hamilton's method that it was finally abandoned and Webster's method readopted, in part due to the efforts of Cornell University philosophy professor and statistician Walter Willcox.

Apportionment for the U.S. House of Representatives requires that each state receives at least one seat. Under Jefferson's method or Webster's method a state may be allotted no seat. The adjustment is easy for these methods, as long as the number of seats is at least the number of states. Simply allot at least one representative to each state in the rounding process, but otherwise continue in the normal way. In the framework of divisors, we simply reset the rounding cutoff between 0 and 1 to be 0, instead of 1 for Jefferson's method or 0.5 for Webster's method.

*E*XERCISES FOR **S**ECTION **2.2**

1. Suppose that Minnesota's nine electoral votes in the 1892 presidential election, won by Grover Cleveland, had been apportioned to the candidates in proportion to the number of votes each received rather than all going to the candidate with the most votes. Allocate Minnesota's nine electoral votes based on the vote totals given here using

  (a) Jefferson's method,   **(b)** Webster's method.

Candidate	Grover Cleveland	Benjamin Harrison	James R. Weaver	John Bidwell
Number of Votes	100,589	122,736	30,399	14,117

2. Suppose that Illinois' 29 electoral votes in the 1912 presidential election, won by Woodrow Wilson, had been apportioned to the candidates in proportion to the number of votes each received rather than all going to the candidate with the most votes. Allocate Illinois' 29 electoral votes based on the vote totals given here using

  (a) Jefferson's method,   **(b)** Webster's method.

  (c) When switching from Jefferson's method to Webster's method, does the shift in seats go from candidates with more votes to those with fewer, or vice versa?

Candidate	Woodrow Wilson	Theodore Roosevelt	William H. Taft	Eugene V. Debs
Number of Votes	405,048	386,478	253,593	81,278

3. The approximate 1997 enrollments in the five high schools of the Lincoln County School District in Oregon are shown in the table. Suppose the district must assign 27 student teachers to the five high schools. Apportion the student teachers based on the student enrollment at the schools using

  (a) Jefferson's method,   **(b)** Webster's method.

  (c) When switching from Jefferson's method to Webster's method, does the shift in seats go from schools with more students to those with fewer, or vice versa?

School Enrollment	Eddyville 60 (estimated)	Newport 647	Taft 704	Toledo 456	Waldport 346

Source: Lincoln County School District.[14]

4. A real estate company employs 42 Realtors. The Realtors are to be apportioned to the six company branch offices based on the total sales made in each office over the past year as shown in the table. Apportion the Realtors using

(a) Jefferson's method,   (b) Webster's method.

Office	Total Sales (in thousands of dollars)
Tanglewood	7871
Park Hill	7122
Cityview	5894
Berkeley	5124
Ridglea	5003
Westcliff	2323

5. Suppose a hypothetical state legislature is to have 42 members apportioned among the five counties in Hawaii based on the 1990 census data shown in the table. Perform the apportionment using

(a) Jefferson's method,   (b) Webster's method,

(c) Hamilton's method,   (d) Lowndes' method.

(e) Which method seems to help the large Honolulu County the most?

(f) Which method seems to help the tiny Kalawao County the most?

County Population	Hawaii 120,317	Honolulu 836,231	Kalawao 130	Kauai 51,177	Maui 100,374

Source: U.S. Bureau of the Census.[15]

6. Suppose Delaware's 21-member state senate is apportioned among its three counties based on the estimates of 1997 population in the table. Perform the apportionment using

(a) Jefferson's method,   (b) Webster's method,

(c) Hamilton's method,   (d) Lowndes' method.

County Population	Kent 122,709	New Castle 474,838	Sussex 134,034

Source: U.S. Bureau of the Census.[16]

7. The number of residents of the six main dormitories at Hendrix College in Conway, Arkansas, is given in the table. Suppose the student government has a house of 23 student representatives, and the house seats are to be apportioned among the six dormitories based on the number of residents. Apportion the 23 house seats using

(a) Jefferson's method,   (b) Webster's method.

Dormitory Number of Residents	Couch 152	Galloway 96	Hardin 147	Martin 130	Raney 96	Veasey 120

Source: Hendrix College.[17]

8. A summer camp for children entering grades 2 through 6 employs 17 camp counselors. Each of the counselors will be in charge of a group of children all in the same grade. The number of children entering each grade is listed in the table. Apportion the camp counselors among the different grades using

(a) Jefferson's method, (b) Webster's method.

Grade	2	3	4	5	6
Number of Children	25	30	41	36	43

In Exercises 9 and 10, consider that elections to Germany's Parliament are on a proportional basis by party in each region of the country, with the number of seats each party wins determined by the percentage of votes the party receives.

9. Apportion the 101 representatives to Germany's Parliament from the Rheinland-Palatinate state among four major political parties using the percentage of the vote each received in the 1996 election results shown in the table as the parties' populations. Use

(a) Jefferson's method, (b) Webster's method.

(c) Do either of these apportionments violate the Quota Property?

(d) Does either apportionment agree with the actual apportionment of Social Democrats 43, Christian Democrats 41, Free Democrats 10, and Greens 7?

Party	Social Democrats	Christian Democrats	Free Democrats	Greens
Percentage of Votes	39.8	38.7	8.9	6.9

Source: The American Institute for Contemporary German Studies, The Johns Hopkins University.[18]

10. Apportion the 75 representatives to Germany's Parliament from the Schleswig-Holstein state among five major political parties using the percentage of the vote each received in the 1996 election results shown in the table as the parties' populations. Use

(a) Jefferson's method, (b) Webster's method.

(c) Do either of these apportionments violate the Quota Property?

(d) Does either apportionment agree with the actual apportionment of Social Democrats 33, Christian Democrats 30, Greens 6, Free Democrats 4, and South Schleswig Voter Union 2?

Party	Percentage of Votes
Social Democrats	39.8
Christian Democrats	37.2
Greens	8.1
Free Democrats	5.7
South Schleswig Voter Union	2.5

Source: The American Institute for Contemporary German Studies, The Johns Hopkins University.[19]

11. Apportion 400 seats among the major political parties to the 1994 National Assembly of South Africa based on the number of votes each received as shown in the table. Use

(a) Jefferson's method, (b) Webster's method,

(c) Hamilton's method, (d) Lowndes' method.

(e) Do any of these apportionments violate the Quota Property?

(f) Do any of these apportionments agree with the actual apportionment of African

National Congress 252, National Party 82, Inkatha Freedom Party 43, Freedom Front 9, Democratic Party 7, Pan Africanist Congress 5, and African Christian Democratic Party 2?

Party	Number of Votes
African National Congress	12,237,655
National Party	3,983,690
Inkatha Freedom Party	2,058,294
Freedom Front	424,555
Democratic Party	338,426
Pan Africanist Congress	243,478
African Christian Democratic Party	88,104

Source: Republic of South Africa Web site.[20]

12. Apportion the 105 seats in the 1792 U.S. House of Representatives based on the 1790 census figures in the table, using

(a) Jefferson's method, (b) Webster's method.

(c) Does either apportionment violate the Quota Property?

(For a comparison with the apportionments obtained using Hamilton's method and Lowndes' method, see Exercise 10 of Section 2.1.)

State	Population	State	Population
Connecticut	237,946	New Jersey	184,139
Delaware	59,094	New York	340,120
Georgia	82,538	North Carolina	393,751
Kentucky	73,677	Pennsylvania	434,373
Maine	96,540	Rhode Island	68,825
Maryland	319,728	South Carolina	249,073
Massachusetts	378,787	Vermont	85,539
New Hampshire	141,885	Virginia	747,610

Source: U.S. Census Office.[21]

13. The estimates for crude oil production, in thousands of 42-gallon barrels, by region for the year 1996 are listed in the table.

Region	Number of Barrels (in thousands) per day
North America	11,156
Central and South America	5,963
Western Europe	6,304
Eastern Europe and former U.S.S.R.	6,918
Middle East	19,195
Africa	7,305
Far East and Oceania	7,212
Total	64,053

Source: Energy Information Administration.[22]

(a) Make a table showing the percentage of the crude oil production by region rounded to the nearest whole number. What is the total of the rounded percentages?

(b) Following the idea of Example 5, use Webster's method to present the data in terms of *whole number* percentages in such a way that the sum of the percentages is exactly 100.

(c) Use Jefferson's method to present the data in terms of *whole number* percentages in such a way that the sum of the percentages is exactly 100.

14. The estimated receipts (in millions of dollars) for the different economic sectors in the United States in the year 1992 are listed in the following table.

Economic Sector	Estimated Receipts (in millions of dollars)
Construction	563,284
Manufacturing	3,083,095
Transportation, communication, utilities	933,516
Wholesale trade	3,222,844
Retail trade	1,973,606
Finance, insurance, real estate	1,827,343
Service	1,803,244
Agriculture, mining, unclassified	198,251
Total	13,605,183

Source: U.S. Bureau of the Census.[23]

(a) Make a table showing the percentage of the economy in 1992 for each of the sectors in the table, rounded to the nearest tenth of a percent (to one decimal place). What is the total of the rounded percentages?

(b) Following the idea of Example 5, use Webster's method to present the data in terms of percentages, to the tenth of a percent, in such a way that the sum of the percentages is exactly 100.

(c) Use Jefferson's method to present the data in terms of percentages, to the tenth of a percent, in such a way that the sum of the percentages is exactly 100.

One variant of the apportionment methods we have studied, which is designed to avoid too much splintering of the vote, is to require an individual or party to receive some minimal number or percentage of the votes, such as 1%, 5%, or 10%, in order to receive any seats or delegates. Exercises 15 and 16 provide examples of this.

15. Oregon selected its delegates to the 1996 Republican Convention in proportion to results in its state primary, with a threshold of 10% of the vote required to earn any delegates. Use the percentages of the vote the candidates received to apportion Oregon's 23 delegates among those candidates eligible to earn delegates using

(a) Jefferson's method, **(b)** Webster's method.

Candidate	Percentage of the Votes
Bob Dole	52
Pat Buchanan	22
Steve Forbes	13
Lamar Alexander	7
Alan Keyes	4
Richard Lugar	1
Phil Gramm	<1
Bob Dornan	<1
Morry Taylor	<1

16. Election to the Great National Assembly of Turkey is by parties on a proportional basis, with the number of seats each party wins determined by the percentage of the votes the party received and with a 10% threshold required to gain any seats. Use the percentages of the vote the parties received to apportion the 1993 Assembly of 450 members among those parties eligible to receive seats using

(a) Jefferson's method, **(b)** Webster's method.

Party	Percentage of Votes
Welfare Party	21.3
Motherland Party	19.5
True Path Party	19.2
Democratic Left Party	14.7
Republican People's Party	10.7
Nationalist Movement Party	8.2
Democratic People's Party	4.2
New Democratic Movement	0.5

Source: Wilfried P. C. G. Derksen Web site.[24]

Sometimes apportionments have the additional requirement that each state is entitled to some minimal number of seats. Exercises 17 and 18 provide examples of this.

17. Apportion the seats in Exercise 5, assuming that each county of Hawaii is entitled to at least one seat, and suitably adapting the methods of

(a) Jefferson,   **(b)** Webster.

18. Apportion the seats in Exercise 5, assuming that each county of Hawaii is entitled to at least three seats, and suitably adapting the methods of

(a) Jefferson,   **(b)** Webster.

19. Explain why a divisor that is too large to give the correct number of seats using Webster's method is also too large to use with Jefferson's method.

20. Explain why a divisor that is too small to give the correct number of seats using Jefferson's method is also too small to use with Webster's method.

21. Why do Webster's method and Hamilton's method agree when there are only two states? (Ignore the possibility that the fractional parts of the natural quotas are exactly 0.5.)

22. Find an example with two states for which Jefferson's method and Webster's method give different apportionments.

2.3   APPORTIONMENT IN TODAY'S HOUSE OF REPRESENTATIVES

Joseph Hill, chief statistician of the Bureau of the Census, proposed a new, more complicated divisor method in 1911. It met with indifferent response until championed by Harvard mathematician Edward V. Huntington beginning in the 1920s. This method, now called the Hill–Huntington method, was adopted in 1941 and remains in effect today. It beat out Webster's method in part because of the professional reputation and persuasive powers of its major proponents, but probably more so simply because of the short-term political interests of its supporters. The sole difference based on the 1940 census was one seat that would go to Arkansas, a Democratic state, under the Hill–Huntington method, or to Michigan, a state with Republican tendencies, under Webster's method. A bill establishing the Hill–Huntington method of apportionment was proposed in the House of Representatives. All Democrats except those from Michigan voted in favor of the bill, whereas every Republican opposed it. After passing in the House, the bill was passed by the Senate and signed into law by President Roosevelt. More recent research by Michel L. Balinski and H. Peyton Young seems to suggest that Webster's method is a fairer apportionment method.

The difference among different divisor methods is in how we round the fractions in the quotas. Recall that with Jefferson's method we round all of the quotas down, whereas with Webster's method we round the quotas in the natural way. With the Hill–Huntington method, we round in yet another way.

The Hill–Huntington Method

Edward V. Huntington (1874–1952).

With Webster's method, the cutoff for rounding up to n seats as opposed to rounding down to $n - 1$ seats is at the average, or arithmetic mean, of $n - 1$ and n, in other words, $n - 0.5$. The Hill–Huntington method sets the cutoff for rounding at the **geometric mean** of $n - 1$ and n, defined as $\sqrt{(n - 1)n}$. For instance, for a quota that lies between 4 and 5, the cutoff for rounding is given by

$$\sqrt{4 \cdot 5} = \sqrt{20} \approx 4.47214.$$

If the quota lies at or above this number, we round it up to 5, and if it lies below this number, we round it down to 4. Similarly, for a quota that lies between 7 and 8, the cutoff for rounding is given by

$$\sqrt{7 \cdot 8} = \sqrt{56} \approx 7.48331.$$

Notice that the fractional part of the cutoff, here 0.48331, is a little larger than it was for a quota between 4 and 5. In general, as n gets larger, the fractional part of the cutoff for rounding gets larger. To illustrate this, in Table 2.25 we have computed the value $\sqrt{(n - 1)n}$ for several values of $n - 1$ and n.

Table 2.25	Cutoffs for the Hill–Huntington Method
$n - 1, n$	**$\sqrt{(n - 1)n}$**
0, 1	0
1, 2	1.41421
2, 3	2.44949
3, 4	3.46410
4, 5	4.47214
9, 10	9.48683
75, 76	75.49834
999, 1000	999.49987

For the Hill–Huntington method, except for the cutoff between 0 and 1, the fractional part of the cutoff for rounding is always between 0.4 and 0.5, and the fractional part of the cutoff gets closer and closer to 0.5 as n increases. *This means that a fractional part less than 0.4 always rounds down (unless the quota is less than 1) and that a fractional part greater than 0.5 always rounds up.* Therefore, we only have to compute the actual cutoff for those quotas with fractional part between 0.4 and 0.5. Observe that the Hill–Huntington method always gives a state at least one representative because quotas less than 1 always round up to 1.

Other than a slightly tricky rounding rule, the **Hill–Huntington method** is the same as other divisor methods and may be summarized as follows.

The Hill–Huntington Method

Find a divisor such that when the allocation is given by rounding all the resulting modified quotas, with the cutoff for rounding between $n - 1$ and n given by the geometric mean $\sqrt{(n - 1)n}$, the correct number of seats is apportioned.

Note that a suitable divisor for the Hill–Huntington method may be higher or lower than the natural divisor, as is the case with Webster's method. In fact, the two methods often give exactly the same apportionment. Although the cutoffs for the Hill–Huntington method may seem unnecessarily complicated, they arise as consequences of certain properties one could want in an apportionment method, just as do Jefferson's and Webster's methods. In particular, the Hill–Huntington method minimizes the largest of the relative differences in population per representative taken over all pairs of states, where the relative difference in population per representative between two states is the difference in population per representative divided by the smaller population per representative. It also minimizes the largest of the relative differences in representation per individual.

The next two examples and the exercises at the end of this section have been specifically chosen to feature the special Hill–Huntington rounding method as well as to continue to show how to adjust the divisor when necessary. In practice, there are many times that the natural divisor works for the Hill–Huntington method, just as for Webster's method, and relatively few times when we must actually compute the rounding cutoff.

Before proceeding to the examples, we will work out a formula for the threshold divisor for the Hill–Huntington method. Recall that the divisor at which the number of seats assigned to a particular state increases from $n - 1$ to n is called a threshold divisor for n seats. Using the same line of reasoning as for the Jefferson's and Webster's threshold divisors, we see that the threshold divisor for the Hill–Huntington method is given by

$$\text{threshold divisor for } n \text{ seats under the Hill–Huntington method} = \frac{\text{population of the state}}{\sqrt{(n - 1)n}}.$$

As always, if we use a divisor greater than a particular state's threshold divisor for n seats, then the state will be assigned fewer than n seats. If we use a divisor less than the state's threshold divisor, the state will be assigned n or more seats.

Example 1   *The Hill–Huntington Method*

Using the Hill–Huntington method, apportion representatives to a hypothetical state legislature with 88 seats among the eight counties in Connecticut, based on the population figures from the 1940 census in Table 2.26.

Table 2.26	**Population of Connecticut in 1940**
County	*Population*
Fairficld	418,384
Hartford	450,189
Litchfield	87,041
Middlesex	55,999
New Haven	484,316
New London	125,224
Tolland	31,866
Windham	56,223

Source: U.S. Bureau of the Census.[25]

Solution: The total population of the eight counties is 1,709,242 and the natural divisor is $1,709,242/88 \approx 19,423.2045$. We next compute the natural quotas for the counties as shown in Table 2.27.

Table 2.27	**Initial Allocation for Connecticut**		
County	*Population*	*Natural Quota D = 19,423.2045*	*Initial Allocation*
Fairfield	418,384	21.5404	22
Hartford	450,189	23.1779	23
Litchfield	87,041	4.4813	5
Middlesex	55,999	2.8831	3
New Haven	484,316	24.9349	25
New London	125,224	6.4471	6
Tolland	31,866	1.6406	2
Windham	56,223	2.8946	3
Total	1,709,242		89

Hartford County's natural quota, with a fractional part less than 0.4, rounds down. The natural quotas of Fairfield, Middlesex, New Haven, Tolland, and Windham counties, with fractional parts greater than 0.5, round up. Because the remaining natural quotas have fractional parts between 0.4 0.5, we have to compute their cutoffs. For Litchfield County, the cutoff is $\sqrt{4 \cdot 5} \approx 4.47214 < 4.4813$, so Litchfield's quota rounds up. The cutoff for New London County is $\sqrt{6 \cdot 7} \approx 6.48074 > 6.4471$, so New London's quota rounds down. The resulting initial allocation, shown in Table 2.27, assigns one seat too many. We need to find a larger divisor to reduce the number of seats allocated.

Because Litchfield's natural quota of 4.4813 is only about 0.009 above its cutoff of 4.47214, it might lose a seat. However, Fairfield's natural quota is 21.5404, and the cutoff between 21 and 22 is $\sqrt{21 \cdot 22} \approx 21.49419$, so Fairfield only beat the cutoff by 0.046. Because Fairfield's population is much larger than Litchfield's, it might lose a seat rather than Litchfield. Let's compute the threshold divisors for both. We have

$$\text{Fairfield's threshold divisor for 22 seats} = \frac{418,384}{\sqrt{21 \cdot 22}} = \frac{418,384}{\sqrt{462}} \approx 19,464.9853,$$

Litchfield's threshold divisor for 5 seats $= \dfrac{87{,}041}{\sqrt{4 \cdot 5}} = \dfrac{87{,}041}{\sqrt{20}} \approx 19{,}462.9593$.

Therefore, Litchfield will be the first to lose a seat when we increase the size of the divisor. Let's choose 19,464 to be our new divisor because it lies well above Litchfield's threshold divisor and well below Fairfield's. We compute the modified quotas and new allocation in Table 2.28 and see that the allocation is correct this time.

Table 2.28	Apportionment for Connecticut				
County	*Population*	*Natural Quota* *D = 19,423.2045*	*Initial* *Allocation*	*Modified Quota* *D = 19,464*	*Final* *Allocation*
Fairfield	418,384	21.5404	22	21.4953	22
Hartford	450,189	23.1779	23	23.1293	23
Litchfield	87,041	4.4813	5	4.4719	4
Middlesex	55,999	2.8831	3	2.8771	3
New Haven	484,316	24.9349	25	24.8827	25
New London	125,224	6.4471	6	6.4336	6
Tolland	31,866	1.6406	2	1.6372	2
Windham	56,223	2.8946	3	2.8886	3
Total	1,709,242		89		88

Example 2 *The Hill–Huntington Method*

A firm has purchased 40 pairs of Super Bowl tickets to award to its top employees. It decides to allocate them to its various divisions in proportion to the number of employees in the divisions as listed in Table 2.29. Use the Hill–Huntington method to allocate the tickets. Is there any reason the firm may want to slightly bias the apportionment in favor of smaller divisions?

Table 2.29	Personnel by Division
Division	*Number of Employees*
Accounting	97
Customer service	196
Personnel	26
Production	1125
Public relations	58
Research & development	142

Solution: The total number of employees of the company is 1644 people and the natural divisor is $1644/40 = 41.1$. We next compute the natural quotas in Table 2.30.

Table 2.30	Initial Allocation of Tickets		
Division	**Size**	**Natural Quota** $D = 41.1$	**Initial Allocation**
Accounting	97	2.3601	2
Customer service	196	4.7689	5
Personnel	26	0.6326	1
Production	1125	27.3723	27
Public relations	58	1.4112	1
Research & development	142	3.4550	3
Total	1644		39

This time, we need only compute the cutoffs for public relations and for research & development. Both fall short of their respective cutoffs of $\sqrt{1 \cdot 2} = \sqrt{2} \approx 1.41421$ and $\sqrt{3 \cdot 4} = \sqrt{12} \approx 3.46410$, respectively, so we round down. We arrive at the initial allocation shown in Table 2.30 and see that it is one seat short of the 40 seats we need to assign. We need to find a smaller divisor to increase the number of seats allocated.

The only divisions that have natural quotas particularly close to the cutoff are public relations and research & development. Public relations falls 0.00301 short of being rounded up, and research & development falls 0.0091 short. We note that the production division is quite large, so even though its natural quota is far away from its cutoff of $\sqrt{27 \cdot 28} = \sqrt{756} \approx 27.49545$, it may be the division that will gain a seat. When we compute the threshold divisors for production, public relations, and research & development, we find that

$$\text{production's threshold divisor for 28 seats} = \frac{1125}{\sqrt{27 \cdot 28}} = \frac{1125}{\sqrt{756}} \approx 40.91585,$$

$$\text{public relation's threshold divisor for 2 seats} = \frac{58}{\sqrt{1 \cdot 2}} = \frac{58}{\sqrt{2}} \approx 41.01219,$$

$$\text{research \& development's threshold divisor for 4 seats} = \frac{142}{\sqrt{3 \cdot 4}} = \frac{142}{\sqrt{12}} \approx 40.99187.$$

We can let our new divisor be 41, so that public relations will gain the extra pair of tickets as shown in Table 2.31. A real squeaker!

Table 2.31	Apportionment of Super Bowl Tickets				
Division	**Size**	**Natural Quota** $D = 41.1$	**Initial Allocation**	**Modified Quota** $D = 41$	**Final Allocation**
Accounting	97	2.3601	2	2.3659	2
Customer service	196	4.7689	5	4.7805	5
Personnel	26	0.6326	1	0.6341	1
Production	1125	27.3723	27	27.4390	27
Public relations	58	1.4112	1	1.4146	2
Research & development	142	3.4550	3	3.4634	3
Total	1644		39		40

One reason the company may wish to favor smaller divisions is to make sure every division receives recognition of its contributions to the firm. It probably does not make much difference psychologically to those in production whether they get 27, 28, or 29 pairs of tickets (except to the few individuals who may think they just missed the cut).

Technology Tip

Commercial software packages are available that carry out all of the apportionment methods we have discussed, along with several other methods.

As with any divisor method, apportionments produced by the Hill–Huntington method may violate the Quota Property. This method also seems to show a slight favoritism toward smaller states.

Other Apportionment Methods

There are three other well-known divisor methods not included here. We have mentioned *Adams' method*, proposed in 1832 by former President John Quincy Adams. Adams' method, in which quotas are rounded up, is the polar opposite of Jefferson's method and has a bias in favor of the smaller states. Thus, the initial allocation will allot too many seats, so that we must find a divisor larger than the natural divisor in order to allocate the proper number of seats. It also follows that, as long as there are at least as many seats as states, every state will receive at least one seat under Adams' method. This would not necessarily happen with Hamilton's or Jefferson's methods, but does happen with Lowndes' method and the Hill–Huntington method. As with all divisor methods, apportionments produced by Adams' method may violate the Quota Property. New York, with its 1820 quota of 32.5031, would only have received 31 seats under Adams' method. Adams' method is the opposite of Jefferson's method in that it rounds up rather than down, so it is not surprising that Adams' method never gives a state more than its natural quota rounded up. We stated that Jefferson's method minimizes the greatest overrepresentation per individual. In its position diametrically opposite Jefferson's method, it is not surprising that Adams' method minimizes the greatest underrepresentation per individual.

Another divisor method, called *Dean's method*, was also proposed around 1832 in a letter to Daniel Webster from James Dean, professor of astronomy and mathematics at the University of Vermont. In Dean's method, the cutoff for determining whether a state receives $n - 1$ or n seats is at $(n - 1)n/(n - 0.5)$. Dean's method minimizes the differences in population per seat among the states. In terms of favoritism toward larger or smaller states, Dean's method falls between Adams' method and the Hill–Huntington method. Dean's method was never adopted by Congress.

A third method omitted here is *Condorcet's method*, named after the eighteenth century Frenchman whom we also encountered in Chapter 1. In this divisor method, the quotas are rounded up at 0.4 rather than at 0.5 as in Webster's method. Condorcet proposed it for use in the government formed in France after the French Revolution.

BRANCHING OUT 2.2

APPORTIONMENT METHODS USED IN OTHER COUNTRIES

Laws designating particular methods of apportionment procedures first appeared in the United States. However, the same procedures were subsequently rediscovered and used in other countries as well. In some cases, the specific method of computation may differ from those used in the United States, while still always yielding the same apportionment as one of our methods. This has led to many different names for the same method and even some confusion about which methods are the same.

In many countries, the legislature is elected by a system known as *proportional representation*. In this system, voters in each electoral district select delegates whose numbers are apportioned to the political parties in proportion to the vote each party receives. Lawyer and professor Victor d'Hondt reinvented and proposed Jefferson's method for electing part of each of Belgian's two houses in 1878. D'Hondt

wrote two papers on the subject. Also known as the method of highest averages, and the method of greatest divisors (a misnomer), Jefferson's method is used to select at least a part of the national legislatures today in Angola, Argentina, Brazil, Bulgaria, Croatia, Dominican Republic, Finland, Germany, Iceland, Israel, Italy, Japan, Mozambique, the Netherlands, Portugal, San Marino, Slovenia, Spain, Switzerland, and Uruguay. Frenchman A. Sainte-Laguë reinvented Webster's method from a different perspective in 1910. Sainte-Laguë's method or Webster's method, or a slight variant thereof, is used to apportion at least part of the legislatures in Denmark, Latvia, Norway, Poland, and Sweden. Hamilton's method, also known as the method of largest or greatest remainders, is used today in Austria, Belgium, El Salvador, Indonesia, Liechtenstein, and Madagascar.

We also want to mention that the bias for or against larger states is not necessarily a bad thing. Elections in other countries often feature more than two major parties or groups. If the rules of election are designed to give representation proportional to the votes received by a group, then using Jefferson's method to allocate seats would encourage the formation of coalitions where two or more parties combine into one. On the other hand, Adams' method would discourage such mergings and actually encourage the disintegration of parties. Either alternative could be desirable, depending on the system.

EXERCISES FOR SECTION 2.3

In Exercises 1–4, find the Hill–Huntington cutoff for rounding a quota that lies between the two given numbers.

1. 6 and 7   *2.* 11 and 12

3. 780 and 781   *4.* 945 and 946

5. Suppose that Pennsylvania's 38 electoral votes in the 1912 presidential election, won by Woodrow Wilson, had been apportioned to the candidates in proportion to the number of votes they received rather than all going to Theodore Roosevelt, who had the most votes. Use the Hill–Huntington method to allocate Pennsylvania's 38 electoral votes based on the vote totals in the table.

Candidate	Woodrow Wilson	Theodore Roosevelt	William H. Taft	Eugene V. Debs
Number of Votes	395,637	444,894	273,360	83,614

6. Suppose the 50 seats of Rhode Island's state senate are apportioned among the five counties in Rhode Island based on the 1990 census figures in the table. Use the Hill–Huntington method to carry out the apportionment.

County	Bristol	Kent	Newport	Providence	Washington
Population	48,859	161,135	87,194	596,270	110,006

Source: U.S. Bureau of the Census.[26]

7. Suppose a hypothetical state legislature of Connecticut is to have 79 seats, apportioned among the eight counties in Connecticut and based on the 1990 census figures in the table. Use the Hill–Huntington method to carry out the apportionment.

County	Fairfield	Hartford	Litchfield	Middlesex	New Haven	New London	Tolland	Windham
Population	827,645	851,783	174,092	143,196	804,219	254,957	128,699	102,525

Source: U.S. Bureau of the Census.[27]

8. A research psychologist has a grant to study 140 families. She wants a representative sample of families of different sizes. Use the Hill–Huntington method to determine the number of families of each size she should include in her study based on the estimated number of families of each size in the United States in 1995 given in the table.

Family Size	2	3	4	5	6	7+
Number of Families (in thousands)	32,736	17,065	15,396	6,774	2,311	1,334

Source: U.S. Bureau of the Census.[28]

9. Suppose a conference of Muslims from the six countries of North Africa is being planned. There are to be 44 official delegates, allocated based on estimated populations of the countries in the year 2000 given in the table.

Country	Algeria	Egypt	Libya	Morocco	Tunisia	Western Sahara
Population (in thousands)	31,743	67,542	6,294	32,189	9,599	245

Source: U.S. Dept. of Commerce.[29]

Apportion the delegates using

(a) the Hill–Huntington method,   (b) Hamilton's method,

(c) Lowndes' method,   (d) Jefferson's method,

(e) Webster's method.

(f) Which method seems to help Egypt, the largest country, the most?

10. Based on the enrollment figures from the previous year, a junior college expects the total enrollment in mathematics courses, broken down by course, to be as given in the table.

Course	Algebra	Trigonometry	Precalculus	Calculus I	Calculus II
Total Enrollment	286	133	678	568	295

The college employs 9 mathematics professors and they each teach 4 classes, so a total of 36 classes will be offered. The mathematics department chair decides to apportion the 36 classes among the 5 courses according to the expected enrollment. Apportion the classes using

(a) the Hill–Huntington method,   (b) Hamilton's method,

(c) Lowndes' method,   (d) Jefferson's method,

(e) Webster's method.

(f) Which method seems to hurt the student–professor ratio for Precalculus, the largest course, the most?

11. A city police department employs 175 patrol officers. The officers are to be apportioned among the 6 precincts based on the number of crimes reported in each precinct over the past year as shown in the table.

Precinct	1	2	3	4	5	6
Number of Crimes	3015	624	1775	1479	1212	921

Apportion the officers using

(a) the Hill–Huntington method,   (b) Hamilton's method,

(c) Lowndes' method,   (d) Jefferson's method,

(e) Webster's method.

12. A bank with six branches wants to set up a total of 11 ATM machines at the branches. The ATMs will be apportioned according to the average number of customers per day at each branch as given in the table.

Branch	Average Number of Customers per Day
Arlington	157
Belmont	298
Churchill	68
Oaklawn	354
Remington	99
Suffolk	167

Apportion the ATMs using

(a) the Hill–Huntington method,   (b) Hamilton's method,

(c) Lowndes' method,   (d) Jefferson's method,

(e) Webster's method.

13. Election to Uruguay's Chamber of Deputies is by parties on a proportional basis, with the number of seats each party wins determined by the percentage of the votes the party receives. Use the Hill–Huntington method to apportion the 1994 Chamber of 99 members among the four major political parties using the percentages of the vote they received in the election results shown in the table as the parties' populations.

Party	Percentage of Votes
Red Party	31.4
National White Party	30.2
Progressive Encounter	29.3
New Space	5.0

Source: Wilfried P. C. G. Derksen Web site.[30]

14. Election to Namibia's National Assembly is by parties on a proportional basis, with the number of seats each party wins determined by the percentage of the votes the party receives. Use the Hill–Huntington method to apportion the 72-seat Assembly among the five major political parties using the percentages of the vote they received in the 1994 election results shown in the table as the parties' populations.

Party	Percentage of Votes
South West African People's Organization	73.9
Democratic Turnhall Alliance	20.8
United Democratic Front	2.7
Monitor Action Group	0.8
Democratic Coalition of Namibia	0.8

Source: Wilfried P. C. G. Derksen Web site.[31]

15. Apportion the 105 seats in the 1792 U.S. House of Representatives based on the 1790 census figures in the table, using the Hill–Huntington method.

State	Population	State	Population
Connecticut	237,946	New Jersey	184,139
Delaware	59,094	New York	340,120
Georgia	82,538	North Carolina	393,751
Kentucky	73,677	Pennsylvania	434,373
Maine	96,540	Rhode Island	68,825
Maryland	319,728	South Carolina	249,073
Massachusetts	378,787	Vermont	85,539
New Hampshire	141,885	Virginia	747,610

Source: U.S. Census Office.[32]

(For a comparison with the apportionments obtained using other methods, see Exercise 10 in Section 2.1 and Exercise 12 in Section 2.2.)

16. Estimates for the total value of goods exported out of the United States in 1993, in millions of dollars, are listed in the table by commodity.

Commodity	Value of Goods (in millions of dollars)
Food and agriculture	30,818
Beverages and tobacco	5,675
Crude materials	22,329
Mineral fuels and related products	62,886
Chemicals and related products	42,153
Machinery and transport equipment	358,049
Manufactured goods and other	248,942
Total	770,852

Source: U.S. Bureau of the Census.[33]

(a) Make a table showing the percentage of each commodity exported, rounded to the nearest whole number. What is the total of the rounded percentages?

(b) Use the Hill–Huntington method to present the data in the table in terms of whole number percentages so that the sum of the percentages is exactly 100.

17. Estimates for the total value of goods imported into the United States in 1993, in millions of dollars, are listed in the table by commodity.

Commodity	Value of Goods (in millions of dollars)
Food and agriculture	44,694
Beverages and tobacco	8,074
Crude materials	34,820
Mineral fuels and related products	10,478
Chemicals and related products	61,755
Machinery and transport equipment	282,861
Manufactured goods and other	142,060
Total	584,742

Source: U.S. Bureau of the Census.[34]

(a) Make a table showing the percentage of each commodity imported, rounded to the nearest tenth of a percent (to one decimal place). What is the total of the rounded percentages?

(b) Use the Hill–Huntington method to present the data in the table in terms of percentages, to the tenth of a percent, so that the sum of the percentages is exactly 100.

In Exercises 18–20, note that Adams' method is a divisor method in which the quotas are rounded up to the nearest whole number.

18. Find an equation for the threshold divisor for n seats for Adams' method.

19. Apportion the seats in Exercise 7 using Adams' method.

20. Apportion the seats in Exercise 6 using Adams' method.

In Exercises 21–23, note that Dean's method is a divisor method in which the cutoff for determining whether a state receives $n - 1$ or n seats is at $(n - 1)n/(n - 0.5)$.

21. Find an equation for the threshold divisor for n seats for Dean's method.

22. Apportion the seats in Exercise 6 using Dean's method.

23. Apportion the seats in Exercise 7 using Dean's method.

In Exercises 24–26, note that Condorcet's method is a divisor method similar to Webster's method except that the quotas are rounded up at 0.4 rather than at 0.5.

24. Find an equation for the threshold divisor for n seats for Condorcet's method.

25. Apportion the seats in Exercise 7 using Condorcet's method.

26. Apportion the seats in Exercise 6 using Condorcet's method.

27. Explain why any divisor that is too small to give the correct number of seats using Webster's method is also too small to use with the Hill–Huntington method.

28. Explain why any divisor that is too large to give the correct number of seats using the Hill–Huntington method is also too large to use with Webster's method.

29. Find an example with just two states for which Webster's method and the Hill–Huntington method give different apportionments.

30. Find an example in which no states have the same population, and yet the Hill–Huntington method cannot give a valid apportionment.

2.4   THE SEARCH FOR AN IDEAL APPORTIONMENT METHOD

Along with looking at five methods of apportionment in the last three sections, we touched on some of the politics involved in the selection of apportionment methods. Members of Congress chose from among competing methods based on the immediate impact of the methods on their state, region, or party. What about the mathematics of apportionment? Are certain methods better than others by some objective, mathematical measures? In this section we briefly discuss a few properties that might be considered in the search for an ideal apportionment method.

We ran into the Quota Property, probably the most natural property we would like an apportionment method to satisfy, in Section 2.1. We recall its definition here.

The Quota Property

An apportionment is said to satisfy the **Quota Property** if each state's allocation is less than one from its natural quota. In other words, the final allocation for each state is equal to its natural quota, rounded either up or down to a whole number. An apportionment method is said to satisfy the Quota Property if *every* apportionment produced by the method satisfies the Quota Property.

Both Hamilton's and Lowndes' methods satisfy the Quota Property. We have seen that Jefferson's and Webster's methods do not satisfy the Quota Property, nor does the Hill–Huntington method. Michel L. Balinski and H. Peyton Young proved the following:

Theorem: No divisor method satisfies the Quota Property.

It is important to note that apportionments produced by different methods violate the Quota Property with different frequency. For example, Balinski and Young slightly varied the state populations from the 1970 census that went into the reapportionment of the House of Representatives in order to estimate the frequency with which the Quota Property would be violated for apportionments produced by the various divisor methods debated over the history of Congress. Based on their study, apportionments produced by Jefferson's method can be expected to violate quota within this fairly narrow range of populations nearly every time. On the other hand, apportionments based on Webster's method might be expected to violate quota only once every one or two thousand apportionments, whereas apportionments produced by the Hill–Huntington method might violate it once every few hundred apportionments. A similar study based on the 1980 census shows apportionments produced by Webster's method violating the Quota Property somewhat more often, though still significantly under 1% of the time. Surely we can live with such slight imperfections. On the other hand, since Hamilton's and Lowndes' methods satisfy the Quota Property, why not use one of them? In Section 2.1, we indicated that Lowndes' method has such a bias toward smaller states that it never had a significant chance of adoption. Hamilton's method does not appear to have any such bias, but it has other flaws, as we will now see.

Until 1921, the size of the House usually increased so that no state was faced with the loss of a seat. (The apportionment for the year 1842, when the size of the House was decreased, was the exception.) The apportionment following the 1880 census revealed a disturbing phenomenon. In those days, the Bureau of the Census prepared apportionments for Congress for a range of potential house sizes based on Hamilton's method, in this case from 275 to 350 seats. With a house size of 299 seats, Alabama was due 8 seats. However, with a house size of 300, it was due only 7! This is counter to common sense, and became known as the **Alabama paradox**. Although the earlier 1870 apportionment over a range of house sizes had Rhode Island losing a seat as the size of the House increased from 270 to 280, this phenomenon attracted little attention at the time. Let's examine what happened in 1882. Table 2.32 shows the natural quotas of the three relevant states.

Table 2.32	**1880 Quotas Leading to the Alabama Paradox**	
State	*Quota at 299 Seats*	*Quota at 300 Seats*
Alabama	7.6459	7.6715
Illinois	18.6400	18.7024
Texas	9.6399	9.6721

With a house size of 299 seats, there are 20 seats left to allot after the initial allocation. Nineteen states have fractional parts larger than that of Alabama, which will then get the twentieth and final extra seat. Illinois and Texas just miss out on an extra seat. With a house size of 300, there are 18 seats left to allot. This time the fractional parts of the quotas of Illinois and Texas slightly surpass that of Alabama, which just misses getting an extra seat. Just as when we decreased divisors, when the house size increases, the natural quotas go up by a fixed percentage, so that the larger states go up by a greater amount. In the end, a house size of 325 was chosen in part because Hamilton's method and Webster's method gave the same allocation for this house size.

Judicious choice of house size prevented major problems following the 1890 census. However, apportionments for house sizes from 350 to 400 following the census of 1900 exhibited six instances of the Alabama paradox, affecting five states. This led to the final downfall of Hamilton's method in the U.S. house of Representatives. The apportionments for the five affected states are shown in Table 2.33.

| Table 2.33 | The 1900 Census and the Alabama Paradox |

State	House Size							
	350–356	357	358–360	361	362	363	364–367	368–372
Colorado	3	2	3	3	3	3	3	3
Connecticut	4	4	4	4	4	4	4	4
Maine	3	3	3	3	3	3	3	3
North Dakota	1	1	1	2	2	2	1	2
Virginia	9	9	9	9	10	9	9	9

State	House Size							
	373	374	375–382	383–384	385	386	387	388–390	391–400
Colorado	3	3	3	3	3	3	3	3	3
Connecticut	5	4	5	5	5	5	5	5	5
Maine	3	3	3	4	4	3	4	3	4
North Dakota	2	2	2	2	2	2	2	2	2
Virginia	9	9	9	9	10	10	10	10	10

You might notice that there is exactly one house size for which Colorado gets 2 seats instead of 3—namely, 357. Populism was a strong political force in Colorado at the time. The chair of the House Select Committee on the Twelfth Census was Albert J. Hopkins of Illinois, known to be hostile to the populist movement. So what house size do you think his committee recommended? Why 357, of course! After long and bitterly personal debate, Hamilton's method was abandoned, and Webster's method reenacted.

It seems desirable to avoid the Alabama paradox if possible. If an apportionment method is not susceptible to the Alabama paradox, we will say it satisfies the House Size Property stated next.

The House Size Property

An apportionment method is said to satisfy the **House Size Property** if, for fixed populations, no state ever loses a seat when the size of the house increases.

What we call the House Size Property is often called *house monotonicity* in the technical literature on apportionment. Whereas neither Hamilton's method nor Lowndes' method satisfies the House Size Property, Jefferson's, Webster's, and the Hill–Huntington methods all do. More generally, Balinski and Young, the main contributors to the general theory of apportionment, proved the following in their 1982 book *Fair Representation.*

> ***Theorem:*** Every divisor method satisfies the House Size Property.

We already saw that none of the five apportionment methods we studied satisfy both the Quota Property and the House Size Property. However, there are complicated methods that satisfy both. One such method was developed by Balinski and Young and is known as the *Quota Method*. However, you might reasonably argue that the House Size Property is unimportant today in the U.S. House of Representatives or in any other body where the house size is fixed.

There are many other properties one might strive for in an apportionment method. We consider one final property.

> ### *The Population Property*
>
> An apportionment method is said to satisfy the **Population Property** if, for a fixed house size, no state whose population increases ever loses a seat to another state whose population decreases.

Note that the sizes of other states may be changing as well. Often called *population monotonicity* in the technical literature, the Population Property is considered one of the most desirable properties. Concisely and very generally speaking, migration patterns in the population should not cause a shift of seats in the opposite direction.

We now come to the big theorem in the theory of apportionment, analogous to Arrow's Impossibility Theorem in the theory of voting methods, which we discussed in Chapter 1. It was also proved by Balinski and Young.

> ***Theorem:*** An apportionment method satisfies the Population Property if and only if it is a "generalized" divisor method, where the rounding method may depend on the number of states and the house size. Consequently, no apportionment method satisfies both the Quota Property and the Population Property.

What are we left with? As in the theory of voting, we just saw that there is no ideal method of apportionment. However, as in the theory of voting, we can discover what problems arise and how often we can expect them to occur. For instance, we saw that Webster's method satisfies the House Size Property and the Population Property, and that apportionments produced by Webster's method satisfy the Quota Property for most situations similar to those arising in the apportionment of the U.S. House of Representatives. The Hill–Huntington method, currently in use in the House, also satisfies the first two properties, and apportionments based on Hill–Huntington do not violate the Quota Property all that much more often than those under Webster's method. The study of these and other properties, spurred by the work of Balinski and Young, continues to be of interest.

EXERCISES FOR SECTION 2.4

1. For each apportionment method in the table, place a check in the corresponding box if the method satisfies the given property.

		Quota Property	*House Size Property*	*Population Property*
Quota Methods	Hamilton			
	Lowndes			
Divisor Methods	Jefferson			
	Webster			
	Hill–Huntington			

In Exercises 2–4, consider all possible apportionment methods, not just the five that we studied in depth.

2. Give an example of an apportionment method satisfying both the Quota Property and the House Size Property or explain why none exists.

3. Give an example of an apportionment method satisfying both the Quota Property and the Population Property or explain why none exists.

4. Give an example of an apportionment method satisfying both the House Size Property and the Population Property or explain why none exists.

5. Show that the apportionment produced by Jefferson's method has two violations of the Quota Property for a 50-member house, with populations given by the following table.

State	A	B	C	D	E	F
Population	14,978	12,991	9,260	5,453	4,624	2,753

6. Show that the apportionment produced by Webster's method violates the Quota Property for a 110-member house and four states, with populations given by the following table.

State	A	B	C	D
Population	10,099	431	332	143

7. Show that Hamilton's method leads to a violation of the House Size Property as the house increases from 24 to 25 seats, with populations given by the following table.

State	A	B	C	D	E
Population	1676	1454	921	778	615

8. Show that Lowndes' method leads to a violation of the House Size Property as the house increases from 26 to 27 seats, with populations given by the following table.

State	A	B	C	D	E
Population	1920	1682	1313	486	241

9. Show that Lowndes' method leads to a violation of the Population Property for a 23-member house, with populations given by the following table.

State	A	B	C	D
Old Population	11,167	7,536	3,356	1,332
New Population	11,160	7,536	3,360	1,200

10. Show that Hamilton's method leads to a violation of the Population Property for a 40-member house, with populations given by the following table.

State	A	B	C	D
Old Population	1865	1269	712	594
New Population	1924	1272	710	594

In doing Exercises 11 and 12, remember that the Quota Property is most easily violated when there are great differences in the sizes of states.

11. Find an example with three states for which Jefferson's method violates the Quota Property.

12. Find an example with three states for which the Hill–Huntington method violates the Quota Property.

13. Explain why Webster's method satisfies the Quota Property when restricted to the case of countries with only two states. (Ignore the possibility that the fractional parts of the quotas are exactly 0.5.)

14. (a) For a country with exactly three states, what are the possible numbers of seats short that an initial apportionment (using the natural divisor) can be from the desired house size when using Jefferson's method?

    (b) For each of the cases in (a), how many states could have fractional part one-half or greater?

    (c) Use these observations, case by case, to explain why Webster's method satisfies the Quota Property when restricted to the case of countries with three states. (Webster's method is the only divisor method for which this is true.)

15. Explain why Hamilton's method satisfies the House Size Property when restricted to the case of countries with only two states.

16. Give an example of a country with three states and a house size that increases from 3 to 4 seats that shows Hamilton's method violates the House Size Property.

17. Give an example of a country with four states that shows Lowndes' method violates the House Size Property.

18. Explain why Hamilton's method must satisfy the Population Property when restricted to the case of countries with only two states.

19. Explain why Hamilton's method must satisfy the Population Property whenever we assume that the total population does not change.

20. Explain why Lowndes' method must satisfy the Population Property when restricted to the case of countries with only two states.

21. Give an example of a country with three states which shows that the Population Property fails to hold for Lowndes' method.

22. Give an example of a country with three states which shows that the Population Property fails to hold for Hamilton's method.

WRITING EXERCISES

1. Explain in your own words why Jefferson's method favors large states and why Lowndes' method favors small states.

2. Suppose the president of a small college forms a housing committee that includes 13 student representatives. The number of representatives allocated to each of the four residence halls on campus is based on the number of residents in each hall and the apportionment is done using Jefferson's method. The resulting allocation is shown in the table.

Residence Hall	Number of Residents	Number of Representatives
Gates	682	3
Moncrief	1045	5
Brown	389	1
Bass	834	4

Suppose you are a resident of Brown Hall and unhappy about the fact that Brown Hall was allowed only one representative. Write a letter of protest to the college president, explaining in full the reasons for your dissatisfaction.

3. Of the apportionment methods we studied in this chapter, which do you feel is the best method overall? Which method do you feel is the weakest? Support your answer with full explanations.

4. In many countries, election to legislative bodies is by parties on a proportional basis, with the number of seats each party wins determined by the percentage of votes the party received. Many of these proportional representation systems use Jefferson's method, which favors larger states, rather than Webster's method or the Hill–Huntington method, which are more neutral, to allocate seats in the legislature to the different political parties. Explain some of the political effects of this choice and why such effects might be desirable.

5. With Jefferson's method, every state will get a number of seats at least as large as its natural quota rounded down to the nearest whole number. However, the Quota Property may be violated so that some states may get more than their natural quota rounded up to the nearest whole number. Do you think that this Quota Property violation is a serious flaw? Support your answer with an argument.

6. For apportioning representatives to the U.S. House of Representatives, do you think it is better to have a method that favors large states or one that favors small states if you must choose between one or the other? Support your answer with an argument.

7. Because all apportionment methods must violate either the Quota Property or the Population Property, which do you think is the best to violate? Support your answer with an argument.

PROJECTS

1. Research and report on some legal challenge to an apportionment. (One such instance is the challenge by the state of Montana of the apportionment of the U.S. House of Representatives based on the 1990 census.)

2. Write a computer or calculator program or create a spreadsheet that will help carry out all the apportionment methods we saw.

3. Many computer programs that create pie charts or bar graphs round percentages so they add up to 100. (Examples are Microsoft Graph and Microsoft Excel.) Test one of these programs with several populations and house sizes for which our methods give different apportionments to see if this program seems to use one of our methods. Before you start, which do you think would be the likeliest methods the programs might use?

4. For most of the 1800s, the house size of the U.S. House of Representatives was chosen so that Hamilton's method and Webster's method gave the same apportionment. Select three of the "countries" with at least five states from the text examples or exercises. For each, apportion houses ranging from 20 to 25 members, inclusive, by both methods to see if they agree. Note any instances where the apportionments using Webster's method violate the Quota Property or those using Hamilton's method violate the House Size Property. What observations or conclusions do these examples suggest? Do they reveal most of the problems we discussed?

5. Generate five random populations between 1 and 9999. You might do this with the random number generator on a calculator or computer. These often provide decimal numbers between 0 and 1; if so, take the first four digits (throwing out 0000 if it occurs) as the population of a state. You could also generate random populations by constructing ten slips of paper with digits 0 through 9 and drawing one of the ten for each digit (again throwing out 0000 if it occurs). With these random populations, apportion a 100-member house using Jefferson's method, Webster's method, and the Hill–Huntington method. Repeat this a total of ten times.

(a) For each method, how many of the apportionments violate the Quota Property?

(b) Based on the results of part (a), how do the different methods compare with regard to violating the Quota Property? Is this what you would expect, and if so, why?

(c) Give some situations in which the method used here to generate populations might give an accurate simulation of the distribution of "populations" in the real world and some situations in which it would not.

KEY TERMS

Alabama paradox, 117
divisor, 84
divisor method, 83
fractional part, 75
geometric mean, 105
Hamilton's method, 75
Hill–Huntington method, 106
house size, 72

House Size Property, 118
Jefferson's method, 84
Lowndes' method, 78
modified quota, 84
natural divisor, 72
natural quota, 72
population, 72
Population Property, 119

Quota Property, 80
relative fractional part, 78
seat, 72
state, 72
threshold divisor for n seats, 87
total population, 72
Webster's method, 93

CHAPTER 2 REVIEW EXERCISES

1. Four Michigan Blue Cross and Blue Shield insurance companies proposed a merger in 1997. Their 35-member board of directors included representatives from the different facets of health care. Suppose they had instead allocated positions on the board to the four different companies based on the number of patients they had. Allocate positions on the board of directors using

(a) Hamilton's method,   **(b)** Lowndes' method,

(c) Jefferson's method,   **(d)** Webster's method,

(e) the Hill–Huntington method.

Company	Number of Patients
Blue Care Network of East Michigan	85,750
Blue Care Network-Great Lakes	142,450
Blue Care Network Mid Michigan	68,170
Blue Care Network of Southeast Michigan	238,280

Source: Blue Cross Blue Shield Blue Care Network of Michigan.[35]

2. Partial results of Greenland's Inatsi-satut election of March 5, 1995 are shown in the table. Apportion 28 seats to the three political parties based on these results, using

(a) Hamilton's method,   **(b)** Lowndes' method,

(c) Jefferson's method,   **(d)** Webster's method,

(e) the Hill–Huntington method.

Party	Siumut	Inuit Ataqatigiit	Atassut
Percentage of Votes	38.5	20.3	29.7

Source: Wilfried P. C. G. Derksen Web site.[36]

3. The 1997 enrollments of the junior high schools in the Chino Valley Unified School District in California are shown in the table. Suppose the district receives a grant to purchase 225 new computers for its five junior high schools. Apportion the computers to the schools based on their enrollments, using

(a) Hamilton's method,   **(b)** Lowndes' method,

(c) Jefferson's method,   **(d)** Webster's method,

(e) the Hill–Huntington method.

(f) Which apportionment is least favorable to the largest school, Canyon Hills?

(g) Do any of these apportionments violate the Quota Property?

School	Canyon Hills	Magnolia	Ramona	Townsend	Woodcrest
Enrollment	1050	924	917	841	502

Source: Chino Valley Unified School District.[37]

4. North Dakota selects its delegates to the Republican Convention in proportion to the results in its state primary. Use the percentages of the vote the major candidates received in the 1996 primary to apportion North Dakota's 18 delegates using

(a) Hamilton's method,   **(b)** Lowndes' method,

(c) Jefferson's method,   **(d)** Webster's method,

(e) the Hill–Huntington method.

(f) How do these apportionments compare to the actual apportionment of Dole 8, Forbes 4, Buchanan 3, Gramm 2, Alexander 1?

Candidate	Percentage of Votes
Bob Dole	42
Steve Forbes	20
Pat Buchanan	18
Phil Gramm	10
Lamar Alexander	6
Alan Keyes	3
Richard Lugar	1

5. Construct an example with four states in which a suitable divisor for Webster's method must be less than the natural divisor, which in turn must be less than a suitable divisor for the Hill–Huntington method.

6. If the natural divisor works for Webster's method, explain why Webster's method and Hamilton's method give the same apportionment.

7. Explain why Jefferson's method will never give a state fewer seats than its natural quota rounded down.

8. Explain why Lowndes' method must satisfy the Population Property whenever we further assume that the total population does not change.

Suggested Readings

Balinski, Michel L., and H. Peyton Young. *Fair Representation*. New Haven, Conn.: Yale University Press, 1982. The definitive historical reference of apportionment methods for the U.S. House of Representatives. Shows choices of method in their political contexts, with technical mathematics gathered in a lengthy appendix.

Olivastro, Dominic. "One Nation, Indivisible." *The Sciences* 32:5 (September/October 1992), 51–53. A concise history of apportionment in the U.S. House of Representatives.

Steen, Lynn A. "The Arithmetic of Apportionment." *Science News* 121:19 (May 8, 1982), 317–318. A brief history of apportionment in the U.S. House of Representatives.

Woodall, D. R. "How Proportional Is Proportional Representation?" *The Mathematical Intelligencer* 8:4 (1986), 36–46. Basic properties of apportionment methods, using the European names for the methods.

BN 0-471-13935-3

Before the development of coinage around 1000 B.C., people made loans and repaid them with interest. Over five thousand years ago, farmers borrowed grain

Babylonian deed of sale in Cuneiform with its envelope.

for planting and repaid it with an extra measure at harvest. In ancient Babylon, financial dealings began with agricultural products such as grain and cattle as the items of exchange. Later, precious metals and then coins were exchanged. The Babylonians used mathematics to record these dealings and to compute interest payments. Many of the surviving Babylonian cuneiform clay tablets are financial records.

Mortgages for the purchase of land and pawning of personal goods in exchange for money to be repaid were established in Greece by 700 B.C. Checks payable to the bearer go back as far as Ptolemaic Egypt over two thousand years ago. Today's financial transactions are more complex, but the idea of money turned over to another in exchange for a promise of repayment with interest is unchanged.

In a complex web of financial obligations, including mortgages, car loans, credit cards, and other loans, a family of four in the United States has an average debt of nearly $80,000 and manages to save an average of $3000 per year. When money changes hands, it is often done through a check or an electronic transfer with no cash involved. The pulse of our society depends on this vast network of obligations governed by these transactions.

In this chapter we will learn how interest is computed. We will see how the rate of interest and the frequency with which it is compounded affects the growth of savings. We will then move from considering the growth of a single, initial deposit with interest to considering a regular series of deposits, such as in some savings plans or retirement accounts. On the other side of the coin, we conclude the chapter with a discussion of loans repaid in installments. We will learn how banks determine the size of payments on a car loan or mortgage and the total cost of paying back such loans.

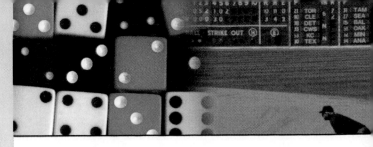

CHAPTER 3

THE MATHEMATICS OF MONEY

Solving interest problems often requires working with exponents. Here, we review the techniques needed to solve equations involving exponents.

The product $2 \cdot 2 \cdot 2 \cdot 2$ can be written as 2^4. In a like manner, for any number x and any positive integer n, the product of n copies of x can be written as x^n. That is,

$$x^n = \underbrace{x \cdot x \cdots x}_{n \text{ times}}.$$

We also have

$$x^{-n} = \frac{1}{x^n}.$$

For instance,

$$2^{-4} = \frac{1}{2^4} = \frac{1}{16} = 0.0625.$$

The values 2^4 and 2^{-4} can be computed using the exponentiation key on your calculator. In many of the applications we will look at, the exponents will be large and a calculator will be necessary. For example, using a calculator we find that

$$(1.04)^{72} \approx 16.84226241,$$

where we have rounded the value to eight decimal places. We will use the approximation symbol $\approx$ whenever values are rounded. *When solving algebraic equations in this chapter, we will round to eight decimal places any numbers that arise in intermediate steps.* The reason for keeping all these digits is that when we round too much in an intermediate step, the error introduced may be magnified by a later step and result in a final answer that is too far off the mark. In this section, where the calculations themselves are what we are studying, we will also retain eight decimal places in our final answers. However, in the mathematics of finance problems in later sections, where we are solving for interest rates, money, or time, we will round our final answer to two decimal places. Two decimal places will normally give a final answer precise enough for business purposes, especially when solving for dollars, because we will be rounding to the nearest penny.

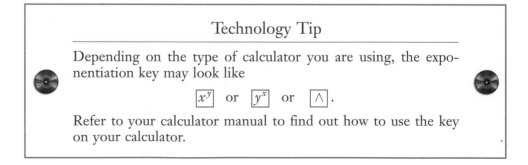

Technology Tip

Depending on the type of calculator you are using, the exponentiation key may look like

$$\boxed{x^y} \quad \text{or} \quad \boxed{y^x} \quad \text{or} \quad \boxed{\wedge}\,.$$

Refer to your calculator manual to find out how to use the key on your calculator.

We will also need to compute roots of numbers. For a positive number x and a positive integer n, the nth root of x is the positive number denoted by $\sqrt[n]{x}$ whose nth power gives x. (Negative numbers have nth roots only for n odd.) We will write the *n*th root of *x* in the fractional form

$$\sqrt[n]{x} = x^{1/n}.$$

For example, because $2^3 = 8$, we see that

$$8^{1/3} = 2.$$

The exponentiation key on the calculator can be used to compute the *n*th root.

In some of the equations that arise in the mathematics of finance, an expression involving the unknown variable will be raised to a power that is a number. For example,

$$(1 + y)^{36} = 1.31$$

is an equation of this type. To solve for the unknown variable in this kind of equation, we will need the following rule, which holds for any positive number *x* and any exponents *a* and *b*:

$$(x^a)^b = x^{ab}.$$

For example,

$$(3^2)^4 = 3^8 = 6561.$$

To understand why we multiply the exponents 2 and the 4 to get the exponent 8, notice that

$$(3^2)^4 = (3^2)(3^2)(3^2)(3^2) = (3 \cdot 3)(3 \cdot 3)(3 \cdot 3)(3 \cdot 3) = 3^8.$$

We will be particularly interested in the special case of the rule that says that for *x* positive and any nonzero number *a*,

$$\boxed{(x^a)^{1/a} = x^{a \cdot 1/a} = x^1 = x.}$$

In the next two examples, we see how this rule of exponents can be used to solve equations.

Example 1

Solve for *x* in the equation $x^{15} = 1325$.

Solution: To solve for *x* we raise both sides to the power 1/15. Then the exponent of *x* will be equal to 1.

$$x^{15} = 1325$$
$$(x^{15})^{1/15} = 1325^{1/15}$$
$$x^1 = x = 1325^{1/15} \approx 1.61490777.$$

■

Example 2

Solve for *y* in the equation $(1 + y)^{36} = 1.31$ where $1 + y > 0$.

Solution: We begin by raising both sides to the power 1/36, and we get

$$(1 + y)^{36} = 1.31$$
$$[(1 + y)^{36}]^{1/36} = (1.31)^{1/36}$$
$$1 + y = (1.31)^{1/36}.$$

Now we solve for y by subtracting 1 from each side of the equation and we see that

$$y = (1.31)^{1/36} - 1 \approx 0.00752895.$$

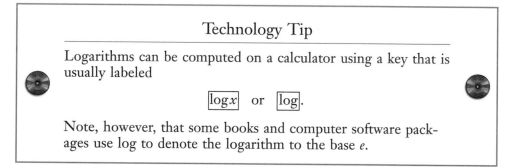

In interest calculations, we will often encounter equations in which we want to solve for a variable that appears in the exponent. For example,

$$(1.025)^x = 3$$

is an equation of this type. To solve for a variable in an exponent we use logarithms. We will use log, the logarithm to the base 10. For any positive number x, log x is the exponent to which 10 must be raised to give x. For instance, $\log 100 = 2$ because $10^2 = 100$. Similarly, $\log(1/10) = -1$ because $10^{-1} = 1/10$. Note that log x is defined only for positive x because no power of 10 is a negative number. For example, using a calculator we find that

$$\log 47 \approx 1.67209786.$$

Technology Tip

Logarithms can be computed on a calculator using a key that is usually labeled

$$\boxed{\log x} \quad \text{or} \quad \boxed{\log}.$$

Note, however, that some books and computer software packages use log to denote the logarithm to the base e.

There are several rules of logarithms, but we need only the following rule, which holds for any positive number a and any real number b:

$$\boxed{\log(a^b) = b \log a.}$$

You may want to verify this rule with a few examples using your calculator. We will now see how this rule can be used to solve equations.

Example 3

Solve for x in the equation $(1.025)^x = 3$.

Solution: We begin by taking the logarithms of both sides of the equation:

$$(1.025)^x = 3$$
$$\log(1.025)^x = \log 3.$$

Next, we apply the rule previously stated and we have

$$x \log(1.025) = \log 3.$$

Finally, dividing both sides of the equation by $\log(1.025)$, we solve for x and evaluate the result on a calculator to get

$$x = \frac{\log 3}{\log(1.025)} \approx \frac{0.47712125}{0.01072387} \approx 44.49151752.$$

In this calculation, we calculated the values of $\log 3$ and $\log(1.025)$ before performing the division. In general, it is faster to carry out the entire calculation at once on a calculator, recording only the final answer. This answer is often more accurate because the calculator may hold more decimal places in the intermediate calculations than you would write down. When the calculation in our example was done directly on a calculator, the final answer was 44.49153708. ∎

Example 4

Solve for t in the equation $4^{5t} + 64 = 4250$.

Solution: We first isolate the term 4^{5t} on one side of the equation:

$$4^{5t} + 64 = 4250$$
$$4^{5t} = 4250 - 64$$
$$4^{5t} = 4186.$$

Next, we take the logarithms of both sides of the equation and apply the logarithm rule to get

$$\log 4^{5t} = \log 4186$$
$$5t \log 4 = \log 4186.$$

Now we can divide both sides of the equation by both 5 and $\log 4$ to solve for t, and we have

$$t = \frac{\log 4186}{5 \log 4} \approx 1.20313566. \qquad ∎$$

Example 5

Solve for x in the equation $54.73 = \dfrac{1 - (1.004)^{-12x}}{0.004}$.

Solution: Before taking logarithms, we must isolate the term $(1.004)^{-12x}$ on one side of the equation. We begin by multiplying each side of the equation by 0.004:

$$54.73 = \frac{1 - (1.004)^{-12x}}{0.004}$$
$$(54.73)(0.004) = 1 - (1.004)^{-12x}$$
$$0.21892 = 1 - (1.004)^{-12x}.$$

Next, we subtract 1 from each side to get

$$0.21892 - 1 = -(1.004)^{-12x}$$
$$-0.78108 = -(1.004)^{-12x}.$$

To solve for x we would like to take the logarithms of both sides of the equation. However, we cannot take the logarithm of a negative number, so we must first multiply both sides of the equation by -1:

$$(-1)(-0.78108) = (-1)(-(1.004)^{-12x})$$
$$0.78108 = (1.004)^{-12x}.$$

Now, because $(1.004)^{-12x}$ is isolated on one side of the equation, we are ready to take the logarithms of both sides of the equation and solve for x:

$$\log(0.78108) = \log(1.004)^{-12x}$$
$$\log(0.78108) = -12x\log(1.004)$$
$$x = \frac{\log(0.78108)}{-12\log(1.004)} \approx 5.15774017.$$

BRANCHING OUT 3.1

THE DEVELOPMENT OF LOGARITHMS

French mathematician and physicist Pierre Simon Laplace (1749–1827) claimed that logarithms so reduced the calculations of astronomy that they effectively doubled the life of an astronomer. Who invented (discovered is probably a better word) such an important concept? Credit goes to the Scotsman John Napier (1550–1617) and the Englishman Henry Briggs (1561–1631). Napier was a well-to-do minor lord, and included the title Master of the Mint in his resume. Briggs was brought up in modest means and continued this lifestyle as a university professor. Napier's published works on the use of logarithms and tables for their use first popularized them. Napier's "logarithm," not quite a logarithm according to today's definition, involved the constant e ≈ 2.71828183 and was somewhat unwieldy to use. Briggs' logarithm was the familiar logarithm to the base 10, for which he contributed extensive tables computed by hand, accurate to 14 decimal places. Because our number system is base 10, Briggs' logarithms were significantly easier than Napier's to use. Seventeenth-century writer William Lilly reported that the first time Napier

and Briggs met, "almost one quarter of an hour was spent, each beholding the other with admiration, before one word was spoken." Logarithms led to the invention of the slide rule (in its modern form) by Englishman William Oughtred in the 1600s. The use of logarithmic tables and slide rules for approximate multiplication continued until they were supplanted by hand-held calculators in the 1970s.

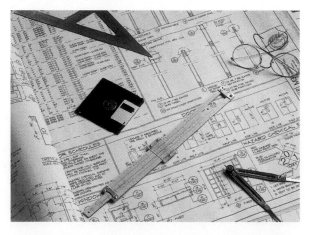

EXERCISES FOR SECTION 3.1

In each of the following exercises, solve for the unknown variable. Round your final answers to eight decimal places.

1. $x^9 = 568$
2. $x^{17} = 39$
3. $y^{50} = 7.23$ where $y > 0$
4. $z^5 = 358$
5. $x^{12} = 1570$ where $x > 0$
6. $t^{72} = 1.35$ where $t > 0$
7. $3^x = 782$
8. $5^x = 6026$
9. $(1.73)^{-z} = 8$
10. $6^{-x} = 0.32$
11. $9^t = 0.5$
12. $(1.047)^x = 9.2$
13. $8^{2x} = 49$
14. $6^{5y} = 0.39$
15. $(7.53)^{-4x} = 0.249$
16. $9^{-8x} = 104$
17. $(1 + x)^{108} = 7.3$ where $1 + x > 0$
18. $(1 + x)^{12} = 68$ where $1 + x > 0$
19. $(3 + t)^4 = 82$ where $3 + t > 0$
20. $(x - 2)^{72} = 4.32$ where $x - 2 > 0$
21. $4 + 2^t = 6520$
22. $(6.4)^x - 4 = 20$
23. $4^{-12z} + 1 = 1.75$
24. $1 + 7^{9z} = 6$
25. $(1.5)^{2x} - 3 = 8.6$
26. $2 + (9.2)^{-8x} = 2.32$
27. $186 = \dfrac{(1.02)^{4x} - 1}{0.02}$
28. $57.5 = \dfrac{(1.003)^{12t} - 1}{0.003}$
29. $72.3 = \dfrac{1 - (1.01)^{-12y}}{0.01}$
30. $2.41 = \dfrac{1 - (1.015)^{-4x}}{0.015}$

3.2 SIMPLE INTEREST

When we invest money in a bank account, the bank pays us interest. The most basic form of interest is simple interest. With **simple interest**, the interest is paid only at the end of a specified period of time and is paid only on the amount initially deposited, called the **principal**. The interest paid depends on the principal, the interest rate, and the length of time the account is held.

Suppose $200 is deposited into a simple interest account at 5% interest per year for 1 year. Then the amount of interest I is 5% of $200, which is given by

$$I = (200)(0.05) = \$10.$$

If the account is held for 6 years, then the interest paid is 6 times the amount of interest paid for 1 year and is given by

$$I = (200)(0.05)(6) = \$60.$$

Following this line of reasoning, we see that in general

simple interest = principal · rate · time.

We can now write a general algebraic formula for simple interest.

Simple Interest Formula

$$I = Prt \qquad \text{where} \quad \begin{cases} I = \text{interest} \\ P = \text{principal} \\ r = \text{annual interest rate} \\ t = \text{time (in years)} \end{cases}$$

The interest rate r we use in the formula is the annual interest rate. It is important to note that the interest rate r must be written as a decimal. If you are given the interest rate as a percentage, you convert it to a decimal by dividing by 100. For example, if the annual rate is 9.6%, then $r = 9.6/100 = 0.096$.

Example 1   *Simple Interest*

If $3500 is invested at the simple interest rate of 7% per year, how much interest will be earned if the investment is held for 4 years?

Solution: Here the principal is $P = 3500$, the annual interest rate is $r = 0.07$, and $t = 4$ years. Using the simple interest formula $I = Prt$, we find that the interest earned is

$$I = (3500)(0.07)(4) = \$980.$$

■

Example 2   *Simple Interest*

Suppose $2700 is deposited into an account earning simple interest of 4.8% annually. If the account is closed at the end of 15 months, how much interest will the bank pay?

Solution: Because the time is given in months rather than years, we first convert the months to years. We do this by dividing by 12, and we find that $t = 15/12 = 1.25$ years. Now we can use the simple interest formula $I = Prt$ with $P = 2700$, $r = 0.048$, and $t = 1.25$, and we see that the interest paid is

$$I = (2700)(0.048)(1.25) = \$162.$$

■

For any kind of investment, we are interested in what it will grow to be worth in the future. Appropriately, this amount is called the **future value** of the investment. The future value of any investment will depend on how much is invested and on the rate and method by which interest is paid. It will also depend on whether the investment is made in one lump sum or by several deposits over time. For a lump-sum investment earning simple interest we can easily compute its future value by adding the interest to be earned to the original principal invested. Using the letter F to denote the future value, we see that the future value of a simple interest investment is given by

$$F = \text{principal} + \text{interest} = \text{principal} + \text{principal} \cdot \text{rate} \cdot \text{time}.$$

Letting P denote the principal, I interest, r the rate, and t time, we see that

$$F = P + I = P + Prt = P(1 + rt).$$

Conversely, if we have a particular future value F in mind, we might ask what amount, P, would have to be deposited today in order to achieve a future value of F after t years of earning interest at a rate r. We call this amount P the **present value** corresponding to the future value F. For a simple interest investment, the present value is simply the principal P that will result in a future value F. Therefore, in the simple interest future value formula $F = P(1 + rt)$, we can think of P as either the principal or the present value depending on the particular viewpoint of the problem we are trying to solve.

Simple Interest Future Value Formula

$$F = P(1 + rt) \qquad \text{where} \quad \begin{cases} F = \text{future value after } t \text{ years} \\ P = \text{principal or present value} \\ r = \text{annual interest rate} \\ t = \text{time (in years)} \end{cases}$$

Example 3 *Finding the Future Value*

If you put $4250 in an account today that earns simple interest at a rate of 6.53% per year, what will the investment be worth 8 years from now?

Solution: We want to find the future value F of the investment after 8 years. Using the simple interest future value formula with the principal $P = 4250$, $r = 0.0653$, and $t = 8$, we see that after 8 years the investment will be worth

$$F = 4250(1 + (0.0653)8) = 4250(1 + 0.5224) = 4250(1.5224) = \$6470.20.$$

Example 4 *Finding the Present Value*

Find the amount of money you must deposit in a simple interest account paying 4.3% annually so that the account will be worth $5000 in 3 years.

Solution: In this example we want to find the present value, P, in the simple interest future value formula given that $F = 5000$, $r = 0.043$, and $t = 3$. Writing the formula and substituting the known values, we have

$$F = P(1 + rt)$$
$$5000 = P(1 + (0.043)3)$$
$$5000 = P(1 + 0.129)$$
$$5000 = P(1.129).$$

Solving for P gives

$$P = \frac{5000}{1.129} \approx \$4428.70.$$

Note that in the last step the value of P was rounded to two decimal places.

■

When a loan specifies simple interest, you can compute the interest due or the balance due on the loan using the formulas for simple interest or future value in the usual way. Such calculations are shown in the next two examples.

Example 5   *Simple Interest Loan*

A man borrows $300 from his parents to be repaid in 100 days at a simple interest rate of 5% annually. How much will he need to repay them 100 days later?

Solution: He will owe his parents the future value of $300 after 100 days, so we use the simple interest future value formula $F = P(1 + rt)$. Because t must be given in years, we have to convert 100 days to years by dividing by 365. We get $t = 100/365 \approx 0.27397260$ after rounding to eight decimal places. We have $P = 300$ and $r = 0.05$, so the future value is given by

$$F \approx 300(1 + (0.05)(0.27397260)) = 300(1 + 0.01369863)$$

$$= 300(1.01369863) \approx \$304.11,$$

and we see that he will owe $304.11 at the end of 100 days.

■

Example 6   *Pawn Shop Loan*

The receipt from a loan from a pawnshop is shown in Figure 3.1. The $130 loan was made on November 6, 1996 and was to be repaid with simple interest on December 6, 1996. At the end of the loan term the borrower had to pay back $156. Verify that the annual interest rate of 240% as stated on the receipt (as the annual percentage rate) is correct.

figure 3.1
Pawn shop receipt.

Solution: In this case we know the present value $P = 130$, the future value $F = 156$, and the time $t = 1$ month. Because t must be given in years, we have to convert 1 month to years by dividing by 12. We get $t = 1/12 \approx 0.08333333$ after rounding to eight decimal places. To find the rate r, we substitute into the simple interest future value formula and get

$$156 \approx 130(1 + r(0.08333333)).$$

To solve for r, we first isolate $r(0.08333333)$:

$$\frac{156}{130} \approx 1 + r(0.08333333)$$
$$1.2 \approx 1 + r(0.08333333)$$
$$1.2 - 1 \approx r(0.08333333)$$
$$0.2 \approx r(0.08333333).$$

Now, solving for r gives

$$r \approx \frac{0.2}{0.08333333} \approx 2.40000010 \approx 240\%.$$

So we see that the stated interest rate of 240% was correct. (The reason for the slight discrepancy in the value of r was due to rounding.)

$\blacksquare$

The interest rate of 240% seen in the last example is typical of the outrageous interest rates charged by pawn shops. Why do the pawn shop owners get away with this? They are not taking on very much risk because the borrowers must leave some kind of item or items of value as collateral for the loan that is being made, and if the loan is not repaid, the pawn shop owner can sell the merchandise. Pawn shops can charge high interest rates because their customers most likely do not have easy access to loans at lower rates, whereas the pawn transaction is easy and convenient for them.

BRANCHING OUT 3.2

THE ORIGIN OF THE WORD AND SYMBOL FOR PERCENT

The word percent, meaning per hundred, goes back to the Latin term per centum. Early twentieth-century scholar D. E. Smith traced the % symbol back to 1425 in Italy, where per cento meant per hundred. The cento, or hundred, was written $\overset{c}{c}$. The unknown author who introduced the precursor of the modern % notation wrote $P\,\overset{\circ}{\frown}{}^{\circ}$. Around 1650, the $\overset{\circ}{\frown}{}^{\circ}$ became $\frac{0}{0}$. When the fraction $\frac{a}{b}$ was typeset as a/b some two hundred years ago, per cent became 0/0 or stylized as %, and the P disappeared along the way. There is a similar, though obscure, notation ‰, meaning per mille, or per thousand.

EXERCISES FOR SECTION 3.2

The interest rates given in the following exercises are annual rates. Round all dollar amounts in the answers to the nearest cent.

In Exercises 1–6, find the simple interest I earned on the given principal for the time period and interest rate specified.

1. $P = \$4000$, $r = 5\%$, $t = 3$ years
2. $P = \$2300$, $r = 4\%$, $t = 5$ years
3. $P = \$752$, $r = 12.1\%$, $t = 8$ years
4. $P = \$20{,}000$, $r = 6.25\%$, $t = 2$ years
5. $P = \$35{,}200$, $r = 9\%$, $t = 7$ months
6. $P = \$85$, $r = 3.5\%$, $t = 14$ months

In Exercises 7–10, find the future value F of the given principal earning simple interest for the time period and interest rate specified.

7. $P = \$320$, $r = 3\%$, $t = 4$ years
8. $P = \$7123$, $r = 10.5\%$, $t = 5$ months
9. $P = \$51{,}225$, $r = 4.35\%$, $t = 50$ days
10. $P = \$8000$, $r = 7\%$, $t = 2$ years

In Exercises 11–14, find the present value P under simple interest given the future value, time period, and interest rate specified.

11. $F = \$2000$, $r = 9\%$, $t = 6$ months
12. $F = \$830$, $r = 8.4\%$, $t = 3$ years
13. $F = \$90{,}000$, $r = 14.2\%$, $t = 4$ years
14. $F = \$4400$, $r = 6\%$, $t = 10$ weeks

In Exercises 15 and 16, find the time t, in years, for the given principal to reach the future value under simple interest for the interest rate specified.

15. $P = \$300$, $F = \$400$, $r = 6.2\%$ 16. $P = \$20{,}150$, $F = \$29{,}000$, $r = 8\%$

In Exercises 17 and 18, find the annual interest rate r, for the given principal to reach the future value under simple interest for the time period specified.

17. $P = \$123$, $F = \$140$, $t = 10$ months   18. $P = \$1500$, $F = \$1850$, $t = 2$ years

19. Six years ago, a woman made an investment paying 7.5% simple interest. If her investment is now worth $3245, how much did she originally invest?

20. At Vanderbilt University in 1996, if a student did not pay tuition and fees on time, the student's account would be assessed a monthly late-payment fee of 18% annual simple interest. If a student owed $13,943 and the payment was two months late, what is the total amount the student would have to pay?

21. A $6000 investment in Intel Corporation stock in 1988 rose in value to $49,375 in 1997. What was the rate of return, figured as an annual simple interest rate, for this investment?

22. The Continental Congress borrowed $3.8 million in 1777 at 6% interest. How much simple interest would it have paid by 1782, when it was unable to continue making payments?

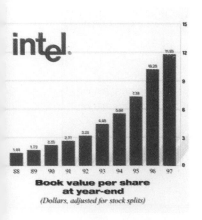

intel

Book value per share at year-end
(Dollars, adjusted for stock splits)

23. A man received a salary bonus of $1000. He needs $1150 for the down payment on a car. If he can invest his bonus at 8.25% simple interest, how long must he wait until he has enough money for the down payment?

24. A woman borrowed $75 from her parents, agreeing to pay them back in 9 months with simple interest. If the woman paid her parents $100 at the end of 9 months, what annual simple interest rate did she pay?

25. In 44 B.C., Brutus, the man who murdered Julius Caesar, made a loan to the city of Salamis. He charged an interest rate of 48%, even though 12% was the maximum legal rate. How much simple interest would Brutus make on a loan of 3000 talents for 3 years?

26. Suppose some municipal bonds pay 6.2% simple interest. How much should you invest in the bonds if you want them to be worth $5000 in 10 years?

27. Of states that limit total pawn interest rates, the state with the highest limit is Georgia at 300% simple interest. Suppose you borrow $50 from a pawn shop charging 300% simple interest. How much would you owe if you paid back the loan after 7 months?

28. If $32,000 is invested at 4.83% simple interest, how long will it take for the investment to be worth $40,000?

29. Suppose a man borrows $500 from a friend agreeing to pay it back in 2 years with 6% simple interest. How much will the man owe his friend at the end of the 2 years?

30. When people arrested for a crime do not have the money to post bail, they often call on a bail bondsman to put up the bail money. Bail bondsmen get their money back once the court proceedings of the arrested person have concluded. Laws in the state of Tennessee are fairly typical, allowing the bondsman to charge a premium of 10% of the bond and an additional $25 fee. If a person's bond is $10,000 and the case is over in 3 months, what is the annual simple interest rate on what is essentially a $10,000 loan for 3 months? Consider the combined amount of the 10% premium and the $25 fee to be the interest paid.

31. In the early to mid 1900s, New York City loan sharks would loan $4 at the beginning of the week to be repaid with $5 at the end of the week. What annual rate of simple interest did the loan sharks charge?

3.3 COMPOUND INTEREST

Many kinds of investments, including a typical savings account in a bank, earn interest that is compounded at regular intervals. The interest is paid on the initial deposit as well as on any previous interest payments credited to the account. This kind of interest is called **compound interest**. If banks only offered simple interest, then depositors could increase their interest earnings by withdrawing the money, drawing interest, and redepositing both the original principal and the interest earned, so that interest is paid on any previous interest earned. This would, at the very least, lead to a lot of extra work for the banks, so they offer compound interest.

As you might guess, the formula for finding the future value of an account with compounded interest is less obvious than for a simple interest account, so we begin by considering a simple example. Suppose you deposit $300 into an account earning 5% interest compounded annually and leave the account open for 4 years without any further deposits. At the end of the first year, interest will be paid on the $300, so the total amount in the account will be

$$300 + 300(0.05) = 300(1 + 0.05) = 300(1.05).$$

Notice that adding 5% interest is the same as multiplying by (1.05). At the end of the second year, 5% interest is paid again. Because the balance at the start of the second year is $300(1.05) = \$315$ and because paying 5% interest is equivalent to multiplying by (1.05), we see that, after interest is paid, the amount in the account at the end of 2 years will be

$$[300(1.05)](1.05) = 300(1.05)^2.$$

At the end of 3 years, interest will be paid on the balance of $300(1.05)^2 = \$330.75$ from the previous year, so the amount in the account will be

$$[300(1.05)^2](1.05) = 300(1.05)^3.$$

Finally, following this line of reasoning, we find that at the end of 4 years the amount in the account will be

$$[300(1.05)^3](1.05) = 300(1.05)^4 \approx \$364.65,$$

where the final amount was rounded to the nearest cent. This example suggests the formula for the future value of an account earning interest compounded annually. If P is the principal invested and r is the annual rate of interest compounded annually, then the future value F after t years, where t is an integer, is given by $F = P(1 + r)^t$.

In the general situation, interest may be credited more often than once a year. The interval of time between interest payments is called a **period**. Let n be the number of periods per year. For example, if interest is compounded monthly, then $n = 12$. The interest that is paid each period is called the periodic interest.

If the annual rate of interest is r and interest is compounded n times per year, then the **periodic interest rate** is given by

$$\text{periodic interest rate} = \frac{r}{n}.$$

For example, if the annual interest rate is 4.8% and the interest is compounded monthly, then the periodic interest rate is given by

$$\frac{r}{n} = \frac{0.048}{12} = 0.004.$$

Suppose P dollars is invested in a compound interest account. As in the case of simple interest, we call this initial amount P the present value or principal. Following the ideas from our example, we see that after one period the amount in the account is given by

$$P\left(1 + \frac{r}{n}\right),$$

and, in general, after k periods the amount in the account is given by

$$P\left(1 + \frac{r}{n}\right)^{k}.$$

Because there are n periods each year, we see that the total number of periods is given by $k = nt$, where t is the number of years the account is held. Combining these ideas, for nt an integer, we have the following general compound interest formula.

Compound Interest Formula

$$F = P\left(1 + \frac{r}{n}\right)^{nt} \quad \text{where} \quad \begin{cases} F = \text{future value after } t \text{ years} \\ P = \text{principal or present value} \\ r = \text{annual interest rate} \\ n = \text{number of periods per year} \\ t = \text{time (in years)} \end{cases}$$

To remember this formula, think of it as

future value = principal · (1 + periodic rate)^total number of periods.

To see the effect of compounding interest, we look at a graph. In Figure 3.2, the amount in an account with an initial deposit of $1000 that earns 8% interest is graphed both for the case of simple interest and for the case where the interest is compounded monthly. The graph for simple interest is a straight line, whereas the graph for the accrued value of the account earning compounded interest is curved upward. This is always true. The increasing slope of the compound interest curve shows us why compound interest always eventually outpaces simple interest, even if the simple interest rate is much higher.

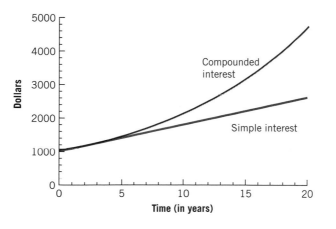

Figure 3.2
Amount in an account with an initial deposit of $1000 at 8% interest with interest compounded monthly and with simple interest.

Example 1   *Finding the Future Value*

Suppose $2400 is deposited in an account earning 5.54% interest compounded quarterly. How much will be in the account after 9 years?

Solution: Because interest is compounded quarterly, $n = 4$. We also have $P = 2400$, $r = 0.0554$, and $t = 9$. Using the compound interest formula, we find that the amount in the account after 9 years is given by

$$F = 2400\left(1 + \frac{0.0554}{4}\right)^{4 \cdot 9} = 2400(1.01385)^{36} \approx \$3937.90,$$

where the final answer has been rounded to the nearest penny.

■

The more frequently interest is compounded, the faster the account grows. This principle is illustrated in Example 2.

Example 2   *Comparing Compounding Periods*

Find the future value of a \$2000 investment after 6 years if it is earning an annual interest rate of 7% compounded

(a) annually,
(b) daily.

Solution:
(a) For interest compounded annually, $n = 1$, so after 6 years

$$F = 2000(1 + 0.07)^6 = 2000(1.07)^6 \approx \$3001.46.$$

(b) When the interest is compounded daily, $n = 365$, and we have

$$F = 2000\left(1 + \frac{0.07}{365}\right)^{365 \cdot 6} \approx 2000(1.00019178)^{2190} \approx \$3043.80.$$

With daily compounding, the future value is \$42.34 more than the future value of the investment when it is compounded annually.

■

If we had rounded the amount $1 + \frac{0.07}{365}$ to 1.000192 in the last calculation, we would have obtained $F \approx \$3045.26$, whereas rounding to 1.0002 would give $F \approx \$3099.07$. You get the most accurate answer by using all the decimal places you have available in each step of the calculations. Recall our conventions in this chapter: When solving algebraic equations, we will round any numbers that arise in intermediate steps to eight decimal places, and we will round our final answers in the mathematics of finance problems—where we are solving for interest rates, money, or time—to two decimal places.

In the last example, we saw that when the interest was compounded more frequently, the interest earned over time was greater. Because the actual interest earned on an account depends on the annual rate and on how often the interest is compounded, it would be nice to have a single number that allows us to compare different accounts. There is such a number and it is commonly called the **annual percentage yield** or **APY**. The APY is the actual percentage increase in the account over a 1-year period. From the compound interest formula, we see

Banks often advertise their interest rates in terms of APY.

that in a 1-year period the principal P is increased by a factor of

$$\left(1 + \frac{r}{n}\right)^{n \cdot 1} = \left(1 + \frac{r}{n}\right)^n.$$

For instance, if the annual rate is 6% and the interest is compounded monthly, the principal P is increased by a factor of

$$\left(1 + \frac{0.06}{12}\right)^{12} = (1.005)^{12} \approx 1.0617.$$

Therefore, we see that the annual percentage yield is 0.0617 or 6.17%. Following this idea, we have the following general formula:

Annual Percentage Yield (APY)

The annual percentage yield of an account with an annual interest rate r compounded n times per year is given in decimal form by

$$\text{APY} = \left(1 + \frac{r}{n}\right)^n - 1.$$

The APY is sometimes referred to as the annual effective yield or annual equivalent yield.

Example 3   *APY*

For an account with an annual interest rate of 8%, find the annual percentage yield when it is compounded

(a) quarterly,

(b) monthly,

(c) daily.

Solution:

(a) To find the APY when the interest is compounded quarterly, we use our formula for the annual percentage yield with $r = 0.08$ and $n = 4$. We see that

$$\text{APY} = \left(1 + \frac{0.08}{4}\right)^4 - 1 = (1.02)^4 - 1 = 0.08243216 \approx 8.24\%.$$

(b) Similarly, when the interest is compounded monthly, we find that

$$\text{APY} = \left(1 + \frac{0.08}{12}\right)^{12} - 1 \approx (1.00666667)^{12} - 1 \approx 0.08299955 \approx 8.30\%.$$

(c) For interest compounded daily,

$$\text{APY} = \left(1 + \frac{0.08}{365}\right)^{365} - 1 \approx (1.00021918)^{365} - 1 \approx 0.08327833 \approx 8.33\%.$$

∎

In Example 3, we saw that going from compounding monthly to daily resulted in a smaller increase in the APY than when we went from compounding quarterly to monthly. It turns out that going from compounding daily to compounding every hour, minute, second, or even "continuously" does not have much effect on the APY.

If you want a single deposit, earning compound interest at a fixed rate, to be worth some particular future value after a certain number of years, how much should you deposit? This is a question of finding the present value P in the compound interest formula. We see this idea in the next example.

A CLOSER LOOK 3.1

THE DIMINISHING EFFECT OF COMPOUNDING MORE FREQUENTLY

What effect does compounding more often than daily have on the APY? Let's see by comparing the APYs corresponding to an annual rate of 10% for interest compounded quarterly, monthly, daily, and every hour, minute, or second. (To avoid rounding problems, all intermediate calculations here were carried to 20 decimal places.)

Length of Period	APY (percent)
Quarter	10.38128906
Month	10.47130674
Day	10.51557816
Hour	10.51702873
Minute	10.51709076
Second	10.51709179

We see that the APY does not increase much when interest is compounded more often than daily. By letting the number of periods n get larger and larger, we have the idea of compounding continuously. The formula for the APY that results in this case is $e^r - 1$, where r is the annual interest rate and $e \approx 2.71828183$ is a special number that arises naturally in many areas of mathematics. In the case of an annual rate of 10%, the APY when interest is compounded continuously is given by

$$e^{0.10} - 1 \approx 0.1051709181 = 10.51709181\%,$$

which is only marginally higher than the APY when compounding every second.

Example 4   *Finding the Present Value*

A couple wants to invest some money for their child's college education. If they want to have $80,000 in 10 years and the investment pays 6.5% compounded monthly, how much should they deposit?

Solution: In this situation, we find the present value P corresponding to future value $F = 80,000$ when $r = 0.065$, $t = 10$, and $n = 12$. Substituting into the compound interest formula gives

$$80,000 \approx P\left(1 + \frac{0.065}{12}\right)^{12 \cdot 10}$$
$$80,000 \approx P(1.00541667)^{120}$$
$$80,000 \approx P(1.91218451),$$

and solving for P gives

$$P \approx \frac{80,000}{1.91218451} \approx \$41,836.97.$$

This is a rather large amount of money to deposit at one time. A more typical way to save for college tuition is to make several smaller deposits over a period of time. We will look at systematic savings plans of this type in the next section.

In some situations, the amount you have to invest is fixed but you may have in mind a goal of how much you want the investment to be worth in the future. The question is to find the interest rate required to reach that goal. In choosing investments, we often have some choice of interest rates, but with higher rates there may be some form of risk assumed by the investor. In the next example, we see how to solve for the interest rate required to meet a financial goal.

Example 5   *Finding the Interest Rate*

If $2500 is invested with interest compounded quarterly, what annual interest rate is needed for the investment to be worth $3000 in 15 months?

Solution: We must write the time t in years, so we divide 15 months by 12 to get

$$t = \frac{15}{12} = 1.25 \text{ years.}$$

We also have $P = 2500$, $F = 3000$, and $n = 4$. Using the compound interest formula, we get

$$3000 = 2500\left(1 + \frac{r}{4}\right)^{4 \cdot 1.25}$$
$$3000 = 2500\left(1 + \frac{r}{4}\right)^{5}.$$

Dividing each side by 2500, we have

$$\frac{3000}{2500} = \left(1 + \frac{r}{4}\right)^5$$

$$1.2 = \left(1 + \frac{r}{4}\right)^5.$$

Now we raise both sides to the power 1/5,

$$(1.2)^{1/5} = \left[\left(1 + \frac{r}{4}\right)^5\right]^{1/5}$$

$$1.03713729 \approx \left(1 + \frac{r}{4}\right),$$

and solving for r gives

$$0.03713729 \approx \frac{r}{4}$$

$$r \approx 4(0.03713729)$$

$$r \approx 0.1485 = 14.85\%.$$

We see that an annual interest rate of about 14.85% will make the investment grow to the desired amount.

Next we will see how we can use logarithms to solve for the time t in a problem involving compound interest.

Example 6   *Finding Time*

If $1600 is deposited into an account earning 6% compounded monthly, how long will it take for the account to be worth $2500?

Solution: In this case, we know that $P = 1600$, $F = 2500$, $r = 0.06$, and $n = 12$. Substituting into the compound interest formula gives

$$2500 = 1600\left(1 + \frac{0.06}{12}\right)^{12t}$$

$$2500 = 1600(1.005)^{12t}.$$

Because t appears in an exponent, we use logarithms to solve for it. We first divide both sides by 1600 to isolate the term $(1.005)^{12t}$:

$$\frac{2500}{1600} = (1.005)^{12t}$$

$$1.5625 = (1.005)^{12t}.$$

Next we take the logarithm of each side of the equation, obtaining

$$\log(1.5625) = \log(1.005)^{12t}$$

$$\log(1.5625) = 12t\log(1.005).$$

Finally, solving for t gives

$$t = \frac{\log(1.5625)}{12\log(1.005)} \approx 7.46 \text{ years.}$$

BRANCHING OUT 3.3

THE HIGH COST OF INTEREST:
THE INTER CHANGE BANK JUDGMENT

Shock waves went through the small Swiss canton of Ticino when, on October 3, 1994, it was ordered by a New York state court to pay more than $125 billion to an American family who claimed to have lost money when the local Inter Change Bank failed in the 1960s. The roots of the case date back to 1961, with the death of Seattle real estate investor Sterling Granville Higgins. In 1989, lawyers for Mr. Higgins' brother Robert Higgins filed suit in New York, claiming that in 1966 the estate of Sterling Higgins deposited $600 million in options on oil and mineral deposits in Venezuela with Inter Change Bank. The plaintiffs claimed that when the bank went bankrupt a year later many of the estate documents were lost. They claimed the bank had agreed to pay an extraordinary interest rate of 1% per week on the investment.

Because the Swiss government bankruptcy agencies that liquidated the bank never showed up in court, state Justice Gerald Held ruled in the estate's favor and ordered the bank to pay $125,444,300,221, corresponding to the value in 1994 of a $600 million investment made in 1966 with an annual percentage yield of about 21%. If the interest had actually been calculated at 1% per week compounded weekly, the judgment would have been for over $1000 trillion.

Ticino, with a population of only about 300,000 and annual government budget of about $2.1 billion, was in no position to pay the judgment and appealed the decision. The circumstances surrounding the case raised many questions. Both of the Higgins brothers had spent time in jail. Sterling died in San Quentin Prison, where he was serving time for writing bad checks. Robert had been convicted of fraud. At the time of Sterling Higgins' death, his net worth according to papers filed with the court was less than $5000. Swiss officials claimed that it was impossible that $600 million was deposited in 1966 because the bank books did not show anything near that in total value. They also claimed they were never told to appear in court and believed they were not subject to a judgment from a U.S. state court. The case was moved to U.S. District Court in Brooklyn, and on February 28, 1996, much to the relief of Ticino and the Swiss government, the judgment was overturned.

In the last example, we found that the account would be worth $2500 in about 7.46 years, which translates to 7 years, 5.52 months. However, if the interest is only paid at the end of the month and only on money left until the end of the month, then we would have to wait 7 years, 6 months to have at least $2500. Because interest is only paid at the end of each period, the formula is only exactly correct at the end of each period in this case. We are not concerned with this when solving for time, because our answers give a reasonable approximation to the answer we are seeking. We will take this approach whenever we are solving for time throughout this chapter. Note, however, that with some investments, if the account is closed in the middle of a period, interest will be paid even for the fraction of the period the money was invested. In such cases, the compound interest formula always gives a future value very close to the closing balance.

Inflation

Inflation is the increase in the prices of goods and services. If a 10% inflation rate holds for 3 years running, the cost of a $10 lunch increases to $11 after 1

BRANCHING OUT 3.4

HYPERINFLATION

Lenin stated in 1910 that "The surest way to overthrow an established social order is to debauch its currency." In a 1919 book, the famous British economist John Maynard Keynes paraphrased Lenin and went on to say "The process [inflation] engages all the hidden forces of economic law on the side of destruction, and does it in a manner which not one man in a million is able to diagnose." Whereas the highest rate of inflation in the United States since 1920 was 14.4% in 1947, other countries have suffered from runaway inflation, or *hyperinflation,* where prices might double or triple in a month, or even worse. If prices triple every month, they would increase by a factor of $3^{12} = 531,441$ (or 53,144,100%) over the course of a year. Such inflation rates rarely last beyond a few months, but even much lower increases take a great toll over the course of several years. For instance, hyperinflation almost inevitably results in shortages of consumer goods. It is most common in developing countries with heavy national debts. Simplistically speaking, the government simply prints the money it needs to remain solvent. Unfortunately, such action diminishes the purchasing power of all the money in

Marks being bailed as waste paper in Berlin. Their value as old paper was thousands of times their worth as marks.

year, then to $12.10 after 2 years, and to $13.31 after the third year. Inflation works just like compounding of interest, and we can use the idea of compound interest to calculate the effect of inflation on prices and salaries. In this case, we consider the annual inflation rate as the rate of interest and view the increases in salaries and prices as being compounded annually. This idea is illustrated in our next two examples.

Example 7   *Inflation*

Suppose the rate of inflation for the next 5 years is projected to be 4% per year. How much would you expect a car that costs $23,000 today to cost in 5 years?

Solution: An inflation rate of 4% per year has the same effect as paying interest of 4% per year compounded annually, where the price of the car today is the present value P and the price it will be in 5 years is the future value F. Therefore, we can use the compound interest formula with $P = 23,000, r = 0.04, t = 5,$

circulation. Weimar Germany, in the aftermath of World War I, provides an early example. Before the war, the combined value of all mortgages in the country was on the order of forty billion marks. When inflation peaked in 1923, forty billion marks might buy a loaf of bread. To combat this problem, billion mark notes were put into circulation.

A League of Nations report on the inflation throughout Europe in the 1920s noted that "Inflation is the one form of taxation which even the weakest government can enforce, when it can enforce nothing else." An inflation rate averaging over 10% *per day* in Hungary in 1945 resulted in what is possibly the largest denomination of currency ever circulated, the 100 quintillion (100,000,000,000,000,000,000) pengo note issued in 1946.

More recently, hyperinflation has occurred in Central and South America and in the countries of Eastern Europe as they make the transition out of the former Soviet bloc. The inflation rate has often been several thousand percent for a year or two. Amazingly, the Serbian 50,000 dinar note of the early 1990s was really worth 50 billion dinars, the other six zeros disappearing for accounting purposes. As a result of raging inflation in the 1980s, Brazilian labor unions pressed for daily payment of wages. It is easy to see why the civil, economic, and political instability caused by severe inflation steers all but the most nearsighted and desperate of politicians away from monetary recklessness.

100 quintillion pengo note.

and $n = 1$. Evaluating F we get

$$F = 23{,}000\left(1 + \frac{0.04}{1}\right)^{1\cdot 5}$$
$$F = 23{,}000(1.04)^5$$
$$F \approx 23{,}000(1.21665290)$$
$$F \approx \$27{,}983.02.$$

Therefore, we should expect the car to cost $27,983.02 or roughly $28,000 in 5 years.

Example 8   *Inflation*

Based on the Consumer Price Index, the rate of inflation in the United States between 1986 and 1996 was about 3.7% per year. What salary in 1986 would be equivalent to a $30,000 salary in 1996?

Solution: To solve this, we view the problem from a 1986 perspective. Then the salary of $30,000 in 1996 is the future value F, and the equivalent salary in 1986 is the corresponding present value P under an interest rate of 3.7% compounded annually. Because there are 10 years between 1986 and 1996, we have $t = 10$. Substituting $F = 30{,}000$, $r = 0.037$, $t = 10$, and $n = 1$ into the compound interest formula and solving for P, we have

$$30{,}000 = P\left(1 + \frac{0.037}{1}\right)^{1\cdot 10}$$
$$30{,}000 = P(1.037)^{10}$$
$$30{,}000 \approx P(1.43809496)$$
$$P \approx \frac{30{,}000}{1.43809496} \approx \$20{,}860.93.$$

EXERCISES FOR SECTION 3.3

In Exercises 1–8, find the future value F of the given principal under compound interest for the time period and interest rate specified.

1. $P = \$6000$, $t = 7$ years, $r = 4\%$ compounded annually
2. $P = \$35{,}000$, $t = 10$ years, $r = 7.3\%$ compounded annually
3. $P = \$7250$, $t = 3$ years, $r = 10\%$ compounded quarterly
4. $P = \$900$, $t = 20$ years, $r = 8\%$ compounded monthly
5. $P = \$2356$, $t = 9$ months, $r = 5.7\%$ compounded monthly
6. $P = \$50{,}000$, $t = 6$ years, $r = 3.5\%$ compounded quarterly
7. $P = \$544$, $t = 15$ years, $r = 5\%$ compounded daily $\ulcorner 151.6 \urcorner$
8. $P = \$7000$, $t = 2$ years, $r = 11.25\%$ compounded daily

In Exercises 9–12, find the present value P given the future value, time period, and interest rate specified.

9. $F = \$5400$, $t = 4$ years, $r = 6.5\%$ compounded annually

10. $F = \$16,200$, $t = 18$ months, $r = 3\%$ compounded quarterly

11. $F = \$185$, $t = 2$ years, $r = 10.3\%$ compounded daily

12. $F = \$14,000$, $t = 24$ years, $r = 6\%$ compounded monthly

In Exercises 13–16, find the time t, in years, for the given principal to reach the given future value at the interest rate specified.

13. $P = \$1200$, $F = \$2000$, $r = 5.8\%$ compounded monthly

14. $P = \$348$, $F = \$500$, $r = 7.53\%$ compounded daily

15. $P = \$10,000$, $F = \$40,000$, $r = 9\%$ compounded quarterly

16. $P = \$960$, $F = \$1400$, $r = 13\%$ compounded annually

In Exercises 17–20, find the annual interest rate r that will yield the given future value starting from the given principal over the specified time period.

17. $P = \$3000$, $F = \$4500$, $t = 4$ years, interest compounded annually

18. $P = \$12,600$, $F = \$20,000$, $t = 8$ years, interest compounded monthly

19. $P = \$65$, $F = \$227$, $t = 19$ years, interest compounded daily   6.58 %

20. $P = \$980$, $F = \$1100$, $t = 9$ months, interest compounded quarterly

In Exercises 21–24, find the annual percentage yield (APY) for the given interest rate compounded as specified.

21. $r = 4.7\%$ compounded monthly

22. $r = 10\%$ compounded quarterly

23. $r = 3.95\%$ compounded daily

24. $r = 6.5\%$ compounded monthly

25. For an account with an annual interest rate of 5%, find the annual percentage yield (APY) when it is compounded

   (a) quarterly,   (b) monthly,   (c) daily.

26. For an account with an annual interest rate of 6%, find the annual percentage yield (APY) when it is compounded

   (a) quarterly,   (b) monthly,   (c) daily.

27. Find the future value of an account after 5 years if \$540 is deposited initially and the account earns 8% interest compounded

   (a) annually,   (b) quarterly,   (c) monthly,   (d) daily.

28. Find the future value of an account after 12 years if \$3920 is deposited initially and the account earns 6.2% interest compounded

   (a) annually,   (b) quarterly,   (c) monthly,   (d) daily.

29. In May 1997, NationsBank was offering an interest rate of 4.71% compounded quarterly on 6-month certificates of deposit, whereas Bank of America was offering 4.70% compounded daily. If you wanted to invest in a 6-month certificate of deposit, which bank should you choose?

30. A woman wants to set up a college savings account for her granddaughter. If the account earns 5.4% compounded monthly, how much should she invest today so that the account will be worth $50,000 in 18 years?

31. A man has set a goal of saving $15,000 to bankroll the launching of his own business in 2 years. If he has $12,000 to invest today, what interest rate, compounded quarterly, would he have to earn on his investment so that it would grow to $15,000 in 2 years?

32. Suppose you lend $850 to a friend, and your friend agrees to pay the loan back in 3 years at 12% interest compounded quarterly. How much will the friend owe you at the end of 3 years?

33. In 1999, Ken Griffey Jr., outfielder for the Seattle Mariners, had a salary contract for $8.5 million, but only $7.25 million was paid in 1999 and the remaining $1.25 million was deferred until the year 2003. Assuming the Mariners could have made an investment at 7.5% interest compounded annually, what amount of money would they have needed to invest in 1999 so that the investment would be worth the $1.25 million they would need to pay Ken Griffey Jr. in 2003?   *930,000.66*

Ken Griffey Jr.

34. A woman who is now 28 years old invests $30,000 in a retirement account. What interest rate, compounded monthly, would she have to earn in order for the account to be worth $500,000 when she retires at age 65?

35. It is quite common for investors to settle for interest rates far below what they could be earning. In June 1997, the United States average interest rate for 5-year certificates of deposit with a minimum deposit of $10,000 was 5.642% compounded monthly, whereas one of the best rates in the country, 6.625% compounded monthly, was being offered by Advanta National Bank USA in Delaware. How much more interest would be earned on a $10,000 5-year certificate if it is invested at 6.625% compounded monthly rather than 5.642% compounded monthly?

36. California savings banks paid 15% interest in 1860, not too long after the Gold Rush. If the interest had been compounded quarterly, how long would it have taken a $4000 deposit to grow to $10,000?

37. A family has set a goal of saving $18,000 to be used toward the purchase of a new swimming pool. If they put $14,000 into an account earning 7% interest compounded daily, how long will they have to wait until the account has $18,000 in it?

38. Based on the Consumer Price Index, the rate of inflation in the United States during the high inflation years of President Jimmy Carter's term, beginning in January 1977 and ending in January 1981, was about 10.7% per year. What salary in 1977 would be equivalent to a $20,000 salary in 1981?

45,993.34 +

39. Tuition at Syracuse University grew by about 7.5% per year between 1986 and 1996. In 1996, the tuition was $16,710. Assuming that it will continue to grow at a rate of about 7.5% per year, what will tuition be at Syracuse in the year 2010?   *45,983.34*

40. Suppose the rate of inflation for the next 8 years is predicted to be 3% per year. How much would a house that costs $75,000 today be expected to cost in 8 years?

49,452.13 +

41. Based on the Consumer Price Index, the rate of inflation in the United States between 1980 and 1996 was about 4.1% per year. What salary in 1996 would be equivalent to a $26,000 salary in 1980?

42. In 1959, the rate for a first-class letter mailed within the United States was 4¢. In 1999, it was 33¢. Find the annual rate of growth (inflation rate) in the cost of mailing a letter from 1959 to 1999.

83,856.13 +

43. Suppose the rate of inflation for the next 4 years is expected to be 4.5% per year and you plan to save up enough money to purchase a $100,000 home 4 years from now. How much would a home that will cost $100,000 in 4 years be worth today?

44. The inflation rate in the Ukraine was 4735% per year in 1993. If this rate had continued compounding annually for five more years, how much would an item that cost 10 Karbovanets (the Ukrainian currency) in 1993 be expected to cost in the year 1998?

45. If a deposit is made into an account earning 3.92% compounded daily, how long will it take the investment to double in value?

46. If a deposit is made into an account earning 6% interest compounded monthly, how long will it take the amount in the account to triple in value?

47. Suppose $950 is deposited into an account earning 4% interest compounded daily. After 2 years, all of the money is withdrawn from the account and reinvested in an account earning 9% interest compounded quarterly for 6 more years. How much will the investment be worth after the entire 8 years? *$1755.47*

48. Suppose $3000 was deposited into a savings account in a bank. For the first 4 years the bank paid interest at a rate of 3.9% compounded monthly. For the next 2 years the bank paid interest at a rate of 4.2% compounded monthly. How much was in the account at the end of the entire 6 years?

49. One-thousand-year railroad bonds were issued in 1861 when interest rates on railroad bonds averaged 6.33%. How much would a bond purchased for $10,000 in 1861 be worth in 2861 assuming

   (a) simple interest of 6.33% is paid?

   (b) the interest is compounded monthly with an annual rate of 6.33%?

50. A woman deposits $500 into an account earning 4% interest compounded monthly. One year later she deposits an additional $300 into the same account. What will be the total amount in the account 4 years after she made her first deposit?

51. A $9200 deposit is made into an account earning 6.4% interest compounded quarterly. Two years later, an additional $3300 is deposited into the same account. What will be the total amount in the account 6 years after the first deposit was made? *17,720.55*

52. In 1997, Colonial Savings Bank offered a money market account with an annual percentage yield (APY) of 4.00%. The interest was compounded monthly. Find the annual interest rate for this account.

53. In 1996, the in-state tuition and fees for 4 years at Florida State University were $7528, a 4.67% increase from the year before. The Florida Prepaid College Program allows tuition and fees to be prepaid. In 1996, one way to prepay the in-state tuition and fees at a 4-year public college for a third-grade student who would enter college 10 years later in 2006 was with a lump sum of $5914.

   (a) If the inflation rate for tuition and fees at Florida State University is 4.67% per year from 1996 until 10 years later in 2006, what would the tuition and fees be in 2006?

   (b) Assuming the lump-sum payment of $5914 in 1996 would be worth the tuition and fees amount found in part (a) in 2006, what interest rate, compounded annually, would have been earned?

54. In the initial Inter Change Bank judgment discussed in Branching Out 3.3, the investment of $600 million made in 1966 was claimed to be worth $125,444,300,221 in 1994.

   (a) Find the interest rate that would give the same growth if simple interest were paid.

   (b) Find the interest rate that would give the same growth if interest were compounded monthly.

   (c) If the interest had been compounded at 1% interest per week, compounded weekly, find the value of the $600 million investment in 1994.

(a) $11,882.34

(b) 7.23%

3.4   THE REWARDS OF SYSTEMATIC SAVINGS

In many cases when we want to save money, we do not have a large amount to deposit at the outset. For instance, when parents try to save for their children's college education, they often do so by setting aside a small amount of money each month to save enough money over a long period of time. We call an account in which a fixed amount of money is deposited at regular intervals a **systematic savings plan**. These kinds of plans are sometimes called annuities, although the word annuity has several different meanings.

To calculate the amount of money that accumulates in a systematic savings plan, we must take into account the fact that because the deposits are made at different times, the amount of interest each deposit earns will be different. Let's see how this situation works by considering a particular example.

Suppose $80 is deposited at the end of each month for 5 months into a savings account earning 6% interest compounded monthly. How much money, including interest, will be in the account at the end of 5 months? This is simply the sum of the future values for the five deposits to be made at the end of each month. To figure this out, we look at each deposit separately. The first deposit is made at the end of the first month, so it will have been in the account earning interest for 4 months. Using the compound interest formula with $P = \$80$, $r = 0.06$, $t = \frac{4}{12}$ year, and $n = 12$, we find that the first deposit will be worth

$$80\left(1 + \frac{0.06}{12}\right)^{12 \cdot \frac{4}{12}} = 80(1.005)^4.$$

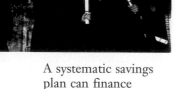

A systematic savings plan can finance college tuition.

The second deposit will only be in the account for 3 months, so $t = \frac{3}{12}$ in the compound interest formula and we see that the second deposit will be worth

$$80\left(1 + \frac{0.06}{12}\right)^{12 \cdot \frac{3}{12}} = 80(1.005)^3.$$

Following this reasoning, we see that the third deposit will be worth

$$80\left(1 + \frac{0.06}{12}\right)^{12 \cdot \frac{2}{12}} = 80(1.005)^2,$$

and the fourth deposit will be worth

$$80\left(1 + \frac{0.06}{12}\right)^{12 \cdot \frac{1}{12}} = 80(1.005)^1.$$

The fifth deposit is made at the very end of the 5 months, so it will have earned no interest and is only worth $80. Adding up the future values of each of the separate deposits, we see that at the end of 5 months the entire account will be worth

$$80(1.005)^4 + 80(1.005)^3 + 80(1.005)^2 + 80(1.005) + 80 \approx \$404.02.$$

Finding the amount in the account this way was not too bad for 5 months, but the work required to find the amount in the account after several years looks overwhelming. This is where a little mathematics goes a long way in simplifying the calculations.

We first notice that the reasoning we used to come up with the sum for the entire amount in the account would work for any number of months. If the deposits are made at the end of each month for k months, then the amount in the

A CLOSER LOOK 3.2

ADDING IT UP: THE SUM OF A GEOMETRIC SERIES

The formula for the sum of a geometric series

$$a^{k-1} + a^{k-2} + \cdots + a + 1 = \frac{a^k - 1}{a - 1},$$

for $a \neq 1$, follows from the identity

$$(a - 1)(a^{k-1} + a^{k-2} + \cdots + a + 1) = a^k - 1.$$

This last equation can be seen to be true by multi-

plying out the left-hand side of the equation and seeing that everything cancels out except $a^k - 1$. For instance, if $k = 4$ we have

$$\begin{aligned}
(a - 1)(a^3 + a^2 + a + 1) \\
= (a^4 + a^3 + a^2 + a) - (a^3 + a^2 + a + 1) \\
= a^4 - 1.
\end{aligned}$$

account after k months is the sum of the future values:

$$\begin{aligned}
80(1.005)^{k-1} + 80(1.005)^{k-2} + \cdots + 80(1.005) + 80 \\
= 80[(1.005)^{k-1} + (1.005)^{k-2} + \cdots + (1.005) + 1].
\end{aligned}$$

Now we can apply the formula for the sum of a geometric series. This formula says that for any number a not equal to 1 we have

$$a^{k-1} + a^{k-2} + \cdots + a + 1 = \frac{a^k - 1}{a - 1}.$$

Applying this formula to the last equation, we find that the amount in the account after k months is given by

$$80\left(\frac{(1.005)^k - 1}{1.005 - 1}\right) = 80\left(\frac{(1.005)^k - 1}{0.005}\right).$$

This formula is easy to use even if the number of months k is very large.

Now let's take these ideas to come up with a formula for the general situation. We notice that the 80 is just the amount of each deposit, the number 0.005 is the periodic interest $\frac{r}{n}$, where r is the annual interest rate, and n is the number of periods each year. Also, the number k is the number of periods the account is held, so $k = nt$, where t is the number of years the account is held. Putting all of this together, for nt an integer, we have the following general formula for the future value of a systematic savings plan.

Systematic Savings Plan Formula

$$F = D\left(\frac{\left(1 + \dfrac{r}{n}\right)^{nt} - 1}{\dfrac{r}{n}}\right) \quad \text{where} \quad \begin{cases} F = \text{future value after } t \text{ years} \\ D = \text{amount of each deposit} \\ r = \text{annual interest rate} \\ n = \text{number of periods per year} \\ t = \text{time (in years)} \end{cases}$$

Deposits are made at the end of each period.

It is important to note that the intervals at which the deposits are made have to be the same as the intervals at which interest is compounded in order for the formula to apply. It is not very difficult to adapt the formula to the situation when this is not the case, but we will not pursue it here, and instead leave the derivation of the formula as an exercise.

In deriving the systematic savings plan formula, we also assumed that the interest rate did not vary and that the deposits were all the same size and at regular intervals. If either of these were not the case, a simple formula like the one we found would not be possible. In most true savings situations, interest rates vary over time and the deposits vary in size and may be irregularly spaced. However, we can still use our systematic savings plan formula to make reasonable estimates. Such estimates are key to long-term financial planning.

Example 1　*Finding the Future Value*

Suppose you deposit $200 at the end of each month into an account earning 5.4% interest compounded monthly. How much will be in the account, how much of the future value will be from deposits, and how much of the future value will be from interest after making deposits for

(a) 15 years?
(b) 40 years?

Solution:

(a) The amount in the account after 15 years of deposits is the future value F given by the systematic savings formula with $D = 200$, $r = 0.054$, $t = 15$, and $n = 12$. Substituting into the formula, we find that

$$F = 200\left(\dfrac{\left(1 + \dfrac{0.054}{12}\right)^{12 \cdot 15} - 1}{\dfrac{0.054}{12}}\right)$$

$$F = 200\left(\dfrac{(1.0045)^{180} - 1}{0.0045}\right)$$

$$F \approx 200(276.40602937)$$

$$F \approx \$55{,}281.21.$$

To find the total amount deposited, we simply multiply the amount of each deposit, $200, by the number of deposits $nt = 12 \cdot 15 = 180$. We find that the total amount deposited is

$$\$200 \cdot 180 = \$36{,}000.$$

Because only $36,000 of the future value of $55,281.01 came from deposits, the rest must have come from interest. So the total amount of interest accumulated is given by

$$\$55{,}281.21 - \$36{,}000 = \$19{,}281.21.$$

(b) The amount F in the account after 40 years is given by

$$F = 200\left(\frac{\left(1 + \dfrac{0.054}{12}\right)^{12\cdot40} - 1}{\dfrac{0.054}{12}}\right)$$

$$F = 200\left(\frac{(1.0045)^{480} - 1}{0.0045}\right)$$

$$F \approx 200(1695.38301294)$$

$$F \approx \$339{,}076.60.$$

Because $12 \cdot 40 = 480$ deposits of \$200 each are made over 40 years, the amount of the future value from deposits is

$$\$200 \cdot 480 = \$96{,}000,$$

and the amount of the future value from interest is given by

$$\$339{,}076.60 - \$96{,}000 = \$243{,}076.60.$$

After 40 years, more of the money in the account is from interest than from deposits. In fact, as we can see in the graph in Figure 3.3, after only about 23 years, the interest accumulated in the account had surpassed the amount deposited. If the account were continued even longer, an even higher percentage of the account would come from interest.

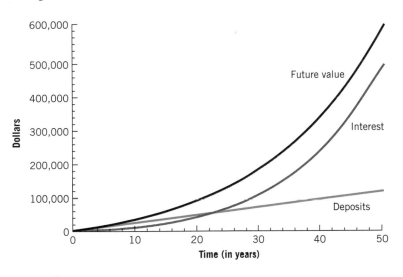

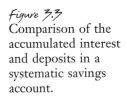

Figure 3.3
Comparison of the accumulated interest and deposits in a systematic savings account.

Example 2   Saving for College

The parents of a kindergarten student want to save \$160,000 to pay for the estimated cost of her college education. They plan to invest their savings in a mutual fund, and based on the past performance of the fund and its projected future performance, they estimate that it will give returns that are equivalent to earning

BRANCHING OUT 3.5

THE TAX ADVANTAGES OF IRAS

Individual retirement accounts (IRAs) are investments with tax advantages for the investor. One kind of IRA, known as a traditional IRA, does not require the payment of taxes on interest earned until the money is withdrawn after retirement. In addition, for an investor who is not covered by an employer-sponsored retirement plan or whose income is below a specified level, the money invested in a traditional IRA is deductible from his or her income. In this case, all the money in the account is taxable on withdrawal. Individuals can invest up to $2000 each year in a traditional IRA.

To see how the tax advantages affect the growth of an investment, consider the example of an individual in the 28% tax bracket who can take full advantage of a traditional IRA's tax breaks. Suppose the person invests $2000 at the end of each year for 30 years in a traditional IRA at 7% interest, compounded annually. The amount in the account after t years is given by

$$F = 2000\left(\frac{(1.07)^t - 1}{0.07}\right).$$

However, we assume that the withdrawals will be taxed at 28% on withdrawal, so the true value of the account to the investor after t years is only

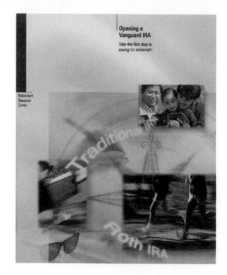

100% − 28% = 72% of the total or

$$F = (0.72)\left(2000\left(\frac{(1.07)^t - 1}{0.07}\right)\right).$$

By comparison, if the individual invested in a regular account, then the $2000 would first be taxed at 28%, leaving only 72% of $2000, or $(0.72) \times \$2000 = \$1440$ to be deposited each year. In addition, because the interest will be taxed at 28%

about 8% interest compounded quarterly. How much should the parents deposit at the end of each quarter into an account paying 8% interest compounded quarterly so that the account will be worth $160,000 in 13 years?

Solution: In this case we have $F = 160,000$, $r = 0.08$, $t = 13$, $n = 4$, and we want to find the amount of each deposit D. Substituting into the systematic savings plan formula and solving for D, we get

$$160,000 = D\left(\frac{\left(1 + \dfrac{0.08}{4}\right)^{4 \cdot 13} - 1}{\dfrac{0.08}{4}}\right)$$

each year, we can consider the true interest rate to be $(0.72) \times 7\% = 5.04\%$. Therefore, after taxes, the amount in the account after t years is given by

$$F = 1440\left(\frac{(1.0504)^t - 1}{0.0504}\right).$$

The withdrawals are not taxable, so this is the true value of the account for the investor after t years. The true value of the accounts to the investor over the 30-year period is shown in the graph.

After 30 years, the traditional IRA account is worth \$136,023.53 and the account with no tax breaks is worth only \$96,331.71. The tax advantages are even greater if the investor is able to avoid Social Security taxes on the amount deposited, as is possible in some cases, or if the investor is in a lower tax bracket after retirement.

Most investors have the option of choosing another kind of IRA, called a Roth IRA. With this kind of IRA, the money invested cannot be deducted but the interest earnings are never taxed. Which kind of IRA is a better option for an investor depends on the investor's tax bracket before and after retirement and on how long the money is invested.

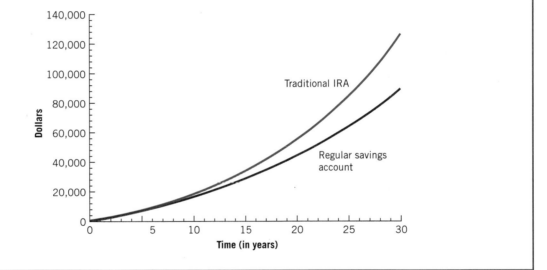

$$160,000 = D\left(\frac{(1.02)^{52} - 1}{0.02}\right)$$

$$160,000 \approx D(90.01640927)$$

$$D \approx \frac{160,000}{90.01640927}$$

$$D \approx \$1777.45.$$

Making a deposit of \$1777.45 each quarter means that the parents will have to budget almost \$600 each month toward the fund—a sobering thought.

Example 3   *Finding Time*

Suppose you want to have $3000 available for the down payment on a car. If you deposit $120 every month, at the month's end, into an account earning 6% compounded monthly, how long will it take until you have the down payment?

Solution: In this situation we know $F = 3000$, $D = 120$, $r = 0.06$, and $n = 12$. We use the systematic savings plan formula to solve for the number of years t it will take to save $3000. Substituting into the formula, we have

$$3000 = 120\left(\frac{\left(1 + \dfrac{0.06}{12}\right)^{12t} - 1}{\dfrac{0.06}{12}}\right)$$

$$3000 = 120\left(\frac{(1.005)^{12t} - 1}{0.005}\right).$$

How do we solve for this t? A systematic approach is the key. We know that in the end we need to use logarithms, but first we need to isolate the term $(1.005)^{12t}$. We use algebra to bring us step by step to our goal.

$$\frac{3000}{120} = \frac{(1.005)^{12t} - 1}{0.005}$$

$$25 = \frac{(1.005)^{12t} - 1}{0.005}$$

$$(25)(0.005) = (1.005)^{12t} - 1$$

$$0.125 = (1.005)^{12t} - 1$$

$$0.125 + 1 = (1.005)^{12t}$$

$$1.125 = (1.005)^{12t}.$$

Now the term $(1.005)^{12t}$ is isolated, so we take logarithms of both sides of the equation and solve for t.

$$\log(1.125) = \log(1.005)^{12t}$$

$$\log(1.125) = 12t\log(1.005)$$

$$t = \frac{\log(1.125)}{12\log(1.005)}$$

$$t \approx 1.97 \text{ years.}$$

The time we solved for in Example 3 is not exactly the end of a period. The next period will end at 2 years or 24 months, so in fact when $t = 1.97$, because the last deposit has not been made and interest has probably not been paid for the last period, the account will not be worth $3000. In fact, it will never be worth exactly $3000, and its value will exceed $3000 when $t = 2$. As with solving for time in the case of compound interest problems, we will not be concerned with this here. So, for instance, in Example 3, we will keep the answer $t = 1.97$. Just keep in mind that the times that we solve for are approximate. They should be sufficiently accurate for financial planning.

EXERCISES FOR SECTION 3.4

In Exercises 1–4, find the future value F for a systematic savings plan with the given deposits made at the ends of the compounding periods for the time period and interest rate specified.

$59561.68 +

1. $D = \$640$, $t = 15$ years, $r = 5.6\%$ compounded quarterly

2. $D = \$25$, $t = 30$ years, $r = 4.8\%$ compounded monthly

$151898.73 +

3. $D = \$700$, $t = 10$ years, $r = 11\%$ compounded monthly

4. $D = \$35$, $t = 50$ years, $r = 9.2\%$ compounded quarterly

In Exercises 5–8, find the amount of each deposit D for a systematic savings plan with the deposits made at the ends of the compounding periods given the future value, time period, and interest rate specified.

$687.42 +

5. $F = \$40,000$, $t = 11$ years, $r = 5\%$ compounded quarterly

6. $F = \$892$, $t = 3$ years, $r = 4\%$ compounded quarterly

$222.61

7. $F = \$5000$, $t = 21$ months, $r = 8\%$ compounded monthly

8. $F = \$3,000,000$, $t = 40$ years, $r = 6.2\%$ compounded monthly

In Exercises 9–12, find the time t, in years, for a systematic savings plan with the given deposits made at the ends of the compounding periods to reach the given future value under the specified interest rate.

9. $D = \$75$, $F = \$1200$, $r = 6\%$ compounded monthly

10. $D = \$800$, $F = \$25,000$, $r = 5.8\%$ compounded monthly

11. $D = \$130$, $F = \$4000$, $r = 12\%$ compounded quarterly

12. $D = \$680$, $F = \$200,000$, $r = 9.3\%$ compounded quarterly

13. A woman has $75 deducted from her paycheck at the end of each month and put into a savings account earning 9% interest compounded monthly. She continues these deposits for 10 years.

   (a) How much will the account be worth after 10 years?

   (b) How much of the future value will be from deposits?

   (c) How much of the future value will be from interest?

14. Suppose a 27-year-old man bought an annuity (a systematic savings plan) through the Jackson National Life Insurance Company in 1997. The plan calls for deposits of $100 at the end of each month and pays an interest rate of 6.25% compounded monthly.

   (a) How much will the annuity be worth when the man is 65 years old in 2035?

   (b) How much of the future value will be from deposits?

   (c) How much of the future value will be from interest?

15. A family wants to save $7000 to use toward the purchase of a piano. If they deposit $600 at the end of each quarter into an account earning 9% interest compounded quarterly, how long will it take for them to save $7000?

16. Suppose you would like to have $12,000 available to be used toward the down payment on a house in 3 years. If you can invest the money in an account earning 6.4% interest compounded quarterly, how much should you deposit at each quarter's end so that you will have $12,000 in 3 years?

17. Suppose a man makes monthly deposits into a retirement savings account earning 7%

compounded monthly. How much will he have in the account when he retires at age 70 if

(a) he makes deposits of $100 at each month's end beginning at age 22?

(b) he makes deposits of $200 at each month's end beginning at age 40?

18. From 1926 to 1996, while inflation averaged 3.12%, U.S. stocks posted an average annual return of 10.5%. During the same time, long-term U.S. bonds gave an average annual return of 5.17%, and Treasury bills earned 3.72% per year on average. Making the simplifying assumption that the interest rates were constant, for each of the three investments, find the amount you would have in 1996 if you had made annual investments of $1000 at the end of each year from 1926 to 1996.

19. A couple predicts that they will need to save $150,000 for their child's college education, and they predict that they will be able to earn about 9% interest, compounded monthly, on their investments.

(a) If they begin the deposits at the end of each month when their child is a newborn, so that they have 18 years of deposits, how large must each deposit be?

(b) If they do not begin making deposits until the child is 10 years old, so that they have only 8 years of deposits, how large must each deposit be?

20. In May 1997, the United States average interest rate for money market accounts was 2.63%, whereas one of the best rates in the country, 5.46%, was being offered by First Deposit National Bank in New Hampshire. If $100 is invested at the end of each month into an account, how long will it take the account to grow to $10,000 if it is invested at

(a) 2.63% compounded monthly?

(b) 5.46% compounded monthly?

21. The Mississippi Prepaid Affordable College Tuition Program (MPACT) is a prepaid tuition program through which the college tuition and fees at any Mississippi public university or college may be paid for in advance. To cover the costs of in-state tuition at a 4-year university for a newborn who will enter college in 18 years, purchasers of the plan have the options of paying one lump-sum payment of $9141 immediately, payments of $185 at each month's end for 5 years beginning immediately, or payments of $80 at each month's end for 18 years. Show that the future values after 18 years of each of these deposits into an account with an interest rate of 8% compounded monthly are fairly close in value.

22. Suppose $600 is deposited at the end of each month into an account earning 10% interest compounded monthly. After 8 years the deposits are discontinued, but the money is left in the account and continues to earn the same interest. How much will be in the account 15 years after it was first opened?

23. A woman pays $130 at each quarter's end into an account earning 5.4% interest compounded quarterly. After 10 years of deposits, she changes the amount of the deposits to $200 per quarter. If the money continues to earn interest at the same rate, how much will be in the account 30 years after it was first opened?

24. A family currently has $10,000 of the $20,000 they want to have available for a down payment on a house. They invest this $10,000 in a savings account paying 5.7% interest compounded monthly, and at the end of each month they deposit another $150 into the account. How long will it take them to save enough for the down payment?

25. It would not be unusual to have a systematic savings plan in which interest is compounded more often than deposits are made. For instance, annual deposits might be made into an account with interest compounded monthly. The systematic savings formula derived in the text does not directly apply to this case, but it can be adapted.

(a) Find a formula for the future value F after t years of a systematic savings plan with an annual interest rate r, compounded n times a year, into which deposits of size D are made at the end of every k periods. (Assume that nt is exactly divisible by k.)

    (b) Use the formula from part (a) to find the future value after 7 years of an account with interest at 6% compounded monthly into which deposits of $50 are made at the end of each quarter.

26. Although it is not possible to algebraically solve exactly for the interest rate r in the systematic savings plan formula, r can be estimated by using a calculator or computer that will perform such estimates or by making a series of guesses for r. When using the guessing method, you should determine if your first guess was too high or too low and adjust accordingly, continuing to adjust your guesses until the value of r you have found gives a reasonably close answer. Suppose deposits of $230 are made at the end of each month into an account earning interest compounded monthly. After 6 years, the account is worth $20,432. Use some method to estimate the annual interest rate r to two decimal places.

27. Derive a general formula for the future value of a systematic savings plan when deposits are made at the beginning, rather than the end, of each period.

3.5   AMORTIZED LOANS

Most of us borrow money many times over the course of our lifetime. We may take loans to pay tuition in college, purchase a car, start a business, or buy a home. Loans are contracts between a lender and a borrower who agrees to pay back the loan under some agreed upon terms. With an **amortized loan**, a lender loans a borrower a lump sum of money, and the borrower pays back the loan by making *equal* payments at some regular intervals. To earn money, the lender charges interest on the balance of the loan. The payments are due at the end of each compounding period, and each payment covers the periodic interest due on the current balance of the loan. In addition, each payment includes some extra money that is used to bring down the balance so that the loan will eventually be paid off. In the early payments, the portion of the loan that is applied to paying the interest is higher than later in the life of the loan when the balance has been reduced. The benefits of setting up a loan with regular equal payments is that it gives the borrower a definite schedule for paying off the loan, it assures the lender of a flow of cash, and it provides a regular check as to whether the borrower is meeting his or her obligations.

    When lenders loan money at an interest rate r, they expect the income stream coming in from the payments to be equivalent in value to investing the principal themselves at the same interest rate r. In other words, the future value of the series of payments must be the same as the future value of investing the principal in an account with the same rate and compounding period. We will now see how this idea can be used to find a formula that allows us to figure out exactly what size the equal payments must be in order for an amortized loan to be paid off in a specified period of time.

    The initial amount of money borrowed is called the **principal** or **present value** of the loan, and we denote it by the letter P. Suppose the interest rate on the loan is r, compounded n times a year. Then the future value F of the initial loan amount P after t years is given by the compound interest formula

$$F = P\left(1 + \frac{r}{n}\right)^{nt}.$$

If the loan is paid off in t years with payments of R dollars at the end of each period, then the future value F of this series of payments is given by the systematic savings plan formula

$$F = R\left(\frac{\left(1 + \frac{r}{n}\right)^{nt} - 1}{\frac{r}{n}}\right).$$

To find a formula that relates the payment amount R to the principal, rate, compounding period, and length of the loan, we use the key fact that these two future values must be equal. Setting them equal to each other, we have

$$P\left(1 + \frac{r}{n}\right)^{nt} = R\left(\frac{\left(1 + \frac{r}{n}\right)^{nt} - 1}{\frac{r}{n}}\right).$$

We can now solve for P as follows:

$$P = \frac{1}{\left(1 + \frac{r}{n}\right)^{nt}}\left[R\left(\frac{\left(1 + \frac{r}{n}\right)^{nt} - 1}{\frac{r}{n}}\right)\right]$$

$$= R\left(\frac{1 - \left(1 + \frac{r}{n}\right)^{-nt}}{\frac{r}{n}}\right).$$

Thus, for nt an integer, we have the following loan formula.

Loan Formula

$$P = R\left(\frac{1 - \left(1 + \frac{r}{n}\right)^{-nt}}{\frac{r}{n}}\right) \qquad \text{where} \quad \begin{cases} P = \text{principal or present value} \\ R = \text{amount of each payment} \\ r = \text{annual interest rate} \\ n = \text{number of periods per year} \\ t = \text{time (in years)} \end{cases}$$

The preceding derivation illustrates a general mathematical technique. When the same quantity can be computed in two different ways, by setting the two expressions equal a potentially new formula or identity often may be discovered.

Example 1   *Cost of Buying a Car*

Suppose you want to purchase a Honda Accord and the added charges for taxes, title, license, and fees bring the total to $16,032.31. You must make a 10% down payment and pay for the remainder through a car loan at an interest rate of 7.9% compounded monthly. You are to repay the loan with monthly payments for 4 years.

(a) What are the monthly payments?
(b) What is the total amount you make in payments over the life of the loan?
(c) How much interest will you pay over the life of the loan?

Solution:

(a) To find the monthly payments, we must first find the principal for the loan that will be the purchase price minus the down payment. The down payment is 10% of $16,032.31 or

$$(0.10)(16,032.31) = 1603.231 \approx \$1603.23,$$

so the principal for the loan will be given by

$$P = 16{,}032.31 - 1603.23 = \$14{,}429.08.$$

Now, to find the monthly payments we substitute $P = 14{,}429.08$, $r = 0.079$, $n = 12$, and $t = 4$ into the loan formula and solve for the monthly payment R:

$$14{,}429.08 = R\left(\frac{1-\left(1+\dfrac{0.079}{12}\right)^{-12\cdot4}}{\dfrac{0.079}{12}}\right)$$

$$14{,}429.08 \approx R\left(\frac{1-(1.00658333)^{-48}}{0.00658333}\right).$$

$$14{,}429.08 \approx R(41.04078199)$$

$$R \approx \frac{14{,}429.08}{41.04078199}$$

$$R \approx \$351.58.$$

(b) You will make $nt = 12 \cdot 4 = 48$ payments. Because the amount of each payment is $R = \$351.58$, the total amount paid over the life of the loan is

$$\$351.58 \cdot 48 = \$16{,}875.84.$$

(c) The interest you will pay over the life of the loan is the difference between the total amount you paid and the initial principal P. Thus the total interest paid is

$$\$16{,}875.84 - \$14{,}429.08 = \$2446.76.$$

The total interest paid over the life of the loan is sometimes called the *finance charge*. The term finance charge may also refer to the monthly interest charged on a revolving credit account such as a credit card account.

Sometimes borrowers know the maximum periodic payments they can afford and they want to know how much money they can borrow. We examine this question in the following example.

Example 2   *Student Loan*

A college senior does not have enough money to cover expenses and plans to get extra money through working and through a student loan that will have quarterly payments for 6 years. She estimates that after graduation she can afford to make quarterly loan payments of $300. If the loan is made at an interest rate of 7% compounded quarterly, how much can she afford to borrow?

Solution: In this situation we know the quarterly payments will be $R = 300$, and we know $r = 0.07$, $n = 4$, and $t = 6$. Substituting into the loan formula, we find the principal P:

$$P = 300\left(\frac{1 - \left(1 + \dfrac{0.07}{4}\right)^{-4 \cdot 6}}{\dfrac{0.07}{4}}\right)$$

$$P = 300\left(\frac{1 - (1.0175)^{-24}}{0.0175}\right)$$

$$P \approx 300(19.46068565)$$

$$P \approx \$5838.21.$$

We see that she can afford to borrow $5838.21.

With most loans, the borrower has the option of paying more than the required payments each period. By doing so, not only is the loan paid off more quickly, but also the total interest paid is reduced, and therefore the total amount paid over the life of the loan is reduced. The savings achieved by such extra payments can be quite striking. We see how this works in the next example.

Example 3   *Saving by Paying Off a Loan Quickly*

A couple purchased a house by taking out a $125,000 mortgage (loan) with monthly payments for 30 years. The interest rate on the loan was 8.25% compounded monthly.

(a) How much would the scheduled payments on the loan be?

(b) If they pay $100 extra each month, how long will it take to pay off the loan?

(c) How much does the couple save in payments over the life of the loan by paying $100 extra each month?

Solution:

(a) We have $P = 125{,}000$, $r = 0.0825$, $n = 12$, and $t = 30$. Using the loan formula to solve for the scheduled payment R, we get

$$125{,}000 = R\left(\frac{1 - \left(1 + \dfrac{0.0825}{12}\right)^{-12 \cdot 30}}{\dfrac{0.0825}{12}}\right)$$

$$125{,}000 = R\left(\frac{1 - (1.006875)^{-360}}{0.006875}\right)$$

$$125{,}000 \approx R(133.10853891)$$

$$R \approx \frac{125{,}000}{133.10853891} \approx \$939.08.$$

(b) By paying \$100 extra each month, the payment is now

$$R = \$939.08 + \$100 = \$1039.08.$$

To find how long it will take to repay the loan with this new payment amount, we substitute $P = 125{,}000$, $R = 1039.08$, $r = 0.0825$, and $n = 12$ into the loan formula, and solve for t.

$$125{,}000 = 1039.08\left(\frac{1 - \left(1 + \dfrac{0.0825}{12}\right)^{-12t}}{\dfrac{0.0825}{12}}\right)$$

$$125{,}000 = 1039.08\left(\frac{1 - (1.006875)^{-12t}}{0.006875}\right).$$

To solve for t, we need to use logarithms, but first we need to isolate the term $(1.006875)^{-12t}$.

$$\frac{125{,}000}{1039.08} = \frac{1 - (1.006875)^{-12t}}{0.006875}$$

$$120.2987258 \approx \frac{1 - (1.006875)^{-12t}}{0.006875}$$

$$(120.2987258)(0.006875) \approx 1 - (1.006875)^{-12t}$$

$$0.82705374 \approx 1 - (1.006875)^{-12t}.$$

Next, we subtract 1 from each side of the equation:

$$0.82705374 - 1 \approx -(1.006875)^{-12t}$$

$$-0.17294626 \approx -(1.006875)^{-12t}.$$

Because we cannot take logarithms of negative numbers, we multiply both sides of the equation by -1 and get

$$(-1)(-0.17294626) \approx (-1)(-(1.006875)^{-12t})$$

$$0.17294626 \approx (1.006875)^{-12t}.$$

Now we take the logarithms of both sides of the equation and solve for t:

$$\log(0.17294626) \approx \log(1.006875)^{-12t}$$
$$\log(0.17294626) \approx -12t\log(1.006875)$$
$$t \approx \frac{\log(0.17294626)}{-12\log(1.006875)}$$
$$t \approx 21.34 \text{ years.}$$

We see that by paying an extra $100 each month, the loan is paid off in 21.34 years rather than the original 30 years. As with solving for time in the previous sections, we must keep in mind that the answer is approximate.

(c) The total amount paid over the life of the loan is the amount of each payment R multiplied by the number of payments nt. For the scheduled payments, this total is given by

$$(\$939.08)(12)(30) = \$338,068.80.$$

When paying $100 extra each month, the payment R is $1039.08 and the time t is roughly 21.34 years, so the total amount paid over the life of the loan is approximately

$$(\$1039.08)(12)(21.34) = \$266,087.61.$$

We see that the total amount the couple saves in interest over the life of the loan by paying $100 extra each month is approximately

$$\$338,068.80 - \$266,087.61 = \$71,981.19.$$

■

It is important to note that in Example 3 we did not consider interest that could be earned by investing the $100 each month. Furthermore, we ignored taxes on earned interest and tax deductions on mortgages. All these factors should be taken into account when making the decision of whether or not to pay off a loan more quickly.

Finding a Loan Balance

When you have a loan, you may want to know how much you still owe on the loan at a given point in time. This amount is called the **loan balance**. The loan balance is the present value of the series of payments that are left to make. In other words, the loan balance is the principal that would have to be borrowed today at the specified rate to yield the series of payments remaining to be paid. To compute the balance of a loan, we use the loan formula in a different way. If we let R be the usual payment amount and t be the amount of time left in the life of the loan, then the present value P given by the loan formula is the current balance of the loan. We now see how this works.

Example 4   *Loan Balance*

Suppose a $12,000 loan is taken out at an interest rate of 11.3% compounded monthly with monthly payments for 5 years. After 2 years of payments, what is the balance of the loan?

BRANCHING OUT 3.6

THE MYSTERIOUS APR

Annual percentage rate, or APR, is to loans as annual percentage yield is to savings. However, because of up-front fees that vary from loan to loan, APR is much harder to calculate. In fact, different banks may compute APR slightly differently on exactly the same loan. Federal laws mandate that lenders provide the annual percentage rate for consumer protection. Specifically, finance charge and annual percentage rate are defined in Sections 1605 and 1606 of Chapter 41, Consumer Credit Protection, of Title 15, Commerce and Trade, of the United States Code. The APR must include costs arising as part of the extension of credit such as application fees, loan fees, or points. The APR does not include charges that would arise in a cash transaction such as title fees, deed preparation, or escrow for taxes and insurance.

Let's illustrate with an example. Suppose you finance the purchase of a home through a $100,000 mortgage with 30 years of monthly payments at 9% interest. From the amortization formula, we find the monthly payment R satisfies

$$100,000 = R\left(\frac{1 - \left(1 + \frac{0.09}{12}\right)^{-30 \cdot 12}}{\frac{0.09}{12}}\right).$$

Solving for R in the usual way, we find $R = \$804.62$.

Suppose the loan requires a $100 application fee and an up-front payment of 1 point. Then you pay the bank the $100 application fee and 1% of $100,000, or $1000, for a total of $1100. Thus, the bank only has to loan you $100,000 − $1100 = $98,900 from its own funds. The interest rate on a loan of $98,900 with monthly payments of $804.62 has an effective interest rate satisfying

$$98,900 = 804.62\left(\frac{1 - \left(1 + \frac{r}{12}\right)^{-30 \cdot 12}}{\frac{r}{12}}\right).$$

We cannot combine terms algebraically in any way that allows us to solve exactly for r. However, using a numerical equation solver available on many calculators or using a similar computer software package, we estimate $r \approx 9.124\%$. The closing statement from the bank is required to state this as the APR.

Solution: First we must figure out the payment amount R. We use the loan formula with $P = 12,000$, $r = 0.113$, $n = 12$, and $t = 5$ and solve for R.

$$12,000 = R\left(\frac{1 - \left(1 + \frac{0.113}{12}\right)^{-12 \cdot 5}}{\frac{0.113}{12}}\right)$$

$$12,000 \approx R\left(\frac{1 - (1.00941667)^{-60}}{0.00941667}\right)$$

$$12,000 \approx R(45.67808441)$$

$$R \approx \frac{12,000}{45.67808441}$$

$$R \approx \$262.71.$$

After 2 years of payments, the amount of time left in the life of the loan is 3 years. To find the balance left after 2 years, we use the loan formula with $R = 262.71$, $r = 0.113$, $n = 12$, and $t = 3$. This gives the present value, P, of the remaining series of payments:

$$P = 262.71 \left(\frac{1 - \left(1 + \frac{0.113}{12}\right)^{-12 \cdot 3}}{\frac{0.113}{12}} \right)$$

$$P \approx 262.71 \left(\frac{1 - (1.00941667)^{-36}}{0.00941667} \right)$$

$$P \approx \$7989.73.$$

Thus the balance of the account after two years is $7989.73.

Amortization Schedules

An **amortization schedule** is a list of payments to be made on a loan that breaks down each payment into principal and interest. Many lenders give borrowers an amortization schedule each year, often in the form of a coupon book. An amortization schedule provides a good record of the total amount of interest paid each year on a loan. These records may be important for tax purposes if the interest is tax deductible, as is the case for home loans and home equity loans. Amortization schedules sometimes include the balance of the loan after each payment, thus providing an easy way to look up the balance of the loan at any time.

In the next example, we see how an amortization schedule is constructed.

$\mathbf{E}$xample 5   *Amortization Schedule*

Suppose a $2758.35 furniture purchase is financed with a loan at 9.6% interest compounded monthly with monthly payments for 4 months. Write an amortization schedule for the loan.

Solution: We first use the loan formula to find the amount of each payment. Substituting $P = 2758.35$, $r = 0.096$, $n = 12$, and $t = 4/12$ into the loan formula, we get

$$2758.35 = R \left(\frac{1 - \left(1 + \frac{0.096}{12}\right)^{-12 \cdot (4/12)}}{\frac{0.096}{12}} \right)$$

$$2758.35 = R \left(\frac{1 - (1.008)^{-4}}{0.008} \right)$$

$$2758.35 \approx R(3.92126231)$$

$$R \approx \frac{2758.35}{3.92126231}$$

$$R \approx \$703.43.$$

Now we are ready to set up an amortization schedule. The schedule lists the payment number, the amount of each payment, the amount of interest and principal in each payment, and the loan balance. We can arrange this information in a table that starts out as follows:

Payment Number	Payment	Interest Paid	Principal Paid	Balance
				2758.35
1				
2				
3				
4				

To fill out the next row in the table, we enter $703.43 in the payment column. To find what portion of this payment is interest, we multiply the balance appearing in the previous row by the interest rate for one period $r/n = 0.096/12 = 0.008$. We find

$$\text{interest paid} = (\$2758.35)(0.008) \approx \$22.07.$$

The remainder of the first payment goes toward paying the principal, so we have

$$\text{principal paid} = \$703.43 - \$22.07 = \$681.36.$$

We compute the new balance by subtracting the principal paid from the previous balance:

$$\text{balance} = \$2758.35 - \$681.36 = \$2076.99.$$

Entering this information into the table, we have

Payment Number	Payment	Interest Paid	Principal Paid	Balance
				2758.35
1	703.43	22.07	681.36	2076.99
2				
3				
4				

To fill in the next row, we first enter the payment and then compute the interest paid. It is given by the new balance multiplied by the monthly interest rate as follows:

$$\text{interest paid} = (\$2076.99)(0.008) \approx \$16.62.$$

The principal paid is then given by

$$\text{principal paid} = \$703.43 - \$16.62 = \$686.81,$$

and the new balance is given by

$$\text{balance} = \$2076.99 - \$686.81 = \$1390.18.$$

Entering this information into our amortization schedule we get

Payment Number	Payment	Interest Paid	Principal Paid	Balance
				2758.35
1	703.43	22.07	681.36	2076.99
2	703.43	16.62	686.81	1390.18
3				
4				

Continuing in this way, we complete the next row of the amortization schedule and we have

Payment Number	Payment	Interest Paid	Principal Paid	Balance
				2758.35
1	703.43	22.07	681.36	2076.99
2	703.43	16.62	686.81	1390.18
3	703.43	11.12	692.31	697.87
4				

The last line of an amortization schedule is computed in a somewhat different way, and the last payment may be slightly different from the other payments. In the last line, the interest paid is computed in the usual way and we find that

$$\text{interest paid} = (\$697.87)(0.008) \approx \$5.58.$$

BRANCHING OUT 3.7

LOTTERY WINNINGS: LUMP SUM VS. ANNUAL PAYMENTS

In 1997, the New Jersey Lottery gave lottery players the option of having any winnings paid immediately in a one-time lump sum or paid in 20 equal annual payments with one payment made immediately and the remaining payments made at the end of each year for the next 19 years. For the June 23, 1997 lottery winner, the cash option was a single payment of $5,217,843.45. For the annual payment option, the winner would receive $9,539,071 in 20 equal payments of about $476,953.55 each. Why was the lump-sum payment only a little more than half of the amount that would be received under the annual payment option?

In this case, the lump sum of $5,217,843.45 is the actual amount of money that the state had available to pay the winner. If the winner had chosen the option of annual payments, the state would have made the first payment and invested the remaining money in some interest-bearing fund from which periodic payments would be paid out. It is similar to the amortized loan situation, where a lump sum of money is transferred from one party to another, and the party receiving the lump sum must make payments to the other party over time. In this particular case, the interest rate that the state is getting on its money is about 7.5262% per year. The cash value of the lottery

The principal paid in the last payment must pay off the balance of the loan, and therefore we simply let the principal paid equal the previous balance. Finally, we compute the amount of the last payment by adding the interest paid and the principal paid and we find that

$$\text{payment} = \$5.58 + \$697.87 = \$703.45.$$

We see that in this case the last payment was $0.02 higher than the other payments. In an amortization schedule, the last payment may be higher, lower, or the same as the other payments. Completing our amortization schedule, we have

Payment Number	Payment	Interest Paid	Principal Paid	Balance
				2758.35
1	703.43	22.07	681.36	2076.99
2	703.43	16.62	686.81	1390.18
3	703.43	11.12	692.31	697.87
4	703.45	5.58	697.87	0

Notice that with each successive payment, the amount of interest paid goes down and the amount of principal paid goes up. This phenomenon always occurs with amortized loans. With a typical 30-year home loan the payments for the first several years consist mostly of interest payments. Because this is the case,

prize is the present value in this situation. The annual payments correspond to the loan payments R in our loan formula except that because one initial payment is made up front, the present value of $5,217,843.45 is given by

$$5{,}217{,}843.45 = R + R\left(\frac{1 - (1.075262)^{-19}}{0.075262} \right).$$

Solving for R, we see that

$$5{,}217{,}843.45 = R\left(1 + \frac{1 - (1.075262)^{-19}}{0.075262} \right)$$

$$5{,}217{,}843.45 \approx R(10.93994182)$$

$$R \approx \frac{5{,}217{,}843.45}{10.93994182}$$

$$\approx \$476{,}953.49 \approx \$476{,}953.55.$$

When should the winner choose the lump sum option? Ignoring taxes, the winner should take the lump sum if it can be invested at better than 7.5262%

interest compounded annually. However, taxes will probably be higher on the large lump sum, and that must be taken into consideration. Of course, there are other factors that come into play in choosing a method of payment, for instance the age or any pressing financial needs of the winner.

Some state lotteries, for instance the New York Lotto and the Colorado Lotto, offer the option of a lump-sum payment or a *graduated* payment plan. With the graduated payments, the payments gradually increase in size. In this case, the state can hold some of the money longer so the cash value of the lottery prize is an even smaller fraction of the sum of the scheduled payments.

Not all state lotteries offer a lump-sum option, and those states that do require the players to select that option when the lottery ticket is purchased. Because of this, there are several investment companies that will pay lottery winners lump-sum payments in exchange for their stream of payments.

one simple way to estimate the size of the payments on a 30-year loan is to compute the monthly interest due on the original balance. The actual payments will be only somewhat higher. This idea will not apply to shorter term loans as you can see in Example 5, where the initial interest is $22.07, but the payments are $703.43 except for the final one.

Technology Tip

The repetitive nature of constructing amortization tables makes their construction an ideal spreadsheet application.

EXERCISES FOR SECTION 3.5

In Exercises 1–4, find the principal P of a loan with the given payment, time period, and interest rate.

1. $R = \$290.62$, $t = 5$ years, $r = 8\%$ compounded monthly
2. $R = \$1427$, $t = 15$ years, $r = 10.7\%$ compounded quarterly
3. $R = \$90$, $t = 6$ years, $r = 7.4\%$ compounded quarterly
4. $R = \$31.49$, $t = 9$ months, $r = 6\%$ compounded monthly

In Exercises 5–8, find the amount of each payment R for a loan with the given principal, time period, and interest rate.

5. $P = \$105,000$, $t = 20$ years, $r = 7.75\%$ compounded monthly
6. $P = \$4623.14$, $t = 8$ years, $r = 11\%$ compounded quarterly
7. $P = \$7430.65$, $t = 18$ months, $r = 13\%$ compounded quarterly
8. $P = \$87,400$, $t = 10$ years, $r = 8.8\%$ compounded monthly

In Exercises 9–12, find the time t, in years, required to pay off a loan with the given principal, payment, and interest rate.

9. $P = \$16,425$, $R = \$500$, $r = 9\%$ compounded monthly
10. $P = \$60.82$, $R = \$10$, $r = 19.5\%$ compounded monthly
11. $P = \$300,000$, $R = \$12,000$, $r = 7.3\%$ compounded quarterly
12. $P = \$8472$, $R = \$350$, $r = 6\%$ compounded quarterly

13. Suppose you can obtain an 8-year bank loan at an interest rate of 9% compounded quarterly and you have determined that you can afford to make payments of $350 each quarter. How much can you afford to borrow?

14. The balance on a Travelers Bank Visa Card account is $827.08. The minimum payment due is $20 and the interest rate is 17% compounded monthly. Suppose only $20 is paid on the account each month.

    (a) How long would it take to pay off the balance?
    (b) How much total interest is paid by the time the balance is paid off?

15. The Federal Perkins Loan is a low-interest student loan available to students with exceptional financial need. The loan has an interest rate of 5% compounded monthly with monthly payments and the borrower has up to 10 years to repay. Suppose a stu-

dent has a $9000 Perkins Loan program and will take 10 years to repay it.

(a) What are the monthly payments?

(b) What is the total amount of payments over the life of the loan?

(c) How much interest is paid over the life of the loan?

16. An Edsel, a classic car from the late 1950s, is offered for $7500. Suppose it is purchased with a 15% down payment and a 3-year loan at 9.5% interest compounded monthly.

(a) What are the monthly payments?

(b) What is the total amount of payments over the life of the loan?

(c) How much interest is paid over the life of the loan?

17. Suppose a student takes out a 10-year $9000 student loan at 7% interest compounded quarterly with quarterly payments. After 8 years of payments, the student decides to pay off the balance of the loan. How much will the student have to pay?

18. A woman took out a 4-year, $15,460.35 car loan at 9% interest compounded monthly with monthly payments. After making payments for 3 years, what is the balance of the loan?

19. To purchase a $23,620 swimming pool, a family makes a 10% down payment and finances the remainder with a loan at 8.3% interest compounded quarterly. If they make loan payments of $2000 per quarter, how long will it take them to pay off the loan?

20. From 1985 to 1986, the average mortgage rate in the United States on a 30-year loan fell from 12.42% to 10.18%. Suppose a family could only afford mortgage payments of $700 per month. How large a loan could they afford if the interest is compounded monthly at a rate of

(a) 12.42%? **(b)** 10.18%?

21. The highest average interest rate on 30-year mortgages in the United States from 1972 to 1998 was 16.63% in 1981. In that year the median house price was about $66,800. Suppose a 30-year mortgage was taken out on a $66,800 home in 1981 at an interest rate of 16.63% compounded monthly. Find

(a) the monthly payments,

(b) the balance of the loan in the year 1986,

(c) the balance of the loan in the year 2000.

22. A small company took out a 5-year $20,000 loan at 9.5% interest compounded quarterly with quarterly payments. After 3 years of payments, the company wants to pay off the balance of the loan. How much will they have to pay?

23. A car dealer offered a Ford Taurus for a total price of $16,738. Assume that a 10% down payment was required. The customer had a choice of taking out a loan for the remaining cost of the car at 0.9% interest compounded monthly or taking a $750 rebate to be used to reduce the size of the loan by $750. If the rebate were taken, then the buyer could only obtain a loan at 7.95% interest compounded monthly. In either case, the loan required monthly payments for 4 years. For which option are the monthly payments the lowest?

24. Suppose $10,000 is borrowed at 8.7% interest compounded monthly.

(a) How long will it take to pay off the loan with monthly payments of $200?

(b) How long will it take to pay off the loan with monthly payments of $100?

(c) What happens if you use the loan formula to find out how long it will take to pay off the loan with monthly payments of $50? Why does the formula not work in this case?

25. A family takes out a $138,542.91 mortgage at 8.4% interest compounded monthly with monthly payments for 30 years.

    (a) How much would the scheduled payments on the loan be?
    (b) If they decide to pay an extra $200 each month on the house payments, how long will it take to pay off the loan?
    (c) How much will they save in payments over the life of the loan by paying $200 extra each month?

26. Suppose you have a student loan of $5700 at 7% interest compounded quarterly with quarterly payments for 10 years.

    (a) How much would the scheduled payments on the loan be?
    (b) If you pay an extra $75 each quarter, how long will it take to pay off the loan?
    (c) How much will you save in payments over the life of the loan by paying $75 extra each quarter?

27. In June 1997, the average annual rate on a $100,000 mortgage for a 30-year fixed interest rate loan in the United States was 7.90%, compounded monthly. The average rate on a 15-year fixed interest rate loan was 7.54%, compounded monthly. Suppose a mortgage is taken out for $100,000 and the interest is at these average rates.

    (a) Find the monthly payment for the 30-year mortgage.
    (b) Find the monthly payment for the 15-year mortgage.
    (c) How much would a borrower save in total payments over the life of the loan by choosing the 15-year mortgage rather than the 30-year mortgage?

28. Don Davis Auto Group dealership had a 1997 Toyota Camry for a total price of $20,291.34. Two choices of financing were offered by the dealer: a 36-month loan at 5.9% compounded monthly and a 48-month loan at 6.9% compounded monthly. Either of the loans would have monthly payments. Suppose the Camry is purchased with a 10% down payment and the remainder is financed with a loan. Find the total amount that would be made in payments and the total interest paid over the life of the loan for each of the two loan choices.

29. An undergraduate takes out an $8000 student loan at 6% interest compounded quarterly with quarterly payments for 10 years. Write an amortization schedule for the first year of the loan.

30. A family borrows $75,350.25 to purchase a home. The loan is at 9.6% interest compounded monthly with monthly payments for 15 years. Write an amortization schedule for the first 6 months of the loan.

31. A man finances the purchase of his living room furniture with a loan of $5429.38 at 8.4% interest compounded monthly with monthly payments for 6 months. Write an amortization schedule for the 6 months of payments.

32. Write out an amortization schedule for a $2500 two-year loan at 7.2% interest compounded quarterly with quarterly payments.

33. Write out an amortization schedule for a $7300 loan at 11.5% interest compounded quarterly with quarterly payments for 9 months.

34. Suppose a $2356.34 computer purchase is financed with a loan at 13.5% interest compounded monthly with monthly payments for 5 months. Write an amortization schedule for the loan.

35. A man borrowed $115,000 to buy a house. The loan he obtained had an interest rate of 9.45% compounded monthly with monthly payments for 30 years. After 10 years of payments, interest rates have dropped and he is considering refinancing by paying off the old loan and taking out a new 20-year loan at 8.5% interest compounded

monthly with monthly payments.

   (a) Find the total amount the man would make in payments over the life of the loan for the original loan.

   (b) Find the principal for the new loan.

   (c) Find the total amount the man will make in payments for both loans if he decides to refinance.

   (d) If the bank will charge the man $3000 to refinance, should he refinance? *(Do not consider any interest that might be earned by investing the refinance charge or investing the difference in monthly payments.)*

36. A couple purchased a house with a $237,400.30 loan. The loan had an interest rate of 8.7% compounded monthly with monthly payments for 30 years. After 15 years of payments, they are considering refinancing by paying off the old loan and taking out a new 15-year loan at 7.75% interest compounded monthly with monthly payments.

   (a) Find the total amount that would be made in payments over the life of the loan for the original loan.

   (b) Find the principal for the new loan.

   (c) Find the total amount that would be made in payments for both loans if they decided to refinance.

   (d) If the bank will charge the couple $1500 to refinance, should they refinance? *(Do not consider any interest that might be earned by investing the refinance charge or investing the difference in monthly payments.)*

For Exercises 37 and 38, consider the following: In purchasing a house, the buyers sometimes have the option of paying some extra money up front in order to lower the interest rate they will pay for their mortgage. The money paid up front is called *points*, and each point costs the buyer 1% of the total amount of the mortgage. *(In all parts of these questions, do not consider any interest that might be earned by investing the point or investing any difference in monthly payments.)*

37. In August 1997, Merit Mortgage offered home mortgages at a rate of 7.625% with no points or 7.25% with one point. The interest in both cases was compounded monthly, and the loan was to be paid back with monthly payments for 30 years. Suppose a man borrows $95,000 from Merit Mortgage to purchase a home when these rates are in effect.

   (a) If the loan is to be held for the entire 30 years, how much will paying the point save in payments over the life of the loan?

   (b) If the loan is to be held for the entire 30 years, should the man pay the point?

   (c) Suppose the man plans to sell the house after only 2 years. How much will paying the point save in payments over the 2 years?

   (d) What is the balance on the loan after 2 years if the point is paid?

   (e) What is the balance on the loan after 2 years if the point is not paid?

   (f) If the man plans to sell the house in 2 years, should he pay the point?

38. In August 1997, Commerce Bank offered 15-year home mortgages at 7.5% with no points and at 7.125% with one point. The interest in both cases was compounded monthly, and the loan was to be paid back with monthly payments. Suppose a woman plans to take out an $83,500 mortgage to purchase a condominium when these rates are in effect.

   (a) If the loan is to be held for the entire 15 years, how much will paying the point save in payments over the life of the loan?

(b) If she plans to hold the loan for the entire 15 years, should she pay the point?

(c) If she plans to sell the condominium in 3 years, how much will paying for the point save her in payments over the 3 years?

(d) What is the balance on the loan after 3 years if she pays the point?

(e) What is the balance on the loan after 3 years if she does not pay the point?

(f) If she plans to sell the condominium in 3 years, should she pay the point?

39. In Exercise 37, we ignored any interest that might have been earned by investing the point or any difference in monthly payments in a savings account. Suppose that in the situation in Exercise 37, either the borrower does not pay the point and invests the money he would have paid for the point in an account earning 3% interest compounded monthly, or he does pay the point and each month he invests the reduction in the monthly loan payment into an account earning 3% interest compounded monthly.

(a) If the borrower plans to hold the loan for the entire 30 years, should he pay the point?

(b) If the borrower plans to hold the loan for only 2 years, should he pay the point?

40. In Exercise 38, we ignored any interest that might have been earned by investing the point or any difference in monthly payments into a savings account. Suppose that if the woman does not pay the point, she invests the money she would have paid for the point in an account earning 5% interest compounded monthly, and if she does pay the point, then each month she invests the reduction in the monthly loan payment into an account earning 5% interest compounded monthly.

(a) If the woman holds the loan for the entire 15 years, should she pay the point?

(b) If the woman holds the loan for only 3 years, should she pay the point?

For Exercises 41–46, consider the following: There are some situations in which a lump sum is invested into an account earning interest and regular payments are made from the account until it is depleted. This kind of account is often called an annuity, although the word annuity has some other alternative meanings. If the lump sum invested is P, interest is compounded n times per year at an annual rate r, and payments of size R are paid out of the account at the end of every period until the account runs out of money at the end of t years, then the loan formula applies to this situation. In this case, we view the bank or holder of the account as the borrower and the person receiving the payments as the lender.

41. Suppose you want to set up an account from which $500 payments can be withdrawn at the end of each month for 15 years. If the account will earn 7.25% compounded monthly, what lump sum would need to be invested at the start?

42. A student will be attending college for the next 4 years, and her parents want to set up an account from which $3000 payments for college expenses can be withdrawn at the end of each quarter of the next 4 years. If the account will earn 6.8% interest compounded quarterly, how much should be invested?

43. Suppose a woman's retirement account currently has $257,548.72 in it and it is earning 6.72% interest compounded monthly. If she plans to withdraw $4000 from the account at the end of each month, how long will it take for the funds in the account to be depleted?

44. A grandmother wants to set up a trust fund account from which payments to her grandson can be withdrawn at the end of each month for the next 20 years. If $80,000 is invested in the account and it earns 7% interest compounded monthly, what size will the payments be?

45. In a "Hole-In-One Shootout" benefiting the Lena Pope Home, the grand prize was $1,000,000 to be paid in equal monthly installments over 30 years. Assuming the payments will be paid at the end of each month out of an account earning 6.5% interest compounded monthly, what lump sum would need to be invested at the start?

46. Dallas Cowboys running back Emmitt Smith signed a contract in 1996 paying a signing bonus of $1.5 million for each of the succeeding 7 years.

   (a) Assuming the Cowboys could have invested the money into an account earning 7% interest compounded annually, what lump sum would they need to have invested into an account in 1996 so that Emmitt Smith's $1.5 million payments could have been withdrawn at the end of each year for the following 7 years?

   (b) Earlier in the contract negotiations the Cowboys had announced a signing bonus of $15 million spread out over "many years." Assuming the lump sum they planned to invest was the one you found in part (a) and the payments would be made out of an account earning 7% interest compounded annually, with payments made to Emmitt Smith at the end of each year, estimate the number of these "many years" if the total of the payments were to come to $15 million. (You cannot solve for the time exactly here; instead, you will have to estimate it by using a calculator or computer that will perform such estimations or by making guesses that will close in on the answer.)

For Exercises 47 and 48, consider the following: As described in Branching Out 3.7, when a person wins a state lottery and the winnings are paid out with equal annual payments over many years (typically, about 20 years), the state does not have the entire jackpot on hand but instead invests a smaller amount of money into a fund earning interest, from which the payments to the winner will be paid. If the lump sum invested is P, interest is compounded n times per year at an annual rate r, and payments of size R are paid out of the account with one payment immediately and then at the end of every period until the account runs out of money at the end of t years, then the loan formula applies to this situation except that, because one initial payment is made up front, the present value P is given by

$$P = R + R \left(\frac{1 - \left(1 + \dfrac{r}{n}\right)^{-nt}}{\dfrac{r}{n}} \right).$$

47. In the Tri-State Megabucks lottery run by the three states Maine, New Hampshire, and Vermont, winners may collect the jackpot in one of two ways: either in 25 equal annual payments or with a one-time lump-sum cash payment up front. If the payment option is selected, the first of the 25 payments is paid immediately and the remaining payments are paid at the end of each year for the next 24 years. Suppose the Tri-State Megabucks lottery now has $3 million available for a lottery prize. Suppose the states can invest this amount into an account earning 6.82% compounded annually, and the money from this account can be used to make the 25 payments if the winner chooses the payment option. For this arrangement, find

   (a) the amount of each annual payment,

   (b) the total amount paid out in payments (this is the amount that the state will announce as the jackpot amount).

48. The Pennsylvania Wild Card Lotto gives the winners the option of choosing a lump-sum payment or collecting the entire jackpot in 21 equal annual payments with the first payment made right away and the remaining 20 payments paid at the end of each year for the next 20 years. One week the projected jackpot was $1.2 million and the lump-sum option was projected to be $600,000. Suppose you could invest the lump sum in an account earning 7% interest compounded annually and then make 21 equal payments to yourself out of the account in the same way as the state's payment plan so that the account would run out of funds with the last payment.

(a) Find the amount of each payment.

(b) If you play, should you choose the lump-sum option or the payment option?

49. An amortized loan may have interest compounded more often than payments are made. For instance, quarterly payments might be made on a loan for which interest is compounded monthly. The amortized loan formula derived in the text does not directly apply to this case, but it can be adapted.

(a) Find an amortized loan formula for a loan with principal P, an annual interest rate r, compounded n times a year, for which payments of size R are made at the end of every k periods for t years. (Assume nt is exactly divisible by k.)

(b) Use the formula from part (a) to find the payment on a $5000 loan at 8% interest compounded monthly with payments made at the end of each quarter for 4 years.

50. Although it is not possible to algebraically solve exactly for the interest rate r in the amortized loan formula, r can be estimated by use of a calculator or computer that will perform such estimates or by making a series of guesses for r. When using the guessing method, you should determine whether your first guess was too high or too low and adjust accordingly, and continue to adjust your guesses until the value for r you have found gives a reasonably close answer. Suppose that the payments on a $20,000 loan with interest compounded monthly and monthly payments for 5 years are $526. Use some method to estimate the annual interest rate r to two decimal places.

WRITING EXERCISES

1. Would you expect the variation in interest rates among different lenders to be greater now or in the past? Give a detailed argument supporting your position.

2. Collateral, credit history of the loan applicant, the length and size of a loan, and the purpose of a loan all affect the interest rates banks charge. Discuss how these factors influence interest rates. Can you think of other factors that influence rates as well?

3. Unlike the fixed-rate mortgages we have been considering, interest rates on an adjustable-rate mortgage (ARM) may change over the life of the loan. Typically, they begin at a rate one or two percentage points below that of a fixed-rate mortgage. Each year the rate is adjusted (up or down) to be some set level above the U.S. government's current prime interest rate. The consumer is usually somewhat sheltered from huge increases in the rate by caps on the increase in any one year and by a ceiling on the highest rate over the life of the loan. Discuss different circumstances in which the initial lower interest rate of an ARM is worth the additional risk of higher rates in the future.

4. Suppose you are a financial planner and are seeking new clients. Write a letter that you will send to families with small children in which you are trying to impress on them the importance of early and careful financial planning in saving for college. Of course, you want to persuade them to come to you for help.

PROJECTS

1. Study the effect of changing interest rates on the affordability of homes. Research the historical relation of interest rates to number of home purchases and housing starts.

2. Compare the traditional IRA and the Roth IRA. Under what circumstances should an investor prefer one over the other? You should take into account the tax bracket of the investor before and after retirement and how long the money will be invested.

3. Choose a state with a lottery in which players can choose to have their winnings paid in a lump sum or in a series of payments. Taking into consideration not only the payments but also state and federal income taxes, analyze the question of which option a player should choose.

4. Compare the financial aspects of leasing a new car versus buying the same car.

5. Create a spreadsheet or computer program that will compute the amortization table for a loan.

KEY TERMS

amortization schedule, 170
amortized loan, 163
annual percentage yield (APY), 142
compound interest, 139
compound interest formula, 141
future value, 134

inflation, 148
loan balance, 168
loan formula, 164
period, 140
periodic interest rate, 140
present value, 135, 163

principal, 133, 163
simple interest, 133
simple interest formula, 134
simple interest future value formula, 135
systematic savings plan, 154
systematic savings plan formula, 155

CHAPTER 3 REVIEW EXERCISES

1. The average rate of interest on a 6-month certificate of deposit in the United States in 1981 was a very high 15.79% and by 1993 it was only 3.28%. How much more interest would be earned on a $10,000 6-month certificate of deposit at 15.79% compounded monthly than one at 3.28% compounded monthly?

2. A loan shark offers to lend $500 on the condition that he is paid back $800 at the end of 9 months. What simple interest rate is the loan shark charging?

3. Suppose a 22-year-old man starts to make $500 deposits at the end of each quarter into an account earning 7% compounded quarterly.

   (a) How much will be in the account when the man is 68 years old?
   (b) How much of the future value will be from deposits?
   (c) How much of the future value will be from interest?

4. Suppose a 7-year loan at 8.2% compounded monthly with monthly payments is used to purchase a 1997 Rolls Royce MPW for $211,100.

   (a) What are the monthly payments?
   (b) What are the total payments over the life of the loan?
   (c) How much interest is paid over the life of the loan?

5. Based on the Consumer Price Index, the rate of inflation in the United States between 1940 and 1990 was about 4.6% per year. What salary in 1940 would be equivalent to a $40,000 salary in 1990?

6. If $3000 is invested at 7% interest compounded quarterly, how long will it take until the investment is worth $10,000?

7. From 1978 to 1979, the average mortgage rate in the United States on a 30-year loan rose from 9.63% to 11.19%. Suppose a family could only afford mortgage payments of $600 per month. How large a loan could they afford if the interest is compounded monthly at a rate of

   (a) 9.63%?   (b) 11.19%?

8. A couple wants to save $15,000 to be used toward the down payment on a home. If they can invest money in an account earning 7.3% interest compounded quarterly, how much do they have to deposit at the end of each quarter in order to save $15,000 in 4 years?

9. Queen Elizabeth I was the first English monarch to enforce the maximum legal interest rate of 10% that had been established 15 years earlier by King Edward VI. In 1561, she borrowed 30,000 pounds in London at 10% interest. How much simple interest would she have paid by 1570?

10. A couple borrows $943.61 to pay for a refrigerator. The loan is at 7.8% interest compounded monthly with monthly payments for 6 months. Write an amortization schedule for the loan.

11. A man takes out a $18,493.05 car loan at 9.9% interest compounded monthly with monthly payments for 5 years. What is the balance of the loan after

   (a) 1 year?   (b) 4 years?

12. On May 12, 1997, the Bank von Ernst in Switzerland offered a 12-month certificate of deposit at an interest rate of 1.68% compounded monthly. If 30,000 Swiss francs were invested in this CD, how much would it be worth after 1 year?

13. Suppose a small company takes out a loan of $14,000 at 8.4% interest compounded monthly with monthly payments for 10 years.

   (a) How much would the scheduled payments on the loan be?
   (b) If the company decides to pay an extra $50 each month on the loan payments, how long will it take to pay off the loan?
   (c) How much will the company save in payments over the life of the loan by paying $50 extra each month?

14. Suppose you invest $500 and would like the investment to grow to $10,000 in 16 years. What interest rate, compounded daily, would you have to earn in order for this to happen?

15. On August 12, 1997, First National Bank and Trust paid 5.75% interest compounded monthly on a 4-year certificate of deposit with a minimum deposit of $5000. Find the annual percentage yield (APY).

16. To save $160 to use toward the purchase of a video game system, a boy deposits $10 at each month's end into a savings account earning 4% interest compounded monthly. How long will it take before the account is worth $160?

17. If $3800 is borrowed at 5% interest compounded quarterly and payments of $200 are made each quarter, how long will it take to pay off the loan?

18. On December 1, 1996, First Federal Bank for Savings in Chicago offered used car loans at 8.09% interest if a 10% down payment was made or at 8.59% interest if no down payment was made. In both cases the interest is compounded monthly and pay-

ments are made each month for 4 years. Suppose a $9000 used car is purchased. Find the total amount paid (including any down payment) for the car on each of these two financing options.

19. A woman deposits $300 at each quarter's end into a savings plan account earning 4.5% compounded quarterly. After 3 years, the interest rate drops to 4% compounded quarterly and she stops making deposits. If she leaves the money in the account, how much will the account be worth 8 years after she opened it?

20. The state of Virginia has a prepaid college tuition program, available for ninth graders and younger children. In 1996, a prepaid in-state tuition contract for an infant, who would be expected to enter college in 18 years, could be purchased for a lump sum of $14,660. For those who chose the option of making monthly payments from the time the child was an infant until he or she entered college 18 years later, the cost was $128 per month. The expected total cost of tuition for a four-year college 18 years later was $79,100.

(a) Assuming tuition will cost $79,100 in 18 years, what is the annual interest rate, compounded monthly, that the purchasers of the lump-sum contract earn on their $14,660 investment?

(b) Find the future value of the $128 monthly payments for 18 years at the interest rate you found in part (a), compounded monthly. Assume the payments are made at the end of each month.

(c) Which of the payment options is a better deal for the purchasers?

ᔑUGGESTED READINGS

Garner, Robert J., et al. *Ernst & Young's Retirement Planning Guide.* New York: John Wiley, 1997. Very detailed personal finance guide, including a thorough coverage of tax issues. One of many such books.

Money Magazine's 100 Steps to Wealth. Libertyville, Ill.: JMJHI, 1997. Comprehensive, three-volume treatise on investment strategies focusing on types of stocks, mutual funds, and bonds. One of many such books.

Wade, William W. *From Barter to Banking: The Story of Money.* New York: Crowell-Collier Press, 1967. Early history, checks, the Federal Reserve, inflation, and the world monetary network.

Williams, Jonathan, et al., eds. *Money: A History.* New York: St. Martin's Press, 1997. Colorful survey of the history of coins and currency.

Nothing is more predictable or definite than $2 + 2 = 4$. By contrast, the flip of a coin to decide which team will kick off in a football game is quite unpredictable, yet probability allows us to understand and use such randomness.

We use randomness when we make decisions based on the flip of a coin and when we deal from a shuffled deck of cards. To make these processes fair or sporting, the outcomes must be as uncertain as possible. Heads should be just as likely as tails; each player must have an equal chance of being dealt any of the myriad of possible hands.

Aristotle described means by which the Athenians selected officials and juries by lottery, recognizing the fairness of arrangements that gave all an equal chance of serving. In the same spirit, a lottery was used in the United States in 1970 to determine the order for being drafted among those eligible for military service.

Probability provides a means for us to understand randomness by giving us a measure of how likely it is that an event will occur. Probability reveals your chances of winning a state lottery. In some situations, such as playing blackjack in a casino, a knowledge of probability can improve your chances of success.

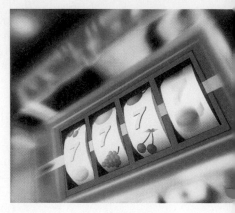

Many human characteristics such as eye color are determined by genes from both parents, so the probability of a child's having blue eyes can be worked out given the eye colors of his or her ancestors. Here, probability is used to answer biological questions. Eye color is not a serious matter, but the same calculations work for many inheritable diseases, and these are serious matters.

In this chapter we will learn how to compute probabilities, beginning with the most intuitive probabilities and building up to more complex situations in which the probabilities involved are far from obvious. We will also see how probability can be used to compute expected value, which is a measure of the average outcome. Through this knowledge, we will have the ability to make reasoned decisions in situations where randomness plays a role.

CHAPTER 4

PROBABILITY

4.1   ELEMENTARY PROBABILITY

We will compute the probability of an event occurring in a particular experiment. An **experiment** is simply a process that can be repeated and may result in different **outcomes**. For example, rolling a single six-sided die is an experiment with six possible outcomes: rolling a 1, 2, 3, 4, 5, or 6. An **event** is some subset of the possible outcomes of an experiment. In the case of rolling a single die, rolling an even number is an example of an event. This event occurs with any of the three outcomes: rolling a 2, 4, or 6. The probability of an event E occurring, which we denote by $P(E)$, is given by the following formula:

Probability of an Event

If all the outcomes of an experiment are equally likely, then

$$P(E) = \frac{\text{number of outcomes that result in event } E}{\text{total number of possible outcomes}}.$$

Because the number of outcomes that result in the event E is less than or equal to the total number of possible outcomes, we know that

$$0 \le P(E) \le 1.$$

Let's now see how to use the formula for the probability of an event in some examples.

Example 1   *Computing Probability*

A box of chocolate candy contains 25 pieces. Seven of the pieces are filled with vanilla cream, 12 are filled with caramel, and the remaining 6 are filled with coconut. If you select a piece of chocolate at random, what is the probability that it will be filled with vanilla?

Solution: Because there are 25 pieces of chocolate from which to choose, there are 25 possible outcomes. Only seven of the pieces are filled with vanilla cream, so there are seven outcomes resulting in a vanilla cream. Therefore,

$$P(\text{vanilla cream}) = \frac{\text{number of outcomes that result in a vanilla cream}}{\text{total number of outcomes}}$$

$$= \frac{7}{25} = 0.28.$$

■

We note here that sometimes probabilities are expressed as a percentage between 0% and 100%. To do this, we simply convert the decimal value of $P(E)$ into a percentage by multiplying by 100. For instance, in Example 1 we found

$P(\text{vanilla cream}) = 7/25 = 0.28$. Writing this probability as a percentage, we find $P(\text{vanilla cream}) = (0.28)(100) = 28\%$.

Example 2. *Rolling a Single Die*

A single die is rolled. Find the probability of rolling

(a) a 3,
(b) an even number,
(c) a 4 or 5,
(d) an 8,
(e) a number greater than 0.

Solution:
(a) When rolling a single die, there are six equally likely outcomes: rolling a 1, 2, 3, 4, 5, or 6. There is only one way the event of rolling a 3 can occur. So the probability of rolling a 3 is given by

$$P(3) = \frac{\text{number of outcomes that result in a 3}}{\text{total number of possible outcomes}} = \frac{1}{6}.$$

(b) There are three outcomes that result in rolling an even number: rolling a 2, 4, or 6. So we have

$$P(\text{even}) = \frac{3}{6} = \frac{1}{2}.$$

(c) There are two outcomes that result in rolling a 4 or 5. Therefore,

$$P(4 \text{ or } 5) = \frac{2}{6} = \frac{1}{3}.$$

(d) The event of rolling an 8 does not occur with any outcome, so we see that

$$P(8) = \frac{0}{6} = 0.$$

(e) Every outcome results in rolling a number greater than 0. Thus,

$$P(\text{number greater than } 0) = \frac{6}{6} = 1.$$

The last two parts of Example 2 illustrate the following important facts:

$$P(\text{an event that is certain to occur}) = 1,$$
$$P(\text{an event that can never occur}) = 0.$$

The more likely it is that an event E will occur, the closer $P(E)$ will be to 1. The less likely it is to occur, the closer $P(E)$ will be to zero.

A CLOSER LOOK 4.1

PLAYING CARDS

A standard deck of playing cards has 52 cards. The cards come in four different suits: spades, hearts, diamonds, and clubs. Each suit has thirteen cards with ranks 2, 3, 4, 5, 6, 7, 8, 9, 10, jack, queen, king, and ace.

There are four cards of each rank, one of each suit. The jacks, queens, and kings are called face cards. The hearts and diamonds are red cards, and the spades and clubs are black cards.

Example 3   *Drawing a Card*

A card is drawn at random from a standard deck of 52 cards. What is the probability that the card is

(a) an ace?
(b) a heart?
(c) a king or a queen?

Solution:

(a) Because four of the cards in a standard deck are aces, the probability of drawing an ace is

$$P(\text{ace}) = \frac{4}{52} = \frac{1}{13}.$$

(b) Thirteen of the 52 cards are hearts, so

$$P(\text{heart}) = \frac{13}{52} = \frac{1}{4}.$$

(c) Because there are four kings and four queens in the deck, 8 of the 52 cards are either a king or queen and we have

$$P(\text{king or queen}) = \frac{8}{52} = \frac{2}{13}.$$

■

Example 4   *Coin Flipping*

Suppose you flip a coin two times in a row. What is the probability that you will get exactly one head?

Solution: There are four equally likely possible outcomes of the experiment of flipping a coin twice. We denote these outcomes by HH, HT, TH, and TT, where, for example, HT stands for the outcome of a head appearing on the first toss and a tail appearing on the second toss. We see that of the four possible outcomes, only the two results HT and TH have exactly one head. So we have

$$P(\text{exactly one head}) = \frac{2}{4} = \frac{1}{2}.$$

Example 5   *Rolling Two Dice*

In many games, such as Monopoly, two dice are thrown and the player moves the number equal to the sum of the two dice. Suppose two dice are thrown. What is the probability that the sum of the two dice will be

(a) 6?

(b) 12?

Solution:

(a) When rolling two dice, there are six possible outcomes for the first die and six possible outcomes for the second die. This gives a total of $6 \times 6 = 36$ possible outcomes for rolling the two dice. It is important to understand that, for instance, the outcome of a 2 on the first die and a 4 on the second die is different from the outcome of a 4 on the first die and a 2 on the second die. We will denote these two outcomes, illustrated in Figure 4.1, by (2, 4) and (4, 2), respectively.

Figure 4.1
Two possible outcomes of rolling two dice.

(2, 4)   (4, 2)

All of the outcomes of rolling two dice are listed in Table 4.1.

Table 4.1	The 36 Possible Rolls of Two Dice				
(1, 1)	(1, 2)	(1, 3)	(1, 4)	(1, 5)	(1, 6)
(2, 1)	(2, 2)	(2, 3)	(2, 4)	(2, 5)	(2, 6)
(3, 1)	(3, 2)	(3, 3)	(3, 4)	(3, 5)	(3, 6)
(4, 1)	(4, 2)	(4, 3)	(4, 4)	(4, 5)	(4, 6)
(5, 1)	(5, 2)	(5, 3)	(5, 4)	(5, 5)	(5, 6)
(6, 1)	(6, 2)	(6, 3)	(6, 4)	(6, 5)	(6, 6)

The outcomes that have a sum of 6 are

$$(5, 1),\quad (1, 5),\quad (4, 2),\quad (2, 4),\quad (3, 3),$$

and we see that there are five such outcomes. Therefore, the probability of rolling a sum of 6 is

$$P(\text{sum of }6) = \frac{5}{36}.$$

(b) In Table 4.1, we see that the only outcome that has a sum of 12 is (6, 6)—in other words, rolling a 6 on both of the dice. Thus,

$$P(\text{sum of }12) = \frac{1}{36}.$$

Sometimes we want to compute the probability that a certain event E will not occur. We denote this probability by $P(\text{not } E)$. Because it is a certainty that the event E will either occur or not occur, we have

$$P(E) + P(\text{not } E) = 1,$$

and therefore we have the following important formula:

$$\boxed{P(\text{not } E) = 1 - P(E)}$$

Example 6   *Freshmen Survey*

A survey of college freshmen conducted by the American Council on Education and UCLA in 1996 found that a record 33.1% of freshmen cited financial aid as a "very important" reason for selecting their college. What is the probability that a random college freshman in 1996 did not consider financial aid very important in selecting his or her college?

Solution: Using the formula $P(\text{not } E) = 1 - P(E)$, we have

$$P(\text{financial aid not very important}) = 1 - P(\text{financial aid very important})$$
$$= 1 - 0.331 = 0.669.$$

A probability computed by using our probability formula is sometimes called a **theoretical probability**. There are many examples in real life where the theoretical probability of an event cannot be determined. Sometimes we have to determine the probability of a certain event by observing the actual outcomes of an experiment when it is repeated many times. This kind of probability is called **empirical probability**. To compute the empirical probability, we simply divide the number of times the event has occurred by the total number of times the experiment has been performed.

Example 7   *Free Throws in Basketball*

Through the games of January 20 in the 1997–98 NBA season, Reggie Miller of the Indiana Pacers made 209 of 239 free throws. What was the empirical probability that he would make a free throw on his next attempt?

Solution: Because the event of making a free throw has occurred 209 times of the 239 times the experiment was performed, the empirical probability of making a free throw is given by

$$\frac{209}{239} \approx 0.874.$$

In instances when we can determine the theoretical probability, how does the theoretical probability compare to the empirical probability of the event? When a coin is flipped, we expect that the probability of getting a head is 1/2 because there is one way to get a head and two equally likely outcomes. However, if you flip a coin exactly twice, you do not know for certain that you will get a head exactly once and a tail exactly once. On the other hand, if the experiment of flipping a coin is repeated a large number of times, we can expect the empirical probability to be close to 1/2. In fact, as the number of times an experiment is repeated grows larger, the empirical probability almost certainly gets closer and closer to the theoretical probability. This property is known as the **Law of Large**

BRANCHING OUT 4.1

A SURPRISING ANSWER TO A PROBABILITY QUESTION

In the *Parade Magazine* column "Ask Marilyn," Marilyn Vos Savant posed the following question: "A woman and a man (unrelated) each have two children. At least one of the woman's children is a boy, and the man's older child is a boy. Do the chances that the woman has two boys equal the chances that the man has two boys?" She answered the question by saying that the probability that the woman has two boys is about 1/3, whereas the probability that the man has two boys is about 1/2. Many of her readers wrote to her objecting to her answer, claiming that the probability that the woman has two boys is 1/2, just as it is for the man. One reader even wrote, "I will send $1000 to your favorite charity if you can prove me wrong. The chances of both the woman and the man having two boys are equal."

Ms. Vos Savant responded to this challenge by providing some empirical evidence to support her claim. She asked her readers to take part in a survey of women with exactly two children, at least one of which is a boy. She published the results of the survey in her article, where she reported that of the

17,946 respondents, 35.9% had two boys, very close to the 1/3 she predicted. To further support her argument, she contacted the U.S. Bureau of the Census. The Census Bureau surveys a random sample of families each year. They reported that from 1987 to 1993, 42,888 families with exactly two children were interviewed. Of these families, 9523 had two girls and the remaining 33,365 had at least one boy. Those families with at least one boy broke down as follows: 11,118 had an older boy and younger girl, 10,913 had an older girl and a younger boy, and 11,334 had two boys. Therefore, from among the families with at least one boy, we see that

$$\frac{11,334}{33,365} \approx 34\%$$

had two boys, very close to the predicted 1/3. Needless to say, the reader who promised to send $1000 to a charity appears to have lost the argument.

Although the data appear to support Ms. Vos Savant's answer, we still have the question of why the probabilities turn out this way. To understand the answer, we will list all of the possible scenarios in each case. Let GB denote the case of the oldest child being a girl and the younger child being a boy, and similarly let BG and BB denote the other possibilities. For the woman with at least one boy, there are three possibilities, all about equally likely: BG, GB, or BB. In the case of the man whose oldest child is a boy, there are only two possibilities, each about equally likely: BB or BG. Therefore, the probability that the woman has two boys is 1/3, and the probability that the man has two boys is 1/2.

Source: Adapted with permission from *Parade*, copyright © 1997.

Numbers. Gambling casinos can be reasonably assured of making a profit because of the Law of Large Numbers. Any individual gambler may make money, but because there are a very large number of gamblers and the probabilities of the games are set up in the casinos' favor, the casinos will profit over the long run.

EXERCISES FOR SECTION 4.1

1. A single die is rolled. Find the probability of rolling

   (a) a 6,   **(b)** an odd number,   **(c)** a number less than 3,   **(d)** a 7.

2. A 12-sided die with sides numbered 1 through 12 is rolled. Assuming that all sides are equally likely to be rolled, what is the probability that the number rolled is

   (a) a 5?   **(b)** an even number?   **(c)** an 11 or 12?

3. A bowl of goldfish has 12 females and 8 males. If you select a fish at random, what is the probability that it will be female?

4. The FBI's *Uniform Crime Reports* for 1996 reported 19,650 deaths in the United States by murder or nonnegligent manslaughter when the U.S. population was estimated to be 265,284,000. What is the probability that a random person in the United States in 1996 was not a murder or nonnegligent manslaughter victim?

5. In 1996, the National Center for Health Statistics reported that 23.4% of deaths in the United States were caused by cancer. What is the probability that a random person who died in 1996 did not die of cancer?

6. A jar of pens has six black pens, four red pens, and seven blue pens. If a pen is selected at random, what is the probability that it will be red?

7. One card is drawn at random from a deck of cards. What is the probability the card will be

   (a) a jack?   **(b)** a face card?

8. One card is drawn at random from a deck of cards. What is the probability the card will be

   (a) a spade?   **(b)** a heart or a club?

9. One card is drawn at random from a deck of cards. What is the probability that the card will not be an ace?

10. Stanford University received 16,842 applications from students wishing to enroll as freshmen for the Fall 1997 semester. Of these, 2596 were accepted for admission. What is the probability that a random applicant received a letter of acceptance?

11. Assuming that the spinner in the figure cannot land on a line, what is the probability that it will land on a 2?

12. Assuming that the spinner in the figure cannot land on a line, what is the probability that it will land on a 1?

13. An experiment consists of flipping a coin and tossing a regular die.

    (a) List all of the possible outcomes of the experiment, where, for example, H2 stands for the outcome that a head comes up on the coin and a 2 on the die.

    (b) What is the probability of getting a tail and a 3?

    (c) What is the probability of getting a head and an even number?

14. An experiment consists of tossing a coin three times in a row.

    (a) List all of the possible outcomes, where, for example, HHT stands for the outcome where a head appears on the first toss, a head appears on the second toss, and the third toss is a tail.

    (b) What is the probability of getting exactly two heads?

    (c) What is the probability of getting no heads?

    (d) What is the probability of getting at least one head?

15. A school is running a raffle for which 870 tickets were sold. One ticket will be drawn and the holder of the ticket will win $500. If you buy six raffle tickets, what is the probability that you will have the winning ticket?

16. If the probability that you will lose a contest is 7/9, what is the probability that you will win the contest?

17. A couple plans to have three children. Assuming that boys and girls are equally likely, find the probability that the couple will have exactly one girl.

18. A custodian has 15 keys on a ring. He needs to open a door but does not know which of the 15 keys is the correct one, so he tries them systematically one after another. What is the probability that he will find the correct key within his first four tries?

19. Two dice are thrown. What is the probability that the sum of the two dice will be 9?

20. Two dice are thrown. What is the probability that the sum of the two dice will be 8?

21. Two dice are thrown. What is the probability that the sum of the two dice will not be 6?

22. Suppose the spinner in the figure is spun two times in a row. Assuming that the spinner cannot land on a line, what is the probability that it will land on 2 at least once?

23. A question on a multiple choice test has five possible answers, only one of which is correct. If you guess at random on the question, what is the probability of guessing the correct answer?

24. Three hundred and sixteen students were enrolled in statistics classes. Eighty-six of these students received A's, 80 received B's, 74 received C's, 51 received D's, and the remaining 25 failed the course. What is the probability that a randomly selected student in a statistics course that semester received a grade of

(a) A?   **(b)** C or better?

25. In the NBA draft lottery, 1000 Ping-Pong balls are placed in a bin. Each of the 13 NBA teams that did not make the playoffs is assigned some of the Ping-Pong balls, with the number of balls dependent on the team's season record. The team with the worst record is assigned 250 balls, whereas the team with the second to worst record receives 200 balls and the one with the third to worst record receives 157 balls. One ball is drawn and the team with that ball gets the number one draft pick. What is the probability that

(a) the team with the worst record will get the number one draft pick?

(b) one of the worst three teams will get the number one draft pick?

26. In the *Sports Illustrated* Baseball, Football, and Golf games there are two cubic dice with faces labeled 0, 1, 2, 3, 4, 5 on one and 0, 0, 1, 2, 3, 4 on the other. When these two dice are rolled and the faces are summed, the sum must be between 0 and 9. By writing out a diagram such as the one for standard dice in Table 4.1, find the probability that the sum of the two rolled dice will be

(a) 0 **(b)** 1 **(c)** 2 **(d)** 3 **(e)** 4 **(f)** 5 **(g)** 6 **(h)** 7 **(i)** 8 **(j)** 9.

27. For a peculiar pair of dice, discovered by Colonel George Sicherman and known as Sicherman Dice, one die has faces labeled 1, 3, 4, 5, 6, and 8 and the other has faces labeled 1, 2, 2, 3, 3, and 4. The remarkable feature of these dice is that the probability of obtaining a particular value for the sum of the two dice is exactly the same probability as for a pair of standard dice. By writing out a diagram such as the one for standard dice in Table 4.1, find the probability that the sum of the two rolled dice will be

(a) 2   **(b)** 3   **(c)** 4   **(d)** 5   **(e)** 6   **(f)** 7

(g) 8   **(h)** 9   **(i)** 10   **(j)** 11   **(k)** 12.

28. According to the National Climatic Data Center, Columbus, Ohio had 204 cloudy days in 1995. On the basis of this information, what is the empirical probability that it will be cloudy in Columbus on a random day?

29. For the 1997 football season, Jeff Jaeger of the Chicago Bears attempted 26 field goals and made 21. On the basis of these statistics, what was the empirical probability that he would make a field goal?

30. Suppose an inspector testing a batch of grass seed found that 392 of the 500 seeds she planted germinated. What is the empirical probability that a random seed from the batch will germinate?

31. The Air Transport Association of America reported that between 1994 and 1996 there were 747 fatalities on scheduled commercial airplanes. These accidents occurred on a total of nine flights. There were 23.8 million departures during this time period with a total of 1,657,822,000 passengers.

(a) Using the number of fatalities, find the empirical probability of a random passenger flying a single time dying in a plane accident over this time period.

(b) Find the empirical probability that a random flight would be in an accident involving fatalities.

(c) Based on the probabilities computed in parts (a) and (b), does it seem likely that there are many survivors of plane accidents involving fatalities? Explain your reasoning.

32. In a variation of the old game show "Let's Make a Deal," a contestant picks one of three doors. Behind one of the doors lies a grand prize. After the contestant initially picks a door at random, the host always opens one of the two remaining doors. The host knows where the grand prize is, and he purposely never opens the door with the grand prize behind it. Assume that if the contestant picks the grand prize door first, then the host is equally likely to open either of the two remaining doors. After opening the second door, the host allows the contestant the option of switching to another door. Should the contestant switch? Explain your answer. (This question has a surprising answer and it has led to many arguments, even among mathematicians.)

33. Two cards are placed in a hat. One of the cards has two blue sides, and the other card has one side blue and one side red. If you draw a card out of the hat and see that the side you are looking at is blue, what is the probability that the other side is also blue? (This question also has a surprising answer.)

4.2   ODDS

In everyday use, probabilities are often expressed in terms of odds. It is a common misconception that odds and probability are the same thing. This is simply not the case. As we will see, if the probability of the event occurring is 1/4, this does not mean that the odds against an event occurring are 4 to 1.

In gambling situations and everyday language, odds are usually given in the form of **odds against** an event occurring. If the odds against an event occurring are a to b, this means that for every a times the event does not occur, we expect the event to occur b times. In other words, if the experiment is run $a + b$ times, then the event is expected to occur b times and not to occur a times.

Example 1   *Rolling a Die*

What are the odds against rolling a 2 on a regular die?

Solution: There is only one way to roll a 2, and five ways not to roll a 2. This tells us that the odds against rolling a 2 on a regular die are 5 to 1.   ∎

Whenever possible, we will write odds in the form a to b, where a and b are whole numbers with no factors in common. If a and b do have a factor in common, we simply cancel out the factor from both a and b. This does not change the ratio of a to b, so the odds have the same meaning. This is illustrated in the next example.

Example 2   *A Raffle*

One hundred and four raffle tickets were sold in a raffle for a new car. A man really wanted the car, so he bought 72 of the 104 tickets. What are the odds against the man winning the car?

Solution: Because the man has 72 tickets, there are 72 ways he can win. There were 104 tickets sold, so there are $104 - 72 = 32$ ways he can lose. Therefore, the odds against him winning the car are 32 to 72. However, because 32 and 72 are both divisible by 8, we divide each of them by 8 to express these odds as 4 to 9.

■

We could write odds in the form of **odds for** an event occurring rather than odds against. If the odds against an event occurring are a to b, then the odds for the event occurring are b to a. Because the odds that are stated in the press and in everyday language are usually odds against, we will only consider odds against.

Suppose we know the probability $P(E)$ of an event E occurring, and we want to find the odds against E occurring. Because the ratio of a to b is the ratio of the number of times E will not occur to the number of times E will occur, we have the following formula:

Converting from Probability to Odds Against

The odds against an event E occurring are a to b, where

$$\frac{a}{b} = \frac{\text{probability } E \text{ will not occur}}{\text{probability } E \text{ will occur}} = \frac{1 - P(E)}{P(E)}.$$

We generally express the odds a to b with a and b as whole numbers with no factors in common. Therefore, when applying the last formula you may have to alter the fraction to make it a ratio of whole numbers with no factors in common.

Example 3   *On-Time Flights*

According to the U.S. Department of Transportation, in 1996 about 68% of TWA flights were on time (within 15 minutes of scheduled times). What were the odds against a TWA flight being on time in 1996?

Solution: The odds against a TWA flight being on time were a to b, where

$$\frac{a}{b} = \frac{1 - 0.68}{0.68} = \frac{0.32}{0.68} = \frac{32}{68} = \frac{8}{17}.$$

So we see that the odds were 8 to 17.

■

At the beginning of this section we noted that if the probability of an event occurring is 1/4, this does not mean the odds against the event occurring are 4 to 1. We see this in Example 4.

Example 4   *Comparison of Odds to Probability*

Suppose the probability of an event occurring is $P(E) = 1/4$. What are the odds against the event occurring?

Solution: The odds against the event occurring are a to b, where

$$\frac{a}{b} = \frac{1 - 1/4}{1/4} = \frac{3/4}{1/4} = \frac{3}{1}.$$

So the odds are 3 to 1.

■

Sometimes you may be given the odds against an event occurring, and you would like to know the probability of the event occurring. We know that if the odds against an event E occurring are a to b, then the event E is expected to occur b times every $a + b$ times the experiment is run. Because $P(E)$ can be thought of as the ratio of successful occurrences of E to the number of times the experiment is run, we have the following formula:

Converting from Odds Against to Probability

If the odds against E occurring are a to b, then

$$P(E) = \frac{b}{a + b}.$$

Example 5 *Sweepstakes*

Suppose a sweepstakes contest states that the odds against winning some prize are 55 to 7. If you enter the sweepstakes, what is the probability that you will win some prize?

Solution: Using the formula for converting from odds against to probability, we see that

$$P(\text{winning some prize}) = \frac{7}{55 + 7} = \frac{7}{62} \approx 0.1129.$$

■

House Odds and Fair Bets

At gambling casinos the payoffs for particular bets are stated in terms of odds. However, the odds that are posted are not the true odds, but rather something called the house odds. The **house odds** tell a gambler what the payoff is on a bet. In particular, if the house odds are a to b, then for every b dollars bet on the event occurring, the gambler would win a dollars in addition to having the b dollar bet returned whenever the event occurred. If the event does not occur, the gambler loses the b dollars bet. In other words, if the player wins, the house pays a/b dollars for every dollar wagered in addition to returning the wager, and the house keeps the wager if the player loses. As you would expect, the house odds at a casino are set in such a way that the casino will almost certainly make a profit over the long term. In horse racing, the house odds are determined by a system called pari-mutuel betting. In this system, the house odds depend on how much the betting public wagers on each horse. However, because the track takes a certain percentage of the total money bet off the top, the posted house odds are usually not favorable to the bettor.

Example 6   *The Preakness Stakes*

When Real Quiet, the winner of the 1998 Kentucky Derby, ran in the 1998 Preakness Stakes, the second race in the Triple Crown, the house odds against Real Quiet winning were 5 to 2. Real Quiet went on to win the Preakness Stakes. If you had bet $8 on Real Quiet, how much would you have won in addition to having your $8 returned?

Solution: Because you win $(5/2) = $2.50 for every $1 bet and you bet $8, you would have won $8 \cdot \$2.50 = \$20$.

∎

Real Quiet at the
Preakness Stakes.

If the house odds are equal to the true odds against the event occurring, we say a wager in this situation is a **fair bet**. The way that casinos make profits is by setting the house odds so that the payoff is less than this ideal. In other words, the house odds are set so that the ratio a/b is smaller than it would be for the true odds against the event occurring. We see how this works in the next example.

Example 7   *Roulette*

In a common form of roulette, a ball is spun around a wheel with 38 slots of equal size and eventually lands in one of the slots. The slots are numbered 0, 00, and 1 through 36. One wager that can be made in roulette is to bet that the ball will land on a single number between 1 and 36. The house odds for betting on a single number are 35 to 1. Is wagering on a single number a fair bet?

Solution: Because there is only 1 way to win and 37 ways to lose, the odds against landing on the number are 37 to 1. The house odds are only 35 to 1, so the ratio of house odds 35/1 is less than the ratio of true odds 37/1. Therefore, wagering on a single number is not a fair bet, but instead is a bet that favors the casino. Exactly how favorable or unfavorable a wager is can be measured by computing a number called its expected value. This topic is covered in Section 4.7.

∎

Roulette wheel.

Example 8   *Horse-Race Betting*

Suppose the house odds on a particular horse in a race are 5 to 8. If you believe the horse has a 60% chance of winning the race, do you think wagering on the horse is a fair bet?

Solution: If you are right and the horse has a 60% chance of winning the race, then the true odds against the horse winning the race are a to b, where

$$\frac{a}{b} = \frac{1 - 0.60}{0.60} = \frac{0.40}{0.60} = \frac{2}{3}.$$

Because the ratio of true odds 5/8 = 0.625 is less than the ratio of the true odds 2/3 ≈ 0.667, you would not consider the wager to be a fair bet, but instead a wager in favor of the track.

Although wagers made at casinos and tracks usually do not favor the bettor, there are some exceptions to this rule. When the house odds are determined by the betting public as in the case of horse racing and betting on sports, the house odds may sometimes be favorable to a knowledgeable bettor. For example, a very good horse may be somewhat unknown to the general horse-racing audience and the resulting house odds against the horse may be higher than they really should be.

BRANCHING OUT 4.2

MAKING A PROFIT WITH SUPER BOWL BETTING

Casinos in Las Vegas post house odds for betting on a team to win the Super Bowl long before the football season even begins. The opening odds for the 1999 Super Bowl are shown in the list.

Although some of these house odds give very attractive payoffs, you must keep in mind that the teams with long odds, such as the New Orleans Saints at 150 to 1, have very little chance of being in the Super Bowl, much less winning it. It turns out that the payoffs are really quite poor. If we convert all the odds to probabilities and then add them up, we would find that they sum to approximately 1.7, whereas the true probabilities should add up to 1. By setting the house odds so low, the casinos stand to make a healthy profit on Super Bowl betting.

Green Bay Packers	2 to 1
San Francisco 49ers	7 to 2
Denver Broncos	8 to 1
Dallas Cowboys	8 to 1
Pittsburgh Steelers	10 to 1
Jacksonville Jaguars	10 to 1
Kansas City Chiefs	12 to 1
New England Patriots	15 to 1
Tampa Bay Buccaneers	18 to 1
Miami Dolphins	20 to 1
Detroit Lions	20 to 1
New York Jets	25 to 1
New York Giants	25 to 1
Washington Redskins	25 to 1
Carolina Panthers	30 to 1
Oakland Raiders	30 to 1
Philadelphia Eagles	30 to 1
Tennessee Oilers	30 to 1
Minnesota Vikings	30 to 1
Seattle Seahawks	30 to 1
Cincinnati Bengals	40 to 1
Buffalo Bills	50 to 1
San Diego Chargers	60 to 1
Baltimore Ravens	60 to 1
Atlanta Falcons	60 to 1
Indianapolis Colts	75 to 1
Chicago Bears	75 to 1
Arizona Cardinals	75 to 1
Saint Louis Rams	100 to 1
New Orleans Saints	150 to 1

John Elway after the Broncos' victory at Super Bowl XXXII.

EXERCISES FOR SECTION 4.2

1. What are the odds against rolling an even number on a regular die?

2. If a card is drawn at random from a standard deck, what are the odds against the card being an ace?

3. If a sportswriter thinks the odds against an AFC team winning the next Super Bowl are 7 to 2, what does he think is the probability that an AFC team will win the next Super Bowl?

4. A survey conducted by the U.S. Bureau of the Census in 1997 showed that 84% of the people in the United States are covered by health insurance. What are the odds against a person in the United States being covered by health insurance?

5. There is an 85% probability that a team will win a game. What are the odds against the team winning the game?

6. According to *The World Book of Odds*, the odds against the home team winning a major league baseball game are 20 to 23. What is the probability of the home team winning a game?

7. The odds against winning a $2 prize in the Lucky Gold game from the Colorado Lottery are 9 to 2. What is the probability of winning $2 in the game?

8. If the probability of guessing correctly on a multiple choice exam question is 1/5, what are the odds against guessing correctly?

9. A coin is tossed two times in a row. What are the odds against getting two heads?

10. There are 18 children on a swim team. Twelve of the children will be selected at random to represent the team at a meet in another town. What are the odds against a particular child being selected?

11. If a game spinner has probability 3/8 of landing on blue, what are the odds against the spinner landing on blue?

12. In the game of Monopoly, players roll two dice and move the number of spaces given by the sum of the dice. Pennsylvania Avenue is five spaces from Boardwalk on the game board. What are the odds against a player on Pennsylvania Avenue landing on Boardwalk on the next play?

13. The *World Book of Odds* states that the odds against a German soldier being killed in World War II were 5 to 1. What was the probability of a German soldier being killed?

14. At a local video store, there is a gumball machine with gumballs of various colors. If you put a quarter in the machine and get a blue gumball, you win a free video rental. If only 2% of the gumballs are blue, what are the odds against getting a blue gumball?

15. The probability that a baby of a random 45-year-old woman will have Down's syndrome is 1/35. What are the odds against the baby having Down's syndrome?

16. If the odds against winning first prize in a sweepstakes contest are 2500 to 1, what is the probability of winning first prize?

17. When Seattle Slew raced in the Belmont Stakes horse race in 1977, the house odds against him winning were 2 to 5. How much would a person who placed a $2 bet on Seattle Slew have won in addition to having the wager returned when Seattle Slew went on to win the race?

18. The winner of the 1953 Kentucky Derby, Dark Star, was truly a dark horse in the race. The house odds against Dark Star were 249 to 10. How much would a person

who placed a $150 bet on Dark Star to win the Derby have won in addition to having the wager returned?

19. On March 10, 1998, two days before the 1998 NCAA Basketball Tournament began, the Las Vegas house odds against Kentucky winning the tournament were 10 to 1. If a gambler had placed a $75 bet on Kentucky, what would the gambler have won in addition to having the wager returned when Kentucky went on to win the tournament?

20. Die-hard football fans can place their bets on the Super Bowl well before the game takes place. On October 27, 1997, the Las Vegas house odds on the Denver Broncos winning the 1998 Super Bowl were 5 to 2. If a gambler had bet $46 on the Broncos, how much would the gambler have won in addition to having the wager returned when the Broncos went on to win it?

21. In a common form of roulette, a ball is spun around a wheel with 38 slots of equal size and eventually lands in one of the slots. Eighteen of the slots are colored red. One wager that can be made in roulette is to bet that the ball will land on the color red.

    (a) What are the odds against winning a bet on red?

    (b) The house odds on a bet on red are typically 1 to 1. Is a bet on red a fair bet?

22. In football betting at some casinos, a bettor can wager $11 on a game with a point spread. If the team the bettor chooses beats the point spread, then the $11 bet is returned to the bettor along with an additional $10. If the team doesn't beat the point spread, the bettor loses the $11 wager. (With a point spread, the favored team must win by more than the point spread in order for a bet on the favored team to be a winning bet. For instance, suppose the Green Bay Packers are favored to beat the Chicago Bears by 10 points. If Green Bay wins by more than 10 points, then a bet on Green Bay is a winning bet. If Green Bay wins by fewer than 10 points, ties, or loses, then a bet on the Chicago Bears is a winning bet. If Green Bay wins by exactly 10 points, then the bet is called off and the gambler's wager is returned.)

    (a) What are the house odds on this kind of bet?

    (b) Suppose the probability of winning the bet is 1/2 and the probability of losing the bet is also 1/2. Is this a fair bet?

23. Suppose the house odds on a particular horse in a race are 4 to 1. If you believe the horse has a 20% chance of winning the race, do you think that wagering on the horse is a fair bet?

24. Suppose the house odds on a particular horse in a race are 7 to 10. If you believe the horse has a 70% chance of winning the race, do you think that wagering on the horse is a fair bet?

4.3   THE ADDITION RULE

Suppose we want to find the probability of event A or event B occurring. For example, in drawing a card at random from a regular deck of 52 cards, we might want to know the probability, $P(\text{ace or heart})$, of drawing an ace or a heart. By an ace *or* a heart, we include the possibility that the card is both an ace and a heart. It is tempting to say that the probability of drawing an ace or a heart is simply the sum $P(\text{ace}) + P(\text{heart})$, but we would be wrong in this case. The reason this method does not work is the ace of hearts. To compute $P(\text{ace or heart})$, we need to count the number of cards that are aces or hearts and then divide this number by 52. To count the number of cards that are aces or hearts, it is helpful to look at the **Venn diagram** in Figure 4.2.

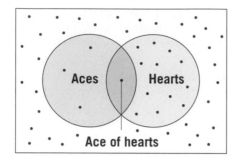

figure 4.2
Venn diagram for a
deck of cards.

In the Venn diagram, the 52 dots in the rectangle represent all 52 cards. The dots in the circle labeled aces represent the 4 "aces," and those in the circle labeled "hearts" represent the 13 hearts. Because the ace of hearts is both an ace and a heart, it lies in the intersection of the two circles. To count the number of cards that are aces or hearts, we add the 4 aces to the 13 hearts and then subtract the 1 ace of hearts because we have counted it twice. This gives us

number of aces or hearts = 4 + 13 − 1 = 16.

To compute $P(\text{ace or heart})$, we divide our calculation through by 52 to get

$$P(\text{ace or heart}) = \frac{4}{52} + \frac{13}{52} - \frac{1}{52} = \frac{16}{52} = \frac{4}{13}.$$

Using these ideas, we can now find a general formula for $P(A \text{ or } B)$, where A and B are two different events. *Note that when we say A or B happens we mean that A happens without B, that B happens without A, or that both happen at the same time.* The Venn diagram for this situation is shown in Figure 4.3.

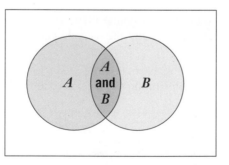

figure 4.3
Venn diagram for
events A and B.

To find $P(A \text{ or } B)$ we add $P(A)$ and $P(B)$ and then subtract $P(A \text{ and } B)$, because we counted the events that lie in A and B twice. This leads to the following formula, which we call the **addition rule**:

The Addition Rule

$$P(A \text{ or } B) = P(A) + P(B) - P(A \text{ and } B)$$

Example 1   *Blood Type and Rh Factor*

Forty-three percent of the world's population have type O blood, 85% of the world's population are Rh-positive, and 37% have type O blood and are Rh-positive. What is the probability that any individual will have type O blood or be Rh-positive?

Solution: Using the addition rule, we have

$$P(\text{type O or Rh-positive}) = P(\text{type O}) + P(\text{Rh-positive}) - P(\text{type O and Rh-positive})$$
$$= 0.43 + 0.85 - 0.37 = 0.91.$$

Notice that in this particular case if we had tried to find the probability by adding $P(\text{type O})$ and $P(\text{Rh-positive})$ without subtracting $P(\text{type O and Rh-positive})$, we would have arrived at a probability larger than 1, which is impossible.

■

There are many instances of events A and B, where $P(A \text{ and } B) = 0$. In other words, it is impossible for both A and B to occur simultaneously. In this case, we say that the events A and B are **mutually exclusive**. The Venn diagram corresponding to this situation is drawn in Figure 4.4. Notice that there is no intersection of the circle representing A and the circle representing B.

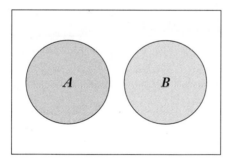

Figure 4.4
Venn diagram for mutually exclusive events A and B.

When two events A and B are mutually exclusive, $P(A \text{ and } B) = 0$, so the addition rule takes on the simplified form $P(A \text{ or } B) = P(A) + P(B)$. Similarly, if more than two events are pairwise mutually exclusive, then to compute the probability of any one of them happening, we simply add all the probabilities. For example, if the events A, B, and C are pairwise mutually exclusive, then A, B, and C would be represented by nonoverlapping circles in a Venn diagram, and $P(A \text{ or } B \text{ or } C) = P(A) + P(B) + P(C)$.

Example 2   *Drawing a Card*

Suppose one card is drawn at random from a deck of cards. Find the probability of drawing

(a) a queen or jack,
(b) a club, heart, or spade.

Solution:

(a) Drawing a queen and drawing a jack are mutually exclusive events, so we have

$$P(\text{queen or jack}) = P(\text{queen}) + P(\text{jack}) = \frac{4}{52} + \frac{4}{52} = \frac{8}{52} = \frac{2}{13}.$$

(b) Because drawing a club, a heart, or a spade are pairwise mutually exclusive events, we can add the probabilities to obtain

$$P(\text{club, heart, or spade}) = P(\text{club}) + P(\text{heart}) + P(\text{spade})$$

$$= \frac{13}{52} + \frac{13}{52} + \frac{13}{52} = \frac{39}{52} = \frac{3}{4}.$$

Another way to answer this question is to notice that

$$P(\text{club, heart, or spade}) = P(\text{not a diamond}) = 1 - P(\text{diamond})$$

$$= 1 - \frac{13}{52} = \frac{39}{52} = \frac{3}{4}.$$

■

Sometimes we need to apply more than one of the rules of probability in order to solve a problem. For instance, in part (b) of the next example we need to use the addition rule and the fact that $P(\text{not } E) = 1 - P(E)$.

Example 3   *Firearm Deaths*

According to the National Safety Council, in 1994 there were 38,166 firearm deaths in the United States. Of these, 32,694 were males, 10,954 were between the ages of 15 and 24, and 9809 were males between the ages of 15 and 24.

(a) What is the probability that a random person killed by firearms in 1994 was male or between the ages of 15 and 24?

(b) What is the probability that a random person killed by firearms in 1994 was neither male nor between the ages of 15 and 24?

Solution:

(a) Applying the addition rule, we get

$P(\text{male or between 15 and 24})$

$\quad = P(\text{male}) + P(\text{between 15 and 24}) - P(\text{male and between 15 and 24})$

$$= \frac{32{,}694}{38{,}166} + \frac{10{,}954}{38{,}166} - \frac{9809}{38{,}166} = \frac{33{,}839}{38{,}166} \approx 0.887.$$

(b) Using the fact that $P(\text{not } E) = 1 - P(E)$, we have

$P(\text{neither male nor between 15 and 24}) = 1 - P(\text{male and between 15 and 24}).$

Therefore, using our answer to part (a), we get

$$P(\text{neither male nor between 15 and 24}) = 1 - \frac{33{,}839}{38{,}166} = \frac{4327}{38{,}166} \approx 0.113.$$

■

Exercises for Section 4.3

1. Twelve percent of the world's population have type B blood, 15% are Rh-negative, and 2% have type B blood and are Rh-negative. What is the probability that a randomly selected individual will have type B blood or be Rh-negative?

2. In one generation of plants, 1/4 of the plants have white flowers, 1/4 of the plants are short, and 1/16 of the plants are short and have white flowers. What is the probability that a random plant from this generation will be short or will have white flowers?

3. In a particular neighborhood of 60 homes, 35 of the homes have children, 22 have pets, and 15 have both. What is the probability that a randomly selected home in the neighborhood will

   (a) have children or pets?   (b) have neither children nor pets?

4. A U.S. National Highway Traffic Safety Administration report on pedestrian injuries in police-reported traffic accidents in 1993 included the following statistics: in 33.7% of the cases the victim was under 16 years old, in 35% the accident occurred at night, and in 8.8% the victim was under 16 and the accident occurred at night. For a random pedestrian injury case in 1993, what is the probability that

   (a) the victim was under 16 or the accident occurred at night?
   (b) the victim was 16 or over and the accident occurred during the day?

5. Students in public schools in Texas take a standardized test called the Texas Assessment of Academic Skills Test (TAAS). For children in grade 5, the TAAS consists of two tests, one in reading and one in math. In 1994, across the state 77.9% of fifth graders passed the reading test, 63.0% passed the math test, and 58.6% passed both. What is the probability that a random fifth grader passed the reading test or the math test?

6. If a single regular die is rolled, what is the probability of rolling an even number or a number greater than 3?

7. Suppose 37% of those polled approve of the Republican candidate for president, 42% approve of the Democratic candidate for president, and 7% approve of both candidates. What is the probability that a randomly selected person approves of neither candidate?

8. If one card is drawn at random from a deck of cards, what is the probability of drawing neither a king nor a spade?

9. If one card is drawn at random from a deck of cards, what is the probability of drawing an ace or a red card?

10. If one card is drawn at random from a deck of cards, what is the probability of drawing a king, queen, or ace?

11. According to the U.S. Bureau of the Census, the total population of the United States in 1990 was 248,709,873. Of this population, 18,354,443 were under 5 years old and 13,135,272 were over 74 years old. What is the probability that a randomly selected person in the United States in 1990 was over 74 or under 5?

12. According to the American Medical Association, in 1996 there were 737,764 physicians in the United States, 157,387 of whom were female. There were 133,005 physicians under 35 years of age, 47,348 of whom were female. What is the probability that a randomly chosen physician in 1996 was female or under the age of 35?

13. If the probability that a spinner will land on the number 1 is 1/6, the probability that it will land on the number 2 is 1/4, and the probability that it will land on the number 3 is 1/8, what is the probability that it will land on a 1, 2, or 3?

14. Suppose that at a college, 55% of the students earn a degree at the end of four years and 25% of the students earn a degree at the end of five years. What is the probability of earning a degree at the end of four or five years?

For Exercises 15 and 16, note that there are some instances where the addition rule

$$P(A \text{ or } B) = P(A) + P(B) - P(A \text{ and } B)$$

can be used to solve for one of the quantities $P(A)$, $P(B)$, or $P(A \text{ or } B)$. For instance, if you are given $P(A)$, $P(B)$, and $P(A \text{ or } B)$, then you can solve for $P(A \text{ and } B)$.

15. One-third of the children in a class like hot dogs, 1/2 like pizza, and 3/4 like hot dogs or pizza. What is the probability that a randomly selected child in the class likes both hot dogs and pizza?

16. Suppose 18% of the students at a college have an academic scholarship that pays partial tuition, 43% of the students at the college have some need-based financial aid, and 52% of the students have need-based financial aid or an academic scholarship. What is the probability that a randomly selected student has both need-based financial aid and an academic scholarship?

4.4 CONDITIONAL PROBABILITY AND THE MULTIPLICATION RULE

There are instances when we want to compute the probability of a particular event occurring when we have further information that may limit the possible outcomes of the experiment. For example, when rolling a single die, we know that the probability of rolling a 4 is 1/6. However, if we are told that the number rolled is an even number, then the only possible outcomes are rolling a 2, 4, or 6, so the probability that the number rolled is a 4 is 1/3. We see that by limiting the number of possible outcomes, the probability of a particular event may be changed.

The probability that an event B will occur given that the event A has occurred is called a **conditional probability** and is denoted by $P(B|A)$. When discussing conditional probability and independence, we will assume throughout that the given events occur with probabilities greater than zero. Before giving a general formula for $P(B|A)$, we will look at an example.

Example 1 *Conditional Probability*

Suppose a college class has 16 members, consisting of 10 women and 6 men. Four of the women are seniors and three of the men are seniors.

(a) What is the probability that a student selected at random is a senior?

(b) What is the probability that a student selected at random is a senior given that the selected student is a woman?

Solution:

(a) There are 16 students and 7 of them are seniors, so we see that

$$P(\text{senior}) = \frac{7}{16}.$$

(b) If we know the selected member is a woman and there are 10 women in the class, we see that there are 10 possible outcomes. Of the 10 women, there are only 4 seniors. Therefore the conditional probability that the selected student is a senior given that the member is a woman is given by

$$P(\text{senior}\,|\,\text{woman}) = \frac{4}{10} = \frac{2}{5}.$$

Using the method of Example 1, we will now find a general formula for the conditional probability $P(B|A)$. If we are given that event A has occurred, then the only possible outcomes are those that result in event A. Also, the only way B can occur in this case is if an outcome that results in both A and B occurs. Therefore, *assuming all the outcomes of an experiment are equally likely*, we see that

$$P(B|A) = \frac{\text{number of outcomes that result in events } A \text{ and } B}{\text{number of outcomes that result in event } A}.$$

We can write this equation in a more compact form by letting $n(A)$ be the number of outcomes that result in event A and letting $n(A \text{ and } B)$ be the number of outcomes that result in both event A and event B. Then we have the following formula for conditional probability:

Conditional Probability

$$P(B|A) = \frac{n(A \text{ and } B)}{n(A)}$$

Example 2   *Drawing a Card*

Suppose one card is drawn at random from a standard deck. What is the probability that the card is

(a) the jack of hearts?
(b) the jack of hearts, given that it is a heart?
(c) a heart, if you know that it is a red card?

Solution:
(a) Because there are 52 cards in the deck and only 1 jack of hearts, we have

$$P(\text{jack of hearts}) = \frac{1}{52}.$$

(b) There are 13 hearts and only 1 card that is a jack and a heart, so

$$P(\text{jack of hearts}\,|\,\text{heart}) = \frac{n(\text{jack and heart})}{n(\text{heart})} = \frac{1}{13}.$$

(c) Twenty-six cards in the deck are red cards and 13 of them are hearts. Therefore, we have

$$P(\text{heart}\,|\,\text{red}) = \frac{n(\text{heart and red})}{n(\text{red})} = \frac{13}{26} = \frac{1}{2}.$$

■

Example 3 *Minimum Wage Bill*

The vote on a bill in the U.S. House of Representatives on May 23, 1996 to increase the minimum wage to $5.15 from $4.25 per hour was as shown in Table 4.2.

Table 4.2	The Vote on the Minimum Wage		
Party	*Yes*	*No*	*Total*
Democrats	187	6	193
Republicans	93	138	231
Independents	1	0	1
Total	281	144	425

(a) What is the probability that a random representative voted yes?

(b) What is the probability that a random representative voted yes, given that the representative is a Republican?

Solution:

(a) There were 425 representatives, 281 of whom voted yes, so we have

$$P(\text{voted yes}) = \frac{281}{425} \approx 0.661.$$

(b) Using the formula for conditional probability, we get

$$P(\text{voted yes}\,|\,\text{Republican}) = \frac{n(\text{Republican and voted yes})}{n(\text{Republican})} = \frac{93}{231} \approx 0.403.$$

■

Sometimes we want to compute a conditional probability $P(B\,|\,A)$ in a situation where the exact counts $n(A \text{ and } B)$ and $n(A)$ are not available. As long as we know $P(A \text{ and } B)$ and $P(A)$, we can easily compute the conditional probability. To see why, notice that

$$P(B\,|\,A) = \frac{n(A \text{ and } B)}{n(A)} = \frac{\left(\dfrac{n(A \text{ and } B)}{\text{total number of possible outcomes}}\right)}{\left(\dfrac{n(A)}{\text{total number of possible outcomes}}\right)} = \frac{P(A \text{ and } B)}{P(A)},$$

and we see that $P(B\,|\,A)$ can be expressed in terms of $P(A \text{ and } B)$ and $P(A)$. Even though we derived this new formula for the case where all outcomes are equally likely, the formula holds in general. We call this the alternate formula for conditional probability.

> ### *Alternate Formula for Conditional Probability*
>
> $$P(B|A) = \frac{P(A \text{ and } B)}{P(A)}$$

Example 4   *A Poll*

Suppose a poll of local voters indicates that 45% favor a property tax increase, 65% favor the establishment of a state lottery, and 12% favor both. If a voter is in favor of the property tax increase, what is the probability that the voter is also in favor of the lottery?

Solution: We use the alternate formula for conditional probability, and we get

$$P(\text{in favor of lottery} | \text{in favor of tax}) = \frac{P(\text{in favor of tax and lottery})}{P(\text{in favor of tax})}$$

$$= \frac{0.12}{0.45} \approx 0.2667 = 26.67\%.$$

Notice that we did not need to use the fact that 65% of the voters favored the lottery to answer the question.   ∎

The Multiplication Rule

Suppose we want to find $P(A \text{ and } B)$, the probability of both event A and event B occurring. We do not have to look far to find a nice way to compute $P(A \text{ and } B)$. We start with the alternate formula for conditional probability,

$$P(B|A) = \frac{P(A \text{ and } B)}{P(A)},$$

and solve for $P(A \text{ and } B)$ to get the following formula, called the **multiplication rule**:

> ### *The Multiplication Rule*
>
> $$P(A \text{ and } B) = P(A) \cdot P(B|A)$$

Example 5   *Using the Multiplication Rule*

(a) If two cards are drawn at random from a standard deck, what is the probability that both cards drawn are kings?

(b) If three cards are drawn at random from a standard deck, what is the probability that all three cards drawn are kings?

Solution:

(a) From the multiplication rule we know that

P(first card a king and second card a king)

$$= P(\text{first card a king}) \cdot P(\text{second card a king} \,|\, \text{first card a king}).$$

Because there are 4 kings in the deck of 52 cards, we have

$$P(\text{first card a king}) = \frac{4}{52}.$$

After the first king is drawn, there are 3 kings in a deck of only 51 cards, and we see that

$$P(\text{second card a king} \,|\, \text{first card a king}) = \frac{3}{51}.$$

Therefore,

$$P(\text{first card a king and second card a king}) = \frac{4}{52} \cdot \frac{3}{51} = \frac{1}{13} \cdot \frac{1}{17} = \frac{1}{221}.$$

(b) Applying the multiplication rule twice, we see that P(all three cards kings) is given by

P(first card a king) $\cdot$ P(second card a king $|$ first card a king)

$$\cdot P(\text{third card a king} \,|\, \text{first and second cards are kings}).$$

Therefore, we have

$$P(\text{all three cards kings}) = \frac{4}{52} \cdot \frac{3}{51} \cdot \frac{2}{50} = \frac{1}{13} \cdot \frac{1}{17} \cdot \frac{1}{25} = \frac{1}{5525}.$$

Example 6   *A Raffle*

A woman bought five raffle tickets. Four hundred raffle tickets were sold in all, and 10 of the tickets will be awarded prizes. What is the probability that the woman will have only winning tickets?

Solution: In this case we apply the multiplication rule five times. Following the same line of reasoning as in part (b) of Example 5, we see that

$$P(\text{five winning tickets}) = \frac{10}{400} \cdot \frac{9}{399} \cdot \frac{8}{398} \cdot \frac{7}{397} \cdot \frac{6}{396} \approx 0.00000000303.$$

Independence and the Multiplication Rule

There are many instances when the occurrence of one event has no effect on the probability of another event. For example, if a coin is flipped two times in a row and the first coin toss results in a head, the probability of a head on the second toss is still 1/2. In this case, we say that the events of getting a head on the first toss and getting a head on the second toss are independent. More generally, two events A and B are **independent** if the occurrence of one has no effect on the probability of the other occurring. We can state this situation more formally in terms of conditional probability.

> The events A and B are **independent** if
> $$P(B|A) = P(B) \quad \text{or} \quad P(A|B) = P(A).$$

The multiplication rule implies that if $P(B|A) = P(B)$, then $P(A|B) = P(A)$, and vice versa. We leave the verification of this as an exercise.

If two events A and B are independent, the multiplication rule takes on the following simplified form:

> ### *Multiplication Rule for Independent Events*
> If A and B are independent events, then
> $$P(A \text{ and } B) = P(A) \cdot P(B).$$

Example 7   *Independent Events*

Suppose the probability of rain on a particular day is 56% and the probability of your favorite basketball team losing is 3/5. What is the probability that you will have a particularly bad day with both rain falling and your favorite team losing?

Solution: It is reasonable to assume that the events of rain falling and your favorite team losing are independent, so we can use the multiplication rule for independent events to find that

$$P(\text{rain and team losing}) = P(\text{rain}) \cdot P(\text{team losing})$$
$$= (0.56)\left(\frac{3}{5}\right) = (0.56)(0.6) = 0.336.$$

∎

Note that in Example 7, if we were to consider the probability of a football team losing rather than a basketball team, it is not clear that we could have assumed independence. Because football is an outdoor sport and some teams are affected more than others by severe weather, it is fairly likely that the probability of rain and the team's chance of losing would not be independent.

When we want to find the probability of more than two independent events occurring, the multiplication rule tells us that the probability of all the events occurring is just the product of the probabilities of each of the events. This is illustrated in the next two examples.

Example 8 *Manufacturing Process*

Suppose a manufacturing process requires three steps. The first step is successful 98% of the time, the second step is successful 95% of the time, and the third step is successful 99% of the time. Assuming that the successes of the three steps are independent, what is the probability that

(a) all three steps of the manufacturing process will be completed successfully?

(b) none of the three steps will be completed successfully?

Solution:

(a) Applying the multiplication rule for independent events, we have

BRANCHING OUT 4.3

The "Hot Hand Theory" in Free Throw Shooting

It is a common scenario: in the final seconds of a game, a basketball player lines up to take two free throw shots. He must make both shots to put his team ahead. He makes the first shot, and the fans of the opposing team take a deep breath. Will he make the next shot? Does the fact that he made the first shot make it more likely that he will make the second one? Those who believe this have faith in the "hot hand theory" of free throw shooting. But is it a theory that holds true in practice? This question was studied by psychologists A. Tversky and T. Gilovich, and it was further analyzed by statistician Robert L. Wardrop in an article published in *The American Statistician*. They looked at data from the nine regular players on the 1980–81 and 1981–82 Boston Celtics basketball team. The data gathered on two of the players, Larry Bird and Rick Robey, are shown in the tables.

Larry Bird				
		Second		
		Hit	Miss	Total
First	Hit	251	34	285
	Miss	48	5	53
	Total	299	39	338

Rick Robey				
		Second		
		Hit	Miss	Total
First	Hit	54	37	91
	Miss	49	31	80
	Total	103	68	171

From these tables, we see that for Larry Bird,

$$P(\text{hit on second}\,|\,\text{hit on first}) = \frac{251}{285} \approx 0.881,$$

$$P(\text{hit on second}\,|\,\text{miss on first}) = \frac{48}{53} \approx 0.906.$$

For Rick Robey we have

$$P(\text{hit on second}\,|\,\text{hit on first}) = \frac{54}{91} \approx 0.593,$$

$$P(\text{hit on second}\,|\,\text{miss on first}) = \frac{49}{80} \approx 0.613.$$

Larry Bird.

$$P(\text{all three steps correct}) = P(\text{1st step correct}) \cdot P(\text{2nd step correct})$$
$$\cdot P(\text{3rd step correct})$$
$$= (0.98)(0.95)(0.99) = 0.92169.$$

(b) Using the multiplication rule and the fact that $P(\text{not } E) = 1 - P(E)$, we have

$$P(\text{all steps fail}) = P(\text{1st step fails}) \cdot P(\text{2nd step fails}) \cdot P(\text{3rd step fails})$$
$$= (1 - 0.98)(1 - 0.95)(1 - 0.99) = (0.02)(0.05)(0.01)$$
$$= 0.00001.$$

If the hot hand theory held true, we would expect that the probability of a hit on the second free throw would be higher after a hit on the first free throw than after a miss on the free throw. For these two players the opposite was true, although the differences in the probabilities were quite small. The hot hand theory is not supported by the separate data on these two players. However, an interesting thing happens when we combine the data for Larry Bird and Rick Robey. The combined data are shown in the following table:

Combined Free Throw Statistics

		Second		
		Hit	Miss	Total
	Hit	305	71	376
First	Miss	97	36	133
	Total	402	107	509

For the combined data we have

$$P(\text{hit on second} \mid \text{hit on first}) = \frac{305}{376} \approx 0.811,$$

$$P(\text{hit on second} \mid \text{miss on first}) = \frac{97}{133} \approx 0.729.$$

We see that the probability of a hit on the second free throw is higher after a hit on the first than after a miss. The hot hand theory seems to be supported by this combined data. How can this be when such a conclusion was not supported by the data on the two individual players? This situation is an example of a phenomenon known as Simpson's Paradox. Simpson's Paradox occurs in many areas. For instance, it may suggest discrimination in the workplace where none exists, or vice versa. In this example of free throws, the reason the combined data support the hot hand theory is that the data for the second shot after a hit on the first come more from Larry Bird than from the inferior shooter Rick Robey, whereas the opposite is true for the data for the second free throw after a miss on the first. When the same probabilities were computed for each of the nine members of the Boston Celtics who played regularly during the 1980–81 and 1981–82 seasons, Tversky and Gilovich found that four players shot better after a hit on the first free throw and five players shot better after a miss, but none of the players had a statistically significant difference in the two probabilities. However, the combined data across all nine players supported the hot hand theory. Robert Wardrop concluded that this may explain why so many basketball fans believe in the hot hand theory even though it does not seem to hold for individual players. Because basketball fans view many games and many players, they probably think of the data in the combined form rather than by individual player. This would lead to the incorrect conclusion that the hot hand theory is true.

Source: Robert L. Wardrop, "Simpson's Paradox and the Hot Hand in Basketball." *The American Statistician,* 49:1 (February 1995).[1]

Example 9   *Gender of Children*

A couple plans to have five children. Assuming that boy and girl babies are equally likely, what is the probability that the couple will have

(a) five boys?
(b) at least one girl?

Solution:

(a) For each baby born, the probability of having a boy is 1/2. Because the gender of each baby is independent, the probability of having five boys is

$$P(\text{five boys}) = \frac{1}{2} \cdot \frac{1}{2} \cdot \frac{1}{2} \cdot \frac{1}{2} \cdot \frac{1}{2} = \frac{1}{32}.$$

(b) We know from the rule $P(\text{not } E) = 1 - P(E)$ that

$$P(\text{at least one girl}) = 1 - P(\text{five boys}) = 1 - \frac{1}{32} = \frac{31}{32}.$$

∎

To solve part (b) of Examples 8 and 9, we had to apply more than one of the rules of probability. This is also the case for the next example.

Example 10   *AIDS Testing*

One of the standard tests for infection by HIV, the virus that causes AIDS, is the Elias test. The precise reliability of the test varies by laboratory and has improved over time. Assume the test returns a positive result for 99.8% of those with the HIV virus and that when the test is given to those not infected it returns a positive result 1.5% of the time. Heterosexuals without specific risk factors for HIV have an infection rate of about 0.02%.

(a) What is the probability that a random person in this group would have a positive Elias test?
(b) If a random person in this group has a positive Elias test, what is the probability the person is infected with HIV?

Solution:

(a) The person with the positive test result is either infected or not infected. Because these two possibilities are mutually exclusive, when we apply the addition rule we get

$$
\begin{aligned}
P(\text{positive}) &= P(\text{infected and positive}) + P(\text{not infected and positive}) \\
&= P(\text{infected}) \cdot P(\text{positive}|\text{infected}) + P(\text{not infected}) \cdot P(\text{positive}|\text{not infected}) \\
&= (0.0002)(0.998) + (1 - 0.0002)(0.015) \\
&= (0.0002)(0.998) + (0.9998)(0.015) \\
&\approx 0.0152 = 1.52\%.
\end{aligned}
$$

(b) Applying the conditional probability formula, we have

$$P(\text{infected}\,|\,\text{positive}) = \frac{P(\text{infected and positive})}{P(\text{positive})}.$$

Now, using the multiplication rule to compute $P(\text{infected and positive})$ and the result of part (a) for $P(\text{positive})$, we get

$$P(\text{infected}\,|\,\text{positive}) = \frac{P(\text{infected}) \cdot P(\text{positive}\,|\,\text{infected})}{P(\text{positive})}$$

$$\approx \frac{(0.0002)(0.998)}{0.0152} \approx 0.0131 = 1.31\%.$$

We see that in this situation most of the positive results will be false positives, even though the test is fairly good. This surprising situation is a result of the fact that a very high percentage of the population is not infected. The calculations we see here certainly support the argument that follow-up tests are absolutely necessary in such situations. ∎

EXERCISES FOR SECTION 4.4

1. The U.S. Bureau of the Census reported that, in 1993, the children between the ages of 3 and 5 living in households that included their mother broke down into the following groups according to their mother's education level and whether or not the children were enrolled in preschool or kindergarten.

Mother's Education Level	Enrolled (in thousands)	Not Enrolled (in thousands)	Total (in thousands)
Elementary: 0 to 8 years	219	337	556
Some high school: 1–3 years	566	712	1278
High school graduate	2090	1897	3987
Some college: less than a BA	1762	1306	3068
BA degree or higher	1476	693	2169
Total	6113	4945	11,058

Source: U.S. Bureau of the Census.[2]

For a random child in 1993 between the ages of 3 and 5 living in a household that included the mother, find

(a) the probability that the child was enrolled in preschool or kindergarten,

(b) the probability that the child was enrolled in preschool or kindergarten given that the child's mother had only an elementary education,

(c) the probability that the child was enrolled in preschool or kindergarten given that the child's mother had a BA degree or higher.

(d) On the basis of your answers to parts (a) through (c), does there seem to be a correlation between the mother's education level and the likelihood that a child of hers is enrolled in preschool or kindergarten?

2. A study done by the Texas Employment Commission in January 1996 resulted in the following employment data for Tarrant County:

Civilian Labor Force	Employment	Unemployment	Total
Black	67,071	7,770	74,841
Hispanic	69,486	5,049	74,535
Other minority	19,067	1,266	20,333
Nonminority	520,557	21,378	541,935
Total	676,181	35,463	711,644

Source: Texas Employment Commission.

(a) What is the probability that a Hispanic in the Tarrant County Civilian Labor Force is unemployed?

(b) What is the probability that a nonminority in the Tarrant County Civilian Labor Force is unemployed?

(c) What is the probability that a random person in the Tarrant County Civilian Labor Force is employed?

(d) What is the probability that an employed person in the Tarrant County Civilian Labor Force is black?

3. Articles submitted to the *Journal of the American Medical Association* are initially reviewed by editors. On receipt of an article, an editor must decide whether to immediately reject the article or send it out to an expert for further review. A study conducted in 1991, involving male and female editors and male and female authors, found the following breakdown of immediate rejections according to whether the author was male or female and whether the editor was male or female.

	Male Author		Female Author	
	Total Papers Submitted	Immediately Rejected	Total Papers Submitted	Immediately Rejected
Male editor	800	308	204	59
Female editor	605	278	242	108

Source: Julie R. Gilbert, Elaine S. Williams, George D. Lundberg, MD, "Is There Gender Bias in *JAMA*'s Peer Review Process?" *Journal of the American Medical Association* 272:2 (July 1994), 139–142. Copyright 1994, American Medical Association.

(a) What is the probability that a random paper submitted by a male author was rejected given that the editor was male?

(b) What is the probability that a random paper submitted by a female author was rejected given that the editor was male?

(c) What is the probability that a random paper submitted by a male author was rejected given that the editor was female?

(d) What is the probability that a random paper submitted by a female author was rejected given that the editor was female?

(e) What observations can you make based on the probabilities you computed in parts (a) through (d)?

4. In an article appearing in *Nature* magazine, Robert A. J. Matthews reported that although the United Kingdom Meteorology Office's 24-hour rain forecasts are quite successful in predicting the absence of rain, its forecasts are much less useful for predicting rainfall during a particular short time period. The following table, showing the outcomes of the Office's forecast and weather over 1000 one-hour walks in London, appeared in the *Nature* article:

	Rain	*No Rain*	*Total*
Forecast of rain	66	156	222
Forecast of no rain	14	764	778
Total	80	920	1000

Source: Reprinted with permission from *Nature*. Robert A. J. Matthews, "Base-rate errors and rain forecasts," *Nature* 382 (August 1996). Copyright 1996 Macmillan Magazines Limited.

(a) What is the probability that a forecast was correct?

(b) If rain was forecast, what is the probability that the forecast was correct?

(c) If no rain was forecast, what is the probability that the forecast was correct?

5. On a multiple choice question with five possible answers, suppose you know that two of the answers could not be right. If you guess at random from among the remaining possible answers, what is the probability that you will guess correctly?

6. Suppose the probability that a child of a particular couple will have blue eyes is 1/4. If the couple plans to have three children, what is the probability that all three will have blue eyes?

7. If a coin is flipped and a die is tossed, what is the probability of getting a tail and a 4?

8. What is the probability that a rolled die is a 3 given that it is

(a) odd? (b) even?

9. What is the probability that the sum of two rolled dice is 9 given that at least one of the dice rolled is a 5?

10. According to the U.S. Center for Health Statistics, in the United States in 1993, 36.3% of hospital patients were over 64 years old. Whereas only 37.8% of all hospital patients used Medicare as the principal means of payment, 90.3% of patients over 64 years old did. What is the probability that a randomly selected patient was over 64 years old and used Medicare as the principal means of payment?

11. A twin engine airplane has two individual engines. Suppose that each of the engines has a 0.00001% chance of failing on a one-hour flight. Assuming the failures of the two engines are independent, what is the probability that both engines will fail on the same one-hour flight?

12. A card is drawn at random from a deck. What is the probability that it is a jack given that it is a face card?

13. A card is drawn at random from a deck. What is the probability that it is the ace of spades given that it is a black card?

14. The probability of winning a bet on a single number in roulette is 1/38. Suppose you play roulette two times in a row.

(a) What is the probability that you win both times?

(b) What is the probability that you lose both times?

(c) What is the probability that you win exactly once?

(d) Add your answers to parts (a), (b), and (c). Is this the answer you should expect? Why?

15. Three dice are rolled.

(a) What is the probability that all are 5's or 6's?

(b) What is the probability that none is a 5 or a 6?

16. According to United Nations estimates, 0.5% of the world's population was infected with the HIV virus in 1997. If 100 people were selected at random in 1997, what is the probability that at least one of them is infected with HIV?

17. A raffle drawing is being held. There were 120 tickets sold and you bought 2. The first prize will be awarded to the first ticket drawn, and second prize to the second ticket drawn. What is the probability that you will win the first and second prizes?

18. A Travelers Bank personal identification number must consist of four digits and the digits may be used more than once. If you do not know the identification number, what is the probability that you can guess it correctly in one try?

19. Suppose four cards are drawn at random from a standard deck. What is the probability that they are all hearts?

20. Suppose three cards are drawn at random from a standard deck. What is the probability that they are all face cards?

21. The U.S. Bureau of Labor Statistics reports that of those unemployed in 1994, 57.1% were male, 13.4% were college graduates, and 7.6% were male college graduates. What is the probability that a random unemployed male in 1994 was a college graduate?

22. Suppose 68% of the citizens of a state approve of the governor, but only 47% approve of both the governor and the lieutenant governor. If a randomly selected citizen approves of the governor, what is the probability that the citizen also approves of the lieutenant governor?

23. In the Delaware Play 3 lottery game, to place a "straight bet" a player must choose a three-digit number (with zeros allowed for any of the digits). To win, the player's number must match the winning number. What is the probability that a Play 3 straight bet will turn out to be a winner?

24. A class has 17 children, of whom 9 are girls and 8 are boys. If two children are selected at random, what is the probability that both are girls?

25. A jar of 25 lollipops in a clinic has 7 that are grape, 10 cherry, and 8 orange. If a nurse selects four lollipops randomly, what is the probability that they will all be grape?

26. Two cards are drawn at random from a standard deck. What is the probability that they are both queens?

27. A survey of college freshmen conducted by the American Council on Education and UCLA in 1996 found 31.7% of women and 5.7% of men plan to become elementary or secondary school teachers. In 1996, women made up approximately 56% of the total freshmen. What is the probability that a randomly selected college freshman in 1996 was a woman planning to become a teacher?

28. If the probability of winning a carnival game is 3/7 and you play the game two times, what is the probability that you will win both times or you will lose both times?

29. According to the U.S. National Center for Health Statistics, the percentage of births done by Caesarean section in the United States was 22.8% in 1993. Based on this data, for five randomly chosen births in the United States, estimate the probability that

(a) all five of the births were Caesareans,

(b) none of the births were Caesareans.

30. One version of a Master Lock® combination lock can be opened by dialing three numbers between 0 and 39, inclusive, in order. The second number must be different from both the first and third numbers. If you try to guess the combination, what is the probability that you will get it right on the first try?

31. According to a tale (probably greatly embellished, but fun anyway) that has been circulating on the Internet, two students enrolled in an organic chemistry class at Duke University were not prepared to take the final exam at the regularly scheduled time due to an ill-advised round of partying at the University of Virginia. They decided to skip the exam, and afterward went to the professor with the lie that they had a flat tire on their way back from Virginia and did not get back to Duke in time. To their relief, the professor agreed that they could take a make-up final on the following day. When the students showed up for the exam, the professor placed them in separate rooms and handed each of them an exam. The first problem on the exam was something simple worth 5 points. Both students did that problem and then turned the page to find the second problem, which was worth 95 points and read "Which tire?"

    (a) Assuming the students are equally likely to name any of the four tires, what is the probability that the two students name the same tire?

    (b) Assume the probability that a student chooses the right front tire is 40%, the right rear tire is 30%, the left front tire is 20%, and the left rear tire is 10%. What is the probability that the two students name the same tire?

32. The probability that a random baby will have Down's syndrome is approximately 1/800. However, if the parents have a child with Down's syndrome, the probability that their next child will have Down's syndrome increases to about 1/100. If a couple is planning to have two children, what is the estimated probability that both of the children will have Down's syndrome?

33. Suppose a company produces machine parts and 2% of the parts produced are defective. A test that the parts are put through detects 92% of the defective parts. What is the probability that a randomly selected part will be found to be defective?

34. About 10% of all men are left-handed and about 8% of all women are left-handed. Approximately 49% of the entire U.S. population are men and 51% are women. What is the probability that a randomly selected person in the United States is a left-hander?

35. The participants of a 20-kilometer race were surveyed about running injuries over the prior year and the results were reported in *Science and Sports*. Fifty-two percent of the 2200 participants responded. Thirty-four percent of the respondents had sustained one or more injuries, whereas 58 runners had sustained exactly two injuries, 9 exactly three injuries, and none had received more than three injuries.

    (a) What is the probability that a respondent who sustained one or more injuries would sustain more than one?

    (b) What is the probability that a respondent who sustained two or more injuries would sustain three?

36. "Random" audits by the Internal Revenue Service are not truly random in the sense that every return has an equal chance of being selected. Instead, returns to be audited are selected in varying proportions from target groups with suspected patterns of noncompliance with tax laws, with returns chosen at random within these groups. A study by the U.S. General Accounting Office of "random" audits completed shortly after the income tax years 1994–1996 found that 47% of such audits occurred in 11 southern states (Alabama, Arkansas, Florida, Georgia, Louisiana, Mississippi, North Carolina, South Carolina, Tennessee, Texas, and Virginia), although these states had only about 29% of the U.S. population. Nationally, approximately 11 returns out of every million went through such a "random" audit.

(a) Find the probability that a random return from a resident of one of the southern states is subjected to such an audit.

(b) Find the probability that a random return from a resident of the rest of the country is subjected to such an audit.

(c) Based on your answers to parts (a) and (b), does it appear that the audits are truly random? Explain your answer.

37. According to U.S. Bureau of the Census estimates, 41,716,000 people in the United States were not covered by health insurance in 1996. Of these, 18,470,000 had household incomes of less than $25,000 per year. The percentage of the total U.S. population of 265,284,000 in 1996 with household incomes of less than $25,000 per year was 27.8%. What is the probability that in 1996 a random person with a household income of less than $25,000 was not covered by health insurance?

38. Suppose a company wants to do random drug testing of its employees and that 1% of the employees use illegal drugs. If the test for detecting drug use is given to people who use illegal drugs, it returns a positive result 98% of the time. When the test is given to people who do not use drugs, it returns a positive result 5% of the time.

(a) What is the probability that a random employee tested for drugs will test positive for drugs?

(b) If an employee has a positive test result, what is the probability that the employee does not use drugs?

39. Emergency room physicians may use limited ED echocardiography to detect pericardial fluid. Approximately 6.5% of patients in a study by Dave Plummer of the Department of Emergency Medicine at Hennepin County Medical Center in Minneapolis, Minnesota, had pericardial fluid. The test detected the fluid in 98.1% of the patients who had it. It reported a false positive, detecting the fluid where there was none, in 0.1% of those without it. For a patient with a negative test, what is the probability that the patient has pericardial fluid?

40. Use the multiplication rule to show that if $P(B|A) = P(B)$, then $P(A|B) = P(A)$.

41. Given that $P(B|A) < P(B)$, explain why $P(B|\text{not }A) > P(B)$. What happens if $P(B|A) > P(B)$?

4.5 COUNTING TECHNIQUES

To win the New York Lotto game, you must pick the six winning numbers between 1 and 54 that come up in the drawing. The chances of winning are 1 in 25,827,165. It is not practical to list and count these possibilities one by one, but the techniques of this section will let you efficiently count the possibilities in such situations.

We begin by looking at a simple question that will lead to the basic counting law. How many outcomes are there to the experiment of flipping a coin and drawing a number between 1 and 3? We can draw a **tree diagram** to systematically list all the outcomes. In a tree diagram, we start on the left. The tree splits into separate branches for each task that is performed. Each branch represents a different outcome of the task. After the tree is constructed, we can count the total number of outcomes of the entire experiment by counting the number of branches on the right. To draw the tree diagram for the experiment of flipping a coin and drawing a number between 1 and 3, we think of the experiment as two separate tasks. The first task has two possible outcomes: H or T. The second task has three outcomes: 1, 2, or 3. The tree diagram for this example is shown in

Figure 4.5, where, for example, H2 stands for the outcome of getting a head and drawing a 2.

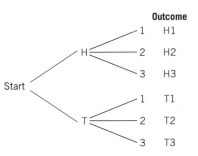

Figure 4.5
Tree diagram for flipping a coin and drawing a number between 1 and 3.

We see in the tree diagram that for each task performed, the number of branches on the tree was multiplied by the number of possible outcomes of the task. Therefore, because there are two possible outcomes to flipping the coin and three possible outcomes to drawing a number between 1 and 3, we see that the total number of possible outcomes for flipping a coin and drawing a number between 1 and 3 is

$$2 \cdot 3 = 6.$$

The idea of using multiplication to count the number of possible outcomes can be generalized, giving the most fundamental law of counting, known as the **basic counting law**, stated next.

> ### Basic Counting Law
>
> If there are M possible ways a task can be performed and, after the first task is complete, there are N possible ways for a second task to be performed, then there are $M \cdot N$ possible ways for the two tasks to be performed in order.

The basic counting law can also be extended to more than two tasks. We simply multiply the number of possible outcomes for each task.

Example 1   *Coin Tossing*

A coin is tossed six times in a row. How many different outcomes are possible?

Solution: Because the coin is tossed six times and there are two possible outcomes to each toss, we see from the basic counting law that the total number of outcomes is given by

$$2 \cdot 2 \cdot 2 \cdot 2 \cdot 2 \cdot 2 = 2^6 = 64.$$

To apply the basic counting law, it is sometimes helpful to draw a slot for each task to be performed. Then, going from the first task to the last task, enter the number of possible outcomes for each task in the slots you have drawn. Last of all, multiply all the numbers you have recorded, and this will give you the total number of outcomes for the series of tasks. This method is illustrated in the next example.

Example 2   *Password*

Suppose a bank account password must consist of four letters followed by two digits. How many passwords are possible if

(a) any letters and digits are allowed?

(b) no repetition of letters or digits is allowed?

(c) no repetition of letters or digits is allowed and the password must start with a T?

Solution:

(a) To choose a password we must fill in the following slots:

$$\underbrace{-\ -\ -\ -}_{\text{letters}}\ \underbrace{-\ -}_{\text{digits}}$$

There are 26 choices for each letter and 10 choices for each digit (0, 1, 2, 3, 4, 5, 6, 7, 8, or 9). Therefore, we see that the number of possible passwords is

$$\underbrace{26 \cdot 26 \cdot 26 \cdot 26}_{\text{letters}} \cdot \underbrace{10 \cdot 10}_{\text{digits}} = 26^4 \cdot 10^2 = 45{,}697{,}600.$$

(b) Because the letters cannot be repeated, once the first letter has been chosen there are only 25 choices for the second letter. Once the first and second letters have been chosen, there are only 24 choices left for the third letter. Continuing, we see that there are only 23 choices left for the fourth letter. Similarly, there are 10 choices for the first digit, but only 9 choices for the second digit. Filling in the slots for letters and digits, we see that the number of passwords possible if no repetition of letters or digits is allowed is given by

$$\underbrace{26 \cdot 25 \cdot 24 \cdot 23}_{\text{letters}} \cdot \underbrace{10 \cdot 9}_{\text{digits}} = 32{,}292{,}000.$$

(c) The answer to this question is similar to the answer to part (b), except that we have only one choice for the first letter because it must be a T. Therefore, the number of passwords possible is

$$\underbrace{1 \cdot 25 \cdot 24 \cdot 23}_{\text{letters}} \cdot \underbrace{10 \cdot 9}_{\text{digits}} = 1{,}242{,}000.$$

∎

Permutations

A **permutation** is an ordering of a group of objects. For example, there are six possible permutations of the objects ▲, ★, and ♥. They are given by

There are many counting problems that involve counting the number of permutations of a group of objects. For example, if we want to count the number of outcomes of an election from first place to last place (ignoring the possibility of ties), we must find the number of ways the candidates in the election can be arranged in a line from first to last. In other words, we must count the number of permutations of the candidates in the election. Listing the permutations as we did with the ▲, ★, and ♥ takes too long if there are many objects, so we need a way to count the number of permutations without actually listing them. Fortunately, this is quite easy if we apply the basic counting law.

To see how to count the number of permutations of a group of objects, let's look at how we would count the number of permutations of ▲, ★, and ♥. For each permutation we have three slots to fill:

$$\underline{\qquad} \qquad \underline{\qquad} \qquad \underline{\qquad}$$

First Second Third

Because there are three objects, we have three choices for the first slot. Once the first slot is filled there are only two choices for the second slot, and once the first two slots are filled there is only one choice left for the third slot. Therefore, we see that the number of permutations of the three objects ▲, ★, and ♥ is given by

$$\underline{\ 3\ } \cdot \underline{\ 2\ } \cdot \underline{\ 1\ } = 6.$$

First Second Third

This idea can be easily generalized to any number of objects. Suppose we have n objects. If we place them in a line, then we have n choices for the first, $n - 1$ choices for the second, and $n - 2$ choices for the third. Continuing this way until we get to the last place where there is only one choice left, we see that the number of permutations of n objects is given by

$$n \cdot (n - 1) \cdot (n - 2) \cdots 3 \cdot 2 \cdot 1.$$

This number is called ***n* factorial** and is denoted by $n!$. For example,

$$3! = 3 \cdot 2 \cdot 1 = 6,$$
$$7! = 7 \cdot 6 \cdot 5 \cdot 4 \cdot 3 \cdot 2 \cdot 1 = 5040,$$
$$1! = 1.$$

Also, by definition

$$0! = 1.$$

This definition makes sense because there is only one way to arrange zero objects.

The preceding discussion is summarized as follows:

Number of Permutations

The number of permutations of n objects is given by

$$n! = n \cdot (n - 1) \cdot (n - 2) \cdots 3 \cdot 2 \cdot 1.$$

Technology Tip
_____

Most calculators have a special key for computing $n!$.

Example 3   *Election Finishes*

There are four candidates running in an election. How many different outcomes from first to last place are possible, assuming there are no ties?

Solution: The number of outcomes is just the number of permutations of the four candidates. Hence, there are

$$4! = 4 \cdot 3 \cdot 2 \cdot 1 = 24$$

different outcomes possible.

∎

Example 4   *Arranging Compact Discs*

How many ways can 12 compact discs be ordered on a shelf?

Solution: The number of orderings possible is equal to the number of permutations of 12 objects, and is therefore given by

$$12! = 479,001,600.$$

∎

Sometimes we want to count the number of ways of making an arrangement, where we use only some of the objects available. For instance, suppose we want to find the number of first-, second-, and third-place finishes possible in a horse race involving eight horses. In this case, each arrangement involves only three horses. To count the number of possibilities, we use the basic counting law. There are eight choices for first place. Once the first-place horse is determined, there are seven choices for second place, and after the first two horses are determined there are only six possible choices for third place. Therefore, the total number of first-, second-, and third-place finishes is given by

$$\underline{8} \cdot \underline{7} \cdot \underline{6} = 336.$$

First   Second   Third

These particular permutations are called permutations of eight objects taken three at a time. In general, an arrangement of r objects selected from a group of n objects is called a permutation of n objects taken r at a time. The number of permutations of n objects taken r at time is denoted by ${}_nP_r$.

To compute ${}_nP_r$, we start with n, then multiply by the next $r - 1$ integers less than n. In other words,

$${}_nP_r = n \cdot (n - 1) \cdot (n - 2) \cdots (n - (r - 1)).$$

It is also possible to express ${}_nP_r$ in terms of factorials. To see how this is done, let's consider again the horse race example. There we saw that ${}_8P_3$, the number of permutations of eight objects taken three at a time, was given by

$$_8P_3 = 8 \cdot 7 \cdot 6.$$

Notice that we began to multiply the way we would in computing 8!, but we stopped before multiplying by $5 \cdot 4 \cdot 3 \cdot 2 \cdot 1$. Therefore, we can think of $_8P_3$ as 8! divided by $5 \cdot 4 \cdot 3 \cdot 2 \cdot 1$. In other words,

$$_8P_3 = 8 \cdot 7 \cdot 6 = \frac{8!}{5!} = \frac{8!}{(8-3)!}.$$

Looking at it this way, we see that we have a formula that depends only on 8, the number of objects we are choosing from, and 3, the number of objects we are arranging at a time. This idea carries over to the general case of n objects taken r at a time, and in the general case we have

$$_nP_r = \frac{n!}{(n-r)!}.$$

We have now derived the following two formulas for $_nP_r$.

Number of Permutations of n Objects Taken r at a Time

$$_nP_r = n \cdot (n-1) \cdot (n-2) \cdots (n-(r-1)) = \frac{n!}{(n-r)!}$$

Notice that the number of permutations of n objects taken n at a time is just

$$_nP_n = \frac{n!}{(n-n)!} = \frac{n!}{0!} = \frac{n!}{1} = n!,$$

and this agrees with our earlier formula for the number of permutations of n when all the objects are arranged.

Technology Tip

Your calculator may have a key for computing $_nP_r$, for which you only have to enter the numbers n and r in some way.

Example 5 *Batting Lineup*

A baseball team has 15 players. The manager has to choose an ordered batting lineup of 9 players. In how many ways can the manager do this?

Solution: Because the manager has 9 ordered spots to fill with 15 players, he is making an arrangement of 15 objects taken 9 at a time. Therefore, the number of ways to do this is

$$_{15}P_9 = \frac{15!}{(15-9)!} = \frac{15!}{6!} = 1,816,214,400.$$

Example 6   *Club Officers*

A club has 92 members. In how many ways can the club select a president and a vice president?

Solution: We can think of the selection of the president and the vice president as a permutation of 92 objects taken 2 at a time, where, for example, the first object will be the president and the second object will be the vice president. Therefore, the number of possibilities is given by

$$_{92}P_2 = 92 \cdot 91 = 8372.$$

If we had used the other formula for $_nP_r$, we would have gotten

$$_{92}P_2 = \frac{92!}{(92 - 2)!} = \frac{92!}{90!}.$$

Some calculators refuse to compute 92! because it is such a large number. If this is the case, we could simplify the factorial expression to get back to the other formula for $_{92}P_2$,

$$_{92}P_2 = \frac{92!}{90!} = \frac{92 \cdot 91 \cdot 90!}{90!} = 92 \cdot 91 = 8372,$$

where the 90! in the numerator and the denominator have canceled out.

■

Combinations

Quite often, when counting the number of ways of selecting a group of objects from a larger set of objects, we do not care about the order in which the objects are selected. For instance, in the New York Lotto game you win by selecting the correct group of six numbers, but the order in which you select the numbers does not matter. A group of objects where order is irrelevant is called a **combination**. When counting the number of combinations, different orderings of the same group of objects are only counted once. For instance, suppose we want to find the number of combinations of the group of objects ▲, ★, ♥, ■, and ♦ taken three at a time. We know that the number of permutations of these five objects taken three at a time is given by

$$_5P_3 = \frac{5!}{(5 - 3)!} = \frac{5!}{2!} = 60.$$

Included in these 60 permutations are several arrangements of the same three objects. For example, the three objects ▲, ■, and ♦ appear in the six permutations

Because the order does not matter in a combination, we consider all these permutations to be the same combination. Every other combination of three objects also appears in exactly six of the permutations. Therefore, the number of permutations of five objects taken three at a time is six times the number of combinations of five objects taken three at a time. Notice that the factor of six came from the number of permutations of three objects, in other words 3!. Therefore, we see that the number of combinations of the objects ▲, ⋆, ♥, ■, and ♦ taken three at a time is given by

$$\frac{_5P_3}{3!} = \frac{60}{6} = 10.$$

This idea can be extended to the case of finding the number of combinations of n objects taken r at a time, which we denote by $_nC_r$. Because we are selecting r objects at a time, by counting the number of permutations of n objects taken r at a time, we have counted each different combination $r!$ times. Therefore, we can find $_nC_r$ by dividing $_nP_r$ by $r!$, and we see that

$$_nC_r = \frac{_nP_r}{r!} = \frac{\left(\dfrac{n!}{(n-r)!}\right)}{r!} = \frac{n!}{r!(n-r)!}.$$

Therefore, we have the following formula:

Number of Combinations of n Objects Taken r at a Time

$$_nC_r = \frac{n!}{r!(n-r)!}$$

A commonly used alternate notation for $_nC_r$ is $\binom{n}{r}$, and it is called n choose r.

Technology Tip

As with $_nP_r$, your calculator may have a key that computes $_nC_r$ for which you only have to enter the numbers n and r in some way.

If your calculator does not have a special key for $_nC_r$, you can just use the formula, but in some cases you will have to do some cancellation as we did in Example 6 to prevent the calculations from becoming too large.

Example 7   *The New York Lotto*

In the New York Lotto game, a player must choose six different numbers between 1 and 54. The order of the numbers is not important. In how many ways can the player do this?

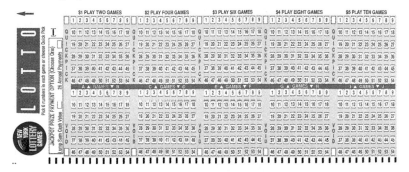

New York Lotto entry form.

Solution: Because the order is not important and the player must choose six numbers from 54 possible choices, we see that the number of ways to do this is given by

$$_{54}C_6 = \frac{54!}{6!(54-6)!} = \frac{54!}{6!48!} = 25{,}827{,}165.$$

We now see how the New York Lotto arrived at the publicly stated probability of winning of 1 in 25,827,165.

Example 8   *Party Invitations*

Alex has 20 classmates, of whom 12 are girls and 8 are boys. He is having a birthday party and is only allowed to invite six classmates. In how many ways can Alex invite

(a) any six classmates?
(b) six boy classmates?
(c) six classmates, four of whom are girls and two of whom are boys?

Solution:
(a) Because the order in which he invites the 6 classmates is unimportant and because he is choosing from 20 classmates, the number of ways Alex can invite 6 classmates is

$$_{20}C_6 = \frac{20!}{6!(20-6)!} = \frac{20!}{6!14!} = 38{,}760.$$

(b) There are only eight boys to choose from, so the number of ways Alex can invite six boy classmates is given by

$$_{8}C_6 = \frac{8!}{6!(8-6)!} = \frac{8!}{6!2!} = 28.$$

(c) In this case, Alex has two tasks: choosing four girls and choosing two boys. According to the basic counting law, the number of ways he can complete both tasks is the product of the number of ways he can complete each separate task. Therefore, the number of ways Alex can invite four girls and two boys is

$$\begin{pmatrix} \text{number of} \\ \text{ways to} \\ \text{invite four girls} \end{pmatrix} \cdot \begin{pmatrix} \text{number of} \\ \text{ways to} \\ \text{invite two boys} \end{pmatrix} = {}_{12}C_4 \cdot {}_8C_2 = 495 \cdot 28 = 13{,}860.$$

Notice that in parts (b) and (c) of the last example, when we computed $_8C_6$ and $_8C_2$, both turned out to be 28. This makes sense because the number of ways to choose six objects from among eight objects should be equal to the number of ways to choose two objects from among eight objects. In either case, you are dividing the eight objects into one group of size 6 and one group of size 2. More generally and by the same reasoning, it is true that $_nC_r = {_nC_{n-r}}$ for $0 \leq r \leq n$. This can also be shown algebraically.

Many interesting counting problems arise in card games. The order in which the cards are dealt is normally not important, so combinations often arise when we count possible card hands. We will see this in the next two examples.

Example 9 *Playing Cards*

Find the number of seven-card hands with

(a) any seven cards,
(b) exactly four hearts.

Solution:
(a) Because there are 52 cards in the deck, we are choosing 7 cards, and order is unimportant, the number of ways to do this is given by

$$_{52}C_7 = 133,784,560.$$

(b) There are 13 hearts in the deck and $52 - 13 = 39$ cards that are not hearts. To draw a seven-card hand with exactly 4 hearts, we must perform the task of choosing 4 hearts from among the 13 hearts in the deck and the task of choosing 3 cards from among the 39 cards that are not hearts. Thus, the number of ways to get a seven-card hand with exactly four hearts is

$$\begin{pmatrix} \text{number of} \\ \text{ways to choose} \\ \text{four hearts} \end{pmatrix} \cdot \begin{pmatrix} \text{number of ways} \\ \text{to choose three cards} \\ \text{that are not hearts} \end{pmatrix} = {_{13}C_4} \cdot {_{39}C_3}$$

$$= 715 \cdot 9139 = 6,534,385.$$

Example 10 *Poker Hands*

Find the number of ways to draw a five-card hand with

(a) three kings and two aces,
(b) three cards of the same rank and one pair of a second rank (called a *full house* in poker),
(c) one pair of cards of one rank, another pair of cards of a second rank, and a fifth card that is of a rank different from that of either of the pairs (called *two pairs* in poker).

Solution:
(a) Because there are four kings and four aces in the full deck, and we have to

BRANCHING OUT 4.4

POKER HANDS

Poker is one of the most popular card games. There are many variations of poker, but in almost all of them the winning player of a hand is determined by comparing five cards held by each of the players. The player with the best hand wins the hand and, in most forms of poker, therefore wins whatever money or chips the other players have put up as wagers. The determination of which hand is best is based on the following rankings of hands from highest to lowest:

Royal Flush:

Ace, king, queen, jack, and ten of the same suit.

Straight Flush:

Five cards in sequence of the same suit, but not a royal flush.

Four of a Kind:

Four cards of the same rank and one other card.

Full House:

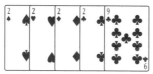

Three cards of one rank and a pair of a second rank.

Flush:

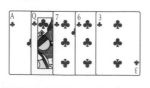

Five cards of the same suit, but not in sequence.

Straight:

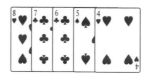

Five cards in sequence, but not of the same suit.

Three of a Kind:

Three cards of one rank, but not a full house or four of a kind.

Two Pairs:

Two pairs of two different ranks and a fifth card of a third rank.

One Pair:

A single pair of the same rank and three other cards of three different ranks.

If two players have hands of the same type, for instance, if both have three of a kind, there are tie-breaking rules based on the ranks of the cards.

Not surprisingly, the probability of being dealt a particular poker hand is smaller for the hands that are ranked higher. For instance, the probability of being dealt a full house is approximately 0.00144, which is smaller than 0.00197, the probability of being dealt a flush. The probability of being dealt any of the very good hands is quite small, but in many types of poker games, players have the opportunity to trade in some of the cards they were first dealt for different ones, making a good hand much more likely.

complete the two tasks of choosing three kings and choosing two aces, we see that the number of ways to do this is

$$\begin{pmatrix} \text{number of} \\ \text{ways to choose} \\ \text{three kings} \end{pmatrix} \cdot \begin{pmatrix} \text{number of} \\ \text{ways to choose} \\ \text{two aces} \end{pmatrix} = {}_4C_3 \cdot {}_4C_2 = 4 \cdot 6 = 24.$$

(b) In this case, we have to choose the rank of the three cards of the same rank and choose the rank of the pair in addition to the tasks of selecting the cards once the ranks are determined. Because there are 13 ranks, there are 13 ways to choose the rank of the three cards of the same rank. Once the rank of the three cards of the same rank is determined, there are only 12 choices left for the rank of the pair. Therefore, the number of ways to do this is given by

$$\begin{pmatrix} \text{number of ways} \\ \text{to choose the rank} \\ \text{of the three cards of} \\ \text{the same rank} \end{pmatrix} \cdot \begin{pmatrix} \text{number of ways to} \\ \text{choose the three cards of} \\ \text{the same rank once} \\ \text{the rank is determined} \end{pmatrix} \cdot \begin{pmatrix} \text{number of} \\ \text{ways to choose} \\ \text{the rank of} \\ \text{the pair} \end{pmatrix} \cdot \begin{pmatrix} \text{number of ways} \\ \text{to choose the pair} \\ \text{once the rank} \\ \text{is determined} \end{pmatrix}$$

$$= 13 \cdot {}_4C_3 \cdot 12 \cdot {}_4C_2 = 13 \cdot 4 \cdot 12 \cdot 6 = 3744.$$

(c) The answer to this question is similar to that of part (b). However, because the order in which we pick the two separate pairs is unimportant, the number of ways to pick the ranks of the two pairs is given by ${}_{13}C_2$. In addition, we also have to complete the task of choosing the fifth card. Because all we know about the last card is that it is of a different rank than either of the two pairs, we choose this card from among the other $52 - (4 + 4) = 52 - 8 = 44$ cards of rank different from the rank of either of the pairs. Putting all this together, we find that the number of ways to get two pairs in a five-card hand is

$$\begin{pmatrix} \text{number of} \\ \text{ways to choose} \\ \text{the ranks of} \\ \text{the two pairs} \end{pmatrix} \cdot \begin{pmatrix} \text{number of ways to} \\ \text{choose the first} \\ \text{pair once the} \\ \text{rank is determined} \end{pmatrix} \cdot \begin{pmatrix} \text{number of ways to} \\ \text{choose the second} \\ \text{pair once the} \\ \text{rank is determined} \end{pmatrix} \cdot \begin{pmatrix} \text{number of} \\ \text{ways to} \\ \text{choose the} \\ \text{last card} \end{pmatrix}$$

$$= {}_{13}C_2 \cdot {}_4C_2 \cdot {}_4C_2 \cdot 44 = 78 \cdot 6 \cdot 6 \cdot 44 = 123{,}552.$$

∎

When solving counting problems that may involve permutations or combinations, it is essential to determine whether or not order is important. We see this in the next example.

Example 11   *Art Contest*

Suppose there are 24 entries in an art contest. First-, second-, and third-place awards will be given along with five honorable mention awards. In how many ways can these awards be made?

Solution: Two tasks must be performed: choosing the first-, second-, and third-place awards from among the 24 entries, and then choosing the five honorable mention awards from among the remaining 21 entries. The order in which the first-, second-, and third-place awards are made does matter, whereas the order in which the honorable mention awards are made does not matter. Therefore, the total number of ways to make the awards is given by

$$
\begin{pmatrix}
\text{number of} \\
\text{ways to award} \\
\text{first, second,} \\
\text{and third}
\end{pmatrix}
\cdot
\begin{pmatrix}
\text{number of ways} \\
\text{to award} \\
\text{honorable mention} \\
\text{to remaining entries}
\end{pmatrix}
= {}_{24}P_3 \cdot {}_{21}C_5
$$

$$
= 12{,}144 \cdot 20{,}349 = 247{,}118{,}256.
$$

∎

EXERCISES FOR SECTION 4.5

1. A cafeteria dinner consists of a salad, an entree, and a dessert. If there are five choices of salad, four choices of entree, and five choices of dessert, how many different dinners are possible?

2. A coin is flipped four times in a row. How many different outcomes are possible?

3. In how many ways can you draw a six-card hand from a standard deck of 52 cards?

4. Suppose there are six horses running in a race. How many different outcomes (orderings from first to sixth place) are possible? Ignore the possibility of ties.

5. A city council has 12 members. Four are Democrats and eight are Republicans.

   (a) In how many ways can a five-member committee be selected?

   (b) In how many ways can a five-member committee be selected if three must be Republicans and two must be Democrats?

   (c) In how many ways can a five-member committee be selected if all must be Republicans?

6. In the Colorado Lotto game, a player must choose six different numbers between 1 and 42. The order of the numbers is not important. In how many ways can a player do this?

7. Suppose a serial number must consist of four different digits, followed by two different letters. How many different serial numbers of this form have first digit 3 and second digit 5?

8. Find the number of ways to draw a four-card hand with two kings and two aces.

9. Find the number of ways to draw a five-card hand consisting of

   (a) all hearts,   (b) cards all of the same suit.

10. A certain Avanti combination lock can be opened by dialing three letters, in order, with the second letter different from both the first and third letters. How many different combinations are possible for a lock of this kind?

11. A student organization has 10 freshman members, 8 sophomores, 12 juniors, and 7 seniors. In how many ways can an eight-member committee be chosen if there must be two members from each class?

12. A terribly absent-minded professor has a class of 11 students, and must assign a grade of A, B, C, D, or F to each student. Unable to locate his grade records, he decides to assign the grades at random. In how many ways can the professor make the grade assignments?

13. A test consists of 15 true-false questions. In how many ways can the test be completed?

14. A multiple choice test consists of 12 questions with four possible answers to each question. In how many ways can the test be completed?

15. Find the number of ways to draw a six-card hand with three cards of one suit and three cards of a second suit.

16. Find the number of ways to draw a six-card hand with four cards of one suit and two other cards of a second suit.

17. Suppose a bank password must consist of five digits or letters (they can be mixed).

    (a) How many such passwords are possible?

    (b) How many such passwords begin with the letter B?

18. Suppose there are 13 children in a class. Eight are boys and five are girls.

    (a) In how many ways can you select seven children from the class?

    (b) In how many ways can you select seven children from the class if four must be boys and three must be girls?

19. A regular six-sided die is rolled three times in a row. How many different outcomes are possible?

20. Referring to meetings of the nine U.S. Supreme Court justices, Justice Sandra Day O'Connor said "Every time we meet, before we go on the bench or before we have a conference, each justice shakes every other justice's hand. That's good, because I think when you touch someone's hand you're less likely to hold a grudge, if you will, should we end up disagreeing the merits about something." In these meetings, how many handshakes take place?

21. License plates for West Virginia have the format of one digit, followed by one letter, and then four more digits. How many different license plates are possible?

22. In California, new license plates either have the format of three letters followed by three digits or they begin with a 1, 2, or a 3, followed by three letters and then three digits. Examples of each of these types of plates are shown here. How many such license plates are possible?

23. In South Carolina, the license plate format is three digits followed by three letters. Suppose a witness to a crime remembers that the criminal drove off in a car with a South Carolina license plate that included the letters S, B, and W. How many different license plates could be of this form?

Justices of the
Supreme Court.

24. The U.S. Senate has 100 members. In how many ways can a committee of five members be formed?

25. The U.S. House of Representatives has 435 members. In how many ways can a committee of three members be formed?

26. Suppose the format for an account number is three different letters followed by three different digits.

   (a) How many different account numbers are possible?

   (b) How many different account numbers of this form can begin with the letter A or the letter B?

27. Suppose there are five candidates running in an election. How many different outcomes (orderings from first to fifth place) are possible? Ignore the possibility of ties.

28. A bowl of fruit contains four apples, five oranges, and three bananas. In how many ways can three oranges be selected?

29. General Motors cars used to have keys with seven segments, with each segment having two possibilities.

   (a) How many such keys are possible?

   (b) General Motors eventually quit using this kind of key because it created security problems. Explain why the relatively small number of possible keys creates a security problem.

30. Find the number of ways to draw a five-card hand with

   (a) exactly three aces,   **(b)** at least three aces.

31. Suppose that six candidates are running for president of a club, and the members are asked to rank their first, second, and third choices on a ballot. In how many different ways can the ballot be filled out?

32. Suppose that six candidates are running for president of a club, and the members are asked to vote for their top two choices but are not asked to rank their two choices. In how many different ways can the ballot be filled out?

33. Find the number of ways to draw a four-card hand with exactly three cards of the same rank and one other card of a different rank.

34. There are 12 children in a preschool class. The classroom has three different activity stations, and four children must be assigned to each station. In how many ways can the assignments be made?

35. Suppose there are 15 mice to be used in an experiment. The mice must be assigned to one of five groups of equal size, with each group receiving a different test drug. In how many ways can the assignments be made?

36. A department has 20 members. They must select a chair, a vice chair, and a six-member advisory committee (neither the chair nor the vice chair can be on the advisory committee). In how many ways can this be done?

37. Several states run a lottery game called Powerball®. To play, a player chooses five different numbers (white balls) from 1 through 49, and one number (the red Powerball) from 1 through 42. The Powerball may or may not repeat a number already picked. The lottery officials also draw five white balls and one red ball. To win the grand prize jackpot, the player must match all the white numbers (the order is unimportant) and the red Powerball number. In how many ways can a player choose the numbers?

38. Show algebraically that $_nC_r = {_nC_{n-r}}$ for $0 \le r \le n$.

4.6 PROBABILITY PROBLEMS USING COUNTING TECHNIQUES

We can now apply the counting techniques we learned in Section 4.5 to problems in probability. To do this we use the basic principle that, when all outcomes are equally likely, the probability $P(E)$ of an event E occurring is given by

$$P(E) = \frac{\text{number of outcomes that result in event } E}{\text{total number of possible outcomes}}.$$

The following examples illustrate the use of counting techniques in probability problems.

Example 1   *Exacta Bet*

In horse race betting, an exacta bet is a bet in which you try to pick the horses that will finish in first and second place, in the correct order. If nine horses are running in a race, what is the probability of winning a random exacta bet?

Solution: There is only one way to correctly guess the exacta, and there are $_9P_2 = 9 \cdot 8$ possible first- and second-place finishes in the race. Therefore, the probability of winning a random exacta bet is

$$P(\text{winning the exacta bet}) = \frac{\text{number of finishes resulting in winning the exacta bet}}{\text{number of possible first- and second-place finishes}}$$

$$= \frac{1}{_9P_2} = \frac{1}{9 \cdot 8} = \frac{1}{72}.$$

Of course, most people placing bets at a race track are not betting at random. An exacta bet in which two very good horses are selected has a much better chance of winning.   ∎

Example 2   *The California Super Lotto*

For each entry in the California Super Lotto game, a player must choose six different numbers between 1 and 51. The lotto officials draw six balls labeled with six different numbers between 1 and 51. To win the big lottery jackpot, the numbers on the player's entry must match all of the state's numbers, but the order is not important. Smaller prizes are awarded if the player matches only three, four, or five of the winning numbers. If a player makes one entry in the California Super Lotto, what is the probability that the player will

(a) match all six winning numbers?

(b) match exactly four of the winning numbers?

Solution:
(a) Because order is unimportant, there are $_{51}C_6$ possible outcomes in the drawing for the California Super Lotto each week. There is only one outcome that will result in a win for the player, namely matching all six winning numbers.

Therefore, the probability that the player will match the six winning numbers is given by

$$P(\text{six winning numbers}) = \frac{1}{{}_{51}C_6} = \frac{1}{18{,}009{,}460} \approx 0.0000000555.$$

(b) To match exactly four winning numbers, the player must choose four numbers from among the six winning numbers and choose two numbers from among the $51 - 6 = 45$ losing numbers. There are ${}_6C_4$ ways to pick the four winning numbers and ${}_{45}C_2$ to pick the two losing numbers. Hence, the probability of matching exactly four winning numbers is

$$P(\text{exactly four winning numbers}) = \frac{{}_6C_4 \cdot {}_{45}C_2}{{}_{51}C_6} = \frac{15 \cdot 990}{18{,}009{,}460}$$

$$= \frac{14{,}850}{18{,}009{,}460} \approx 0.000825.$$

∎

Example 3   *Gender of Children*

A couple plans to have five children. Assuming boy and girl babies are equally likely, what is the probability that the couple will have

(a) five boys?

(b) exactly two boys?

Solution:

(a) We answered this question in an earlier section using the multiplication rule for independent events. It is also possible to answer this question using our counting techniques. We can think of each planned child as a slot to be filled with either a boy or a girl:

$$\underline{\hspace{2em}} \quad \underline{\hspace{2em}} \quad \underline{\hspace{2em}} \quad \underline{\hspace{2em}} \quad \underline{\hspace{2em}}$$

$$\text{1st} \qquad \text{2nd} \qquad \text{3rd} \qquad \text{4th} \qquad \text{5th}$$

Because there are two choices for each slot, we see that there are

$$2 \cdot 2 \cdot 2 \cdot 2 \cdot 2 = 2^5$$

possible outcomes. There is only one outcome resulting in all boys, so we see that the probability of having five boys is given by

$$P(\text{five boys}) = \frac{1}{2^5} = \frac{1}{32}.$$

(b) We did not answer this question using the multiplication rule because it would not have provided us with the answer. In this case, we really need our counting techniques. To compute the probability, the number of possible outcomes is still 2^5, so the denominator is the same as in part (a). To find the numerator, we need to count the number of ways the couple can have exactly two boys. This can be done by counting the number of ways two of the slots for the children can be chosen. We think of the chosen slots as being filled with boys and the remaining slots filled with girls. So we see that counting the number of ways of choosing two slots is the same as counting the number of outcomes resulting in exactly two boys. Because the order in which we choose the two slots is unimportant, we see that there are $_5C_2$ ways to choose them. We can now conclude that the probability of the couple having exactly two boys is given by

$$P(\text{exactly two boys}) = \frac{_5C_2}{2^5} = \frac{10}{32} = \frac{5}{16}.$$

Example 4   *Card Hands*

What is the probability that a five-card hand dealt from a standard deck will be

(a) all clubs?
(b) five cards of the same suit?

Solution:

(a) There are $_{52}C_5$ possible five-card hands. Because there are 13 clubs in a deck, there are $_{13}C_5$ ways to get a hand consisting of all clubs. Therefore, the probability of being dealt five clubs in a five-card hand is

$$P(\text{five clubs}) = \frac{_{13}C_5}{_{52}C_5} = \frac{1287}{2,598,960} \approx 0.000495.$$

(b) To choose five cards of the same suit, we must perform the task of choosing the suit followed by the task of choosing the five cards once the suit is determined. There are four ways to choose the suit and $_{13}C_5$ ways to choose the five cards once the suit is determined, so there are $4 \cdot {_{13}C_5}$ ways to get five cards of the same suit. Therefore,

$$P(\text{five cards of the same suit}) = \frac{4 \cdot {_{13}C_5}}{_{52}C_5} = \frac{4 \cdot 1287}{2,598,960} = \frac{5148}{2,598,960} \approx 0.00198.$$

Example 5   *Maryland License Plates*

Maryland license plates have three letters followed by three digits, as in the one shown in Figure 4.6. What is the probability that a Maryland license plate will have no repeated numbers or digits?

Figure 4.6
Maryland license plate.

Solution: Using the basic counting law, we can see that there are $26 \cdot 26 \cdot 26 \cdot 10 \cdot 10 \cdot 10$ possible license plates of the form of three letters followed by three digits. If the letters and digits cannot be repeated, then there are

$26 \cdot 25 \cdot 24 \cdot 10 \cdot 9 \cdot 8$ possibilities. Therefore,

$$P(\text{no repeated letters or digits}) = \frac{26 \cdot 25 \cdot 24 \cdot 10 \cdot 9 \cdot 8}{26 \cdot 26 \cdot 26 \cdot 10 \cdot 10 \cdot 10} = \frac{11{,}232{,}000}{17{,}576{,}000} \approx 0.639.$$

∎

In the next two examples, we need to apply some of our earlier rules of probability in addition to our counting techniques.

Example 6   *Jelly Beans*

A candy jar contains 45 jelly beans, 20 of which are red, 15 green, and 10 purple. If six jelly beans are picked from the jar at random, what is the probability all six beans will be the same color?

Solution: Because choosing all red or all green or all purple beans are mutually exclusive events, from the addition rule for mutually exclusive events we have

$$P(\text{same color}) = P(\text{all red}) + P(\text{all green}) + P(\text{all purple}) = \frac{_{20}C_6}{_{45}C_6} + \frac{_{15}C_6}{_{45}C_6} + \frac{_{10}C_6}{_{45}C_6}$$

$$= \frac{_{20}C_6 + \,_{15}C_6 + \,_{10}C_6}{_{45}C_6} = \frac{38{,}760 + 5005 + 210}{8{,}145{,}060} = \frac{43{,}975}{8{,}145{,}060}$$

$$\approx 0.00540.$$

∎

Example 7   *Defective Disks*

A carton contains 30 computer disks, 3 of which are defective. If five disks are selected at random, what is the probability that at least one will be defective?

Solution: To answer this question, it is easiest to apply the rule $P(\text{not } E) = 1 - P(E)$ to get $P(\text{at least one is defective}) = 1 - P(\text{none are defective})$. Because 27 of the 30 disks are not defective, and we are selecting 5 disks, we see that

$$P(\text{none are defective}) = \frac{_{27}C_5}{_{30}C_5} = \frac{80{,}730}{142{,}506} \approx 0.567.$$

Therefore,

$$P(\text{at least one is defective}) = 1 - P(\text{none are defective}) = 1 - \frac{_{27}C_5}{_{30}C_5}$$

$$\approx 1 - 0.567 = 0.433.$$

∎

The method we used to compute the probability of at least one disk being defective in the last example can be applied in general to any situation where we want to find the probability of *at least one is "x,"* where x is some condition. In particular,

$$P(\text{at least one is ``}x\text{''}) = 1 - P(\text{none are ``}x\text{''}).$$

We now use this idea to solve a very famous probability problem known as the birthday problem.

The Birthday Problem

Simply stated, the **birthday problem** asks: In a room full of people, what is the probability that at least two of the people in the room have the same birthday? For simplicity, we ignore the possibility of birthdays falling on February 29 in a leap year and we assume that all birthdays are equally likely. To solve the birthday problem, it is easiest to use the formula $P(\text{not } E) = 1 - P(E)$. In the case of the birthday problem, we have

$$P(\text{at least two people have same birthday}) = 1 - P(\text{no people have same birthday}).$$

Let's consider the case when there are only four people in the room. In this case,

$P(\text{no people have same birthday})$

$$= \frac{\text{number of ways four people can have four different birthdays}}{\text{number of ways four people can have any four birthdays}}.$$

To count the number of ways the four people can have birthdays, we can think of each person as a slot to which we assign a birthday:

————   ————   ————   ————

1st person   2nd person   3rd person   4th person

If each person can have any birthday, then there are 365 choices for each slot, and we see that

$$\text{number of ways four people can have any four birthdays} = 365 \cdot 365 \cdot 365 \cdot 365$$
$$= 365^4.$$

Similarly, if each person must have a different birthday, then there are 365 choices for the first person, 364 choices for the second person, 363 choices for the third person, and only 362 choices for the fourth person. Therefore,

$$\text{number of ways four people can have four different birthdays} = 365 \cdot 364 \cdot 363 \cdot 362$$
$$= {}_{365}P_4.$$

Combining this information, we have

$$P(\text{no people have the same birthday}) = \frac{{}_{365}P_4}{365^4},$$

and therefore,

$$P(\text{at least two people have the same birthday}) = 1 - \frac{{}_{365}P_4}{365^4} \approx 1 - 0.9836$$

$$= 0.0164 = 1.64\%.$$

So we see that if four people are in a room, then the probability that at least two of the people have the same birthday is about 1.64%. This is not too surprising because we would not expect there to be a repeated birthday with so few people in the room. However, you may be surprised at how quickly this probability rises as the number of people in the room gets larger.

Now let's answer the birthday problem in the general case of n people in the room. The ideas we used in the case with $n = 4$ are exactly the same in the general case and it is easy to see that

$$P(\text{at least two people have same birthday}) = 1 - P(\text{no people have same birthday})$$

$$= 1 - \frac{_{365}P_n}{365^n}.$$

The probabilities that arise from this formula are quite fascinating. For instance, suppose there are 30 people in the room. Then we see that

$$P(\text{at least two people have the same birthday}) = 1 - \frac{_{365}P_{30}}{365^{30}} \approx 1 - 0.2937$$

$$= 0.7063 = 70.63\%.$$

So we see that if there are 30 people in a room, then more likely than not there will be at least two people with the same birthday. The probabilities that arise for some other possibilities for the number of people in the room are listed in Table 4.3. We see that when only 80 people are in the room, it is highly probable that at least two people have the same birthday. The graph in Figure 4.7 illustrates how the probability that at least two people have the same birthday rises quickly as the number of people in the room increases and is very close to one for 100 people in the room.

Table 4.3	The Birthday Problem
Number of People in the Room	**Probability that at Least Two People Have the Same Birthday (in percent)**
10	11.69
20	41.14
30	70.63
40	89.12
50	97.04
80	99.99

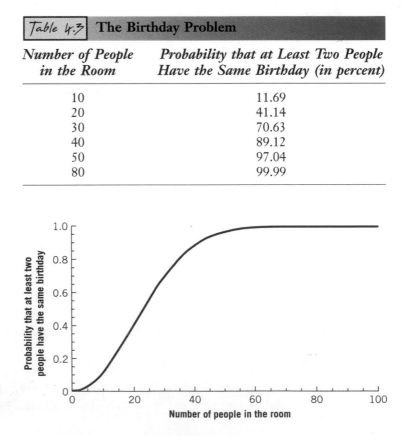

figure 4.7
The probability that at least two people in a room have the same birthday increases as the number of people in the room increases.

BRANCHING OUT 4.5

AMAZING COINCIDENCES ARE TO BE EXPECTED

A story from the *New York Times* in 1986 with the headline "Odds-Defying Jersey Woman Hits Lottery Jackpot 2d Time" quoted a Rutgers University professor as saying that the chances of winning the top prize twice in a lifetime were 1 in about 17.3 trillion. In true tabloid style, a *National Enquirer* headline in 1990 read "4 Sisters Beat 1 in 17 Billion Odds—They All Share the Same Birthday." It is these kinds of stories that make some of us question the laws of probability or believe in miracles, but should we be so amazed that such oddities occur? What we sometimes fail to remember in these situations is that given enough chances for odd things to occur, we should, in fact, expect them to occur occasionally. We look at the lottery story more closely to illustrate this point.

The *New York Times* article told the story of a New Jersey woman, Mrs. Adams, who won the top prize in the New Jersey lottery two times. As statistician James Handley pointed out in his article "Jumping to Coincidences: Defying Odds in the Realm of the Preposterous," the calculation that resulted in the answer of 1 in about 17.3 trillion for someone winning the top prize twice in a lifetime was flawed in two respects: first, in the calculation of the probability that Mrs. Adams would win the lottery twice, and second, in failing to consider that there were many other people besides Mrs. Adams who could have won the lottery twice in a lifetime. The answer of 1 in about 17.3 trillion was based on the assumption that she played only two times, once when the lottery required picking 6 out of 39 numbers and once when 6 out of 42 numbers were required, and it was also assumed that she bought only one ticket each of the two times she played. Based on these assumptions, the probability of winning twice would be

$$\frac{1}{_{39}C_6 \cdot {}_{42}C_6} \approx \frac{1}{(3.3 \times 10^6)(5.25 \times 10^6)} \approx \frac{1}{17 \text{ trillion}}.$$

However, the assumptions made were incorrect because, in fact, Mrs. Adams had bought many tickets each week over several years. As James Handley noted, with multiple purchases over a long period of time, it is certainly more likely that a single player

would win once or even twice. For instance, in the case of a lottery requiring choosing 6 out of 42 numbers, it can be shown that the probability of a particular player who purchases five tickets per week for 4 years winning twice is about 1 in 50 million. If five tickets are purchased each week over a "lottery playing lifetime" of 40 years, then the probability of winning twice improves to 1 in 500,000. Although probabilities of 1 in 50 million or 1 in 500,000 may still qualify as a "miracle" for any particular player, it certainly is not so surprising when we take into account all of the people in New Jersey who play the lottery each week. For instance, if a million people in New Jersey play the way Mrs. Adams plays, we should expect that one of them will hit the jackpot twice over a "lifetime" of 40 years. When we consider further the fact that there are many different lotteries in various states and countries, we would find that the probability of multiple winners popping up somewhere is quite high. So we see that amazing coincidences may not be so amazing after all.

Odds-Defying Jersey Woman Hits Lottery Jackpot 2d Time

By ROBERT D. McFADDEN

Defying odds in the realm of the preposterous — 1 in 17 trillion — a woman who won $3.9 million in the New Jersey state lottery last October has hit the jackpot again and yesterday claim with her fiancé to an a $1.4 million prize.

"Shocking — definitely said 32-year-old Evelyn the manager of a 7-E' store in Point Ple she redeemed h in last Mond otto game.

"They ngs come in threes, s

me First

Actually, latest prize *was* Mrs. Adams's third, counting $500 she won in the state's instant lottery last month, between the big ones. But she said she would probably not go for a third million-dollar coup after last night's drawing, for which she had already bought a batch of tickets.

"I'm going to quit playing," she said. "I'm going to give everyone else a chance."

She was the first two-time million-dollar winner in the history of New Jer-

Associated Press

Evelyn Marie Adams with check.

C.I.A. Accused

By JAMES LeMOYNE

EXERCISES FOR SECTION 4.6

1. In horse race betting, a trifecta bet is one in which you try to pick which horses will finish first, second, and third, in the correct order. If 11 horses are running in a race and you randomly place a trifecta bet, what is the probability of winning the bet?

2. A box of chocolates has 32 pieces, 14 of which contain nuts. If you select six pieces of the chocolates at random, what is the probability that all of the pieces you pick will have nuts?

3. To play the Georgia Fantasy 5 lottery game, a player must choose five numbers between 1 and 35. The order is not important. Five numbers between 1 and 35 will turn out to be winning numbers for a particular game. If you play the Georgia Fantasy 5, what is the probability of choosing the five winning numbers and winning the lottery?

4. To win a Pick-Six bet at a race track, the bettor must pick the winning horse in six different races. If there are 12 horses running in the first four races and 10 in the last two and you randomly place a Pick-Six bet, what is the probability that you will win?

5. A toddler knocked all of his daddy's carefully arranged CDs off the shelf, and then he put them back in a random order. If there are 10 CDs, what is the probability that the toddler will put them back in the correct order?

6. If a six-card hand is dealt from a deck of cards, what is the probability that the hand will be four aces and two kings?

7. A security lock has 12 buttons numbered 1 through 12. To open the lock you must push three particular buttons simultaneously. What is the probability that you will open the lock if you push three buttons at random?

8. If a couple has five children, what is the probability that they have one or two girls? (Assume that girls and boys are equally likely.)

9. If you are dealt five cards from a standard deck, what is the probability of being dealt two hearts and three spades?

10. There are 10 puppies in a cage, of which 6 are male and 4 are female. If you select three puppies at random, what is the probability that they will all be the same gender?

11. If you guess at random on each question of a true-false test with 10 questions, what is the probability that you will get

    (a) all of the answers correct?

    (b) exactly 5 answers correct?

    (c) at least 1 answer correct?

12. A four-card hand is dealt from a standard 52-card deck. What is the probability the hand contains one card of each suit?

13. In the Kansas Cash 4 Life lottery game, players choose four numbers between 0 and 99. The order is not important. Four numbers between 0 and 99 will turn out to be winning numbers for a particular game. If you play the Kansas Cash 4 Life game, what is the probability of matching exactly three of the winning numbers?

14. To play the Minnesota Gopher 5 lottery game, you must choose five numbers between 1 and 39. The order is not important. Five numbers between 1 and 39 will turn out to be winning numbers for a particular game. If you play the Minnesota Gopher 5 game, what is the probability of matching exactly two of the winning numbers?

15. Four children throw their Batman lunch boxes in a pile. If you hand back the boxes at random, what is the probability that each child will get back his or her own lunch box?

16. If you draw three cards at random from a deck, what is the probability that exactly two of the three are of the same suit?

17. A 7-digit phone number cannot begin with a 0 or 1. What is the probability that a random 7-digit phone number of this form has one or more repeated digits?

18. There are 19 people in a room. If you select four people at random, what is the probability that you will select the four tallest people?

19. There are 24 popsicles in a cooler. Ten are grape, six are orange, and eight are cherry. If you grab three popsicles from the cooler, what is the probability that they will

    (a) all be grape?   **(b)** all be the same flavor?   **(c)** be one of each flavor?

20. If a coin is flipped 10 times, what is the probability that you will get

    (a) 10 heads?   **(b)** at least 1 head?   **(c)** exactly 6 heads?

21. If a six-card hand is dealt from a deck, what is the probability that the hand will have two cards of one rank, two cards of a second rank, and two cards of a third rank?

22. If you draw five cards at random from a deck, what is the probability that exactly two of the cards will be aces?

23. License plates in Ohio are of the form of three letters followed by four digits. What is the probability that in such a license plate, chosen at random, all three letters are the same?

24. An art collection of 18 paintings has 14 authentic paintings and 4 forgeries. If you select three paintings from the collection, what is the probability that at least one will be a forgery?

25. There are 20 grapefruit in a basket. Eleven are sweet and nine are sour. If you select three grapefruit at random, what is the probability that they will all be sweet?

26. From a bag containing four red, five white, and six blue balls, three are drawn at random. What is the probability that at least two of the three are the same color?

27. In the game of blackjack, a player can win the hand by receiving a *blackjack*, which consists of an ace and a 10-point card in a two-card hand. (A 10-point card is a king, queen, jack, or ten).

    (a) Find the probability of getting a blackjack when dealt two cards from a standard deck.

    (b) Find the probability of getting a blackjack when dealt two cards from two standard decks that have been shuffled together.

28. A meat processor has packed 50 packages of meat, 5 of which contain contaminated meat. If an inspector checks only 4 of the 50 packages, what is the probability that she will not check any of the contaminated packages?

29. Forty balls numbered 1 to 40 are placed in a bag and four are drawn at random. What is the probability that the first ball drawn is between 1 and 10 (inclusive), the second is between 11 and 20, the third is between 21 and 30, and the last is between 31 and 40?

30. In a matching test, there are six questions with the six answers listed out of order. To complete the test, you must match each answer with the correct question. If you do this at random, what is the probability that you will get 100% on the test?

31. Forty-five schoolchildren must be assigned to three different teachers with classes of 15 students each. Fifteen of the children are girls. If the classes are picked at random, what is the probability that there will be five girls in each class?

32. Six cards are dealt from a standard deck. What is the probability the hand contains three cards of one rank, two cards of a second rank, and one card of a third rank?

33. There are 15 people in a room. What is the probability that at least two have the same birthday?

34. There are 25 people in a room. What is the probability that at least two have the same birthday?

35. There are seven people in a room. What is the probability that at least two of them were born in the same month? (For simplicity, assume that all birth months are equally likely.)

36. There are five people in a room. What is the probability that at least two of them were born in the same month? (For simplicity, assume that all birth months are equally likely.)

In Exercises 37–41, consider the following: Several states run a lottery game called Power-ball®. To play, a player chooses five different numbers (white balls) from 1 through 49, and one number (the red Powerball) from 1 through 42. The Powerball may or may not repeat a number already picked. The lottery officials also draw five white balls and one red ball. To win the grand prize jackpot, the player must match all the white numbers (but the order is unimportant) and the Powerball number.

37. If you play the game, what is the probability of matching all the numbers and winning the grand prize jackpot in Powerball®?

38. In Powerball®, a player can win $5000 by matching exactly four of the white balls and the red Powerball. If you play the game, what is the probability of doing this?

39. In Powerball®, a player can win $7 by matching exactly two of the white balls and the red Powerball. If you play the game, what is the probability of doing this?

40. In Powerball®, a player can win $100,000 by matching all five of the white balls but missing the red Powerball. If you play the game, what is the probability of doing this?

41. In Powerball®, a player can win $7 by matching exactly three of the white balls but missing the red Powerball. If you play the game, what is the probability of doing this?

4.7 EXPECTED VALUE

Sometimes you would like to know the average outcome of an experiment if it is repeated a large number of times. For instance, in gambling situations, you may want to know how much you should expect to lose or win on a bet on average, for then you can figure out the total amount you can expect to lose or gain if you repeat the bet many times. We call the average outcome of an experiment its expected value.

To see how we can compute the expected value of an experiment, let's consider the game of roulette. Recall that in one form of roulette, a ball is spun around a wheel with 38 slots of equal size and eventually lands in one of the slots. The slots are numbered 0, 00, and 1 through 36. One possible wager is to bet that the ball will land on a single number between 1 and 36. You must pay $1 to place the bet. If your selected number is the winner, then you win $35 and your $1 bet is returned to you. Otherwise you lose your $1 bet. Because there are 38 slots on the wheel, the probability of winning a bet on a single number is 1/38, and the probability of losing the bet is 37/38. Therefore, over the long run we can expect that the average return of the $1 bet will be

$$P(\text{winning}) \cdot (\$35) + P(\text{losing}) \cdot (-\$1) = \left(\frac{1}{38}\right) \cdot (\$35) + \left(\frac{37}{38}\right) \cdot (-\$1)$$

$$= -\$\left(\frac{2}{38}\right) \approx -\$0.0526.$$

The ideas used in the roulette example can be generalized to any experiment in which each of the possible outcomes has a numerical value of some kind. This value may be a monetary amount, as in gambling situations, or it may be something other than money. In any case, the expected value of the experiment is given by the following definition:

Expected Value

The **expected value** of an experiment is computed by multiplying the probability of each outcome of the experiment by its value and summing the results.

It is important to understand that the expected value does not tell you what you can expect to happen with any individual trial of an experiment, but rather, it tells you what you can expect to happen *on average*. In the case of the roulette example we just examined, the expected value turned out to be about −$0.0526. On any individual bet you will not lose this amount because you will either win $35 or lose $1. However, if you play the game a large number of times, you should expect your average loss to be about $0.0526 or 5.26¢ per game. For instance, if you play the game 400 times, then you should expect to lose about $400 \cdot \$0.0526 = \$21.04$.

Expected values can come up in situations that do not involve money, as we see in Example 1.

Example 1   *The SAT*

The Scholastic Assessment Test, more commonly known as the SAT, is used as an aid in determining college admissions. This test is a multiple choice test. To discourage random guessing, points are subtracted for wrong answers. Each question has five possible answers, and the test taker must pick one answer or choose not to answer the question. One point is awarded for each correct answer, and for each wrong answer 1/4 point is subtracted.

(a) What is the expected point value of a random guess?

(b) What is the expected point value of a guess if the test taker knows with certainty that one of the answers can be eliminated as a possibility and has no preference among the other four?

Solution:

(a) When guessing from among five possible answers, the probability of guessing correctly is 1/5 and the probability of guessing incorrectly is 4/5. Therefore, the expected point value of a random guess is

$$P(\text{correct}) \cdot 1 + P(\text{incorrect}) \cdot \left(-\frac{1}{4}\right) = \left(\frac{1}{5}\right) \cdot 1 + \left(\frac{4}{5}\right) \cdot \left(-\frac{1}{4}\right)$$

$$= \frac{1}{5} - \frac{1}{5} = 0.$$

Because the expected value is 0, there is nothing to gain or lose on average by guessing.

(b) If the test taker can eliminate one of the possible answers, then the probability of guessing correctly is 1/4 and the probability of guessing incorrectly is 3/4. In this case, we see that the expected point value of a guess is

$$P(\text{correct}) \cdot 1 + P(\text{incorrect}) \cdot \left(-\frac{1}{4}\right) = \left(\frac{1}{4}\right) \cdot 1 + \left(\frac{3}{4}\right) \cdot \left(-\frac{1}{4}\right)$$

$$= \frac{1}{4} - \frac{3}{16} = \frac{4}{16} - \frac{3}{16} = \frac{1}{16}.$$

Because the expected value is 1/16, test takers should expect to earn 1/16 of a point on average for guessing when they can eliminate one of the possible answers. Therefore, in terms of expected value, it is advantageous to guess in this situation.

∎

In many gambling situations, the bettor pays some amount up front to play the game, and the money is not returned even if the bettor wins. In this case, it is often easier to compute the expected payoff and then subtract the amount paid to play to find the expected value of the game. This idea is illustrated in the next two examples.

Example 2   *Raffle*

Suppose a school is holding a raffle, and 1200 raffle tickets are sold for $5 each. One of the tickets will be selected to win a first prize of $2000. Six other tickets will be selected to win prizes of $300 each. How much should you expect to win or lose on average if you buy one raffle ticket?

Solution: We first find the expected payoff for one raffle ticket by multiplying the probability of each possible outcome by the corresponding payoff:

expected payoff $= P(\text{first place}) \cdot \$2000 + P(\text{second place}) \cdot \$300 + P(\text{losing}) \cdot \$0$

$$= \left(\frac{1}{1200}\right) \cdot \$2000 + \left(\frac{6}{1200}\right) \cdot \$300 + \$0$$

$$= \$\left(\frac{2000 + 6 \cdot 300}{1200}\right) = \$\left(\frac{3800}{1200}\right) \approx \$3.17.$$

Notice that we did not need to compute $P(\text{losing})$ because it was multiplied by 0. From now on, if a probability in an expected value calculation is multiplied by 0, we just leave the term out entirely.

Because you paid $5 for the ticket, the expected value of the game for you is given by expected value ≈ $3.17 − $5.00 = −$1.83,

so you should expect to lose $1.83 on average if you purchase a raffle ticket.

Example 3   *Keno*

Keno									
Mark Number of Spots or Ways Played				Account Number			Price Per Game $1.00		
4				Number of Games *1*			Total Price $1.00		
1	2	3	4	5	6	7	8	✗9	10
11	✗12	13	14	15	16	17	18	19	20
21	22	23	24	25	26	27	28	29	30
31	32	33	34	35	36	37	38	39	✗40
We pay on machine-issued tickets-tickets with errors not corrected before start of game will be accepted as issued									
41	✗42	43	44	45	46	47	48	49	50
51	52	53	54	55	56	57	58	59	60
61	62	63	64	65	66	67	68	69	70
71	72	73	74	75	76	77	78	79	80
We are not responsible for Keno numers tickets not validated before start of next game.									

To play the game 4-spot Keno at a particular casino, a player pays $1 to play and must select four numbers between 1 and 80. The order in which the numbers are selected is not important. After the player has made his or her selection, the casino holds a drawing and 20 of the 80 numbers between 1 and 80 are declared "winning numbers." The remaining 60 numbers are losing numbers. The amount of money the player wins depends on how many of his or her selected numbers turn out to be winning numbers. The payoffs in each case are listed in Table 4.4. The $1 bet is not returned to the player.

Table 4.4   **Payoffs in 4-Spot Keno**

Outcome	*Payoff*
0 or 1 winning numbers	$0
2 winning numbers	$1
3 winning numbers	$4
4 winning numbers	$112

What is the expected value of a $1 bet?

Solution: To compute the expected payoff, we need to find the probability of each of the outcomes that results in a payoff. To compute these probabilities, we first note that because order is unimportant, there are $_{80}C_4$ ways to select four numbers.

To select 2 winning numbers, the player must select 2 winning numbers from among the 20 winning numbers and 2 losing numbers from among the 60 losing numbers. Therefore, the number of ways to select two winning numbers is $_{20}C_2 \cdot {}_{60}C_2$, and we see that the probability of selecting two winning numbers is

$$P(\text{two winning numbers}) = \frac{_{20}C_2 \cdot {}_{60}C_2}{_{80}C_4} \approx 0.212635.$$

Similarly, we see that

$$P(\text{three winning numbers}) = \frac{_{20}C_3 \cdot {}_{60}C_1}{_{80}C_4} \approx 0.043248,$$

$$P(\text{four winning numbers}) = \frac{_{20}C_4}{_{80}C_4} \approx 0.003063.$$

We can now compute the expected payoff on a $1 bet, and we find that

$$\begin{aligned}
\text{expected payoff} &= P(\text{two winning numbers}) \cdot \$1 + P(\text{three winning numbers}) \cdot \$4 \\
&\quad + P(\text{four winning numbers}) \cdot \$112 \\
&\approx (0.212635) \cdot \$1 + (0.043248) \cdot \$4 + (0.003063) \cdot \$112 \\
&\approx \$0.729.
\end{aligned}$$

Because it costs $1 to play the game and the $1 is not returned, the expected value of the game is given by

$$\text{expected value} \approx \$0.729 - \$1 = -\$0.271 = -27.1¢$$

Therefore, on a $1 bet a player should expect to lose 27.1¢ on average.

■

We should not be surprised that the expected value of the 4-spot Keno game in Example 3 turned out to be negative. In order for the casino to make money on a game, the expected value of a bet must be in favor of the casino. For each $1 bet, the expected value from the casino's viewpoint is a 27.1¢ gain. In general, the expected value of a $1 bet from the casino's point of view is called the **house edge**, and it is often expressed as a percentage of $1. For instance, because the house expects to win about 27.1¢ on each $1 bet in 4-spot Keno, we say the house edge for this game is 27.1%. Similarly, for roulette we saw that the expected value from the casino's point of view is $0.0526 = 5.26¢ on a $1 bet, so the house edge is 5.26%. In this sense, the game of roulette is more favorable to the bettor than the Keno game we just looked at. By comparison, the house edge for a typical state lottery is around 50%!

Sometimes, expected values can be used to help make business decisions. One instance of this is in the setting of insurance rates. This is illustrated in the next example.

Example 4   *Life Insurance*

A 30-year-old man wants to purchase a $100,000 1-year life insurance policy. On the basis of mortality rates for men of his age and background, the insurance company determines that the probability of the man dying in the next year is 0.00177. What should the insurance company charge the man for the policy, if it would like to make an expected profit of $50?

Solution: To see how much the company should charge for the policy, we first compute the amount the company should expect to pay out in benefits on average for this type of policy. The company should then charge this amount plus an additional $50 in order to have an expected profit of $50.

The company has to pay a benefit only if the man dies. Therefore, the expected benefit the company has to pay is

$$\text{expected benefit paid} = P(\text{dying}) \cdot \$100,000 = 0.00177 \cdot \$100,000 = \$177.$$

Therefore, to make a profit of $50 on average, the company should charge $177 + $50 = $227 for the policy.

■

EXERCISES FOR SECTION 4.7

1. The grading procedure for the SAT exam is described in Example 1. What is the expected point value of a guess if the test taker knows with certainty that two of the answers can be eliminated as possibilities?

2. Suppose a multiple choice exam has three possible answers to each question. You get 5 points for each correct answer, but you lose 3 points for each incorrect answer. No points are gained or lost if you leave the question blank.

   (a) What is the expected point value of a random guess?

   (b) What is the expected point value of a guess if the test taker knows with certainty that one of the answers can be eliminated as a possibility and has no preference among the other two?

   (c) Under what conditions should you answer the question?

3. If you consider the *value* of a roll of a single die to be the number that is rolled, what is the expected value of the roll of a single die?

4. Suppose a contestant on the game show *Jeopardy!* has hit a Daily Double and decides to wager $500. If the contestant answers the question correctly, she wins $500, and if she answers incorrectly, she loses $500. Based on the category of the question, the contestant thinks she has an 80% chance of answering correctly. What is the expected value of her $500 wager?

5. Consider a game that consists of rolling a regular die. You pay $5 to play the game, and the $5 is not returned. If you roll a 1, 2, or 3, you win nothing. If you roll a 4 or 5, you win $6; if you roll a 6, you win $10. What is your expected value for this game?

6. Suppose 320 raffle tickets are sold for $5 each. A prize of $1000 will be awarded to one of the ticketholders. Two other ticketholders will be awarded prizes of $100 each. How much should you expect to win or lose on average if you purchase one raffle ticket?

7. A bag contains 1 slip of paper with $1000 written on it, 3 slips with $100, 5 slips with $20, 10 slips with $5, and 1981 blank slips. For a $1 fee, you may draw a random slip of paper and claim the amount on it as a prize. What is your expected value for the game?

8. Consider a game that consists of drawing a single card at random from a deck. You pay $3 to play the game, and the $3 is not returned. If you draw an ace, you win $10. If you draw a king or queen, you win $5. How much should you expect to win or lose on average if you play this game?

9. A game at a fund-raiser carnival involves throwing a dart at a wall of balloons. If the balloon that is popped has money inside, the player wins the money. Suppose 80% of the balloons have no money inside, 15% have $5 inside, and the remaining 5% have $50 bills inside. If the carnival organizers want to make an average profit of 50¢ per player, what should a player have to pay to play the game? (Assume that each player pops exactly one balloon.)

10. The state of Virginia has an instant game called Double Bonus Bingo. To play, you buy a game card for $2 and then see what you have won. The payoffs and the percentage of cards with these payoffs are as listed in the table.

Payoff	Percentage of Cards
$20,000	0.0001
$5000	0.0005
$1000	0.001
$50	0.125
$20	1
$10	2
$5	3
$4	4
$3	5
$2	10

The $2 you pay for the card is not returned. What is your expected gain or loss when you buy one of these cards?

11. In a common form of roulette, a ball is spun around a wheel with 38 slots of equal size and eventually lands in one of the slots. Eighteen of the slots are colored red. One wager that can be made in roulette is to bet that the ball will land on the color red. The house odds for betting on red are 1 to 1, meaning that if the ball lands on red, then for every $1 wagered, the bettor will win $1 in addition to having the $1 wager returned. If the ball does not land on red, the bettor loses the wager.

   (a) What is the expected value of a $1 wager on red?

   (b) What is the house edge, expressed as a percentage of $1?

12. In football betting at some casinos, a bettor can wager $11 on a game with a point spread. If the team the bettor chooses beats the point spread, then the $11 bet is returned to the bettor along with an additional $10. If the team does not beat the point spread, the bettor losses the $11 wager. (With a point spread, the favored team must win by more than the point spread in order for a bet on the favored team to be a winning bet. For instance, suppose the Green Bay Packers are favored to beat the Chicago Bears by 10 points. If Green Bay wins by more than 10 points, then a bet on Green Bay is a winning bet. If Green Bay wins by fewer than 10 points, ties, or loses, then a bet on the Chicago Bears is a winning bet. If Green Bay wins by exactly 10 points, then the bet is called off and the gambler's wager is returned.) Assuming, for simplicity, the probability of winning the bet is 1/2 and the probability of losing the bet is also 1/2, what should a bettor expect to win or lose on average on an $11 wager?

13. Although the expected value of a gambling game is usually negative, there are some special instances in which this is not the case. Consider a bingo parlor that holds bingo games each night with a prize of $500 going to the winner, and you must pay 25¢ for each card you play. Each card sold has an equal chance of being the winner.

   (a) Suppose it is a busy night when 2500 cards are sold for one game. If you purchase a card and play, what is your expected gain or loss?

   (b) Suppose it is a night when the crowd is very small (perhaps due to bad weather), and there are only 800 cards sold for one game. If you purchase a card and play, what is your expected gain or loss?

14. The house odds for a particular horse in a race are 9 to 5, meaning that if the horse wins, then for every $1 wagered on the horse a bettor will win $(9/5) = $1.80 in addition to having the $1 wager returned. If the horse loses, the bettor loses the wager. If you believe the horse has a 30% chance of winning the race, what is the expected value of a $12 wager?

15. The game of 5-spot Keno is similar to 4-spot Keno as described in Example 3. Eighty numbers are displayed on a board, and 20 of these numbers will be chosen to be the winning numbers. To play, a player selects five numbers on his or her playing card. Suppose a player pays $1 to play, and the $1 is not returned. The payoffs made by the Dunes Casino in Las Vegas for each of the possible outcomes are listed here:

Outcome	*Payoff*
0, 1, or 2 winning numbers	$0
3 winning numbers	$2
4 winning numbers	$20
5 winning numbers	$480

   (a) What is the expected value of this game?

   (b) What is the house edge for this game, expressed as a percentage of $1?

16. The game of 3-spot Keno is similar to 4-spot Keno as described in Example 3. Eighty numbers are displayed on a board, and 20 of these numbers will be chosen to be the winning numbers. To play, a player selects three numbers on his or her playing card. Suppose a player pays $1 to play, and the $1 is not returned. The payoffs made by the Dunes Casino in Las Vegas for each of the possible outcomes are listed here:

Outcome	Payoff
0 or 1 winning numbers	$0
2 winning numbers	$1
3 winning numbers	$42

    (a) What is the expected value of this game?

    (b) What is the house edge for this game, expressed as a percentage of $1?

17. An entry to the Ben & Jerry's® and Yahoo!® "Lids for Kids™" Sweepstakes could be sent in by mail or over the Internet. The prizes and the number awarded are shown here:

Prize	Value	Number Awarded
Lifetime supply of Ben & Jerry's ice cream	$9360	1
Trip to Stowe, Vermont	$2000	5
Year's supply of Ben & Jerry's ice cream	$156	50
Ben & Jerry's/Yahoo! T-shirt	$16	500
Free pint of Ben & Jerry's ice cream	$3	5000
Ben & Jerry's/Yahoo! sticker	$0.50	20,000

Suppose that 100,000 people enter the sweepstakes.

    (a) What is the expected value of a sweepstakes entry that is sent in for free over the Internet?

    (b) What is the expected value of a sweepstakes entry sent in through the mail, taking into account the 40¢ cost of a postage stamp and envelope?

    (c) Some would consider the Ben & Jerry's/Yahoo! sticker to be worthless. Assuming that it has no value, what is the expected value of a sweepstakes entry sent in through the mail, taking into account the 40¢ cost of a postage stamp and envelope?

18. In the Delaware Play 3 lottery game, to place a "straight bet" a player must choose a three-digit number (with zeros allowed for any of the digits). To win the prize of $500, the player's number must match the winning number with all digits in the correct order. It costs $1 to play the game. What is your expected gain or loss when you play this game?

19. The *Fort Worth Star Telegram* newspaper runs a contest called Pigskin Payday. To play the game, a player must try to choose the winners versus a point spread of some number of games from among 20 different college football games listed. We assume there is a 50% chance of choosing the "winner" of a single game. The player has several choices of how to play Pigskin Payday. One choice is to try to pick the winners of all 20 games. A player who chooses all 20 winners wins $100,000 and otherwise wins nothing. A player who does not want to try to pick all 20 winners can try to pick a smaller number of winners. Of course, the fewer winners a player tries to pick, the smaller the payoff. All of the possible ways to play and the payoffs awarded in the case that the player gets all of his or her choices correct are listed in the table. No matter how you play, if you do not get all of your choices correct, you win nothing. It costs nothing to play.

Number of Games	Payoff
20	$100,000
16	$10,000
12	$500
10	$50
8	$15
5	$5

(a) Compute the expected value of the game for each of the six different ways it can be played.

(b) If you want to maximize your expected payoff, how many games should you try to pick?

(c) If you have to mail the contest form in by mail and the 40¢ cost of a stamp and envelope is taken into account, is your expected value ever positive?

20. The jackpot in most state lotteries is dependent on the amount of money that has been bet on the lottery. If there is not a winner one week, the money is added to the jackpot for the next week. If several weeks pass without a winner, then the jackpot may become quite large. Suppose the California Super Lotto has a jackpot of $45 million. To win the lottery a player must choose the six winning numbers from the numbers between 1 and 51, inclusive. Order does not matter. It costs $1 to play.

(a) Ignoring the possibility of multiple winners and any prizes for matching fewer than six winning numbers, what is your expected value if you play this game?

(b) The expected value found in part (a) does not take into account prizes given for matching fewer than six winning numbers or the fact that when jackpots are large many people play and the chance of multiple winners who must split the jackpot increases. What effect, if any, do you think each of these two considerations has on the expected value?

21. Suppose that on a rainy day, a small store in Atlanta expects to get 200 customers on average, whereas on a clear day they average about 300 customers. According to the National Oceanic and Atmospheric Administration, it rains 115 days of the 365 days in an average year in Atlanta. What is the expected number of customers in the store on a random day?

22. An insurance company plans to sell a $150,000 one-year life insurance policy to a 38-year-old woman. On the basis of mortality rates for women of her age and background, the insurance company determines that the probability of the woman dying in the next year is 0.00104. What should the insurance company charge the woman for the policy if it would like to make an expected profit of $45?

23. A man wants to hire a lawyer to handle a $100,000 lawsuit for him. The man must decide between two lawyers. The first lawyer will charge a flat fee of $12,000. The second lawyer will take 30% of the $100,000 if the lawsuit is successful and otherwise get nothing. With either lawyer, the man thinks he has a 75% chance of winning the lawsuit and collecting $100,000.

(a) If the man decides to choose the lawyer for which the expected value of the amount he will have to pay is the smallest, which lawyer should he choose?

(b) Why might the man not choose the lawyer you found in part (a)?

24. Suppose you have $1000 to invest and you have two options to choose between. The first option is a guaranteed gain of $50. The second option is somewhat risky, and if it succeeds it will result in a gain of $400. However, there is only a 70% chance it will succeed, and failure will result in a $700 loss. Which investment option should you choose if you want to maximize the expected value of your investment?

4.8 GENETICS

Using empirical probability, Gregor Mendel (1822–1884), an Austrian monk, was the first person to discover some of the basic laws of heredity. In a monastery garden, Mendel conducted breeding experiments with garden peas. He ran experiments to study the occurrence of seven characteristics that show up in one of two forms. One of the characteristics he studied was seed color, which is either green or yellow. In his experiments he first pollinated the peas by hand until he produced a population of pure green peas, in other words, peas that would produce only green-seeded offspring. Likewise, he produced a population of pure yellow peas. He then crossbred pure green plants with pure yellow plants and he found that in the first generation the resulting offspring always had yellow seeds. Mendel next cross-fertilized this first generation of plants and obtained a second generation in which approximately one-quarter of the plants had green seeds and the remaining three-quarters of the plants had yellow seeds.

Mendel explained the results of his seed-color experiment by suggesting that each inherited characteristic is determined by the interaction of two hereditary factors, one from each parent. We now call these hereditary factors **genes**. Each pea plant has two seed-color genes. If a plant inherits a green color gene from each parent, it will be green. Similarly, if a plant inherits a yellow color gene from each parent, it will be yellow. However, if the plant inherits a green color gene from one parent and a yellow color gene from the other parent, then it will be yellow. The yellow color gene is an example of a dominant gene because it overrides the green color gene. In general, a gene is called **dominant** if it produces the same trait whether paired with a similar or dissimilar gene. The green color gene is an example of a recessive gene. A trait corresponding to a **recessive** gene is produced only when two recessive genes are paired. If a parent pea plant has one gene for each color, there is a 50% probability of passing on each gene to an offspring.

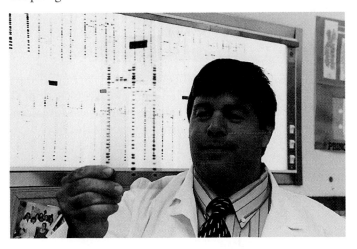

Dr. Arthur Eisenberg, chair of the FBI's DNA Advisory Board, examines DNA test results.

A pair of genes for a trait can be described by a pair of letters. A capital letter is conventionally used to represent a dominant gene, and a small letter is used for a recessive gene. We use a Y to stand for the yellow seed color gene and the letter g for the green seed color gene. A pure yellow plant would have the gene pair YY, whereas a pure green plant would have the pair gg. A plant with one yellow gene and one green gene would have the gene pair Yg or gY. However, because the order of a gene pair does not matter, we will always signify a mixed gene pair with the dominant gene first. Therefore, the yellow-green pair will be described by Yg.

The results of Mendel's crossbreeding experiments can be illustrated by the use of a table, called a **Punnett square** after geneticist Reginald C. Punnett (1875–1967), which shows the results of each of the gene pairings. The Punnett square for crossbreeding a pure yellow plant (YY) with a pure green plant (gg) is given in Table 4.5.

Table 4.5	**Punnett Square for First-Generation Yellow and Green Seeds**	
	Pure Yellow Seed Parent	
	Y	**Y**
Pure Green **g**	Yg	Yg
Seed Parent **g**	Yg	Yg

All of the offspring have the same gene pair Yg, so they will all have yellow seeds. When two plants from this first generation are crossbred, the resulting Punnett square is given in Table 4.6.

Table 4.6	**Punnett Square of Second-Generation Yellow and Green Seeds**	
	Yg Parent	
	Y	**g**
Yg Parent **Y**	YY	Yg
g	Yg	gg

We see that the resulting offspring will have one of the three gene pairs: YY, Yg, or gg. The YY pair is expected to occur one-fourth of the time, the Yg pair two-fourths or half of the time, and the gg pair the remaining one-fourth of the time. The only outcome that results in a green-colored plant is the gg pair, so the probability of a second-generation plant being green is 1/4. The gene pairs YY and Yg both result in yellow plants, and because they occur three-quarters of the time in the second generation, the probability of a second-generation plant being yellow is 3/4.

Example 1   *Pea Plants*

Suppose that a particular yellow-seed-colored pea plant has produced both yellow and green offspring. If this plant is crossbred with a green-seed-colored pea plant, what is the probability that an offspring will be green?

Solution: Because the yellow parent has produced both yellow and green offspring, we know that it is of type Yg. The green plant must be of type gg, because even one yellow gene would result in a yellow plant. Therefore, the Punnett square corresponding to crossbreeding the two plants is given in Table 4.7.

Table 4.7	Punnett Square for Yg and gg	
	Y	**g**
g	Yg	gg
g	Yg	gg

We see that 50% of the offspring will be of type Yg and the other 50% will be of type gg. So the probability that an offspring will be green is 1/2.

Many, but not all, inherited diseases are caused by recessive genes. Only a person with a pair of the recessive disease genes will develop the disease. A person with one of the disease genes and one normal gene will not develop the disease. We call such a person a **carrier**. We now look at these kinds of diseases.

Example 2   *Tay–Sachs Disease*

Tay–Sachs disease is a devastating genetic disease that results in neurologic deterioration and eventually death in early childhood. It is caused by a recessive gene, and only children inheriting two of the genes will develop the disease. There are now carrier detection tests that allow potential parents to learn if they are carriers of the Tay–Sachs gene. Find the probability that a child will have Tay–Sachs disease if

(a) both parents are carriers,

(b) only one of the parents is a carrier.

Solution:

(a) We denote the recessive Tay–Sachs gene by the lowercase letter t, and let T denote the normal gene. Because both parents are carriers, they both have the gene pair Tt, and the Punnett square for this situation is given in Table 4.8.

Table 4.8	Punnett Square for Two Tay–Sachs Carriers	
	T	**t**
T	TT	Tt
t	Tt	tt

The tt pair appears in only one of the four possible outcomes, and therefore the probability of the child inheriting Tay–Sachs disease is 1/4.

(b) If only one of the parents is a carrier, then the Punnett square is given in Table 4.9.

Table 4.9	Punnett Square for One Tay–Sachs Carrier	
	T	**T**
T	TT	TT
t	Tt	Tt

We see that the probability of the child inheriting the disease is zero.

■

Carrier detection tests are not available for all genetic diseases caused by a recessive gene. In this case, the family history of the disease may enable prospective parents to determine the probability that their child will have the disease. This kind of situation is illustrated in the next example.

Example 3   *Genetic Disease*

Suppose that a husband and wife each had a sibling with a genetic disease caused by a recessive gene, although both have parents without the disease. Neither the husband nor the wife has the disease, and they do not know whether they are carriers. If they have a child, what is the probability that the child will have the disease?

Solution: This situation is more complex than those we have considered so far because we cannot know with certainty the gene pairs of the husband and wife. However, we can still answer the question if we use some of our rules of probability. We first note that because neither the husband nor the wife has a parent with the disease and each has a sibling with the disease, all four of the potential grandparents must have the gene pair Dd, where d denotes the disease gene and D the normal gene. Therefore, the Punnett square illustrating the possible gene pairs of the husband and wife are the same and it is given in Table 4.10.

Table 4.10	Punnett Square for Two Carriers	
	D	**d**
D	DD	Dd
d	Dd	dd

Because neither the husband nor the wife has the disease, we know they cannot have gene type dd, so we can eliminate this possibility. From the remainder of the Punnett square, we see that the probability that the husband is a carrier is 2/3 and the probability that he is not a carrier is 1/3. The same probabilities hold for the wife. The child can only inherit the disease if both parents are carriers, and using the multiplication rule for independent events we see that

$$P(\text{husband a carrier and wife a carrier}) = P(\text{husband a carrier}) \cdot P(\text{wife a carrier})$$
$$= \frac{2}{3} \cdot \frac{2}{3} = \frac{4}{9}.$$

Now, if both the wife and husband are carriers, the Punnett square for their child would be the same as the one in Table 4.10. Therefore, the probability that the child would inherit the disease if both parents were carriers is 1/4. Combining all this information and applying the multiplication rule which says that $P(A \text{ and } B) = P(A) \cdot P(B|A)$, we have

$$P(\text{child inherits disease}) = P(\text{parents both carriers and child inherits disease})$$
$$= P(\text{parents both carriers}) \cdot P(\text{child inherits disease} \,|\, \text{parents both carriers})$$
$$= \frac{4}{9} \cdot \frac{1}{4} = \frac{1}{9}.$$

■

Some genetic diseases, such as Huntington's disease, are caused by a dominant gene. We investigate this case in the exercises.

We now consider the probabilities involved in blood types of humans. *When calculating probabilities that offspring have certain traits, we will ignore the small possibility of twins and assume that both sexes are equally likely.* There are four major blood types: A, B, AB, and O. Which blood type a person has is determined by a pair of genes, one coming from the mother and one from the father. The O gene is recessive, so the only way a person can have type O blood is if they have the gene pair oo. The A and B genes are both dominant over o. A person will have type A blood if they have either of the gene pairs AA or Ao. Similarly, the gene pairs BB or Bo result in type B blood. If a person has the gene pair AB, then he or she will have type AB blood.

Example 4   *Blood Type*

Mr. Molina has type A blood, and Mrs. Molina has type O blood. Furthermore, Mr. Molina's mother has type B blood.

(a) If Mr. and Mrs. Molina have one child, what is the probability that the child will have type A blood?

(b) If Mr. and Mrs. Molina have three children, what is the probability that all three children will have type A blood?

Solution:

(a) We know that Mrs. Molina has the gene pair oo. Because Mr. Molina has type A blood, he has either the gene pair Ao or the gene pair AA. However, because Mr. Molina's mother has type B blood, the only gene she could have contributed to him was an o gene. Therefore, we know with certainty that Mr. Molina has the gene pair Ao. The Punnett square for Mr. and Mrs. Molina is given in Table 4.11.

Table 4.11	Punnett Square for Ao and oo Blood Types	
	A	**o**
o	Ao	oo
o	Ao	oo

From the Punnett square, we see that the probability that the child will have type A blood is 1/2.

(b) The blood types of the three children are independent, and the probability that each child will have type A blood is 1/2. Therefore, by the multiplication rule for independent events, we see that the probability that all three will have type A blood is given by

$$\frac{1}{2} \cdot \frac{1}{2} \cdot \frac{1}{2} = \frac{1}{8}.$$

Another feature that distinguishes types of blood is the Rh factor. Human blood is either Rh-positive or Rh-negative depending on a particular gene pair. We let R denote the dominant Rh-positive gene and r denote the recessive Rh-negative gene. A person will be Rh-positive if he or she has either the gene pair RR or Rr, and a person will be Rh-negative only if he or she has the gene pair rr.

Example 5   *Blood Type and Rh Factor*

A married couple is expecting a baby. Both the husband and wife have Rh-positive blood, but each of them has a parent who is Rh-negative. The husband has type AB blood and the wife has type O blood. What is the probability that the baby will have type B, Rh-negative blood? (Assume that the blood type and Rh factor are independent.)

Solution: The Punnett square for the blood type of the baby is given in Table 4.12.

Table 4.12	Punnett Square for AB and oo Blood Types	
	A	**B**
o	Ao	Bo
o	Ao	Bo

Thus, the probability that the baby will have type B blood is 1/2. Because both the husband and wife have a parent who is Rh-negative, they each must have inherited one Rh-negative gene. Therefore, we know that the parents must both have the gene pair Rr, and the Punnett square for the Rh factor of the baby is given in Table 4.13.

Table 4.13	Punnett Square for Rh Factor	
	R	**r**
R	RR	Rr
r	Rr	rr

We see that the probability that the baby will be Rh-negative is 1/4. Now, combining these results and using the multiplication rule for independent events, we have

$$P(\text{type B and Rh-negative}) = P(\text{type B}) \cdot P(\text{Rh-negative}) = \frac{1}{2} \cdot \frac{1}{4} = \frac{1}{8}.$$ ∎

Some inherited traits or diseases, such as hemophilia, colorblindness, and muscular dystrophy, are **sex-linked traits**; that is, they are carried on the same chromosomes that determine a person's gender. Females have two X chromosomes, whereas males have one X chromosome and one Y chromosome. In male offspring, the X chromosome must come from the mother and the Y chromosome must come from the father. Hemophilia is a recessive sex-linked genetic disease. The gene for hemophilia is carried on the X chromosome. The only way a female can inherit hemophilia is if the gene is carried on both of her X chromosomes. In a carrier mother, only one of her X chromosomes carries the gene for hemophilia. However, because males have only one X chromosome, they will have the disease if just this one chromosome carries the defective gene. Because a male's X chromosome comes from his mother, a male can inherit hemophilia only from his mother. A female can inherit hemophilia only from a hemophiliac father and a mother that has at least one defective gene, an extremely unlikely occurrence.

Example 6 *Hemophilia*

Suppose a carrier mother and a normal father are expecting a child.

(a) What is the probability that the child will be a hemophiliac?

(b) If they know the child is a male through prenatal testing, what is the probability that he will be a hemophiliac?

Solution:

(a) We let X_h denote an X chromosome with a defective gene and let X denote a normal X chromosome. The Punnett square for a carrier mother and a normal father is shown in Table 4.14.

Table 4.14	Punnett Square for Hemophilia	
	Carrier Mother	
	X	**X$_h$**
Normal Father $\begin{matrix} X \\ Y \end{matrix}$	XX XY	XX$_h$ X$_h$Y

Only the gene pair X$_h$Y will result in a hemophiliac child, so the probability that they will have a hemophiliac child is 1/4.

(b) Only the outcomes X$_h$Y and XY will result in a male child, and because one of these will result in a hemophiliac child, we see that the probability that the child will be a hemophiliac given that it is a male is 1/2.

■

EXERCISES FOR SECTION 4.8

1. Gregor Mendel investigated the trait of tallness versus shortness in pea plants and found that the trait was determined by a gene pair with the tallness gene T being dominant and the shortness gene s being recessive. Suppose a pea plant with the gene pair Ts is crossbred with a plant with the gene pair ss.

(a) Construct the Punnett square corresponding to this situation.

(b) What is the probability that an offspring of the two plants will be tall?

(c) What is the probability that an offspring of the two plants will be short?

2. The seed shape of pea plants is determined by a gene pair. The two possible genes are the dominant gene R for round seeds and the recessive gene w for wrinkled seeds. Suppose a pea plant with the gene pair Rw is crossbred with another plant with the gene pair Rw.

(a) Construct the Punnett square corresponding to this situation.

(b) What is the probability that a seed of the offspring will be round?

(c) What is the probability that a seed of the offspring will be wrinkled?

In Exercises 3–5, consider the following: Eye color in humans is determined by a gene pair with the gene for brown eyes B being dominant over the gene for blue eyes b. Therefore, a person will have brown eyes if he or she is of type BB or Bb, and will have blue eyes only if he or she is of type bb.

3. If two blue-eyed parents have a child, what is the probability that the child will have blue eyes?

4. Suppose a woman with blue eyes marries a man with brown eyes. If the man's mother had blue eyes, what is the probability that an offspring of the couple will have brown eyes?

5. The first child of a brown-eyed man and a brown-eyed woman has blue eyes.

(a) What is the probability that their second child will have brown eyes?

(b) If this second child has brown eyes and marries someone with blue eyes, what is the probability that a child of theirs will have brown eyes?

In Exercises 6 and 7, consider the following: As explained in Example 2, Tay–Sachs disease is caused by a recessive gene and potential parents can find out if they are carriers of the gene through genetic testing.

6. Suppose two potential parents are both carriers of Tay–Sachs disease, and they plan to have three children. What is the probability that

   (a) all three of the children will inherit Tay–Sachs disease?

   (b) none of the children will inherit Tay Sachs disease?

   (c) at least one of the children will inherit Tay–Sachs disease?

7. Suppose two parents are both carriers of Tay–Sachs disease and they have a child who does not have Tay–Sachs disease. What is the probability that the child is a carrier?

In Exercises 8–10, consider the following: Huntington's disease is characterized by involuntary jerky movements, dementia, and eventual death. The symptoms usually do not show up until after 35 years of age. The disease is caused by a dominant gene H so that persons of type HH or Hh, where h is the normal gene, will have the disease. Because Huntington's disease does not manifest itself until late in life, many people with the disease have children before they realize they are afflicted.

8. If two people with gene type Hh have a child together, what is the probability that the child will develop Huntington's disease?

9. If a person of gene type Hh has four children with a person of gene type hh, what is the probability

   (a) that all four children will develop the disease?

   (b) that at least one of the children will develop the disease?

10. Suppose a woman develops Huntington's disease late in life, and only one of her parents had the disease. If she had a child with a Huntington's disease-free man, what is the probability that their child will develop the disease?

In Exercises 11–13, consider the following: For some types of plants, when plants with different colored flowers are crossbred, one color does not dominate over the other. Instead, the offspring have flowers that are a blend of the colors of the two parent plants. Mendel found that when he crossed a snapdragon with red flowers with a snapdragon with white flowers, the resulting offspring had pink flowers. In other words, when a red snapdragon of type RR was crossbred with a white one of type WW, the resulting offspring of type RW had pink flowers.

11. Suppose a red snapdragon is crossbred with a pink snapdragon. What is the probability that a resulting offspring will be

    (a) red?   (b) white?   (c) pink?

12. Suppose a white snapdragon is crossbred with a pink snapdragon. What is the probability that a resulting offspring will be

    (a) red?   (b) white?   (c) pink?

13. Suppose a pink snapdragon is crossbred with another pink snapdragon. What is the probability that a resulting offspring will be

    (a) red?   (b) white?   (c) pink?

14. If a hemophiliac father and a noncarrier mother are going to have a child, what is the probability that the child will

    (a) be a hemophiliac?   (b) be a carrier?

15. If a nonhemophiliac father and a mother who is a carrier of hemophilia are expecting a child and they know the child is female, what is the probability that the child will be a carrier?

16. Mr. and Mrs. Wyatt are expecting a child. Mrs. Wyatt has a hemophiliac brother and neither of her parents has hemophilia, but she does not know if she is a carrier. Mr.

Wyatt does not have hemophilia. What is the probability that the child will have hemophilia?

In Exercises 17 and 18, use the fact that colorblindness, like hemophilia, is a recessive sex-linked trait and is carried on the X chromosome.

17. If a colorblind father and a carrier mother have a child, what is the probability the child will

(a) be colorblind?   (b) be a carrier?

18. If a noncolorblind father and a mother who is a carrier of colorblindness are expecting a child and through prenatal testing they know the child is male, what is the probability that the child will be colorblind?

In Exercises 19 and 20, consider the following: Some sex-linked genetic diseases, such as vitamin D-resistant rickets, which may produce severe bowing of the legs, are caused by a dominant gene that is carried on the X chromosome. If X_R denotes an X chromosome carrying the rickets trait and X and Y are normal chromosomes, then any female with gene type X_RX or X_RX_R and males with gene type X_RY will have the disease.

19. Suppose a woman who does not have vitamin D-resistant rickets is married to a man who has the disease. If they are expecting a child, what is the probability that the child will have the disease?

20. Suppose a woman who does not have vitamin D-resistant rickets is married to a man who has the disease. If they are expecting a child and through prenatal testing they know the child is a boy, what is the probability that the child will have the disease?

21. A baby has a father with type AB blood and a mother with type O. What is the probability the child has blood type

(a) A?   (b) B?   (c) O?   (d) AB?

22. Mrs. Jung has type A blood, her father has type A, and her mother has type O. If Mr. Jung has type AB blood, what is the probability that a child of Mr. and Mrs. Jung will have blood type

(a) A?   (b) B?   (c) O?   (d) AB?

23. A man has blood type B. The man's mother was type AB and his father was type A. If his wife has blood type O and is expecting a baby, what is the probability the baby will have blood type

(a) A?   (b) B?   (c) O?   (d) AB?

24. A man has type B blood. The man's wife is type AB and is expecting a baby.

(a) Explain why the probability that the baby will have type A blood cannot be determined without further information.

(b) Explain why the probability that the baby will have type B blood can be determined and compute this probability.

25. A man has Rh-negative, type AB blood and his wife is Rh-positive, type A. They are expecting a baby. If they already have a child with Rh-negative, type B blood, what is the probability the baby will be Rh-negative, type A?

26. Suppose a woman has Rh-negative, type O blood and her husband is Rh-positive, type B. Furthermore, the husband's father has Rh-negative, type O blood. If the woman is expecting a baby, what is the probability that the baby will be Rh-negative, type O like its mother?

27. Mrs. Rosen has a genetic disease that is caused by a recessive gene. Mr. Rosen and his parents do not have the disease, but Mr. Rosen's sister has the disease. If the Rosens have a child, what is the probability that the child will inherit the disease?

Mr. and Mrs. Doss each have a sibling with a genetic disease that is caused by a recessive gene. Neither Mr. Doss, Mrs. Doss, nor any of their parents are afflicted with the disease. If Mr. and Mrs. Doss have a child who does not have the disease, what is the probability that the child is a carrier?

WRITING EXERCISES

1. When Skylab dropped out of orbit, a NASA representative estimated the chance that anyone on earth would be hit at 1 in 150. He then observed that, because the earth's population was roughly 4 billion, the chance that a particular individual would be hit would be

$$\frac{1}{150} \cdot \frac{1}{4,000,000,000} = \frac{1}{600,000,000,000}.$$

What is wrong with this reasoning?

2. Police and the FBI have data banks of fingerprints of criminals used in the search for suspects in crimes. Many states have begun compiling data banks of DNA from criminals as well. However, close biological relatives of a criminal share many DNA patterns, so that a high degree of similarity, but not a perfect match, of the DNA from a criminal in the data bank to the DNA obtained from a crime scene would focus attention on a criminal's relatives, who may have no prior criminal history. Comment on whether this is a fair use of the data bank or an improper invasion of privacy.

3. Tests for the presence of disease, drugs, and so on are subject to two types of errors. They may fail to detect presence in a person's system, a "false negative," or may incorrectly report its presence in a person without the disease or drug use, a "false positive." Two different tests for the same disease or substance may vary in the most likely type of error. One test may have almost no false negatives but a significant percentage of false positives, whereas another test may have false positives several percent of the time but relatively few false negatives. Describe how the seriousness of the disease or the reason for drug testing would influence the decision on which test to use, taking the varied points of view of a patient, an insurance company, a potential employer, and a prospective employee.

4. In a letter to the *New York Times*, reader Bill Grassman recounts one version of a popular story, "When the calculations of flights per day, when and where the bombings had occurred and the normal flying patterns of the executive disclosed that the odds of his being on a plane with a bomb were 1 in 13 million, he asked for the probability of his being on a plane with two bombs. On learning that this increased the odds to 1 in 42 billion, he always carried a bomb with him." What is the error in the executive's reasoning?

5. The U.S. Bureau of the Census interviews a random sample of families each year for the National Health Interview Survey. As described in Branching Out 4.1, from 1987 to 1993 it interviewed 42,888 families with exactly two children. The breakdown in family type by the genders of the children, where, for instance, BG means the first child is a boy and the second child a girl, was as follows:

Family Type	Number of Families
BB	11,334
BG	11,118
GG	9,523
GB	10,913

(a) There are 106 boys born to every 100 girls. Based on this fact, what would you expect the breakdown into family type to be?

(b) Notice that the numbers in part (a) do not agree with the census figures. Notice, however, that the census figures are only for families with exactly two children. How do you think that might affect the probabilities? If instead, the statistics had been based on the first two children of any size family, would you expect a different outcome?

6. In defending the safety of large trucks, Thomas Donohue, president and CEO of the American Trucking Association, said "In 1995, 41,798 people died on our nation's highways. Most of those fatalities—88%—did not involve trucks." Explain why this statement is probably a weak defense.

7. In a letter to the editor published in the *Fort Worth Star Telegram*, a man complained that his 24-year-old-son had neither had a traffic accident nor received a ticket for 4 straight years, and yet the son had recently received a notice in the mail that his auto insurance policy would not be renewed because statistics showed that he was overdue for an accident. Was the man correct to be outraged at the insurance company? Explain your answer.

8. During the criminal trial of O. J. Simpson, Alan Dershowitz, one of Mr. Simpson's lawyers, commented on television that only one-tenth of 1% of men who abuse their wives go on to murder them. Mr. Dershowitz probably wanted the audience to conclude that therefore, there was only a 1 in 1000 chance that O. J. Simpson, who had previously admitted to abusing her, murdered his ex-wife, Nicole Brown. Explain what is wrong with this conclusion. What would be a more valid conditional probability to consider here?

Projects

1. Pennsylvania license plates consist of either three letters followed by three numbers or three letters followed by four numbers. The witness to a crime remembers an X and a 7 from the getaway car, but does not remember their position or even whether there was more than one X or 7. How many different license plates could be of this form?

2. In a study of possible gender bias in graduate admissions to the University of California at Berkeley, researchers P. J. Bickel, E. A. Hammel, and J. W. O'Connell considered the outcome of the 12,763 complete applications for admission to graduate programs for the fall of 1973. They considered only the question of *bias*, that is, whether or not women and men were admitted at noticeably different rates. They did not consider the much more complex question of *discrimination*, which would have to take into account the relative strengths of the applicants. The most simplistic way to study the question of bias is to look at the aggregate data across all the graduate programs at Berkeley. These data are shown in the table.

Applicants	Admitted	Not Admitted
Men	3738	4704
Women	1494	2827

Source: Reprinted with permission from P. J. Bickel, E. A. Hammel, and J. W. O'Connell, "Sex Bias in Graduate Admissions Data from Berkeley," *Science* 187 (February 1975). Copyright 1975 American Association for the Advancement of Science and the authors.

(a) Find the probability that a random male applicant would be admitted.

(b) Find the probability that a random female applicant would be admitted.

(c) On the basis of the answers to parts (a) and (b), one might conclude that bias in favor of male applicants existed. However, the researchers in this study found that when the data were looked at department by department, no such bias was evident. The key fact that is left out when looking at the aggregate data is that not all departments are equally easy to enter. To illustrate this, construct an example of a university with four graduate departments for which the aggregate data on admissions are the same as shown in the table and for which every department has a bias in favor of women applicants.

3. Stanford statistician Bradley Efron created the following example of "nontransitive" dice. Consider the following six-sided dice: die A has four faces labeled 4 and two labeled 0, die B has all six faces labeled 3, die C has two faces labeled 6 and four labeled 2, and die D has three faces labeled 5 and three labeled 1. The claim is that, if someone chooses a die first, then it is possible for you to select one of the three remaining dice in such a way that when the dice are rolled, you will roll the higher number more often than the other person. For a single roll of each die, find the probability that the total

(a) on die A is greater than that on die B;

(b) on die B is greater than that on die C;

(c) on die C is greater than that on die D;

(d) on die D is greater than that on die A.

Explain why the claim holds true.

4. Assume that all of the families in a certain society want to produce a male heir, so couples continue to have children until they have a boy, at which point they stop having children.

(a) If boy and girl offspring are equally likely, do you predict that the society will end up with more boys or girls?

(b) Run an experiment with 100 simulated "families." For each family flip a coin to determine the genders of the children born to the family, letting heads be "boys" and tails be "girls," and continue until a boy is born. Record the number of boys and girls born to each family, and then add all the data together to find the total number of boys and girls born. Does this experiment support your prediction?

(c) Repeat the experiment in part (b) assuming that 50 families have boys with probability 2/3 and 50 families have girls with probability 2/3. Specifically, use a standard die to determine the genders of the children born to the family, letting a 1 through 4 on the die be a boy and a 5 or 6 be a girl for 50 families, and a 1 through

4 be a girl and a 5 or 6 be a boy for the other 50. As before, each family should continue to have children until a boy is born. How do the data in this case compare to the data from part (b)? Is this what you would have predicted?

5. Many states have lotteries, like the California Super Lotto game, in which players must choose some numbers within a certain range, where order is not important. To win the big lottery jackpot, the numbers on the player's entry must match all of the numbers drawn by the state. The range of numbers from which to pick and the number of numbers to pick varies from state to state. For instance, in the California Super Lotto game, players choose six numbers between 1 and 51, whereas in the Minnesota Gopher 5 lottery game, players choose five numbers between 1 and 39.

   (a) Find the range of numbers and the number of numbers to pick for all U.S. states that have lotteries of this type. (The search for these data is probably best done over the Internet.)

   (b) Compare the probabilities of winning each of these lotteries.

   (c) Find the populations of each of these states.

   (d) How does the probability of winning a state's lottery relate to the population of the state?

   (e) Explain why states with larger populations might want to have a lottery that is harder to win. What are some of the advantages in terms of publicity for the lottery in having a lottery that players are less likely to win?

KEY TERMS

CHAPTER 4 REVIEW EXERCISES

1. Two dice are thrown. What is the probability that the sum of the two dice will be 10?

2. In a box of machine parts, 4% of the parts are defective. If one part is selected at random, what is the probability that it will not be defective?

3. Three hundred kernels from a batch of popcorn were cooked and only 253 of the kernels popped. What is the empirical probability that a random popcorn kernel from a batch will pop?

4. According to *The World Book of Odds*, the odds against being killed in a canoe accident for every 100 hours spent in the canoe are 6000 to 1. What is the probability of being killed in a canoe accident if you spend 100 hours in a canoe?

5. If the probability of winning a game is 1/3, what are the odds against winning the game?

Secretariat on his way to a 31-length win in the 1973 Belmont Stakes.

6. Secretariat, considered by many to be the greatest racehorse that ever lived, won the 1973 Belmont Stakes by an incredible 31 lengths. Having already won the Kentucky Derby and the Preakness Stakes, the first two races in the Triple Crown, Secretariat was highly favored to win at the Belmont and the house odds were 1 to 10. If a bettor had wagered $50 on Secretariat to win the Belmont, how much would the bettor have won in addition to having the wager returned?

7. Suppose the house odds for a casino game are 19 to 4. If the probability of winning the game is 15%, is it a fair bet?

8. According to a 1995 U.S. Bureau of the Census report, 33% of households in the United States had one wage earner, 36% had two, 10% had three or more, and the remaining 21% had none. What is the probability that a random household in 1995 had one or two wage earners?

9. On May 31, 1997, there were 487,297 personnel on active duty in the U.S. Army. Of these 70,690 were women, 67,986 were commissioned officers, and 9716 were women commissioned officers. What is the probability that a randomly selected member of the U.S. Army on May 31, 1997 was neither a woman nor a commissioned officer?

10. One would expect that the performance of a student in a second-semester calculus class would depend on how well the student had done in the first semester of calculus. Thirty-two students enrolled in Calculus II in the spring semester of 1997 at TCU had also taken Calculus I at TCU. The breakdown of the Calculus I and II grades for these students is shown in the table.

		Grade for Calculus II				
		A	B	C	D	F
	A	4	4	1	0	0
Grade for Calculus I	B	0	2	2	2	1
	C	2	3	6	2	2
	D	0	0	0	0	1

For a random student included in this group,

(a) find the probability that the student got an A in Calculus II given that the student received an A in Calculus I,

(b) find the probability that the student got an A in Calculus II given that the student did not receive an A in Calculus I,

(c) find the probability that the student received a better grade in Calculus II than in Calculus I.

11. If two cards are drawn at random from a deck, what is the probability that they will both be diamonds?

12. Suppose a survey of parents indicates that 53% favor school uniforms for children, 42% favor year-round school, and 20% favor both. If a randomly selected parent favors year-round school, what is the probability that the parent also favors school uniforms?

13. According to the U.S. National Center for Health Statistics, 45.3% of hospital emergency room visits in 1993 were urgent and 25.1% of the patients were under 15 years old. Of the visits by patients under 15 years old, 39.4% were urgent. For a random urgent visit to an emergency room in 1993, what is the probability that the patient was under 15?

14. Rhode Island license plates have two letters followed by a hyphen and then three digits. How many different license plates of this form begin with a C followed by an N?

15. A soccer league consists of seven teams. If each team must play every other team exactly once, how many games must be played?

16. Nine guests leave their coats in a room. If you hand back the coats at random, what is the probability that each guest will get back his or her own coat?

17. There are 25 people in a room. Fourteen are women and 11 are men. If you select eight people at random, what is the probability that you will select four women and four men?

18. There are 28 pens in a box, 12 of which do not work. If you select six pens at random, what is the probability that at least one of them will work?

19. If a coin is flipped seven times, what is the probability that you will get exactly three heads?

20. There are 18 people in a room. What is the probability that at least two have the same birthday?

21. Suppose seven cards are dealt from a standard deck. What is the probability that the hand will have four cards of one suit, two cards of a second suit, and one other card of a third suit?

22. A bowl contains 17 peppermint, 22 butterscotch, and 15 cherry-flavored candies, all of the same size and shape. If you pick 4 pieces of candy at random from the bowl, what is the probability that all will be the same flavor?

23. Suppose you just inherited $100,000 and you are trying to decide how to invest it for the next year. You have narrowed it down to two choices. The first choice is to invest it in a bond with a guaranteed return of 5% interest at the end of the year. The second choice is to bet it all on the Super Bowl with a 51% chance of doubling your money and a 49% chance of losing it all. Which investment option has the highest expected value?

24. To play the Cash 5 game from the Colorado Lottery, a player picks five different numbers between 1 and 32 (the order does not matter). Five of the numbers between 1 and 32 will be chosen to be the winning numbers. It costs $1 to play and the $1 is not returned. The payoffs, shown in the table, are based on the number of winning numbers matched.

Outcome	Payoff
5 winning numbers matched	$20,000
4 winning numbers matched	$200
3 winning numbers matched	$10
2 winning numbers matched	$1
0 or 1 winning numbers matched	$0

  (a) What is the expected value of this game?
  (b) What is the house edge for this game, expressed as a percentage of $1?

25. A man and woman both have brown eyes and each of them had a blue-eyed parent.

  (a) If they have one child, what is the probability that the child will have blue eyes?
  (b) If they have four children, what is the probability that at least one of them will have blue eyes?

26. A husband and wife both have Rh-positive, type A blood and they have a child with Rh-negative, type O blood. They are expecting a second child. What is the probability that the second child will have Rh-positive, type A blood?

27. Colorblindness is a recessive sex-linked trait carried on the X chromosome. If a non-colorblind father and a mother who is a carrier of colorblindness are expecting a child, what is the probability the child will

(a) be colorblind?   **(b)** be a carrier?

∫UGGESTED READINGS

Berry, Donald A., and Seymour Geisser. "Inference in Cases of Disputed Paternity," with comment by Donald Ylvisaker. In *Statistics and the Law*, edited by Morris H. DeGroot, Stephen E. Fienberg, and Joseph B. Kadane. New York: John Wiley, 1986, 353–390. Genetics involving blood, conditional probability, and legal issues.

Gardner, Martin. *Wheels, Life and Other Mathematical Amusements.* New York: W. H. Freeman, 1983. Chapter 5 is a fun look at nontransitive dice and some other probability paradoxes.

Handley, James A. "Jumping to Coincidences: Defying Odds in the Realm of the Preposterous." *The American Statistician* 46:3 (August 1992), 197–202. A nontechnical discussion of seven cases of erroneous probability calculations appearing in the press.

Peterson, Ivars. *The Jungles of Randomness.* New York: John Wiley, 1998. An elementary and enjoyable look at many situations involving randomness, including weird dice, coincidences, and chaos, to name just a few.

Singer, Barry. "Probabilities, Meeting, and Mating." In *The Fascination of Statistics*, edited by Richard J. Brook, Gregory C. Arnold, Thomas H. Hassard, and Robert M. Pringle. New York: Marcel Dekker, 1986, 3–12. An extended example estimating the probabilities of meeting an acceptable mate. Interesting blend of empirical and theoretical probability.

Wardrop, Robert L. "Simpson's Paradox and the Hot Hand in Basketball." *The American Statistician* 49:1 (February 1995), 24–28. Somewhat technical analysis of the probabilities that arise from the free throw data presented by A. Tversky and T. Gilovich.

You encounter statistics daily. This season's temperatures or rainfall compared to past seasons, batting averages and team performances, the movements of financial markets or the Consumer Price Index—

these are all statistics. As you drive along the highway, your own car's statistic—the mileage it gets—determines how often you must purchase gas. Statistics help us to see where we are and to chart our own course. They measure public opinion and predict the outcome of elections. Health statistics guide government and businesses in providing for and controlling the costs of health care, and they are a guide for individual choices on health care.

Statistics are developed by collecting, organizing, and analyzing data in an area of particular interest. If you want to know your car's gas mileage, you record the odometer readings and the amounts of gas bought with successive fill-ups. These are your data. Next come the organization and analysis, which is what this chapter is about: organization, presentation, and analysis of data using statistical tools.

We will divide statistics into two categories. In the first part of this chapter we discuss *descriptive statistics*, ways to present information in a user-friendly fashion. The second part of the chapter is an introduction to *inferential statistics*, where we infer statistical properties based on samples. We conclude the chapter with a brief discussion of how, either accidentally or through deliberate misrepresentation, statistics can be misleading. We also discuss ways to avoid being misled or misleading others.

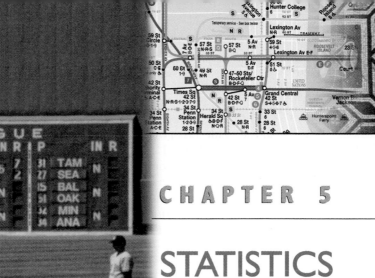

CHAPTER 5

STATISTICS

5.1   ORGANIZING AND PRESENTING DATA

Suppose the grades assigned to the students in a calculus class are F, B, B, A, C, A, A, C, B, C, C, B, C, D, C, B, C, C, and F. A collection of information such as this is called **data**. Grouping like data gives a sense of the distribution of the data. However, by organizing the data into tables or charts, we can get an even better understanding of the distribution.

A **frequency distribution** is a table showing the number of occurrences of each of the different data values. Sometimes, we might prefer to give the fraction of the time each possibility occurs. This is called a **relative frequency distribution**. The relative frequencies are computed by dividing the frequencies by the total number of pieces of data, and are most often given approximately as decimals or as percentages. Frequencies and relative frequencies may be computed for **categorical data**, such as the grades in a class or colors of registered automobiles, as well as for **numerical data**, such as the prices of a set of items.

Example 1   *Frequency Distribution and Relative Frequency Distribution*

Compute the frequency distribution and relative frequency distribution for the following grades assigned to the students in a calculus class:

F, B, B, A, C, A, A, C, B, C, C, B, C, D, C, B, C, C, F.

Solution: There are two ways to find the frequencies of the possible grades. We could scan the list counting the number of A's, then the number of B's, and so forth, or we could go through the list just once, keeping a tally of which values occur as we go along. The method chosen is largely a matter of taste. The former method is probably preferable for short lists and the latter method for longer lists. Using either method, the frequency distribution is shown in Table 5.1.

Table 5.1 **Grade Distribution**	
Grade	*Frequency*
A	3
B	5
C	8
D	1
F	2

As a check on our work, observe that the total of the frequencies is 19, which is equal to the number of grades listed. To obtain the relative frequencies, divide the frequencies by the total of 19. To four decimal places, the relative frequency distribution is given in Table 5.2.

Table 5.2	Grade Distribution	
Grade	*Frequency*	*Relative Frequency*
A	3	0.1579
B	5	0.2632
C	8	0.4211
D	1	0.0526
F	2	0.1053

Relative frequencies should add up to 1, up to a possible slight round-off discrepancy.

Technology Tip

Statistical software, spreadsheets, and calculators that handle lists can be used to make computation of relative frequencies more automatic. Be aware that some round each relative frequency in the standard way, whereas others round so that the relative frequencies sum to exactly 1.

When the data consist of a wide range of numbers or of numbers computed to several decimal places, we often group the data into **classes**, as in the following example.

Example 2   *Grouping Data into Classes*

A 1998 *Consumer Reports* survey[1] of several brands of refrigerators found the following cost per year for running them: $64, $60, $84, $82, $47, $52, $55, $58, $61, $61, $64, $63, $78, $73, $73, $78, $85, $93, $93, $95, $96, $94, $57, $63. Compute the frequency distribution and relative frequency distribution, grouping the data into four approximately equally spaced classes.

Solution: Looking at the list of values, we see that they range from $47 to $96. There are many different values that occur, so rather than counting the frequency of each individual value, it is more reasonable to group the data. Because $96 - 47 = 49$ and $49/4 = 12.25$, we will group the data into four equally spaced classes, covering a range of 13 dollar values each, starting at $46 and ending at $97 as in Table 5.3.

Table 5.3	Cost to Run Refrigerators
Cost per Year (dollars)	*Frequency*
46–58	5
59–71	7
72–84	6
85–97	6

Because there are 24 pieces of data, we obtain the relative frequencies upon dividing the frequencies by 24 as in Table 5.4.

Table 5.4	Cost to Run Refrigerators	
Cost per Year (dollars)	**Frequency**	**Relative Frequency**
46–58	5	0.2083
59–71	7	0.2917
72–84	6	0.2500
85–97	6	0.2500

When grouping data, you have several choices. First is deciding the classes. In the previous example, we could equally well have chosen to group the data into equally spaced classes of width 15 starting at $45 and ending at $104. This division is no better or worse than the one we chose. We have chosen classes of approximately equal size or width. This is usually a good idea, unless you have a good reason not to. One example of where you might not want classes of equal width is in computing the frequency distribution for household income across the United States. For instance, the household income in the United States in 1995 may be broken down into classes as shown in Table 5.5.

Table 5.5	Household Income in 1995
Household Income (dollars)	**Number of Households (in thousands)**
Under 5000	3,651
5000–9999	8,539
10,000–14,999	8,716
15,000–24,999	15,848
25,000–34,999	14,167
35,000–49,999	16,877
50,000–74,999	17,038
75,000–99,999	7,677
100,000 and over	7,114

Source: U.S. Bureau of the Census.[2]

Notice that the first three classes each cover a range of $5000, whereas the remaining classes cover progressively wider ranges of income. If the classes all covered equal ranges of income, the number of households in the classes for the upper income levels would be quite a bit smaller than for the lower ones. More important, whereas a $5000 difference in income would have a huge impact on the life of a household in the lowest income classes, it would have a fairly small impact on the general lifestyle of a household in the upper income classes.

Bar Graphs

One way to illustrate a frequency distribution or relative frequency distribution is with a bar graph. In a **bar graph**, the different categories or classes are indicated on the horizontal axis and either the frequency or the relative frequency is recorded on the vertical axis. The "graph" consists of bars above each category with height equal to the frequency or relative frequency of that category. Generally, the bars are of equal width and separated by a small space, although when we consider histograms, a special kind of bar graph, there will be no separation.

Example 3   *Bar Graphs*

In Example 1, we found the frequency and relative frequency distributions of grades in a calculus class as given in Table 5.6.

Table 5.6	**Grade Distribution**	
Grade	*Frequency*	*Relative Frequency*
A	3	0.1579
B	5	0.2632
C	8	0.4211
D	1	0.0526
F	2	0.1053

Construct a bar graph representing

(a) the frequency distribution,

(b) the relative frequency distribution.

Solution:

(a) We first set the vertical scale for the graph. Looking at the frequencies, we see that the largest frequency is 8, so it is reasonable to set the vertical scale from 0 to 8 in our graph. Putting tick marks on the vertical scale and drawing bars of equal width and spacing, we make the bar graph in Figure 5.1 for the frequency distribution.

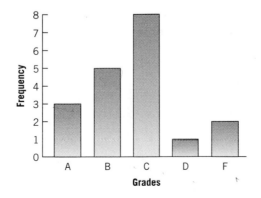

Figure 5.1
Bar graph of grades by frequency.

(b) The relative frequencies go up to 0.4211, so we elect to set the vertical scale from 0 to 0.5 in our graph in Figure 5.2.

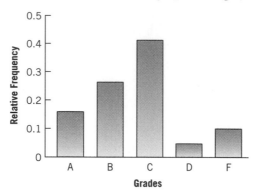

figure 5.2
Bar graph of grades by
relative frequency.

Notice that the bar graphs for the frequency and relative frequency distributions always have the same qualitative shape. With suitable scaling, they may be drawn to have exactly the same shape.

Example 4   *Bar Graph of Data Grouped into Classes*

In Example 2, we grouped the cost per year of running various brands of refrigerators into four classes, arriving at the frequency and relative frequency distributions in Table 5.7.

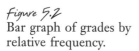

Table 5.7	**Cost to Run Refrigerators**	
Cost per Year (dollars)	**Frequency**	**Relative Frequency**
46–58	5	0.2083
59–71	7	0.2917
72–84	6	0.2500
85–97	6	0.2500

Construct a bar graph representing

(a) the frequency distribution,

(b) the relative frequency distribution.

Solution:

(a) A vertical scale from 0 to 8 for the bar graph of the frequency distribution seems appropriate and we get the bar graph in Figure 5.3.

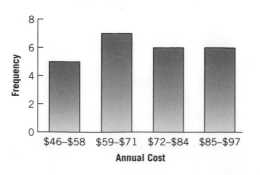

figure 5.3
Bar graph of cost by
frequency.

(b) In the bar graph for the relative frequency distribution in Figure 5.4, we elected to have the vertical scale range from 0 to 0.3.

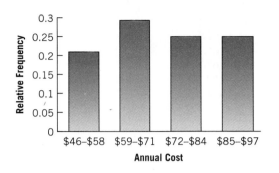

Figure 5.4
Bar graph of cost by relative frequency.

Histograms

Suppose a woman says she is 62 inches tall. Because this number has been rounded to the nearest inch, her actual height is somewhere in the interval from 61.5 inches to 62.5 inches. The frequency or relative frequency distributions for such types of data are commonly illustrated by a special type of bar graph known as a **histogram**. Histograms are also commonly used when the data values include a very large number of different values, even if they have not been rounded. In a histogram, the base of each bar extends over the entire range of values for that bar and there is no separation between the bars. Intervals with frequency 0 must be included. It is preferable to choose the endpoints of the intervals so that none of the data values fall on the endpoints. We will restrict consideration to histograms where the widths of all the bars are equal.

Example 5   *Histograms*

The heights, in inches rounded to the nearest inch, of the female students in a college class are

$$64, 62, 65, 65, 70, 66, 60, 65, 65, 66, 61, 67, 61, 70, 63, 62,$$
$$67, 67, 64, 66, 66, 61, 64, 64, 63, 62, 63, 64, 65, 63, 60.$$

Construct a histogram representing the relative frequency distribution for these data values,

(a) grouping the data into classes of width 2 starting at 59.5,
(b) grouping the data into classes of width 4 starting at 59.5.

Solution:
(a) We begin by computing the relative frequency distribution shown in Table 5.8.

Table 5.8	Height of Women in a Class	
Height (inches)	**Frequency**	**Relative Frequency**
59.5–61.5	5	0.1613
61.5–63.5	7	0.2258
63.5–65.5	10	0.3226
65.5–67.5	7	0.2258
67.5–69.5	0	0
69.5–71.5	2	0.0645

In Figure 5.5, we have the histogram for the relative frequencies. Notice that there are no gaps between the bars and that the bars are drawn over the ranges of values they represent.

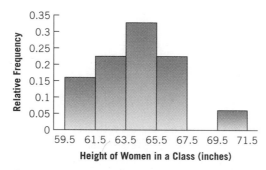

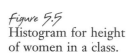
Figure 5.5
Histogram for height of women in a class.

(b) Grouping the data into classes of width 4 gives the relative frequency distribution in Table 5.9.

Table 5.9	Height of Women in a Class	
Height (inches)	**Frequency**	**Relative Frequency**
59.5–63.5	12	0.3871
63.5–67.5	17	0.5484
67.5–71.5	2	0.0645

The corresponding histogram is shown in Figure 5.6.

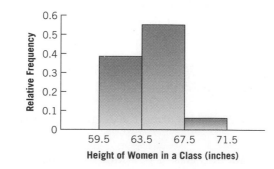

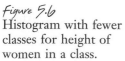
Figure 5.6
Histogram with fewer classes for height of women in a class.

Notice that the histogram in Figure 5.5 is similar in shape to the one in Figure 5.6, only with finer divisions.

Histograms are sometimes "normalized" so that the areas of all the bars in the histogram sum to 1. When dealing with classes of different widths, histograms should always be normalized. We will look at normalized histograms when we study the normal distribution, or "bell curve," in Section 5.4.

Pie Charts

A **pie chart** or circle chart represents categories as pieces of pie or sectors of a circle, whose sizes are proportional to the frequencies and relative frequencies of the classes. Recall that there are 360 degrees in a full circle. Because the relative frequencies represent the fraction of the whole for each category, we can obtain the angle for each sector of the pie chart by multiplying the relative frequencies by 360°:

$$\text{angle} = \text{relative frequency} \times 360°.$$

The sectors are then drawn, often with the help of a protractor, which is an instrument used to measure angles, such as those shown in Figure 5.7.

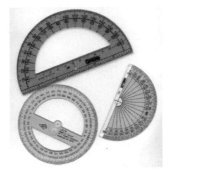

Figure 5.7
Protractors.

There are many stylistic options in labeling a pie chart. We will label within the sectors when possible, and at their edges when not. We will use either the frequency or the relative frequency expressed as a percentage for our labels. We will also include the total frequency as a caption to our chart.

Example 6   *Pie Chart*

In Example 1, we found the frequency and relative frequency distributions of grades in a calculus class as given in Table 5.10.

Table 5.10	**Grade Distribution**	
Grade	*Frequency*	*Relative Frequency*
A	3	0.1579
B	5	0.2632
C	8	0.4211
D	1	0.0526
F	2	0.1053

Draw a pie chart representing the distribution, labeling each sector with its percentage.

Solution: We first multiply the relative frequencies by 360° to find the angle measures. For instance, the angle measure for the sector for the grade of A is $0.1579 \cdot 360° \approx 56.8°$. The other angles are computed in the same way, and the resulting angle measures are recorded in Table 5.11.

Table 5.11	Relative Frequency for Grade Distribution		
Grade	*Frequency*	*Relative Frequency*	*Angle Measure (degrees)*
A	3	0.1579	56.8
B	5	0.2632	94.8
C	8	0.4211	151.6
D	1	0.0526	18.9
F	2	0.1053	37.9

Making the arbitrary choice to begin at the "12 o'clock" position, we mark out the angle measures in succession. A protractor is a great help here; otherwise we can only roughly approximate the angles. We complete our pie chart, shown in Figure 5.8, by labeling each sector with its category and percentage (from the relative frequencies).

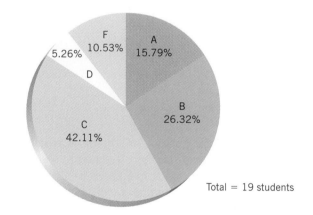

Figure 5.8
Pie chart of grade distribution.

Example 7 *Pie Chart of Data Grouped into Classes*

The number of people in the United States living below the poverty level in 1995 is broken down into classes by age group in Table 5.12.

Table 5.12	Persons Below the Poverty Level
Age (years)	*Frequency (in thousands)*
Under 18	14,665
18–34	9,749
35–54	6,534
Above 54	5,477

Source: U.S. Bureau of the Census.[4]

Compute the relative frequency distribution and draw a pie chart, labeling each sector with its frequency.

Solution: We first compute the relative frequencies and then multiply them by 360° to find the angles of the sectors. The calculations are recorded in Table 5.13.

Table 5.13	**Persons Below the Poverty Level**		
Age (years)	*Frequency (in thousands)*	*Relative Frequency*	*Angle Measure (degrees)*
Under 18	14,665	0.4026	144.9
18–34	9,749	0.2676	96.3
35–54	6,534	0.1794	64.6
Above 54	5,477	0.1504	54.1

The resulting angles add up to only 359.9° due to round-off error. When drawing pie charts by hand, this error will not be significant because there is already some inaccuracy due to hand measurements. If you use some kind of computer software to draw the pie charts, the frequencies will be entered as data, and the software will deal with any rounding issues internally. To draw the pie chart for this case, we again make the arbitrary decision to begin constructing our pie chart at the "12 o'clock" position. We label the slices with frequencies and get the pie chart shown in Figure 5.9.

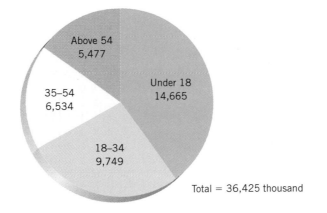

Total = 36,425 thousand

Figure 5.9
Persons below the poverty level in 1995.

Pie charts are also commonly used to illustrate budgets. For this application, frequencies are represented by dollars or some other monetary unit.

Technology Tip

Statistical software, spreadsheets, word processing programs, and statistical graphing calculators can often construct bar graphs, histograms, and pie charts from either frequency or relative frequency data. The most difficult part of using them usually is trying to get the exact appearance you want.

BRANCHING OUT 5.1

THE BIBLE CODES

From finding evidence in Shakespeare's work that suggests they were in fact written by Francis Bacon to repeated number patterns in the New Testament and Al Quran, the holy scripture of Islam, people have found "hidden" words embedded in texts. This came again into the public's eye through the 1994 publication of "Equidistant Letter Sequences in the Book of Genesis" by Doron Witztum, Eliyahu Rips, and Yoav Rosenberg in the journal *Statistical Science*. The authors found names of famous Israeli rabbis in the Torah, the bible of Judaism. Although no explicit statement was made by the authors, many attribute the embeddings found by Witztum, Rips, and Rosenberg to the hand of God. Soon after, Michael Drosnin reported on prophecies he found in the Torah in his 1997 bestseller, *The Bible Code*. Drosnin suggests that perhaps the prophecies he discovered were placed by God or extraterrestrials.

Both works described searching for words embedded in biblical passages (written in Hebrew). The letters of the words must be equally spaced. For the authors of the works, this means that there are the same number of letters between each letter of the embedded word or phrase; spaces and punctuation are ignored. The word may appear normally or backwards. To illustrate, we have searched the following passage from the 25th Psalm in the King James bible for words with mathematical meaning.

25 ... [3]Yea, let none that wait on thee be ashamed: let them be ashamed which transgress without cause. [4]Shew me thy ways, O LORD; teach me thy paths. [5]Lead me in thy truth, and teach me: for thou [art] the God of my salvation; on thee do I wait all the day. [6]Remember, O LORD, thy tender mercies and thy loving kindnesses; for they [have been] ever of old. [7]Remember not the sins of my youth, nor my transgressions: according to thy mercy remember thou me for thy goodness' sake, O LORD.

Embedded in the text are the following:

1. Beginning with the "s" in the first "ashamed" in verse 3 and taking every 9th letter spells out "stats," highlighted in blue.

2. Proceeding backwards and selecting every 14th letter from the first "m" in "Remember" in verse 6 spells "math," highlighted in red.

3. Intermixed with "math" is "line," which starts with the "l" in "salvation" and continues every 4th letter. This is highlighted in green.

4. Also appearing backwards and highlighted in pink, starting with "n" in the "nor" of verse 7 and skipping over three letters, is the suggestive "No fit, redo," perhaps referring to some sort of statistical line graph.

Spacings in the bible codes works can be much longer. For instance, in 1994 Drosnin found the Hebrew version of former Prime Minister Yitzhak Rabin's name with a gap of 4771 intermediate letters. This was "near" a word for "kill." "Near" means that both words appear in the same "small" rectangle when the text is arranged with the same, fixed number of letters appearing in every line. For instance, in the picture the words Chanukah and Hashmonai (the dynasty that led the revolt Chanukah celebrates) appear "near" each other in Genesis. Drosnin interpreted this as predicting Rabin's assassination. When Rabin was assassinated in 1995, Drosnin states, "Suddenly, brutally, I had absolute proof that the Bible code was real." Witztum, Rips, and Rosenberg found the 66 names that had the longest articles in the *Encyclopedia of Great Men in Israel* in Genesis, along with associated dates. They calculate the probability that all 66 pairs of men and dates would appear in as close proximity (according to their particular definition of distance) in a random work the length of Genesis at 1 in 50,000.

Many are skeptical of attributing special significance to findings in either work. One objection is to the specific calculations of probabilities by Witztum, Rips, and Rosenberg or to less precise claims by Drosnin that meaningful equidistant letter sequences, abbreviated ELSs, are unusual. For instance, based on the letter frequencies in the English language given in the following table, the probability that four letters chosen randomly spell "math" is roughly 1/100,000. However, taking into account all possible skips of length 1,000 or less in the 178,149 letters in the King James version of the book of Psalms, one would expect to find "math" a few thousand times. The authors of this text counted almost one hundred occurrences with skips at most 15 letters. As California Institute of Technology Professor Barry Simon expressed it, "So when you search for an ELS of a relatively short word, you are far from searching a needle in a haystack—rather you are searching for a blade of hay."

Letter Frequencies in the English Language

Letter	Relative Frequency	Letter	Relative Frequency
E	0.127	M	0.024
T	0.091	W	0.023
A	0.082	F	0.022
O	0.075	Y	0.020
I	0.070	G	0.020
N	0.067	P	0.019
S	0.063	B	0.015
H	0.061	V	0.010
R	0.060	K	0.008
D	0.043	J	0.002
L	0.040	Q	0.001
C	0.028	X	0.001
U	0.028	Z	0.001

Source: H. Becker and F. Piper, *Cipher Systems: The Protection of Communications.*[3]

Another objection is that the studies allow what Simon and others call "wiggle room." The nature of the Torah, written in Hebrew consonants without vowels, is one source of wiggle room. Do you interpret the English "mth" as "math," "moth," or "mouth"? Another occurs when researchers allow for alternate spellings. For instance, in English the Quran is also written as Quraan, Koran, and Alkoran. Of course, flexibility in what one searches for is another source of wiggle room. A leading skeptic, Australian mathematician Brendan McKay, has found assassination references in *Moby Dick* and technological phrases clustered around an embedding of Microsoft's Bill Gates in the book of Revelations. Although mathematical doubters abound, it also seems reasonable to ask why a superior being would go through such an intricate and convoluted encoding procedure when its discovery would clearly lead to so many interpretations.

EXERCISES FOR SECTION 5.1

1. The grades assigned to the students in an English composition class were as follows:

C, C, B, F, A, A, D, B, C, A, B, C, C, D.

(a) Compute the frequency distribution and relative frequency distribution.

(b) Draw a bar graph representing the frequency distribution.

(c) Draw a bar graph representing the relative frequency distribution.

(d) Draw a pie chart representing the distribution, labeling each sector with its frequency.

2. The Olympic gold medal in women's 100-meter freestyle swimming has been awarded to athletes from only seven different countries as shown here.

1912	Australia	1964	Australia
1920	United States	1968	United States
1924	United States	1972	United States
1928	United States	1976	East Germany
1932	United States	1980	East Germany
1936	Netherlands	1984	United States
1948	Denmark	1988	East Germany
1952	Hungary	1992	China
1956	Australia	1996	China
1960	Australia		

(a) Compute the frequency distribution and relative frequency distribution.

(b) Draw a bar graph representing the frequency distribution.

(c) Draw a bar graph representing the relative frequency distribution.

(d) Draw a pie chart representing the distribution, labeling each sector with its frequency.

3. The number of members on the highest criminal court in the state for each of the 50 states is shown here.

Alabama	9	Louisiana	7	Ohio	7
Alaska	5	Maine	7	Oklahoma	5
Arizona	5	Maryland	7	Oregon	7
Arkansas	7	Massachusetts	7	Pennsylvania	7
California	7	Michigan	7	Rhode Island	5
Colorado	7	Minnesota	7	South Carolina	5
Connecticut	7	Mississippi	9	South Dakota	5
Delaware	5	Missouri	7	Tennessee	5
Florida	7	Montana	7	Texas	9
Georgia	7	Nebraska	7	Utah	5
Hawaii	5	Nevada	5	Vermont	5
Idaho	5	New Hampshire	5	Virginia	7
Illinois	7	New Jersey	7	Washington	9
Indiana	5	New Mexico	5	West Virginia	5
Iowa	9	New York	7	Wisconsin	7
Kansas	7	North Carolina	7	Wyoming	5
Kentucky	7	North Dakota	5		

Source: 1998 Information Please® Almanac.[5]

(a) Compute the frequency distribution and relative frequency distribution.

(b) Draw a bar graph representing the frequency distribution.

(c) Draw a bar graph representing the relative frequency distribution.

(d) Draw a pie chart reresenting the distribution, labeling each sector with its percentage.

(e) Why do you think the data are all odd numbers?

4. The number of children of each of the first 41 presidents of the United States is listed in the table.

Washington	0	Buchanan	0	Harding	0
J. Adams	5	Lincoln	4	Coolidge	2
Jefferson	6	A. Johnson	5	Hoover	2
Madison	0	Grant	4	F. Roosevelt	6
Monroe	2	Hayes	8	Truman	1
J. Q. Adams	4	Garfield	7	Eisenhower	2
Jackson	0	Arthur	3	Kennedy	3
Van Buren	4	Cleveland	5	L. Johnson	2
W. H. Harrison	10	B. Harrison	3	Nixon	2
Tyler	14	Cleveland	given above	Ford	4
Polk	0	McKinley	2	Carter	4
Taylor	6	T. Roosevelt	6	Reagan	4
Fillmore	2	Taft	3	Bush	6
Pierce	3	Wilson	3	Clinton	1

Source: 1998 Information Please® Almanac.[6]

(a) Compute the frequency distribution and relative frequency distribution.

(b) Draw a bar graph representing the frequency distribution.

(c) Draw a bar graph representing the relative frequency distribution.

(d) Draw a pie chart representing the distribution, labeling each sector with its percentage.

5. The natural gas production in 1995 by world region, in billion cubic feet, is given in the following table.

Region	Production (billion cubic feet)
North America	25,177
Central and South America	2,581
Western Europe	8,755
Eastern Europe	25,925
Middle East	4,987
Africa	3,006
Far East and Oceania	7,489

Source: U.S. Energy Information Administration.[7]

(a) Compute the relative frequency distribution.

(b) Draw a bar graph representing the relative frequency distribution.

6. In 1996, General Motors Corporation produced seven different brands of cars. The brands and the numbers produced in 1996 are shown in the table.

Brand	Number Produced
Chevrolet	537,711
Pontiac	541,844
Oldsmobile	300,032
Buick	342,538
Cadillac	162,249
Saturn	313,937
Toyota/Cavalier	11,701
Total	2,210,012

Source: 1998 Information Please® Almanac.[8]

  (a) Compute the relative frequency distribution.

  (b) Draw a bar graph representing the relative frequency distribution.

7. The number of livestock on farms in the United States in 1997 is listed in the table.

Type	Number (*in thousands*)
Cattle	101,460
Swine	59,920
Sheep	7,937
Chickens	403,495
Turkeys	301,378

Source: U.S. Department of Agriculture.[9]

  (a) Draw a bar graph representing the frequency distribution.

  (b) Compute the relative frequency distribution.

  (c) Draw a pie chart, labeling each sector with its percentage.

8. Tourism in 1996 by region of the world, measured by the number of arrivals, is shown in the table.

Region	Number of Arrivals (*in thousands*)
Europe	347,437
Americas	115,511
East Asia/Pacific	90,091
Africa	19,454
Middle East	15,144
South Asia	4,485

Source: World Tourism Organization.

  (a) Draw a bar graph representing the frequency distribution.

  (b) Compute the relative frequency distribution of arrivals by region.

  (c) Draw a pie chart labeled with the percentage in each region.

9. The percentage distribution of means of transportation for getting to work in Charlotte, North Carolina, and New York City in 1990 is recorded in the table.

	Percentage of Total			
	Drove Alone	**Carpool**	**Public Transit**	**Other Means**
Charlotte	80.3	14.8	1.9	3.0
New York	53.5	10.5	28.5	7.5

Source: Federal Highway Administration.[10]

  (a) Find the relative frequency distribution for the Charlotte data.

  (b) Find the relative frequency distribution for the New York data.

  (c) Draw a bar graph representing the relative frequency distribution for the Charlotte data.

  (d) Draw a bar graph representing the relative frequency distribution for the New York data.

10. The marital status of the female population, 18 and older, in the United States in 1970 and in 1995 is shown here.

Female Population (in millions)

Year	Married	Never Married	Widowed	Divorced
1970	47.9	9.6	9.7	2.7
1995	58.9	19.3	11.1	10.3

Source: U.S. Bureau of the Census.[11]

(a) For the year 1970, compute the relative frequency distribution of the female population by marital status.

(b) For the year 1995, compute the relative frequency distribution of the female population by marital status.

(c) Draw a bar graph representing the relative frequency distribution for the 1970 data.

(d) Draw a bar graph representing the relative frequency distribution for the 1995 data.

(e) By comparing the bar graphs for 1970 and 1995, what observations can be made?

11. The distribution of family structure in the United States in 1970 and in 1996 is shown in the table.

	Percent of Total	
Family Type	*1970*	*1996*
Two parents	85	68
Mother only	11	24
Father only	1	4
No parent	3	4

Source: U.S. Bureau of the Census.[12]

(a) Draw a pie chart labeled with percentages for the 1970 data.

(b) Draw a pie chart labeled with percentages for the 1996 data.

(c) By comparing the pie charts for 1970 and 1996, what observations can be made?

12. The relative frequencies of various causes of death in 1970 and in 1993 are given in the table.

	Relative Frequency	
Cause of Death	*1970*	*1993*
Heart disease	0.383	0.326
Cancer	0.172	0.234
Cerebrovascular disease	0.108	0.066
Accidents	0.060	0.039
Pneumonia and influenza	0.033	0.036
Chronic lung disease	0.016	0.045
Other	0.229	0.254

Source: U.S. Bureau of the Census.[13]

(a) Draw a pie chart labeled with percentages for the 1970 data.

(b) Draw a pie chart labeled with percentages for the 1993 data.

(c) What observations can be made from the pie charts about the shifts in the causes of death from 1970 to 1993? What are some possible explanations for the shifts?

13. It is well known that Theodore Roosevelt was the youngest president ever inaugurated and that John F. Kennedy was the youngest ever elected. (Roosevelt was not elected for his first term; he became president on the death of President McKinley.) The ages at inauguration for all U.S. presidents up through Bill Clinton are listed here.

Washington	57	Buchanan	65	Harding	55
J. Adams	61	Lincoln	52	Coolidge	51
Jefferson	57	A. Johnson	56	Hoover	54
Madison	57	Grant	46	F. Roosevelt	51
Monroe	58	Hayes	54	Truman	60
J. Q. Adams	57	Garfield	49	Eisenhower	62
Jackson	61	Arthur	50	Kennedy	43
Van Buren	54	Cleveland	47	L. Johnson	55
W. H. Harrison	68	B. Harrison	55	Nixon	56
Tyler	51	Cleveland	given above	Ford	61
Polk	49	McKinley	54	Carter	52
Taylor	64	T. Roosevelt	42	Reagan	69
Fillmore	50	Taft	51	Bush	64
Pierce	48	Wilson	56	Clinton	46

Source: 1998 Information Please® Almanac.[14]

(a) Group the data into four equally spaced classes, and compute the frequency distribution and relative frequency distribution based on these classes.

(b) Draw a bar graph representing the frequency distribution.

(c) Draw a pie chart representing the distribution, labeling each sector with its percentage.

14. A 1993 *Consumer Reports* survey[15] of leading brands of chicken noodle soup reveals the following calorie counts per 8-ounce serving: 105, 60, 75, 60, 170, 80, 80, 190, 200, 100, 60, 60, 190, 60, 60, 110, 105, 110, 70, 105, 65, 120, 80, 80, 125, 95.

(a) Group the data into four equally spaced classes, and compute the frequency distribution and relative frequency distribution based on these classes.

(b) Draw a bar graph representing the relative frequency distribution.

(c) Draw a pie chart representing the distribution, labeling each sector with its percentage.

15. According to *The Survey of Buying Power Data Service,*[16] the amounts of retail sales in each state and Washington, D.C., in dollars per household, in 1993 were 25,407; 30,056; 23,382; 21,505; 19,970; 23,621; 19,079; 23,626; 21,007; 21,970; 21,346; 22,274; 22,624; 22,140; 23,020; 22,324; 22,154; 24,294; 23,330; 21,424; 21,861; 26,263; 22,091; 15,839; 24,096; 16,484; 20,621; 20,955; 21,250; 23,648; 19,267; 21,399; 18,911; 16,306; 18,295; 20,457; 16,740; 22,076; 22,354; 20,256; 20,098; 23,437; 21,099; 22,441; 22,995; 22,643; 21,332; 23,336; 21,515; 26,088; 34,454.

(a) Group the dollar amounts into five equally spaced classes, and compute the frequency and relative frequency distributions based on these classes.

(b) Draw a bar graph representing the relative frequency distribution.

(c) Draw a pie chart representing the distribution, labeling each sector with its frequency.

16. Many complain that Super Bowls are too often blow-outs. The final scores of Super Bowls I through XXXIII are shown in the table.

I	Green Bay 35	Kansas City 10	XVIII	L.A. Raiders 38	Washington 9
II	Green Bay 33	Oakland 14	XIX	San Francisco 38	Miami 16
III	N.Y. Jets 16	Baltimore 7	XX	Chicago 46	New England 10
IV	Kansas City 23	Minnesota 7	XXI	N.Y. Giants 39	Denver 20
V	Baltimore 16	Dallas 13	XXII	Washington 42	Denver 10
VI	Dallas 24	Miami 3	XXIII	San Francisco 20	Cincinnati 16
VII	Miami 14	Washington 7	XXIV	San Francisco 55	Denver 10
VIII	Miami 24	Minnesota 7	XXV	N.Y. Giants 20	Buffalo 19
IX	Pittsburgh 16	Minnesota 6	XXVI	Washington 37	Buffalo 24
X	Pittsburgh 21	Dallas 17	XXVII	Dallas 52	Buffalo 17
XI	Oakland 32	Minnesota 14	XXVIII	Dallas 30	Buffalo 13
XII	Dallas 27	Denver 10	XXIX	San Francisco 49	San Diego 26
XIII	Pittsburgh 35	Dallas 31	XXX	Dallas 27	Pittsburgh 17
XIV	Pittsburgh 31	L.A. Rams 19	XXXI	Green Bay 35	New England 21
XV	Oakland 27	Philadelphia 10	XXXII	Denver 31	Green Bay 24
XVI	San Francisco 26	Cincinnati 21	XXXIII	Denver 34	Atlanta 19
XVII	Washington 27	Miami 17			

    (a) Group the games into five equally spaced classes based on the margin of victory (the winning score minus the losing score), and compute the frequency distribution and relative frequency distribution based on these classes.

    (b) Draw a bar graph representing the frequency distribution.

    (c) Based on your bar graph, is the complaint well founded that too many Super Bowls are blow-outs?

17. The heights, in inches rounded to the nearest inch, of the male students in a college class are 72, 72, 70, 72, 71, 69, 67, 68, 70, 71, 69, 71, 68, 70, 67, 72. Draw a histogram representing the relative frequency distribution with

    (a) classes of width 1 and the first interval starting at 66.5,

    (b) classes of width 2 and the first interval starting at 66.5.

18. According to the *Office Market Data Book*, the vacancy rates in percent for office buildings in a selection of major U.S. cities in 1994 were 13.0, 15.5, 13.3, 10.0, 18.7, 21.7, 12.8, 19.7, 24.7, 18.4, 11.5, 19.6, 15.4, 17.6, 16.3, 16.3, 15.8, 18.8, 11.7, 14.7, 18.1, 13.4. Draw a histogram representing the frequency distribution with

    (a) classes of width 3 and the first interval starting at 9.75,

    (b) classes of width 5 and the first interval starting at 9.75.

19. A 1992 sampling of different brands of toothpaste by *Consumer Reports*[17] yielded the following set of weights in ounces for the "regular" size: 6.0, 7.0, 5.0, 6.4, 6.4, 6.4, 3.0, 6.4, 6.0, 6.4, 6.4, 6.4, 6.4, 6.0, 6.4, 6.4, 6.4, 6.0, 6.0, 6.4, 6.4, 6.4, 4.5, 3.0, 6.3, 4.6, 6.4, 6.4. Draw a histogram representing the frequency distribution with classes of width 1 and the first interval starting at 2.75.

20. The home run leaders in Major League Baseball from 1970 to 1998 and the number of home runs they hit are shown in the table. Draw a histogram representing the relative frequency distribution with classes of width 6 and the first interval starting at 30.5.

Player	Year	Home Runs	Player	Year	Home Runs
Johnny Bench	1970	45	Darrell Evans	1985	40
Willie Stargell	1971	48	Jesse Barfield	1986	40
Johnny Bench	1972	40	Andre Dawson/Mark McGwire	1987	49
Willie Stargell	1973	44	Jose Canseco	1988	42
Mike Schmidt	1974	36	Kevin Mitchell	1989	47
Mike Schmidt	1975	38	Cecil Fielder	1990	51
Mike Schmidt	1976	38	Cecil Fielder/Jose Canseco	1991	44
George Foster	1977	52	Juan Gonzalez	1992	43
Jim Rice	1978	46	Barry Bonds/Juan Gonzalez	1993	46
Dave Kingman	1979	48	Matt Williams	1994	43
Mike Schmidt	1980	48	Albert Belle	1995	50
Mike Schmidt	1981	31	Mark McGwire	1996	52
Gorman Thomas	1982	39	Mark McGwire	1997	58
Mike Schmidt	1983	40	Mark McGwire	1998	70
Tony Armas	1984	43			

21. The 1990 unemployment rates, in percent, in the 50 states and the District of Columbia are shown in the table. Draw a histogram representing the relative frequency distribution with classes of width 2. Choose the starting point so that no data values fall on the endpoints of the intervals.

Alabama	6.8	Kentucky	5.8	North Dakota	3.9
Alaska	6.9	Louisiana	6.2	Ohio	5.7
Arizona	5.3	Maine	5.1	Oklahoma	5.6
Arkansas	6.9	Maryland	4.6	Oregon	5.5
California	5.6	Massachusetts	6.0	Pennsylvania	5.4
Colorado	4.9	Michigan	7.1	Rhode Island	6.7
Connecticut	5.1	Minnesota	4.8	South Carolina	4.7
Delaware	5.1	Mississippi	7.5	South Dakota	3.7
Dist. of Columbia	6.6	Missouri	5.7	Tennessee	5.2
Florida	5.9	Montana	5.8	Texas	6.2
Georgia	5.4	Nebraska	2.2	Utah	4.3
Hawaii	2.8	Nevada	4.9	Vermont	5.0
Idaho	5.8	New Hampshire	5.6	Virginia	4.3
Illinois	6.2	New Jersey	5.0	Washington	4.9
Indiana	5.3	New Mexico	6.3	West Virginia	8.3
Iowa	4.2	New York	5.2	Wisconsin	4.4
Kansas	4.4	North Carolina	4.1	Wyoming	5.4

Source: U.S. Bureau of the Census.[18]

22. The grade point averages, rounded to the nearest tenth, of the students in a class are as follows: 3.4, 2.3, 2.7, 3.1, 2.9, 3.4, 3.8, 2.5, 2.5, 3.6, 3.1, 3.4, 2.7, 3.3, 3.4, 2.8, 2.3, 4.0, 3.6. Draw a histogram representing the relative frequency distribution with classes of width 0.5. Choose the starting point so that no data values fall on the endpoints of the intervals.

5.2   TYPICAL AND CENTRAL VALUES

In Section 5.1, we looked at different ways to illustrate a given set of data. A loss of information occurred only if we grouped the data into classes. One way to briefly summarize the data is to give a typical or central value. By summarizing a large quantity of data into one value, much information is lost. However, the simplicity of a single value that gives some measure of the data can be very helpful and informative. For instance, a student's grade point average provides potential employers or graduate programs with a concise measure of the student's overall academic record. The average monthly electric bill for a house over the period of a year is helpful in budgeting. Some measure of the typical selling price of the homes in a neighborhood may provide guidance to a real estate agent in setting the price of a home. The batting average of a baseball player gives an opposing team some knowledge of his ability. This might be useful in deciding whether or not to walk the player in a crucial game situation.

In this section, we consider three different measures of typical or central values: the mode, the median, and the mean.

The Mode

We first consider the value, class, or category that occurs most often, called the mode. This can be thought of as a typical value. The advantage of the mode is that it is defined for categorical data, such as colors of automobiles, as well as for numerical data. When data are numerical, we can use the mode, but two other numerical measures called the median and mean are generally more representative of a central value for the distribution.

> The **mode** of a set is the most frequently occurring category or categories.

A distribution may have more than one mode. A distribution with a single mode is called **unimodal**, and one with exactly two modes is called **bimodal**. A distribution may have more than two modes, although if all the categories occur with the same frequency, we often say the distribution has no mode.

Example 1   *Finding the Mode*

In the first example of Section 5.1 we looked at the following distribution of grades from a college calculus course:

F, B, B, A, C, A, A, C, B, C, C, B, C, D, C, B, C, C, F.

Find the mode of these data.

Solution: The frequency distribution for the data is shown in Table 5.14.

Table 5.14	Grade Distribution
Grade	**Frequency**
A	3
B	5
C	8
D	1
F	2

The mode is C, because it is the most frequently occurring grade. Note that the mode corresponds to the tallest bar in a bar graph of the data shown in Figure 5.10.

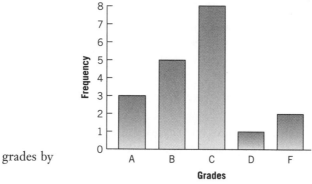

Figure 5.10
Bar graph of grades by frequency.

Example 2   *A Bimodal Distribution*

The number of college football bowl games on each day from December 19, 1996 through January 2, 1997 were 1, 0, 0, 0, 0, 0, 1, 0, 3, 1, 1, 1, 3, 6, 1. Find the mode of these data.

Solution: We compute the frequency distribution shown in Table 5.15.

Table 5.15	Bowl Games per Day
Number of Games	**Frequency**
0	6
1	6
3	2
6	1

We see that 0 and 1 both occur 6 times, so the distribution is bimodal with modes 0 and 1. The bar graph for the data is shown in Figure 5.11, where we see that the bars for 0 and 1 are the tallest.

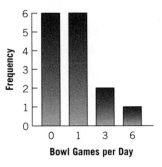

Figure 5.11
Bar graph of bowl
games per day.

The Median

The **median** of a set of numerical data is the center of the data in the sense that a data value is equally likely to lie above or below it. To find the median, we first order the data from the smallest to largest value. It is possible that some values may be repeated. If that is the case, these values must be listed as many times as they occur. If the number of pieces of data is odd (of the form $2n + 1$), then the median will be the middle value in the ordered list. If the number of pieces of data is even (of the form $2n$), then the median will be the average of the two values in the middle of the ordered list. A more formal definition for the median of an ordered list of data is as follows:

The **median** of a nondecreasing list of $2n + 1$ numbers is the middle number, the $(n + 1)$st number.

The **median** of a nondecreasing list of $2n$ numbers is the number halfway between, or "average" of, the middle pair of numbers, the nth and $(n + 1)$st numbers.

(Note that an easy way to compute $n + 1$ from an odd number of the form $2n + 1$ is to add 1 and then divide by 2.)

The next two examples illustrate how we compute the median of a set of data.

Example 3   *Median of an Odd Number of Data Values*

The approximate land area of each of the seven continents is shown in Table 5.16. Find the median land area.

Table 5.16	**Land Areas of the Continents**
Continent	**Approximate Land Area** (in thousand square miles)
Africa	11,724
Antarctica	5,500
Asia	17,226
Australia	2,966
Europe	4,000
North America	9,355
South America	6,878

Source: Encyclopedia Britannica Online.

Solution: In increasing order, the land areas of the continents are 2966, 4000, 5500, 6878, 9355, 11,724, 17,226. The median is the middle of the seven numbers, or the fourth number in our list, which is 6878 thousand square miles.

■

Example 4   *Median of an Even Number of Data Values*

The U.S. National Oceanic and Atmospheric Administration reported that the numbers of lives lost in floods in the United States in the years 1986 through 1993 were 80, 82, 29, 81, 147, 63, 87, and 101. Find the median number of annual fatalities due to floods over this period of years.

Solution: Written in increasing order, the lives lost in floods were 29, 63, 80, 81, 82, 87, 101, 147. We have an even number of pieces of data, 8. Therefore, we look at the middle numbers, 81 and 82, those in the fourth and fifth positions, and the median is their average: $(81 + 82)/2 = 81.5$.

■

Sometimes you may want to find the mode or median when you are already given the frequency distribution. The mode will simply be the value with the highest frequency. To find the median, you must find the total number of values, and then start counting from the smallest until you reach the middle value or values. Example 5 illustrates these ideas.

Example 5   *Finding the Mode and Median Given the Frequency Distribution*

Among each of the 50 states, the number of women holding statewide, elective executive office in 1995 is given by the frequency distribution in Table 5.17.

Table 5.17 Women Holding Statewide Office	
Number of Women	*Frequency*
0	8
1	17
2	14
3	6
4	4
5	1

Source: Center for the American Woman and Politics.[19]

(The one state with five women officeholders was North Dakota.)

(a) Find the mode number of women in statewide, elective executive positions.

(b) Find the median number of women in statewide, elective executive positions.

Solution:
(a) The value 1 is the mode because it has the highest frequency.
(b) Because there are 50 states, the median will be halfway between the values for the 25th and 26th states. There are $8 + 17 = 25$ states with at most one female executive, so that the 25th state has one, and the 26th state has two. Thus the median is $(1 + 2)/2 = 1.5$.

■

You may have heard the term *50th percentile* in the context of standardized test scores. The 50th percentile is just another term for the median. At least half of the data lie at or above the 50th percentile and at least half of the data lie at or below the 50th percentile. It is the midpoint of the data.

The Mean

If you hear someone speaking of the "average" of a set of values, they are most likely referring to the mean. For instance, to compute the average monthly electric bill for a house over a period of a year, you would add all of the monthly bills and then divide by 12. When setting up an annual budget, this average or mean would tell you how much to budget per month for electricity. To write a formula for the mean of a set of data, we let x_i stand for the ith data value. We often abbreviate the sum $x_1 + x_2 + \cdots + x_n$ with the shorthand $\sum_{i=1}^{n} x_i$. The capital Greek letter Σ, called sigma, stands for "sum" or "add." This notation means that we are to sum x_i from $i = 1$ up through $i = n$. We let the Greek letter μ, called mu, stand for mean.

> The **mean** μ of a set of values is the sum of the values divided by the number of values. If $x_1, x_2, \ldots, x_n$ denote the values, then the mean is
> $$\mu = \frac{\sum_{i=1}^{n} x_i}{n} = \frac{x_1 + x_2 + \cdots + x_n}{n}.$$

The values need not be in increasing order to use the formula for the mean.

Example 6   Computing the Mean

In Example 4, we saw that the number of lives lost in floods in the United States in the years 1986 through 1993 were 80, 82, 29, 81, 147, 63, 87, and 101. Find the mean number of annual fatalities due to floods over this period of years.

Solution: There are 8 years worth of data. The mean is
$$\mu = \frac{80 + 82 + 29 + 81 + 147 + 63 + 87 + 101}{8} = \frac{670}{8} = 83.75.$$

■

In Example 4, we saw that the median number of annual fatalities due to floods was 81.5, which is close to, but not the same as, the mean of 83.75. The median and mean are often close, but in some instances they vary greatly. Is the median or the mean the better representative value of a set of data? It depends on the use. Suppose we are considering the incomes of a group of people. If we are talking about pooling the incomes, as in the case of a family, then the mean income is most relevant. If we are talking about providing government services to a small town, the median income is probably more relevant because a few individuals with extraordinarily high incomes can raise the mean but not the median, and their wealth may be largely irrelevant to the rest of the town. In general, the question is whether a relatively small number of very large or very small values are relevant to the typical value. If so, the mean may be more appropriate. If not, the median may be more appropriate.

If a data set is presented in terms of a frequency distribution, our calculations can be condensed. In this case, let f_i denote the frequency of the value x_i. Then the total number of pieces of data is $\sum_{i=1}^{n} f_i$, and the sum of all the data is given by the sum of each x_i multiplied by f_i. Therefore, in this case, we have the following formula for computing the mean:

Mean of Data Presented by a Frequency Distribution

The **mean** μ of the values $x_1, x_2, \ldots, x_n$ occurring with frequencies $f_1, f_2, \ldots, f_n$, respectively, is given by

$$\mu = \frac{\sum\limits_{i=1}^{n} f_i x_i}{\sum\limits_{i=1}^{n} f_i} = \frac{f_1 x_1 + f_2 x_2 + \cdots + f_n x_n}{f_1 + f_2 + \cdots + f_n}.$$

Notice that in this formula for the mean, the x_i's represent all the different values that occur rather than individual values as in our original formula for the mean.

Example 7   *Finding the Mean Given the Frequency Distribution*

As we saw in Example 5, among each of the 50 states, the number of women holding statewide, elective executive office in 1995 is given by the frequency distribution in Table 5.18.

Table 5.18	Women Holding Statewide Office
Number of Women	*Frequency*
0	8
1	17
2	14
3	6
4	4
5	1

Source: Center for the American Woman and Politics.[20]

Find the mean number of women in statewide, elective executive positions.

Solution: Because the value 0 occurs 8 times, 1 occurs 17 times, and so forth, using the formula for the mean of data presented by a frequency distribution, we have

$$\mu = \frac{8 \cdot 0 + 17 \cdot 1 + 14 \cdot 2 + 6 \cdot 3 + 4 \cdot 4 + 1 \cdot 5}{8 + 17 + 14 + 6 + 4 + 1} = \frac{84}{50} = 1.68.$$

By comparison, in Example 5 we found that the mode was 1 and the median was 1.5.

Estimating the Mean of Grouped Data

When the data are already grouped into intervals, we may not be able to compute the mean exactly. However, if we pick a representative value for each class, usually the midpoint of the interval, we can get a reasonable estimate for the mean. We find the midpoint of an interval by averaging its endpoints. Occasionally a class will not be over a specific, finite interval—for instance, a class with a label such as "greater than 100." In this kind of case, we have to make our best guess of a representative value for the class.

Example 8   *Estimating the Mean*

The percentages of residents 65 and older of the 50 states and the District of Columbia in 1994 are given by the frequency distribution in Table 5.19.

Table 5.19	**Percentage of Residents 65 and Older**
Residents 65 and Over (percent)	*Frequency*
4–4.9	1
8–8.9	1
10–10.9	4
11–11.9	10
12–12.9	11
13–13.9	12
14–14.9	7
15–15.9	4
18–18.9	1

Source: U.S. Bureau of the Census.[21]

Estimate the mean of the distribution. (Can you guess the states with percentages in the 4–4.9 and 18–18.9 classes?)

Solution: For the first interval, 4–4.9, we use the representative value $(4 + 4.9)/2 = 4.45$. The representative value for the interval 8–8.9 is $(8 + 8.9)/2 = 8.45$. Similarly, we compute the remaining representative values as shown in Table 5.20.

Table 5.20	Percentage of Residents 65 and Older		
Residents 65 and Over (percent)	**Representative Value**	**Frequency**
4–4.9	4.45	1
8–8.9	8.45	1
10–10.9	10.45	4
11–11.9	11.45	10
12–12.9	12.45	11
13–13.9	13.45	12
14–14.9	14.45	7
15–15.9	15.45	4
18–18.9	18.45	1

Now using our estimated values and the formula for the mean of data presented by a frequency distribution, our estimate for the mean is

$$\mu \approx \frac{1 \cdot 4.45 + 1 \cdot 8.45 + 4 \cdot 10.45 + 10 \cdot 11.45 + 11 \cdot 12.45 + 12 \cdot 13.45 + 7 \cdot 14.45 + 4 \cdot 15.45 + 1 \cdot 18.45}{1 + 1 + 4 + 10 + 11 + 12 + 7 + 4 + 1}$$

$$= \frac{648.95}{51} \approx 12.72.$$

(In 1994, only 4.6% of Alaska's residents were age 65 and over, whereas 18.4% of Florida's were.)

■

Technology Tip

All but the most basic calculator can calculate the mean of a list of data. Most use the notation $\bar{x}$ for the mean instead of μ. More advanced ones can compute medians as well as means and medians from a frequency or relative frequency distribution. Spreadsheets and statistical software nearly always have such features built in.

EXERCISES FOR SECTION 5.2

1. Find the mode of the following grades, which were assigned to the students in an English composition class: C, C, B, F, A, A, D, B, C, A, B, C, C, D.

2. Find the mode of the letters that appear in the following quote by President Franklin D. Roosevelt at his 1933 inaugural address: "The only thing we have to fear is fear itself."

3. The number of wins by the 30 different National Football League Teams in the 1996 season were 10, 10, 9, 7, 6, 13, 9, 7, 6, 5, 12, 12, 6, 3, 3, 11, 10, 9, 8, 1, 10, 9, 8, 8, 4, 13, 9, 8, 7, 7.

   (a) Find the mode number of wins.   (b) Find the median number of wins.

   (c) Find the mean number of wins.

4. A 1994 *Consumer Reports* survey[22] found that the rates in dollars for a single room at a selection of budget hotels were 42, 40, 40, 40, 40, 50, 38, 48, 45, 40, 32, 45, 49, 34.

    (a) Find the mode of prices.   **(b)** Find the median price.

    (c) Find the mean price.

5. The monthly normal daily high temperatures for Chicago, Illinois, in degrees Fahrenheit, are as shown in the table.

Jan.	Feb.	Mar.	Apr.	May	June	July	Aug.	Sept.	Oct.	Nov.	Dec.
29.0	33.5	45.8	58.6	70.1	79.6	83.7	81.8	74.8	63.3	48.4	34.0

Source: U.S. National Oceanic and Atmospheric Administration.[23]

    (a) Find the median of these temperatures.

    (b) Find the mean of these temperatures.

6. The average attendance, in thousands, at NCAA college football games for the 1987 to 1993 seasons, given in order, was 35,008; 34,324; 35,116; 35,330; 35,528; 35,525; 34,871.

    (a) Find the median attendance.   **(b)** Find the mean attendance.

7. According to the U.S. Bureau of the Census, from 1985 through 1995, the percentages of students in grades 10 through 12 who dropped out in a single year in the United States were 5.2, 4.3, 4.1, 4.8, 4.5, 4.0, 4.0, 4.3, 4.2, 5.0, 5.4 given in order by years.

    (a) Find the median dropout rate.   **(b)** Find the mean dropout rate.

8. The number of persons viewing prime-time television during an average minute for each weekday in May 1997 is shown in the table.

Day of Week	Total Persons Viewing (in millions)
Monday	95.3
Tuesday	94.9
Wednesday	93.2
Thursday	93.7
Friday	80.9
Saturday	78.9
Sunday	91.6

Source: Nielsen Media Research.

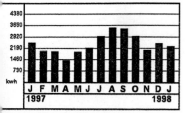

Your electricity use

The average daily high temperature during this billing period was 54°; average daily high temperature during the same billing period last year was 55°.

Your average daily electrical cost was $4.73.

P0030062   012198   00000000000

TUELECTRIC

    (a) Find the median viewing audience over an entire week.

    (b) Find the mean viewing audience over an entire week.

9. Suppose five homes sold in a particular neighborhood one year at the following prices: $123,900; $95,200; $98,620; $210,000; $105,250.

    (a) Find the median sale price.   **(b)** Find the mean sale price.

    (c) How representative of the data do the median and mean values appear to be?

10. The bar chart from a Texas Utilities Electric bill shows the kilowatt-hours used by one household over the course of a year. The bills for each of the months January 1997 through December 1997 for the household were $170.72, $140.25, $137.50, $101.41, $134.74, $188.60, $249.43, $238.84, $287.73, $250.35, $141.62, and $168.54.

    (a) Find the median electric bill.   **(b)** Find the mean electric bill.

    (c) Is the mean or the median a better number to use when budgeting the amount of money spent on electricity?

11. The U.S. National Center for Health Statistics reported that the birth rates per 1,000 women between 15 and 44 years old in the United States from 1983 through 1992 were 65.7, 65.5, 66.3, 65.4, 65.8, 67.3, 69.2, 70.9, 69.6, 68.9, given in order by years.

    (a) Find the median birth rate.   (b) Find the mean birth rate.

    (c) Do you see any general trend in the data?

12. In a 1993 *Consumer Reports* survey[24] of frozen lasagna, the number of milligrams of cholesterol per serving in several brands were 70, 45, 77, 20, 40, 5, 15, 46, 15, 25, 20, 30, 55, 20, 48, 25.

    (a) Find the median cholesterol level.   (b) Find the mean cholesterol level.

    (c) Why do you think the mean is higher than the median for these data?

13. The percentage of the voting-age population turning out to vote in each of the federal elections from 1980 through 1996 is shown in the table.

Year	1980	1982	1984	1986	1988	1990	1992	1994	1996
Percentage Turnout	52.6	39.8	53.1	36.4	50.1	36.5	55.1	38.8	49.1

Source: Federal Election Commission.[25]

    (a) Find the mean percentage turnout over this time period.

    (b) Find the mean percentage turnout over the years 1980, 1984, 1988, 1992, and 1996.

    (c) Find the mean percentage turnout over the years 1982, 1986, 1990, and 1994.

    (d) Explain why the mean you found in part (b) is so different from the one found in part (c).

14. The average starting salary for those with a bachelor's degree in eight different branches of engineering in 1990 is as shown in the table.

Branch	*Average Starting Salary (dollars)*
Civil	28,136
Chemical	35,122
Computer	31,490
Electrical	31,778
Mechanical	32,064
Nuclear	31,750
Petroleum	35,202
Engineering technology	29,318

Source: Reprinted from the *Salary Survey*, with permission of the National Association of Colleges and Employers, copyright holder.

    (a) Find the median salary.   (b) Find the mean salary.

    (c) Will these necessarily be the median and mean rates for the starting salary of all engineers with a bachelor's degree? Explain.

15. A survey of several brands of raisin bran cereals listed the following prices per ounce for the cereals.

Price per Ounce (cents)	10	11	12	13	14	16	22	23	25
Frequency	4	4	1	1	3	2	1	1	1

Source: Consumer Reports, October 1996.[26]

    (a) Find the mode price per ounce.   (b) Find the median price per ounce.

    (c) Find the mean price per ounce.

16. The frequency distribution of Nolan's golf scores on par 3 holes is given here.

Score	3	4	5	6	7	8	9	10
Frequency	3	12	23	35	9	5	1	2

(a) Find the mode score.   **(b)** Find the median score.
(c) Find the mean score.

17. The frequency distribution of the minimum age, rounded to the nearest year, for receiving an unrestricted driver's license in each of the 50 states and the District of Columbia is shown in the table.

Minimum Age (years)	15	16	17	18
Frequency	3	28	12	8

Source: Insurance Institute for Highway Safety.[27]

(a) Find the mode minimum driving age.
(b) Find the median minimum driving age.
(c) Find the mean minimum driving age.

18. The number of triples for each of the 19 nonpitcher players on the 1995 World Series Champion Atlanta Braves is given in the following frequency table.

Number of Triples	0	1	2	3	4	5
Frequency	8	3	4	2	1	1

(a) Find the mode number of triples.   **(b)** Find the median number of triples.
(c) Find the mean number of triples.

19. Thirty of the states with the death penalty allow it to be applied only if the criminal was above a certain legally specified minimum age at the time the crime was committed. The frequency distribution of minimum ages for the death penalty in these states is shown in the table.

Minimum Age (years)	14	16	17	18
Frequency	2	10	4	14

Source: U.S. Bureau of Justice.[28]

(a) Find the mode of the minimum age.
(b) Find the median of the minimum age.
(c) Find the mean of the minimum age.

20. The members of the Supreme Court of the United States are nominated for life terms by the president. The distribution of the number of years of service, rounded to the nearest integer, of all Supreme Court justices who retired before 1998 is shown in the table. Using this information, estimate the mean years of service.

Years of Service	0–5	6–10	11–15	16–20	21–25	26–30	31–35	36–40
Frequency	21	17	15	16	10	11	8	1

Source: 1998 Information Please® Almanac.[29]

21. The energy crisis of the early 1970s prompted legislation aimed at improved fuel economy. One way to achieve such goals is to lighten vehicles. The table shows the number of cars sold, by weight, in the United States in 1975 and 1990.

Weight Class (pounds)	Number of Cars (in thousands)	
	1975	1990
1625–1875	0	1
1875–2125	105	109
2125–2375	375	107
2375–2625	406	1183
2625–2875	281	999
2875–3250	828	3071
3250–3750	1029	2877
3750–4250	1089	1217
4250–4750	1791	71
4750–5250	1505	0
5250–5750	828	1

Source: U.S. Environmental Protection Agency.[30]

(a) Estimate the mean weight of cars sold in 1975.

(b) Estimate the mean weight of cars sold in 1990.

(c) Do the estimated mean weights indicate a significant change from 1975 to 1990 in the weight of vehicles?

22. The projected age breakdown of the population of the United States in the year 2025 is given in the table.

Age (years)	Frequency (in thousands)
Under 5	22,498
5–13	40,413
14–17	17,872
18–24	30,372
25–34	43,119
35–44	42,391
45–54	36,890
55–64	39,542
65–74	35,425
75–84	19,481
85 and over	7,046

Source: U.S. Bureau of the Census.[31]

(a) Estimate the mean age of the projected population using 2 for the representative "Under 5" age and 85 for the representative "85 and over" age.

(b) Estimate the mean age of the projected population using 2 for the representative "Under 5" age and 90 for the representative "85 and over" age.

(c) Estimate the mean age of the projected population using 2 for the representative "Under 5" age and 95 for the representative "85 and over" age.

(d) What age would you choose to represent the "85 and over" category? Explain.

(e) How critical is it to have an accurate representative for the "85 and over" category? Explain.

23. The age breakdown of murder victims in the United States in 1995 is given in the table.

Age (years)	Number
0–12	866
13–19	3069
20–24	3559
25–29	2814
30–34	2526
35–39	1966
40–44	1517
45–49	993
50–54	645
55–74	1337
75 and over	414

Source: U.S. Federal Bureau of Investigation.[32]

(a) Estimate the mean age of murder victims using 75 for the representative "75 and over" age.

(b) Estimate the mean age of murder victims using 80 for the representative "75 and over" age.

(c) Estimate the mean age of murder victims using 85 for the representative "75 and over" age.

(d) What age would you choose to represent the "75 and over" category? Explain.

(e) How critical is it to have an accurate representative for the "75 and over" category? Explain.

24. The ages and average stays of hospital patients in the United States in 1994 are summarized in the table.

Age (years)	Number of Patients (in thousands)	Average Stay (days)
Under 1	730	5.8
1–4	683	3.4
5–14	836	5.1
15–24	3154	3.6
25–34	4377	3.8
35–44	3425	5.1
45–64	6311	5.9
65–74	4894	6.9
75 and over	6433	7.7

Source: U.S. Bureau of the Census.[33]

(a) Estimate the mean stay of a hospital patient.

(b) Estimate the mean age of a hospital patient.

25. In Garrison Keillor's fictional town of Lake Wobegon, "all the women are strong, all the men are good looking, and all the children are above average." Although it is not possible that *all* of the children in a population are above the average of that population, if the mean is used as the measure of "average," then it is possible that most of them could be. To illustrate this, construct an example of a class where each student has an exam score between 0 and 100, and where at least 90% of the students have an above average score.

5.3   MEASURES OF SPREAD

Suppose the exam scores for a small class are 75, 79, 80, 73, 83, and 81. Then the mean of the exam scores is 78.5. If instead, the exam scores had been 46, 100, 98, 94, 92, and 41, the mean would also have been 78.5. However, the numbers are distributed very differently, the latter set being much more spread out. Whenever we condense the description of a set of data into one or two numbers, it is inevitable that we lose information. However, by finding a number representing the spread of a distribution as well as a number that represents the center of the distribution, such as the median or mean, we may more accurately convey the nature of the data. In this section we learn to compute two different measures of spread: the range and the standard deviation.

The Range

The range of a distribution measures the largest difference in values of a distribution.

> The **range** of a set of data is the largest value of the set minus the smallest value of the set.

Example 1   *The Range*

According to the U.S. National Oceanic and Atmospheric Administration, the numbers of tornadoes in the United States in each of the years from 1984 to 1995 were 907, 684, 764, 656, 702, 856, 1133, 1132, 1303, 1173, 1082, 1235. Find the range of these data.

Solution: The smallest and largest of these values are 656 and 1303, respectively. Thus the range is $1303 - 656 = 647$. (Can you think of reasons the numbers could have gone up over the 10-year period?)

The Standard Deviation

Where the range measures the *maximum* spread of the distribution, the standard deviation gives a *representative* spread for a distribution. It is the most common measure of spread and is generally used when the mean is used as a measure of the center of the distribution. You may have had a professor announce the mean and standard deviation for the exam scores in a class.

In measuring the spread of a distribution, we begin by computing the quantity $x_i - \mu$ for each i, which is called the **deviation from the mean**. If we either add or average the deviation from the mean over all i, we get 0. (Convince yourself of this.) A reasonable choice for a representative spread would be the average of the distance from the mean. This can be expressed algebraically as the average of the absolute value of the deviation from the mean. However, there is a better choice, the standard deviation, which we will now define. You might have learned in other math classes that the distance between points (x_1, y_1) and (x_2, y_2) is given by

$$\sqrt{(x_1 - x_2)^2 + (y_1 - y_2)^2}.$$

The standard deviation is somewhat analogous to an n-dimensional distance formula, with each data point corresponding to a dimension. Because of this, the standard deviation has many desirable properties.

We use the standard notation, a lowercase Greek letter σ, called sigma, to denote the standard deviation.

> The **standard deviation**, denoted σ, is the square root of the mean of the squares of the deviations from the mean. If the set of values is $x_1, x_2, \dots, x_n$, then σ is given by
>
> $$\sigma = \sqrt{\frac{\sum_{i=1}^{n}(x_i - \mu)^2}{n}} = \sqrt{\frac{(x_1 - \mu)^2 + (x_2 - \mu)^2 + \cdots + (x_n - \mu)^2}{n}}.$$

Note that the standard deviation can never exceed the range. In fact, it cannot exceed half the range. For accuracy, when computing the standard deviation, keep all of the decimals on your calculator for quantities such as the mean that come up in the middle of the computation.

In the next two examples we compute the standard deviations for the exam scores we considered at the beginning of this section.

Example 2   *Standard Deviation*

Suppose the exam scores for a small class are 75, 79, 80, 73, 83, and 81. Find the mean and standard deviation of the exam scores.

Solution: We begin by computing the mean. It is

$$\mu = \frac{75 + 79 + 80 + 73 + 83 + 81}{6} = \frac{471}{6} = 78.5.$$

Using the formula, we find that the standard deviation is

$$\sigma = \sqrt{\frac{(75 - 78.5)^2 + (79 - 78.5)^2 + (80 - 78.5)^2 + (73 - 78.5)^2 + (83 - 78.5)^2 + (81 - 78.5)^2}{6}}$$

$$= \sqrt{\frac{(-3.5)^2 + (0.5)^2 + (1.5)^2 + (-5.5)^2 + (4.5)^2 + (2.5)^2}{6}} = \sqrt{\frac{71.5}{6}} \approx \sqrt{11.916667} \approx 3.45.$$

■

Example 3   *Comparing Standard Deviations*

Suppose the exam scores for a class were 46, 100, 98, 94, 92, and 41. Find the mean and the standard deviation of the exam scores. How do they compare to the mean and standard deviation of the exam scores in Example 2?

Solution: The mean of the exam scores is

$$\mu = \frac{46 + 100 + 98 + 94 + 92 + 41}{6} = \frac{471}{6} = 78.5.$$

Using the formula, we find that the standard deviation is

$$\sigma = \sqrt{\frac{(46 - 78.5)^2 + (100 - 78.5)^2 + (98 - 78.5)^2 + (94 - 78.5)^2 + (92 - 78.5)^2 + (41 - 78.5)^2}{6}}$$

$$= \sqrt{\frac{(-32.5)^2 + (21.5)^2 + (19.5)^2 + (15.5)^2 + (13.5)^2 + (-37.5)^2}{6}}$$

$$= \sqrt{\frac{3727.5}{6}} = \sqrt{621.25} \approx 24.92.$$

Comparing the solutions to Examples 2 and 3, we see that although the six exam scores in each case had a mean of 78.5, the standard deviations were quite different. As expected, the standard deviation for the exam scores that were more spread out had the larger standard deviation. ∎

Technology Tip

Most software, spreadsheets, and calculators that have built-in capabilities to calculate the mean can also calculate standard deviation automatically.

We sometimes want to compute the standard deviation in a situation in which the data are given by a frequency distribution. Suppose our data are the set of values $x_1, x_2, \ldots, x_n$, with frequencies $f_1, f_2, \ldots, f_n$. Then the total number of pieces of data is the sum of the frequencies, and the sum of the squares of the deviations from the mean is given by

$$\sum_{i=1}^{n} f_i (x_i - \mu)^2.$$

Therefore, the formula for the standard deviation can be written as follows:

Standard Deviation of Data Presented by a Frequency Distribution

The **standard deviation** σ of the values $x_1, x_2, \ldots, x_n$ occurring with frequencies $f_1, f_2, \ldots, f_n$, respectively, is given by

$$\sigma = \sqrt{\frac{\sum_{i=1}^{n} f_i (x_i - \mu)^2}{\sum_{i=1}^{n} f_i}} = \sqrt{\frac{f_1(x_1 - \mu)^2 + f_2(x_2 - \mu)^2 + \cdots + f_n(x_n - \mu)^2}{f_1 + f_2 + \cdots + f_n}}.$$

Example 4 *Finding the Standard Deviation from a Frequency Distribution*

The distribution of maximum Interstate speed limits for cars in 1998 for each of the states except Montana, which has no speed limit, is shown in Table 5.21. Find the mean and the standard deviation.

Table 5.21	Interstate Speed Limits
Speed Limit (mph)	*Frequency*
55	3
65	23
70	13
75	10

Source: U.S. Department of Transportation.[34]

Solution: Using the formula for the mean of data presented by a frequency distribution, we find that the mean speed limit is

$$\mu = \frac{3 \cdot 55 + 23 \cdot 65 + 13 \cdot 70 + 10 \cdot 75}{3 + 23 + 13 + 10} = \frac{3320}{49} \approx 67.76 \text{ mph.}$$

We have converted the mean to a two-digit decimal to fit our upcoming calculations onto one line, but in general as many decimal places as possible should be kept for all intermediate parts of calculations. In fact, you should try to use the numbers directly from your calculator rather than writing them down and reentering them into the calculator later. How you do this will depend on the particular calculator you use. By doing this, the resulting answers will be sufficiently accurate for nearly any practical application. Moving on to compute the standard deviation, we find

$$\sigma \approx \sqrt{\frac{3 \cdot (55 - 67.76)^2 + 23 \cdot (65 - 67.76)^2 + 13 \cdot (70 - 67.76)^2 + 10 \cdot (75 - 67.76)^2}{49}}$$

$$\approx \sqrt{\frac{1253.06}{49}} \approx \sqrt{25.57} \approx 5.06 \text{ mph.}$$

Estimating the Standard Deviation of Grouped Data

When data are grouped into classes, we can only estimate the mean and standard deviation. In this case, we make the approximation that all values in a class are at the midpoint of the respective interval.

Example 5 *Estimating the Standard Deviation*

A *Consumer Reports* survey of several brands of peanut butter reported the frequency distribution of the cost, in cents, per 3-tablespoon serving shown in Table 5.22. Estimate the mean and the standard deviation.

Table 5.22	Peanut Butter Cost
Cost per Serving (cents)	**Frequency**
15–19	20
20–24	7
25–29	6
30–34	0
35–39	2

Source: *Consumer Reports*, September 1995.[35]

Solution: The midpoint of 15 and 19 is $(15 + 19)/2 = 17$, which we use as the representative value for the first class. Dropping the class 30–34 because it has frequency 0, we similarly compute the representatives for the remaining classes as recorded in Table 5.23.

Table 5.23	Peanut Butter Cost	
Cost per Serving (cents)	**Representative Value**	**Frequency**
15–19	17	20
20–24	22	7
25–29	27	6
35–39	37	2

Our estimate for the mean is

$$\mu \approx \frac{20 \cdot 17 + 7 \cdot 22 + 6 \cdot 27 + 2 \cdot 37}{20 + 7 + 6 + 2} = \frac{730}{35} \approx 20.86¢.$$

Our estimate for the standard deviation is

$$\sigma \approx \sqrt{\frac{20 \cdot (17 - 20.86)^2 + 7 \cdot (22 - 20.86)^2 + 6 \cdot (27 - 20.86)^2 + 2 \cdot (37 - 20.86)^2}{20 + 7 + 6 + 2}}$$

$$= \sqrt{\frac{1054.286}{35}} \approx \sqrt{30.12} \approx 5.49¢.$$

Once you have computed a median or mean, a range, or a standard deviation, you should always look back at the data and ask yourself whether your answer seems reasonable. This step will catch a substantial portion of potential errors.

EXERCISES FOR SECTION 5.3

1. The land areas, in square miles, of the five counties of Rhode Island are 25, 170, 104, 413, and 333.

(a) Find the mean. (b) Find the range. (c) Find the standard deviation.

2. According to the U.S. National Center for Education Statistics, the student teacher ratios in elementary and secondary schools in the United States for each of the years 1989 through 1994 were approximately 17.1, 16.9, 16.9, 17.1, 17.0, and 17.2, in order by years.

(a) Find the mean. (b) Find the range. (c) Find the standard deviation.

3. The Congressional Budget Office reported that from 1984 through 1993, the surpluses (or deficits) in the U.S. federal government's unemployment trust fund, in billions of dollars, were 4, 5, 4, 7, 8, 7, 6, −3, −12, 1.

(a) Find the mean. (b) Find the range. (c) Find the standard deviation.

4. A 1998 *Consumer Reports* survey[36] reported the following stopping distances of several brands of tires, in feet, for a car traveling 60 miles per hour on dry pavement: 138, 134, 136, 136, 141, 142, 136, 144, 144, 150, 142.

(a) Find the mean. (b) Find the range. (c) Find the standard deviation.

In Exercises 5–7, consider the given roster of the 1996 Olympics version of the United States basketball "Dream Team."

1996 Olympics U.S. men's basketball team.

Player	Height	Weight (pounds)	Years in the NBA
Charles Barkley	6′6″	252	11
Anfernee Hardaway	6′7″	195	2
Grant Hill	6′8″	225	1
Gary Payton	6′4″	190	5
Scottie Pippen	6′7″	225	8
Hakeem Olajuwon	7′0″	255	11
Shaquille O'Neal	7′1″	301	3
Karl Malone	6′9″	256	10
Reggie Miller	6′7″	185	8
Mitch Richmond	6′5″	215	7
David Robinson	7′1″	225	6
John Stockton	6′1″	175	11

5. For the weights of the players on the Dream Team, find

(a) the mean, (b) the range, (c) the standard deviation.

6. For the number of years in the NBA of the players on the Dream Team, find

(a) the mean, (b) the range, (c) the standard deviation.

7. For the heights of the players on the Dream Team, find

(a) the mean, (b) the range, (c) the standard deviation.

(d) The mean height of American males is approximately 5′10″. Compare this to the mean height of the Dream Team.

8. The National Hurricane Center reported that the number of North Atlantic tropical storms reaching the U.S. coast in each of the years 1986 through 1995 were 6, 7, 12, 11, 14, 8, 7, 8, 7, 19, in order by years.

(a) Find the mean. (b) Find the range. (c) Find the standard deviation.

(d) Do your calculations suggest that the number of storms reaching the coast is variable or stable? Explain your reasoning.

9. Individual federal income taxes in the United States, as a percentage of gross domestic product, are given for selected years in the table.

Year	1965	1970	1975	1980	1985	1990	1995
Percentage	7.3	9.2	8.1	9.2	8.5	8.6	8.5

Source: Congressional Budget Office.[37]

(a) Find the mean.   (b) Find the range.   (c) Find the standard deviation.

(d) Based on your calculations, do you think the tax burden has varied much over this period? Explain your reasoning.

10. (a) Find the mean and standard deviation of the following set of exam scores:
$$83, 91, 70, 75, 68, 93, 68, 59, 81, 73.$$

(b) Find the mean and standard deviation of the following set of exam scores:
$$98, 42, 37, 95, 86, 89, 45, 100, 96, 52.$$

(c) Explain why the standard deviation of the exam scores in part (b) was much larger than the standard deviation of the exam scores in part (a).

11. The normal monthly precipitation, in inches, for Seattle and Pittsburgh is shown in the table.

Normal Monthly Precipitation (inches)

	Jan.	Feb.	Mar.	Apr.	May	June	July	Aug.	Sept.	Oct.	Nov.	Dec.
Seattle	5.38	3.99	3.54	2.33	1.70	1.50	0.76	1.14	1.88	3.23	5.83	5.91
Pittsburgh	2.54	2.39	3.41	3.15	3.59	3.71	3.75	3.21	2.97	2.36	2.85	2.92

Source: U.S. National Oceanic and Atmospheric Administration.[38]

(a) Find the mean monthly precipitation for Seattle.

(b) Find the mean monthly precipitation for Pittsburgh.

(c) Find the range and standard deviation for the monthly precipitation for Seattle.

(d) Find the range and standard deviation for the monthly precipitation for Pittsburgh.

(e) Even though the mean monthly precipitations of Seattle and Pittsburgh are quite close, the ranges and standard deviations are not. What does this say about precipitation in Seattle and Pittsburgh?

12. The daily mean temperatures, in degrees Fahrenheit, for San Diego, California, and Jackson, Mississippi, are shown in the table.

Daily Mean Temperature (°F)

	Jan.	Feb.	Mar.	Apr.	May	June	July	Aug.	Sept.	Oct.	Nov.	Dec.
San Diego	57.4	58.6	59.6	62.0	64.1	66.8	71.0	72.6	71.4	67.7	62.0	57.4
Jackson	44.1	47.9	56.7	64.6	72.0	78.8	81.5	80.9	75.9	64.7	55.8	47.8

Source: U.S. National Oceanic and Atmospheric Administration.[39]

(a) Find the mean of the temperatures for San Diego.

(b) Find the mean of the temperatures for Jackson.

(c) Notice that the means of the temperatures from the two cities are about equal. From looking at the data, which city do you expect to have the largest standard deviation in temperatures? Explain your answer.

(d) Find the standard deviation for the temperatures for San Diego.

(e) Find the standard deviation for the temperatures in Jackson.

(f) Would you expect heating and cooling costs to be greater in San Diego or in Jackson?

13. The 1994 population density by region in the United States, as reported by the U.S. Bureau of the Census, is listed in the table.

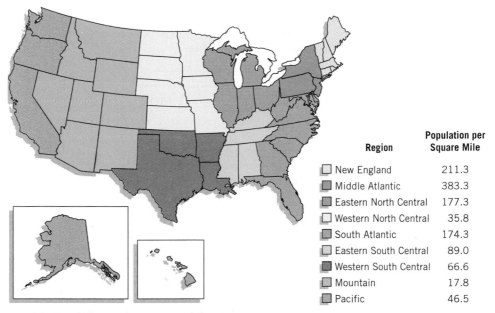

Region	Population per Square Mile
New England	211.3
Middle Atlantic	383.3
Eastern North Central	177.3
Western North Central	35.8
South Atlantic	174.3
Eastern South Central	89.0
Western South Central	66.6
Mountain	17.8
Pacific	46.5

(a) Find the median regional density.

(b) Find the mean of the regional densities.

(c) The density for the United States as a whole was 73.6 persons per square mile. Explain how this can be so much lower than your answers to (a) and (b).

(d) Find the range of the regional densities.

(e) Find the standard deviation of the regional densities.

14. The age distribution of the students in a college class (taught by one of the authors) is shown in the table.

Age	17	18	19	20	21	22	23	25
Frequency	1	7	11	4	4	1	2	2

(a) Find the mean.   (b) Find the range.   (c) Find the standard deviation.

15. The frequency distribution of annual figures for 1985 through 1996 for aircraft accidents worldwide involving one or more passenger fatalities is shown in the table.

Annual Number of Accidents	22	23	25	26	27	28	29	30	34
Frequency	2	1	1	2	1	2	1	1	1

Source: U.S. Bureau of the Census.[40]

(a) Find the mean.   (b) Find the range.   (c) Find the standard deviation.

16. There will be 67 total solar eclipses on Earth from 2001 to 2100. The duration (length of time at the optimal viewing area) of total solar eclipses varies considerably. The frequency distribution of the durations of the 67 total solar eclipses, rounded to the nearest minute, is shown in the table. Estimate the mean and standard deviation of the duration.

Duration (rounded to nearest minute)	1	2	3	4	5	6	7
Frequency	1	18	19	14	8	6	1

Source: NASA/Goddard Space Flight Center.[41]

17. The frequency distribution of the U.S. annual death rates (deaths per 1000 population) for the years 1977–1996 is given in the table. Find the mean and standard deviation of the death rate.

Annual Death Rate	8.5	8.6	8.7	8.8	8.9
Frequency	3	6	3	7	1

Source: National Center for Health Statistics.[42]

18. The distribution of public colleges in the United States by undergraduate tuition rates for in-state students for the 1993–94 academic year is given in the table. Estimate the mean and standard deviation of the tuition at U.S. public colleges.

Range of Tuition (dollars)	Number of Colleges
6000–6999	1
5000–5999	7
4000–4999	31
3000–3999	87
2000–2999	181
1000–1999	207
0–999	12

Source: U.S. Bureau of the Census.[43]

19. The criminal offense rate for 70 large U.S. cities in 1993 is given in the frequency distribution table. Estimate the mean and standard deviation of the criminal offense rate.

Offenses per 100,000 Population	Frequency
4000–5999	2
6000–7999	18
8000–9999	21
10,000–11,999	21
12,000–13,999	2
14,000–15,999	2
16,000–17,999	3
18,000–20,000	1

Source: U.S. Federal Bureau of Investigation.[44]

20. The frequency distribution of ages of those arrested for violent crimes in the United States in 1995 is given in the table. Choosing reasonable class representatives for the "Under 15" and "65 and over" groups (and using class midpoints for the other classes), estimate the mean and standard deviation of the ages.

Age (years)	Number of Arrests
Under 15	34,947
15–17	80,645
18–24	167,223
25–34	182,619
35–44	108,475
45–64	41,278
65 and over	4,043

Source: U.S. Federal Bureau of Investigation.[45]

21. The percentage distribution of the average number of hours worked per week by U.S. workers in 1994 is given in the table.

Hours Worked	Frequency
1–4	1,271
5–14	4,992
15–29	15,115
30–34	9,473
35–39	8,684
40	40,587
41–48	14,075
49–59	13,366
60 and over	9,878

Source: U.S. Bureau of the Census.[46]

Choosing a reasonable class representative for the "60 and over" group (and using class midpoints for the other classes), estimate the mean and standard deviation of the number of hours worked.

THE NORMAL DISTRIBUTION

In the first three sections of this chapter, we looked at ways of describing a set of data, either in its entirety or through measures of center and spread. In this section we study a particular way that data may be distributed called a normal distribution or, more commonly, a bell curve. Most of us have heard the term bell curve in connection with the distribution of grades on an exam or scores on the Scholastic Assessment Test (SAT). The histogram of verbal scores on the 1997 Scholastic Assessment Test shown in Figure 5.12 has a bell-curve shape.

Figure 5.12
Histogram of 1997 verbal SAT scores. (*Source:* The College Board.)

For data that are distributed in this way, we will see how to find the probability that a data point will fall within a given range. The techniques we learn in this section will be used in making estimates based on sampling in the following two sections. In the final section of this chapter, we will discuss the reliability of samples and the ways that erroneous results can arise, either by accident or deliberately.

In the first section of this chapter we looked at histograms for which the height of a bar was the relative frequency of the class. We now consider normalized histograms. In **normalized histograms** the height of the bar of a class is given by the relative frequency of the class divided by the width of the class. The areas of the bars of a normalized histogram are therefore equal to the relative frequencies. As a result, the areas of the bars of a normalized histogram sum to 1. When the widths of the classes are all equal, the normalized histogram has the same shape as the nonnormalized histograms we considered earlier; the only change is in the vertical scale.

We will consider only normalized histograms from now on. If we begin with a histogram representing some frequency distribution and make the class intervals smaller and smaller, the histograms (often) gradually approach the shape of some smooth curve. This process is illustrated in Figure 5.13. You should think of the resulting smooth curve as a histogram with infinitesimally small class intervals. The normalization convention brings a tangible benefit. The relative frequency for any interval is just the area under the curve for that interval!

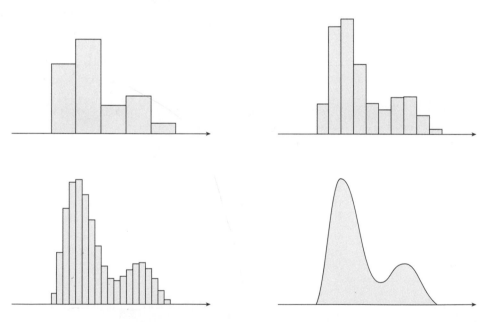

Figure 5.13
Limiting case of histograms with decreasing widths.

One particularly important distribution that gives a good approximation to many types of real-world data is called a **normal distribution** or **bell curve**. Normal distributions are bell-shaped like the curve in Figure 5.14. The center of the bell is at the mean μ of the distribution, and the standard deviation, σ, of the distribution determines how spread out the bell is in shape. For larger σ the curve is more spread out and flat, whereas for smaller σ the curve is narrower and taller.

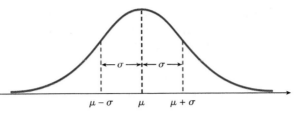

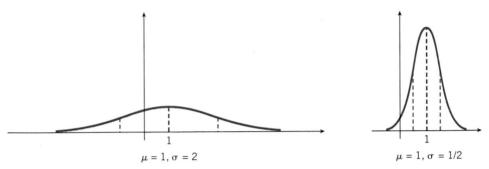

Figure 5.14
The normal distribution.

These ideas are illustrated in the two normal distributions in Figure 5.15.

Figure 5.15
Normal distributions with mean 1 and varying standard deviation.

The algebraic formula for the curve of the normal distribution is

$$y = \frac{1}{\sqrt{2\pi}\sigma} \, e^{-\frac{(x-\mu)^2}{2\sigma^2}},$$

where μ is the mean, σ is the standard deviation, and $\pi \approx 3.14$ and $e \approx 2.72$ are constants. We will not need to use this formula directly. To determine whether or not a normalized histogram has roughly a normal distribution, we first compute the mean μ and standard deviation σ of the data it represents. We then simultaneously plot the curve for the normal distribution with mean μ and standard deviation σ on the same graph as the histogram to see if it gives a reasonable fit. We can see this for scores on the verbal portion of the 1997 Scholastic Assessment Test, which has mean 505 and standard deviation 111, in Figure 5.16. In this case we see that the curve for the normal distribution fits the histogram quite well.

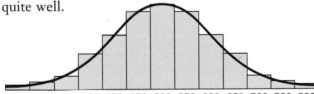

Figure 5.16
Histogram of 1997 verbal SAT scores, $\mu = 505$ and $\sigma = 111$. (*Source:* The College Board.)

The important properties of the normal distribution are listed next.

Properties of a Normal Distribution

1. The total area under the curve is equal to 1.

2. The curve is symmetric about μ. In particular, the area under the curve to the right of μ and the area under the curve to the left of μ are both equal to 0.5.

3. The curve extends infinitely in both directions, getting closer to, but not touching, the horizontal axis.

For any normal distribution, approximately 68% of the data values lie within 1 standard deviation of the mean, 95% lie within 2 standard deviations, and 99.7% lie within 3 standard deviations. So even though the distribution extends infinitely in both directions, most of the data are clustered about the mean.

We will use the normal distribution curve to find the probability that a random data value will lie within a certain range of values. Let X denote a random data value from a set of data whose relative frequencies fall into a normal distribution. Formally, X is called a **random variable**. The probability that X lies between the values a and b is equal to the fraction of values of the distribution that fall between a and b. However, this fraction is just the area under the normal distribution curve between a and b, shown in Figure 5.17. We denote this probability by $P(a \leq X \leq b)$.

figure 5.17
The area of the shaded region is $P(a \leq X \leq b)$.

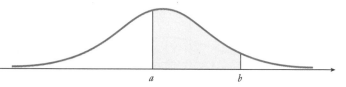

Similarly, the probability that X is at least a, denoted by $P(a \leq X)$ is equal to the area under the normal distribution curve to the right of a, and the probability that X is at most b, denoted by $P(X \leq b)$ is equal to the area under the normal distribution curve to the left of b. Both cases are depicted in Figure 5.18.

(a) The area of the shaded region is $P(a \leq X)$

(b) The area of the shaded region is $P(X \leq b)$

figure 5.18

Summarizing, we have the following important facts:

The probability that a random variable X from a normal distribution lies between the values a and b is given by

$P(a \leq X \leq b) =$ the area under the normal distribution curve between a and b.

Similarly,

$P(a \leq X) =$ the area under the normal distribution curve to the right of a,

and

$P(X \leq b) =$ the area under the normal distribution curve to the left of b.

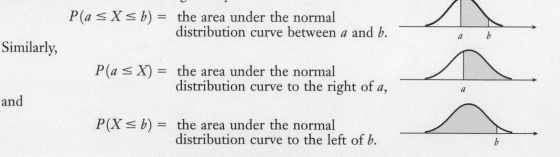

The probability that X is exactly equal to any particular value is equal to the area under the normal distribution curve over an interval of width 0. Because the width is 0, the area must also be 0. Therefore, the probability that X is exactly any single value is 0. This may seem counterintuitive. However, you should remember that any measurement of physical data has been rounded, and hence is only narrowed down to an interval of values. For instance, if heights are rounded to the nearest inch, then the group of people of height $6'$ will include all people between $5'11\frac{1}{2}''$ and $6'\frac{1}{2}''$. Therefore, the probability that a person would give a *stated* height of $6'$ would not be 0. Furthermore, the sample space for real data is finite, so that the normal distribution, with infinitely many possible values, can at best be only a very good approximation of the actual distribution. Because the probability that X is exactly equal to any particular value is 0, we have the following:

$$P(a \leq X \leq b) = P(a \leq X < b) = P(a < X \leq b) = P(a < X < b),$$
$$P(a \leq X) = P(a < X),$$
$$P(X \leq b) = P(X < b).$$

To find the probability that a random variable X lies within a certain range of values, we need to compute the area under a normal distribution curve over that range of values. We could give a rough estimate of this just from an accurate graph of the curve. Methods for more precise estimates rely on calculus. Many of the more advanced scientific calculators can do these calculations. However, normal distributions are so important that tables, such as the table in Appendix A, that can be used to compute these areas were prepared long before the development of calculators. We will first learn how to use this table to compute the areas in the special case of the normal distribution with mean $\mu = 0$ and standard deviation $\sigma = 1$. This particular normal distribution is called the **standard normal distribution**. For the standard normal distribution, we will call the random variable Z. Once we have learned to compute the probabilities for the standard normal distribution, we will learn how to compute the probabilities for any normal distribution.

Unfortunately, because of space limitations it is not feasible to have tables approximating the area between a and b under the standard normal distribution curve for all possible values of a and b. Instead, the table in Appendix A is restricted to the case $a = 0$ and b is positive. However, it is fairly easy to use only this information to deduce the areas for all a and b. We will see how to do this through the next series of examples.

Example 1   *Probability for the Standard Normal Distribution*

Find $P(0 \leq Z \leq 1.37)$ for the standard normal distribution.

Solution: The shaded area in Figure 5.19 is equal to $P(0 \leq Z \leq 1.37)$. To compute the area in this case, we can go right to Appendix A and look up the 1.3 row and the 0.07 column. We find 0.4147, hence $P(0 \leq Z \leq 1.37) \approx 0.4147$.

Figure 5.19
The region representing $P(0 \leq Z \leq 1.37)$.

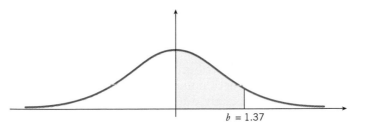

$b = 1.37$

Example 2   *Probability for the Standard Normal Distribution*

Find $P(0.61 \leq Z \leq 2.04)$ for the standard normal distribution.

Solution: The area equal to $P(0.61 \leq Z \leq 2.04)$ is shaded in Figure 5.20.

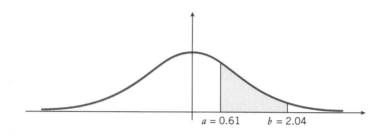

figure 5.20
The region representing $P(0.61 \leq Z \leq 2.04)$.

$a = 0.61$ $b = 2.04$

This area is not given directly in Appendix A but is the difference of the area under the curve from 0 to 2.04 and the area under the curve from 0 to 0.61. Therefore,

$$P(0.61 \leq Z \leq 2.04) = P(0 \leq Z \leq 2.04) - P(0 \leq Z < 0.61).$$

Now, using Appendix A to compute $P(0 \leq Z \leq 2.04)$ and $P(0 \leq Z < 0.61)$, we have

$$P(0.61 \leq Z \leq 2.04) \approx 0.4793 - 0.2291 = 0.2502.$$

Example 3   *Probability for the Standard Normal Distribution*

Find $P(Z < 0.75)$ for the standard normal distribution.

Solution: Once again, it is best to begin with a picture of the area we are trying to find. For $P(Z < 0.75)$, this area is shown in Figure 5.21.

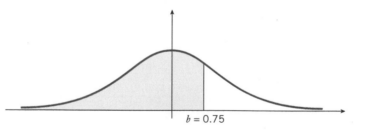

figure 5.21
The region representing $P(Z < 0.75)$.

$b = 0.75$

Recall that it does not matter whether the condition is $Z < 0.75$ or $Z \leq 0.75$. This time we must include the area of the region where $Z < 0$. By the symmetry of the normal distribution, this is just half of the total area under the curve. Because the total area under the curve is 1, we know that the area of the region where $Z < 0$ is exactly 0.5. Thus,

$$P(Z < 0.75) = P(Z < 0) + P(0 \leq Z < 0.75) \approx 0.5 + 0.2734 = 0.7734.$$

Example 4   *Probability for the Standard Normal Distribution*

What is $P(-2.59 \leq Z \leq -1.1)$ for the standard normal distribution?

Solution: From Figure 5.22 and the symmetry of the standard normal distribution, we see that $P(-2.59 \leq Z \leq -1.1)$ is the same as $P(1.1 \leq Z \leq 2.59)$.

figure 5.22
The region representing $P(-2.59 \leq Z \leq -1.1)$.

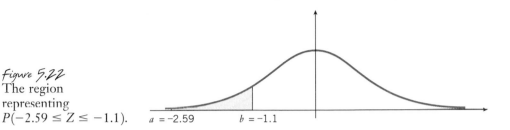

$a = -2.59$ $b = -1.1$

It follows that

$$P(-2.59 \leq Z \leq -1.1) = P(1.1 \leq Z \leq 2.59)$$
$$= P(0 \leq Z \leq 2.59) - P(0 \leq Z < 1.1)$$
$$\approx 0.4952 - 0.3643 = 0.1309.$$

By now you see how to use the table for the standard normal distribution. Remember that the table only applies directly to the standard normal distribution, that is, the normal distribution with mean 0 and standard deviation 1. Although it seems reasonable to think that one would need a different table for every different mean and standard deviation, remarkably this is not the case! To compute probabilities for a random variable X from any normal distribution, we can translate the problem to the case of the normal standard distribution by the use of z-scores, which we examine next.

Suppose we are given a value x from any distribution with mean μ and standard deviation σ. If we want to measure how far x is from the central value, we should compute the difference $x - \mu$, which will be negative if x is smaller than the mean and positive if x is larger than the mean. This difference is very dependent on the units we use. For example, if x represents time, this difference may seem huge if measured in seconds, yet small if measured in years. It makes sense to compare it to a typical spread, which we have represented by the standard deviation. For this reason, we divide the difference by the standard deviation σ to get $(x - \mu)/\sigma$. We call this quantity the z-score associated to x.

For an x-value in a distribution with mean μ and standard deviation σ, the associated **z-score** is given by

$$z = \frac{x - \mu}{\sigma}.$$

The z-score is equal to the number of standard deviations x is from the mean, along with a plus or minus sign indicating whether it is larger or smaller than the mean.

Just as we can convert inches to and from centimeters or degrees Fahrenheit to and from degrees Celsius, we can convert x, in whatever units it is measured, to and from z-scores, measured in standard deviation units. Scores on the verbal portion of the 1997 Scholastic Assessment Test have an approximately normal distribution with mean 505 and standard deviation 111. This is illustrated with a normal distribution curve above the "ruler" in Figure 5.23. The curve is centered at SAT score 505. An SAT score of 600 corresponds to a z-score of $(600 - 505)/111 \approx 0.86$, as we see on the ruler in Figure 5.23. Similarly, an SAT score of 350 corresponds to a z-score of $(350 - 505)/111 \approx -1.40$.

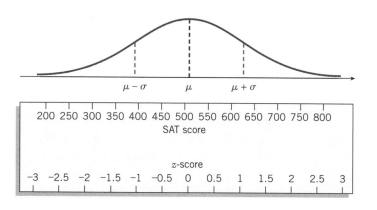

Figure 5.23
A normal distribution with an SAT z-score "ruler," $\mu = 505$ and $\sigma = 111$.

Returning to the problem of computing probabilities, suppose that X is a random variable from a normal distribution with mean μ and standard deviation σ. To compute the probability that X lies within some interval of values, we translate the x-values of the endpoints of the interval to their corresponding z-scores. The probability that X lies within some interval of values is equal to the probability that a random variable Z from the standard normal distribution lies within the translated interval. Summarizing this, we have the following:

If X is a random variable from a normal distribution with mean μ and standard deviation σ, then

$$P(a \le X \le b) = P\left(\frac{a - \mu}{\sigma} \le Z \le \frac{b - \mu}{\sigma}\right),$$

$$P(a \le X) = P\left(\frac{a - \mu}{\sigma} \le Z\right),$$

$$P(X \le b) = P\left(Z \le \frac{b - \mu}{\sigma}\right),$$

where Z is a random variable from the standard normal distribution.

Once we have made the translation from the x-values of the endpoints of the interval to their corresponding z-scores, we proceed as we did for the standard normal distribution and use Appendix A. Again, draw a sketch to decide how to compute the area under the curve.

Example 5 *Probability for a General Normal Distribution*

Suppose X is a random variable from a normal distribution with mean 3 and standard deviation 6. Find $P(0 < X < 5)$.

Solution: We begin by translating the interval $0 < X < 5$ to its equivalent Z-interval. When $a = 0$, we have $z = \frac{0-3}{6} = -0.5$. Similarly, when $b = 5$, we have $z = \frac{5-3}{6} \approx 0.33$. In this last case, we have rounded the z-score to the two decimal places required to use Appendix A. Therefore, $P(0 < X < 5) \approx P(-0.5 < Z < 0.33)$. From our diagram of the standard normal distribution for the translated interval in Figure 5.24, we see that our answer is the sum of two values from the table. (We will omit the values of a and b from the diagram in future examples.)

figure 5.24
The region representing $P(-0.5 < Z < 0.33)$.

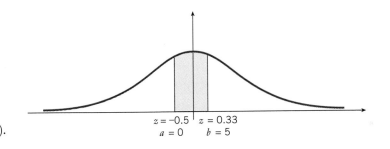

To find this area, add the area below the curve from $Z = -0.5$ to $Z = 0$ to that below the curve from $Z = 0$ to $Z = 0.33$. Then use the symmetry of the normal distribution to see that the area from $Z = -0.5$ to $Z = 0$ is equal to the area from $Z = 0$ to $Z = 0.5$. Doing this, we get

$$P(0 < X < 5) \approx P(-0.5 < Z < 0.33)$$
$$= P(-0.5 < Z < 0) + P(0 \le Z < 0.33)$$
$$= P(0 < Z < 0.5) + P(0 \le Z < 0.33)$$
$$\approx 0.1915 + 0.1293 = 0.3208.$$

Example 6 *Probability for a General Normal Distribution*

If X is a random variable from a normal distribution with mean -4.2 and standard deviation 3.1, find $P(1 \le X)$.

Solution: Converting the endpoint $a = 1$ to its z-score, we find that

$$P(1 \le X) = P\left(\frac{1 - (-4.2)}{3.1} \le Z\right) \approx P(1.68 < Z).$$

The accompanying diagram is shown in Figure 5.25.

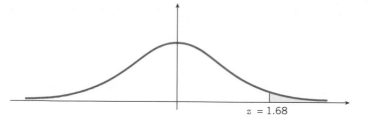

figure 5.25
The region
representing
$P(1.68 \leq Z)$.

Because the graph is symmetrical about its center at $Z = 0$ and the area below the entire graph is 1, we know that $P(0 \leq Z) = 0.5$. The area under the graph where $Z \geq 1.68$ is the area below the graph where $0 \leq Z$, minus that from $Z = 0$ to $Z = 1.68$. Therefore, we have

$$P(1 \leq X) \approx P(1.68 \leq Z) = P(0 \leq Z) - P(0 \leq Z < 1.68)$$
$$\approx 0.5 - 0.4535 = 0.0465.$$

Example 7   *Probability for a General Normal Distribution*

Suppose X is a random variable from a normal distribution with mean 105.3 and standard deviation 21.7. Find $P(X \geq 100)$.

Solution: We have

$$P(X \geq 100) = P\left(Z \geq \frac{100 - 105.3}{21.7} \right) \approx P(Z \geq -0.24),$$

and from Figure 5.26, we see that

$$P(X \geq 100) \approx P(Z \geq -0.24) = P(-0.24 \leq Z < 0) + P(0 \leq Z)$$
$$= P(0 < Z \leq 0.24) + P(0 \leq Z) \approx 0.0948 + 0.5 = 0.5948.$$

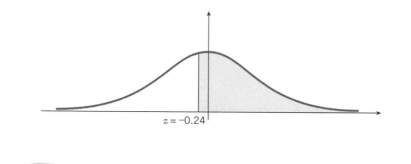

figure 5.26
The region
representing
$P(Z \geq -0.24)$.

Example 8   *The SAT*

The scores on the verbal portion of the 1997 Scholastic Assessment Test were approximately normally distributed with mean 505 and standard deviation 111.

(a) Estimate the percentage of scores falling between 550 and 650.

(b) Estimate the percentage of scores that are at least 700.

Solution:

(a) Converting the SAT scores of 550 and 650 to their corresponding z-scores and referring to Figure 5.27, we obtain

$$P(550 \leq X \leq 650) = P\left(\frac{550 - 505}{111} \leq Z \leq \frac{650 - 505}{111}\right)$$
$$\approx P(0.41 \leq Z \leq 1.31)$$
$$= P(0 \leq Z \leq 1.31) - P(0 \leq Z < 0.41)$$
$$\approx 0.4049 - 0.1591 = 0.2458 = 24.58\%.$$

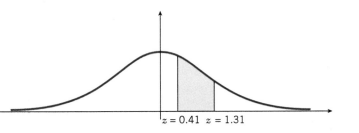

figure 5.27
The region
representing
$P(0.41 \leq Z \leq 1.31)$.

$z = 0.41$ $z = 1.31$

(The actual percentage is 26.57%. Our estimate is a little low primarily because SAT scores can only be multiples of 10. Thus, it matters whether or not we wish to include the scores 550 and 650. The percentage of scores between 550 and 650, noninclusive, is 21.74%. Our estimate is close to the average of these two actual percentages. A slightly more advanced treatment would account for whether the inequalities are strict or not by what is called a "continuity correction.")

(b) We convert an SAT score of 700 to its corresponding z-score and refer to Figure 5.28, and we find

$$P(700 \leq X) = P\left(\frac{700 - 505}{111} \leq Z\right) \approx P(1.76 \leq Z)$$
$$= 0.5 - P(0 \leq Z < 1.76) \approx 0.5 - 0.4608 = 0.0392 = 3.92\%.$$

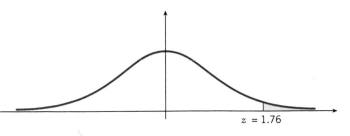

figure 5.28
The region
representing
$P(1.76 \leq Z)$.

$z = 1.76$

(The actual percentage is 4.7%. Our estimate is a little low because SAT scores are not precisely normally distributed.)

BRANCHING OUT 5.2

Don't Try to Cheat a Mathematician

Geometer Henri Poincaré, the most famous of French mathematicians of the latter nineteenth century, once said, "The scientist does not study nature because it is useful to do so, he studies it because he takes pleasure in it, and he takes pleasure in it because it is beautiful." You would think such a person might have his head in the clouds, and be unconcerned and unaware of such mundane matters as the weight of a loaf of bread. The following bit of folklore about Poincaré, if true, suggests otherwise. The story is as follows: French law required that loaves of bread weigh 1 kilogram. However, Poincaré complained to police that loaves purchased daily from his local bakery averaged only 0.9 kilogram. For the next six months, *all* of his daily loaves weighed at least 1 kilogram. When asked if the baker had stopped cheating, Poincaré replied that the baker was still cheating, but had merely stopped cheating Poincaré by reserving one of the heavier loaves for him. How did he deduce this? It seems reasonable that the weight of a random loaf would be distributed in a roughly normal manner. However, Poincaré had kept track of the weights he received and found their distribution remarkably close to that of those loaves weighing at least 1 kilogram from a normal distribution with a mean of 0.9 kilogram. In other words, Poincaré had fairly compelling evidence that the baker weighed loaves until finding one weighing at least 1 kilogram, which he then reserved for Poincaré.

Henri Poincaré (1854–1912).

Percentiles

You may have come across percentiles when looking at a rating of how you performed on a standardized test such as the SAT. If you scored in the 84th percentile, it means that your score was as least as good as that of 84% of the other test takers, and no better than the scores of the remaining 16%. The median is the 50th percentile, a value that separates the lower 50% of values from the upper 50% of values. More generally, we have the following definition:

> The *p*th **percentile** is a number that divides the lower *p* percent of the values of a distribution from the upper $100 - p$ percent. For a normally distributed random variable, the *p*th percentile is a unique number.

When we want to find percentiles for a normal distribution, we reverse the procedure we used in finding the probabilities for a random variable X. In other words, we translate from a probability or area back to a corresponding z-score, and then to the corresponding x-value. We will first look at how this is done in the case of the standard normal distribution, where this final step is unnecessary, and then we will see the technique used to find percentiles for any normal distribution.

Example 9   *Percentiles for the Standard Normal Distribution*

For the standard normal distribution, find

(a) the 80th percentile,
(b) the 17th percentile.

Solution:

(a) To find the 80th percentile, we want to find the z-score for which the probability $P(Z \le z)$ is exactly 0.8. In other words, we want to find the z-score so that the area of the shaded region in Figure 5.29 is 0.8. In drawing Figure 5.29, we recall that the axis of symmetry of the distribution is the 50th percentile.

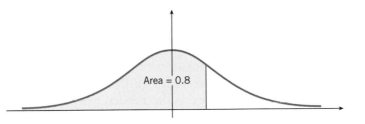

Area = 0.8

figure 5.29
The 80th percentile.

Because the region to the left of the vertical axis has area 0.5, we want to find the z-score so that the area under the curve from 0 to z is $0.8 - 0.5 = 0.3$. Thus, we look for the value of z corresponding to 0.3 in Appendix A. The entry in the table closest to 0.3 is 0.2995 with a corresponding z-score of 0.84. Some statisticians prefer to interpolate, obtaining a value between 0.84 and 0.85, but we will do this only when two consecutive entries are equal distances from the desired value. Thus, the 80th percentile corresponds to a z-score of (approximately) 0.84. Such values will be sufficiently accurate for our purposes.

(b) To compute the 17th percentile, we want to find that value of z for which the area of the shaded region in Figure 5.30(a) is 0.17. The area between the desired z-score and $z = 0$ is $0.5 - 0.17 = 0.33$, as shown in Figure 5.30(b). If we now look up the z-score in Appendix A corresponding to equivalent area 0.33, as illustrated in Figure 5.30(c), our desired z-score is the negative of this number. We find a z-score of 0.95, from which we conclude that the 17th percentile for the standard normal distribution is -0.95.

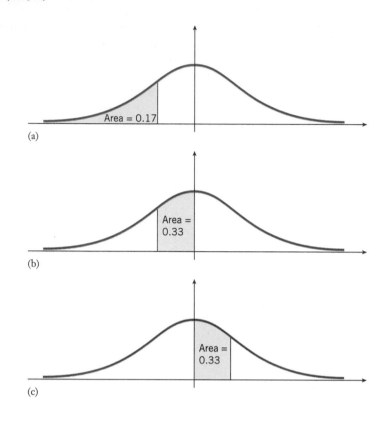

Area = 0.17

(a)

Area = 0.33

(b)

Area = 0.33

(c)

figure 5.30
Computing the 17th
percentile.

In brief, to find percentiles for the standard normal distribution, we find the area of the region between the needed z-score and the $z = 0$ axis. We look up the z-score corresponding to this area in Appendix A; this z-score will always be a nonnegative number. If the percentile exceeds the 50th, we are done. If the percentile is less than the 50th, we take the negative of the z-score found in Appendix A to get the needed z-score.

When the distribution is a more general normal distribution, we first find the relevant z-score and then translate to the corresponding x-value. If we solve for x in terms of z in the formula for z-scores in terms of x-values, we see that

$$x = \mu + z\sigma.$$

It is easiest to remember this formula by keeping in mind that the z-score measures how many standard deviations above or below the mean a score lies. Notice that in the formula $x = \mu + z\sigma$, x is greater than μ when z is positive and x is less than μ when z is negative.

Example 10   *Percentiles for a General Normal Distribution*

For a normal distribution with mean 4 and standard deviation 3, find

(a) the 35th percentile,
(b) the 88th percentile,
(c) the percentile to which the value 6.2 corresponds.

Solution:

(a) We begin as in Example 9 by finding the *z*-score of the 35th percentile from the standard normal distribution. The corresponding graph is shown in Figure 5.31.

Figure 5.31
The 35th percentile.

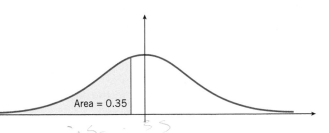

Area = 0.35

To find the *z*-score, we first look up the *z*-score in Appendix A corresponding to the area $0.5 - 0.35 = 0.15$. We find a corresponding *z*-score of 0.39. Because the 35th percentile is to the left of the 50th percentile or mean, we see that $z = -0.39$. Therefore, the *x*-value of the 35th percentile of a normal distribution with mean 4 and standard deviation 3 is given by $4 - 0.39 \cdot 3 = 2.83$.

(b) We again draw a sketch, shown in Figure 5.32, to help us find the *z*-score. For the 88th percentile, we must look up area $0.88 - 0.5 = 0.38$. The value 0.38 is exactly halfway between 0.3790 and 0.3810, with corresponding *z*-scores of 1.17 and 1.18. Thus, we use $z = 1.175$. Therefore, the 88th percentile of a normal distribution with mean 4 and standard deviation 3 is given by $4 + 1.175 \cdot 3 = 7.525$.

Figure 5.32
The 88th percentile.

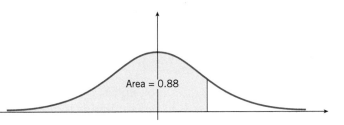

Area = 0.88

(c) We need to find the percentage of values below 6.2. Converting to a *z*-score and referring to the corresponding sketch in Figure 5.33, we find

$$P(X \le 6.2) = P\left(Z \le \frac{6.2 - 4}{3}\right) \approx P(Z \le 0.73)$$

$$= 0.5 + P(0 \le Z \le 0.73) \approx 0.5 + 0.2673 \approx 0.7673.$$

Rounding, we see that 6.2 corresponds to the 77th percentile.

Figure 5.33
The region
representing
$P(Z \le 0.73)$.

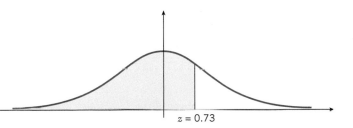

$z = 0.73$

Example 11   *Percentiles and the SAT*

Scores on the verbal portion of the 1997 Scholastic Assessment Test were approximately normally distributed with mean 505 and standard deviation 111.

(a) Estimate the SAT score that falls at the 75th percentile.

(b) Estimate the SAT score that falls at the 40th percentile.

(c) To what percentile does a score of 600 correspond?

Solution:

(a) We begin by sketching the graph in Figure 5.34.

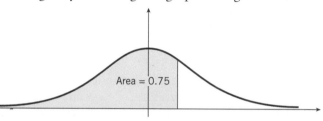

Figure 5.34
The 75th percentile.

We need to look up area $0.75 - 0.5 = 0.25$ in Appendix A. We find that it corresponds most closely to $z = 0.67$. Thus, the SAT score that falls at the 75th percentile is approximately $505 + 0.67 \cdot 111 = 579.37 \approx 579$.

(b) In this case our sketch is the one shown in Figure 5.35.

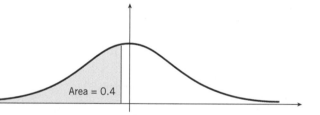

Figure 5.35
The 40th percentile.

We see that we must look up the z-score in Appendix A corresponding to the area $0.5 - 0.4 = 0.1$, and we find it is $z = 0.25$. Because the 40th percentile is to the left of the 50th percentile or mean, we have $z = -0.25$. So we see that the SAT score that falls at the 40th percentile is approximately $505 - 0.25 \cdot 111 = 477.25 \approx 477$.

(c) Translating from SAT scores to the corresponding z-score and referring to the corresponding sketch in Figure 5.36, we find

$$P(X \le 600) = P\left(Z \le \frac{600 - 505}{111} \right) \approx P(Z \le 0.86)$$

$$= 0.5 + P(0 \le Z \le 0.86) \approx 0.5 + 0.3051 \approx 0.8051.$$

We see that a score of 600 corresponds approximately to the 81st percentile.

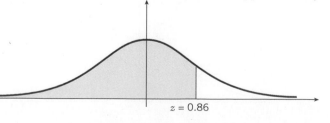

Figure 5.36
Region for SAT score of at most 600.

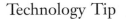

Technology Tip

A statistical or programmable calculator or computer software can be used as an alternative to tables when computing probabilities from z-scores or when finding z-scores for percentiles. The region used to find z for a percentile may vary from program to program. For instance, the table in Appendix A is based on the area under the curve for the interval from 0 to z, but programs may also use the intervals z to ∞ or $-\infty$ to z.

EXERCISES FOR SECTION 5.4

In Exercises 1–10, find the requested probabilities for the standard normal distribution.

1. $P(0 \le Z \le 1.62)$ **2.** $P(-0.37 \le Z \le 0)$

3. $P(-2.1 < Z < -1.2)$ **4.** $P(0.09 < Z < 1.44)$

5. $P(-0.66 \le Z \le 1.32)$ **6.** $P(-1.23 \le Z \le -0.25)$

7. $P(Z > 3.04)$ **8.** $P(-2.2 \le Z)$

9. $P(Z \le 2.82)$ **10.** $P(Z < -1.72)$

In Exercises 11–20, for X normally distributed with mean μ and standard deviation σ, find the indicated probability.

11. $\mu = 6, \sigma = 2, P(0 < X < 8)$ **12.** $\mu = 2, \sigma = 6, P(0 < X < 8)$

13. $\mu = 3.4, \sigma = 1.2, P(2 \le X)$ **14.** $\mu = -1.1, \sigma = 2.1, P(X \ge 0.5)$

15. $\mu = -5, \sigma = 8, P(-3 \le X \le 0)$ **16.** $\mu = 208.3, \sigma = 17.4, P(175 \le X \le 225)$

17. $\mu = 541.2, \sigma = 32.7, P(X < 500)$ **18.** $\mu = 6.7, \sigma = 0.97, P(X < 8.3)$

19. $\mu = 2.88, \sigma = 0.55, P(X \ge 2)$ **20.** $\mu = 3.18, \sigma = 1.28, P(4.1 < X)$

21. Find the 60th percentile for the standard normal distribution.

22. Find the 97th percentile for the standard normal distribution.

23. Find the 15th percentile for the standard normal distribution.

24. Find the 30th percentile for the standard normal distribution.

25. Find the 94th percentile for the standard normal distribution.

26. Find the 5th percentile for the standard normal distribution.

27. Find the 55th percentile for a normal distribution with mean 2 and standard deviation 5.

28. Find the 21st percentile for a normal distribution with mean 10.2 and standard deviation 1.4.

29. Find the 6th percentile for a normal distribution with mean -3 and standard deviation 7.

30. Find the 78th percentile for a normal distribution with mean 5 and standard deviation 0.5.

31. Find the 24th percentile for a normal distribution with mean 100.9 and standard deviation 12.7.

32. Find the 66th percentile for a normal distribution with mean -7 and standard deviation 6.

33. For a normal distribution with mean 38 and standard deviation 7, to what percentile does 40 correspond?

34. For a normal distribution with mean 25 and standard deviation 5, to what percentile does 17 correspond?

35. Scores on the 1997 verbal Scholastic Assessment Test (SAT) had a mean of 505 and a standard deviation of 111. However, all the scores were between 200 and 800 and their distribution was not precisely normal. Suppose the test scores actually did fall into a normal distribution with mean 505 and standard deviation 111. What percentage of scores would you expect to exceed 800?

36. Scores on the 1997 mathematics Scholastic Assessment Test (SAT) had a mean of 511 and a standard deviation of 112. However, all the scores were between 200 and 800 and their distribution was not precisely normal. Suppose the test scores actually did fall into a normal distribution with mean 511 and standard deviation 112. What percentage of scores would lie in the range 200–800?

Exercises 37–44 refer to one form of IQ test for which the scores are approximately normally distributed with mean 100 and standard deviation 16.

37. Estimate the percentage of scores between 110 and 120.

38. Estimate the percentage of scores between 80 and 90.

39. Estimate the percentage of scores below 120.

40. Estimate the percentage of scores above 140.

41. Estimate the score that falls at the 72nd percentile.

42. Estimate the score that falls at the 10th percentile.

43. To what percentile does a score of 130 correspond?

44. To what percentile does a score of 90 correspond?

For Exercises 45–50, use the fact that the height of American males between 25 and 34 years old in the late 1970s was approximately normally distributed with mean 5'10" and standard deviation 2". (In these exercises you should convert all measurements to inches.)

45. Estimate the percentage of men from this group whose heights fall between 5'6" and 5'8".

46. Estimate the percentage of men from this group whose heights are under 6'.

47. Estimate the percentage of men from this group whose heights are over 6'4".

48. Estimate the height of a man from this group at the 90th percentile.

49. Estimate the height of a man from this group at the 10th percentile.

50. NBA star Grant Hill is 6'8". To what percentile does his height correspond? (You will need to refer to the z-scores given below the table in Appendix A.)

For Exercises 51–56, use the fact that the height of American females between 25 and 34 years old in the late 1970s was approximately normally distributed with mean $5'4\frac{1}{2}''$ and standard deviation 2". (In these exercises, convert all measurements to inches.)

51. To what percentile does a woman with height 5' correspond?

52. Estimate the percentage of women from this group whose heights are between $5'4\frac{1}{2}''$ and 5'6".

53. Estimate the percentage of women from this group whose heights are between 5'2" and 5'6".

54. Estimate the percentage of women from this group whose heights are over 5'4".

55. Estimate the height of a woman from this group at the 5th percentile.

56. Estimate the height of a woman from this group at the 62nd percentile.

Exercises 57 and 58 consider grading an exam "on a curve" where it is decided ahead of time that a certain percentage of the class will earn A's, a certain percentage B's, and so on.

57. Suppose the exam scores for a large chemistry class are normally distributed with a mean of 68 and a standard deviation of 14.

    (a) If the top 10% of the scores are given a grade of A, what is the minimum score required to earn an A?

    (b) If the next 25% of the scores are given a grade of B, what is the minimum score required to earn a B?

    (c) If only the bottom 10% of the scores are given a failing grade, what is the minimum score required to pass?

58. Suppose the exam scores for a large mathematics class are normally distributed with a mean of 71 and a standard deviation of 11.

    (a) If the top 15% of the scores are given a grade of A, what is the minimum score required to earn an A?

    (b) If the next 15% of the scores are given a grade of B, what is the minimum score required to earn a B?

    (c) If only the bottom 10% of the scores are given a failing grade, what is the minimum score required to pass?

For Exercises 59–62, consider that on the basis of U.S. Bureau of the Census statistics, the ages of women who bore a child in 1992 were roughly normally distributed with mean 27.5 years old and a standard deviation of 6 years.

59. Of the women who bore a child in 1992, estimate the percentage that were between the ages of 18 and 22.

60. Of the women who bore a child in 1992, estimate the percentage that were over 36 years old.

61. Estimate the age that falls at the 60th percentile of the ages of the women who bore children in 1992.

62. Estimate the percentile for a woman who bore a child at age 40.

For Exercises 63–66, consider that 185 American League baseball players had at least 150 at-bats in the 1997 season. Their batting averages were approximately normally distributed, with a mean of .269 and a standard deviation of .031.

63. The league's leading hitter, Frank Thomas of the Chicago White Sox, hit .347. If the batting averages were actually normally distributed, what percentage of hitters would hit .347 or better?

64. About what percentage would you expect hit between .250 and .300? (One hundred four batters were actually in this range.)

65. The lowest average among these hitters was .197. If the batting averages were actually normally distributed, what percentage of hitters would hit .197 or worse?

66. At what percentile is the magic .300 average?

For Exercises 67–70, consider that a 1991 survey done by the U.S. Bureau of Justice shows that the age of inmates in state prisons was approximately normally distributed with mean 32.4 years and standard deviation 9.9 years.

67. Estimate the percentage of inmates over 25 years old.

68. Estimate the percentage of inmates between the ages of 50 and 60.

69. At approximately what percentile is a 30-year-old inmate?

70. Estimate the age of an inmate at the 20th percentile.

For Exercises 71–74, consider that the weights of light trucks sold in the United States in 1990 were approximately normally distributed, with a mean of 4016 pounds and a standard deviation of 654 pounds.

71. Approximately what percentage of light trucks sold in 1990 weighed between 5000 and 6000 pounds?

72. Estimate the percentage of light trucks sold in 1990 weighing more than 4500 pounds.

73. Estimate the percentile of a 3000-pound truck sold in 1990.

74. Approximately what percentage of light trucks sold in 1990 weighed between 3000 and 4000 pounds?

For Exercises 75 and 76, use the fact that gestation periods (length of time from conception to birth) for humans are approximately normally distributed with a mean of 272 days and a standard deviation of 17 days.

75. What is the percentage of babies born within 2 weeks of the mean?

76. A premature infant is defined as one with a gestation period of less than 253 days. What is the percentage of babies born prematurely?

77. Seven percent of the scores on an exam are 90 or above. Twelve percent are 50 or below. If the scores on the exam are approximately normally distributed, estimate the mean and standard deviation of the exam scores.

5.5   ESTIMATING THE MEAN

In the last section, we determined how to compute the probability that a value taken at random from a normal distribution falls within a certain distance from the mean. In the next two sections, we turn this around and seek to use a sample of values to obtain an interval of values that is very likely to contain the unknown mean of a distribution, which may or may not be normal. Such an interval is called a confidence interval. In this section, we find confidence intervals for a general mean such as income. In the next section, we look at the special case of finding confidence intervals for a proportion, such as the percentage of voters supporting a particular candidate or issue.

We begin by fixing the notation we will need in this section. As before, μ and σ will denote the mean and standard deviation of some distribution. Unfortunately, in this section we are not given the values of either μ or σ. It may be that the population is so large that we cannot survey it all. This would be the case in looking at the heights of all the people in the United States. Or it may be impractical to determine the mean. This is the case with the lifetimes of a production lot of lightbulbs. If we left them burning until all failed, we would be left with no bulbs to sell. We will call μ and σ the **population mean** and **population standard deviation**, respectively, to distinguish them from the mean and standard deviation of the sample data we use to estimate them. We will let n denote the size of a representative random sample and $\bar{x}$ (pronounced "x bar") denote the mean of this sample or **sample mean**. You should think of n as fixed and $\bar{x}$ as variable, depending on the particular sample. Thus, two or more pollsters would likely obtain different values of $\bar{x}$. For large populations, the distribution of sample means is nearly the same whether or not members of the population are permitted to appear more than once in the same sample. Certainly $\bar{x}$ seems a reasonable estimate for the true mean μ. In fact, the mean of $\bar{x}$ over all possible samples is μ. The major issue to be resolved is how much the estimate $\bar{x}$ is likely to differ from μ. More precisely, we are interested in the distribution of $\bar{x}$, because $\bar{x}$ can vary for different samples. We know that the distribution of $\bar{x}$ has mean μ, but we also need further information such as its standard deviation. For this we turn to one of the most important theorems of probability and statistics.

Theorem 5.1

Consider n independent, random samples from a population with mean μ and standard deviation σ.

  (i) When all possible samples of size n are considered, the distribution of sample means $\bar{x}$ has mean μ and standard deviation $\sigma/\sqrt{n}$. If the original population has a normal distribution, so does the distribution of sample means $\bar{x}$.
 (ii) (Central Limit Theorem) As n increases, the distribution of $\bar{x}$, over all possible samples of size n, approaches a normal distribution.

A common practice, which we will follow, is to approximate the distribution of $\bar{x}$ with a normal distribution whenever $n \geq 30$. In other words, we will then use the table in Appendix A to estimate the probabilities for the possible values of $\bar{x}$.

The term *independent* in Theorem 5.1 simply means that each single value sampled from the population has no influence on the rest of the sample. (It is similar to the idea of independent events from Chapter 4.) Part (i) of Theorem 5.1 is not too hard to prove directly from the definitions for mean and standard deviation. Part (ii) is very difficult to prove.

In Figures 5.37 and 5.38, we show the distribution of $\bar{x}$ as n increases for two different population distributions. Observe that the data cluster more and more about μ as n increases, showing that the standard deviation $\sigma/\sqrt{n}$ of $\bar{x}$ decreases with n. Thus, $\bar{x}$ is usually a reasonably accurate estimate of μ for large n. Also notice how the distributions evolve into the familiar shape of the normal distribution as n increases.

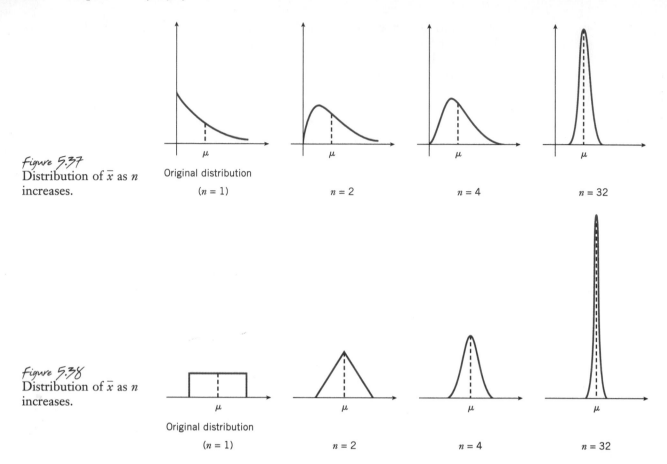

figure 5.37
Distribution of $\bar{x}$ as n increases.

Original distribution

$(n = 1)$ $n = 2$ $n = 4$ $n = 32$

figure 5.38
Distribution of $\bar{x}$ as n increases.

Original distribution

$(n = 1)$ $n = 2$ $n = 4$ $n = 32$

Confidence Intervals

To see how we will use Theorem 5.1, we begin by computing an interval about the mean that will contain the value of $\bar{x}$ obtained from our sample approximately 90% of the time. We say "approximately" because our calculations are based on the distribution of $\bar{x}$ being normal, and although the distribution is approximately normal for large n, it is only guaranteed to be normal in the case that the population distribution is normal. We first find the z-score that will correspond to a total area of 0.90 placed symmetrically about the mean. This would mean we want an area of $0.90/2 = 0.45$ on each side of the mean, as illustrated in Figure 5.39.

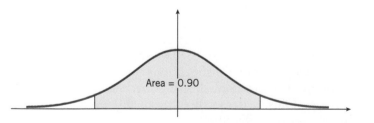

figure 5.39
The 90% confidence interval.

Area = 0.90

Therefore, we need to estimate the z-score that corresponds to a probability or area of 0.45. Because 0.4495, corresponding to a z-score of 1.64, and 0.4505, corresponding to a z-score of 1.65, are equal distances from 0.45, we split the difference, obtaining $z = 1.645$. Theorem 5.1 tells us that the distribution of sample means $\bar{x}$ has mean μ and standard deviation $\sigma/\sqrt{n}$. Therefore, the z-score of $\bar{x}$ is

$$z = \frac{\bar{x} - \mu}{\sigma/\sqrt{n}}.$$

Assuming that $n \geq 30$ so that the distribution of $\bar{x}$ is approximately normal, we know that the z-score of $\bar{x}$ should lie between -1.645 and 1.645 approximately 90% of the time. Therefore,

$$-1.645 \leq \frac{\bar{x} - \mu}{\sigma/\sqrt{n}} \leq 1.645$$

for approximately 90% of all random samples of size n. Now we use this information to find an interval about $\bar{x}$ that will contain the mean μ for approximately 90% of all random samples of size n. We may rewrite the inequality as

$$\frac{-1.645\sigma}{\sqrt{n}} \leq \mu - \bar{x} \leq \frac{1.645\sigma}{\sqrt{n}},$$

or

$$\bar{x} - \frac{1.645\sigma}{\sqrt{n}} \leq \mu \leq \bar{x} + \frac{1.645\sigma}{\sqrt{n}}.$$

We have shown that μ is between $\bar{x} - 1.645\sigma/\sqrt{n}$ and $\bar{x} + 1.645\sigma/\sqrt{n}$ in approximately 90% of all samples of size n. The interval from $\bar{x} - 1.645\sigma/\sqrt{n}$ to $\bar{x} + 1.645\sigma/\sqrt{n}$ is called the 90% confidence interval for the mean μ. There is one obvious problem. Because we do not know the true mean μ, it is almost certain that we do not know the standard deviation σ either.

Sample Standard Deviation and Confidence Intervals

In general, a simple estimate of σ is to use the standard deviation of the values of the sample. Unfortunately, this estimate turns out to be too small on average, but a slight modification makes it workable. Instead of an n in the denominator of our formulas for standard deviation, we will use $n - 1$, which gives a better estimate. The result is called the sample standard deviation and is often denoted by an s.

The **sample standard deviation** s of a sample $x_1, x_2, \ldots, x_n$ is given by

$$s = \sqrt{\frac{\sum_{i=1}^{n}(x_i - \bar{x})^2}{n - 1}} = \sqrt{\frac{(x_1 - \bar{x})^2 + (x_2 - x)^2 \cdots + (x_n - \bar{x})^2}{n - 1}},$$

where $\bar{x}$ denotes the sample mean. When the data are presented by a frequency distribution with the values $x_1, x_2, \ldots, x_n$ occurring with frequencies $f_1, f_2, \ldots, f_n$, respectively, the sample standard deviation is given by

$$s = \sqrt{\frac{\sum_{i=1}^{n} f_i(x_i - \bar{x})^2}{\left(\sum_{i=1}^{n} f_i\right) - 1}} = \sqrt{\frac{f_1(x_1 - \bar{x})^2 + f_2(x_2 - \bar{x})^2 + \cdots + f_n(x_n - x)^2}{f_1 + f_2 + \cdots + f_n - 1}}.$$

The sample standard deviation is a better estimate for σ because the mean value for s^2, taken over all possible samples of size n, is exactly σ^2. For large n, there is very little difference between the sample standard deviation and the standard deviation.

Just as the 90% confidence interval reflects an interval with a 90% probability of containing the mean μ, the $c\%$ confidence interval has a $c\%$ probability of containing μ. Because we have chosen an interval that is symmetric about the sample mean or, equivalently, because when our interval does not contain μ, the mean μ is equally likely to be too big or too small, the confidence interval is said to be **two-tailed**. Although all of the confidence intervals in this book will be two-tailed, there are circumstances where a one-tailed confidence interval is more appropriate, such as when using samples to determine if pollution level is under some acceptable threshold on average.

Confidence Intervals and Margin of Error for Means

Consider a given confidence level $c\%$ and the z-score from the table for the standard normal distribution that corresponds to a probability or area of $c\%/2$. Let $\bar{x}$ be the sample mean of a sample of size n, with $n \geq 30$, and let s be its sample standard deviation. Then a (two-tailed) $c\%$ **confidence interval** for the true mean μ is the interval from $\bar{x} - \dfrac{zs}{\sqrt{n}}$ to $\bar{x} + \dfrac{zs}{\sqrt{n}}$.

The quantity $\dfrac{zs}{\sqrt{n}}$ is called the **margin of error** at the $c\%$ confidence level. The estimate $\bar{x}$ will be within the margin of error of the true population mean μ in about $c\%$ of all samples.

You can see that increasing the confidence level increases the probability that the interval will contain the mean but, of course, simultaneously increases the margin of error. A 90% confidence interval implies that there is a 10% probability that the true mean will lie outside the confidence interval due to the random nature of sampling. *However, it does not address the possibility of errors due to nonrandom sampling, confusing questions, or inaccurate responses.* We will return to such issues in Section 5.7. Furthermore, our confidence interval depends on the distribution of $\bar{x}$ being close to normal.

Example 1   *Margin of Error and Confidence Interval*

A 1992 study by Katherine A. Dettwyler, published in the *American Journal of Physical Anthropology*, of 320 women living in rural Mali yielded a mean height of 63.1 inches with a sample standard deviation of 2.2 inches.

(a) Find the margin of error at the 95% confidence level if we estimate the mean height of rural Mali women to be 63.1 inches.

(b) Find a 95% confidence interval for the mean height of rural Mali women.

Solution:

(a) Looking up the z-score corresponding to an area of $0.95/2 = 0.475$, we find $z = 1.96$. Thus, the margin of error is $1.96 \cdot 2.2/\sqrt{320} \approx 0.241048 \approx 0.2$ inch. Note that we have rounded the margin of error so that its accuracy

BRANCHING OUT 5.3

THE CENSUS CONTROVERSY: HOW SHOULD WE COUNT PEOPLE?

The population counts determined by the U.S. census have significant implications. The most obvious are the apportionment of the 435 seats of the House of Representatives and the allocation of federal funds based on population counts. Because of this, politicians have a great interest in the outcome of a census count. The U.S. Constitution mandates that the population shall be determined by an "actual enumeration" of people, but it does not specify how this enumeration is to be done. It is decidedly impractical to actually count every person. Even in 1790, when the population was less than 4 million, when Thomas Jefferson delivered the first U.S. census to President George Washington, he wrote the official results in black but also included his estimate of the real count in red. Since then, there has always been a discrepancy between the official census and the true population count. It is estimated that the 1990 census missed over 10 million people, while counting some people twice or in the wrong location.

Because attempting to count every person is expensive and inaccurate, the U.S. Bureau of the Census proposed that the census in the year 2000 be based on statistical sampling techniques rather than a strict count. For Census 2000, the bureau planned to attempt to have all households report their data through the mail or via the telephone or possibly the Internet. However, instead of attempting to seek out *every* nonresponding household, as it had in the past, the bureau planned to contact directly at least 90% of nonresponding households and use sampling techniques to make statistical estimates of the remaining households. To double-check the resulting population estimates and make any necessary adjustments, the bureau planned to conduct a second, independent sample of 25,000 blocks about the size of a city block containing a total of 750,000 households and 1.7 million people.

The proposal to use sampling for Census 2000 did not go unnoticed by politicians. When the census attempts to count every person, the people most often missed are urban minorities, who traditionally vote Democratic. As a result, Republicans stand to lose House seats if statistical sampling is used. Not surprisingly, the sampling method has solid support from Democrats and strong opposition from the Republican Party. In February 1998, the House Republican leadership filed a lawsuit seeking to prevent the use of sampling for the census. In a 5–4 vote in January 1999, the U.S. Supreme Court ruled that statistical sampling could not be used to determine the population count used for apportioning the House. However, it left open the possibility of using sampling to arrive at the population counts used in allocating federal funds. In regard to the census controversy, an editorial in the Fort Worth *Star Telegram* observed, "It is well to remember that the politicians who decry using a scientific sampling based on 10 percent of the uncounted homes are happy to stake their political futures on polls that are based on much smaller samplings."

is consistent with that of the mean and sample standard deviation of the sample.

(b) The 95% confidence interval is from $63.1 - 1.96 \cdot 2.2/\sqrt{320} \approx 63.1 - 0.2 = 62.9$ inches to $63.1 + 0.2 = 63.3$ inches. With probability approximately 95%, the mean height of rural Mali women is between 62.9 and 63.3 inches.

There are more advanced methods than those we consider to compare two populations. However, even with the techniques we have, we can draw rough conclusions, as illustrated in the next example.

Example 2   *Weights of High School Football Players*

A study by J. F. Hickson, Jr., et al., published in 1987 in the *Journal of the American Dietetic Association* and involving the diets of 88 high school football players, found a mean body weight of 167.3 pounds with a sample standard deviation of 29.8 pounds, somewhat higher than a mean of 146.6 pounds with a sample standard deviation of 24.9 pounds found in a sample of 1394 high school boys.

(a) Find the 99% confidence interval for the mean weight of a high school football player.

(b) Find the 99% confidence interval for the mean weight of a high school boy.

(c) One would expect high school football players to be bigger than the average boy. Do the data support this?

Solution: For the 99% confidence level, we look up the z-score corresponding to an area of $0.99/2 = 0.495$. Because 0.495 falls exactly midway between 0.4949 and 0.4951 with corresponding z-scores of 2.57 and 2.58, we will use $z = 2.575$.

(a) The confidence interval for the mean weight of high school football players is from $167.3 - 2.575 \cdot 29.8/\sqrt{88} \approx 167.3 - 8.2 = 159.1$ pounds to $167.3 + 8.2 = 175.5$ pounds. With probability approximately 99%, the mean weight of high school football players is between 159.1 pounds and 175.5 pounds.

(b) The confidence interval for the mean weight of high school boys is from $146.6 - 2.575 \cdot 24.9/\sqrt{1394} \approx 146.6 - 1.7 = 144.9$ pounds to $146.6 + 1.7 = 148.3$ pounds. With probability approximately 99%, the mean weight of high school boys is between 144.9 pounds and 148.3 pounds. Notice that the increase in sample size results in a much smaller interval than we found in part (a).

(c) Because the confidence intervals are far apart, the data strongly support a conclusion that high school football players are bigger than the average boy.

Example 3   *Scottish Militiamen*

A study by A. Quetelet of the chest sizes of 5738 Scottish militiamen done over 150 years ago is summarized in Table 5.24.

Table 5.24	Chest Sizes of Militiamen		
Chest Size (inches)	Frequency	Chest Size (inches)	Frequency
33	3	41	934
34	18	42	658
35	81	43	370
36	185	44	92
37	420	45	50
38	749	46	21
39	1073	47	4
40	1079	48	1

Source: A. Quetelet, "Lettres à S.A.R. le Due Régnant de Saxe–Cobourg et Gotha, sur la Théorie des Probabilités," *Appliquée aux Sciences Morales et Politiques.* Brussels: M. Hayez, 1846.

Find a 99.5% confidence interval for the mean chest size for Scottish militiamen from that period.

Solution: First we must compute the mean and sample standard deviation for the militiamen data. Using the formula for the mean of data presented by a frequency distribution, we have

$$\bar{x} = \frac{3 \cdot 33 + 18 \cdot 34 + 81 \cdot 35 + \cdots + 21 \cdot 46 + 4 \cdot 47 + 1 \cdot 48}{5738}$$

$$= \frac{228{,}555}{5738} \approx 39.8318229 \approx 39.83.$$

Using the formula for the sample standard deviation of data presented by a frequency distribution, we have

$$s \approx \sqrt{\frac{3 \cdot (33 - 39.83)^2 + 18 \cdot (34 - 39.83)^2 + \cdots + 1 \cdot (48 - 39.83)^2}{5738 - 1}} \approx 2.05.$$

For the 99.5% confidence level, we look up the z-score corresponding to an area of $0.995/2 = 0.4975$, which leads to $z = 2.81$. We find that the 99.5% confidence interval is from $39.83 - 2.81 \cdot 2.05/\sqrt{5738} \approx 39.83 - 0.08 = 39.75$ inches to $39.83 + 0.08 = 39.91$ inches. With probability approximately 99.5%, the mean chest size of Scottish militiamen at the time was between 39.75 inches and 39.91 inches.

∎

Perhaps the most commonly used confidence level is 95%, especially for polls, followed by the 99% level, which may be more appropriate for medical studies and other studies where acting on the results may have serious repercussions.

EXERCISES FOR SECTION 5.5

1. A sample has sample standard deviation 10. Fill in the margin of error in the table for the given sample size n and confidence level.

	Confidence Level			
n	*60%*	*70%*	*80%*	*90%*
50				
100				
500				
1000				

2. A sample has sample standard deviation 25. Fill in the margin of error in the table for the given sample size n and confidence level.

		Confidence Level		
n	*96%*	*97%*	*98%*	*99%*
100				
300				
900				
2500				

3. In a study by R. Perez-Olmos et al. published in 1990 in the *Belgian Journal of Food Chemistry and Biotechnology*, a sample of 50 Spanish wines had a mean fluoride concentration of 0.20 milligram per liter with a sample standard deviation of 0.078 milligram per liter.

(a) Find the margin of error at the 94% confidence level if one estimates the mean fluoride concentration to be 0.20 milligram per liter.

(b) Find a 94% confidence interval for the mean fluoride concentration of these wines.

4. In a Stanford study by W. Otts, P. Switzer, and K. Willits published in 1994 in the *Journal of the Air Waste Management Association*, a sample of 88 trips on a fixed section of the California Highway El Camino Real in the early 1990s yielded a mean carbon monoxide concentration of 9.8 parts per million with a sample standard deviation of 5.8 parts per million.

(a) Find the margin of error at the 99% confidence level if one estimates the carbon monoxide concentration to be 9.8 parts per million.

(b) Find a 99% confidence interval for the mean carbon monoxide level.

5. In a 1995 study (by P. Jones, L. M. Collins, K. T. Ingram in *Transactions of the ASAE*) of an open top chamber system that was designed to analyze the growth of rice when temperature and carbon dioxide levels are elevated, 288 readings had a mean carbon dioxide level of 660.5 parts per million with a sample standard deviation of 26.6 parts per million. Find a 96% confidence interval for the true mean level.

For Exercises 6–9, consider that the densities of rocks can be estimated by measuring many samples. The data for these exercises are given in the *Handbook of Physical Properties of Rocks*, vol. III, Boca Raton, Fl.: CRC Press, 1984.

6. Find the 98% confidence interval for the mean density of granite, given that 334 samples yield a mean of 2.66 grams per cubic centimeter with a sample standard deviation of 0.06 gram per cubic centimeter.

7. Find the 85% confidence interval for the mean density of basalt, given that 323 samples yield a mean of 2.74 grams per cubic centimeter with a sample standard deviation of 0.47 gram per cubic centimeter.

8. Find the 94% confidence interval for the mean density of quartz porphyry, given that 76 samples yield a mean of 2.62 grams per cubic centimeter with a sample standard deviation of 0.06 gram per cubic centimeter.

9. Find the 97% confidence interval for the mean density of sandstone, given that 107 samples yield a mean of 2.22 grams per cubic centimeter with a sample standard deviation of 0.23 gram per cubic centimeter.

Exercises 10–12 are based on a study by L. A. Jarosz published in 1990 in the *Journal of the American Dietetic Association* on infant feeding and on communication with nutritionists in Liberia.

10. Based on a sample of 74 infants who drank milk-based liquids, the mean age for introducing such liquids was 1.5 months with a sample standard deviation of 1.6 months. Find a 75% confidence interval for this mean.

11. Based on a sample of 113 infants who received any breast-milk foods, the mean age for introducing such foods was 1.9 months with a sample standard deviation of 2.1 months. Find an 80% confidence interval for this mean.

12. Based on a sample of 110 infants who ate nonmilk foods, the mean age for introducing such foods was 2.9 months with a sample standard deviation of 2.5 months. Find an 88% confidence interval for this mean.

For Exercises 13 and 14, consider that fish is often contaminated with mercury, which is toxic at fairly low levels. A study in the early 1990s (by A.-H. Bou-Olayan and S. N. Al-Yakoob in the *Journal of Environmental Science Health, Part A: Environmental Science Engineering*) measured the level of mercury in hair samples of residents of Kuwait. The Swedish limit for mercury is 6 micrograms per gram.

13. The mean level for the 68 female study participants was 4.05 micrograms of mercury per gram of hair, with a sample standard deviation of 4.43 micrograms per gram. Find a 92% confidence interval for the mean mercury level in hair samples from Kuwaiti females.

14. The mean level for the 38 male study participants was 5.52 micrograms of mercury per gram of hair, with a sample standard deviation of 5.37 micrograms per gram. Find a 94% confidence interval for the mean mercury level in hair samples from Kuwaiti males.

15. A 1984 study of anemia in inner city children by Pablo Vazquez-Seoane, Robert Windom, and Howard A. Pearson published in the *New England Journal of Medicine* in 1985 compared hemoglobin blood counts, measured in grams per deciliter (g/dl), for those children enrolled in a particular HMO and given iron supplements, to hemoglobin counts from a similar 1971 study, where supplements were not given.

    (a) The 1984 study consisted of 324 children who were given iron supplements and had a mean hemoglobin count of 11.8 g/dl with a sample standard deviation of 0.9 g/dl. Find the 99% confidence interval for the mean hemoglobin count of children similar to those in the study.

    (b) The 1971 study consisted of 258 children who were not given iron supplements and had a mean hemoglobin count of 11.1 g/dl with a sample standard deviation of 1.7 g/dl. Find the 99% confidence interval for the mean hemoglobin count of children similar to those in the study.

    (c) Does it seem that iron supplements help prevent anemia? Explain.

    (d) If one assumes that the distribution of hemoglobin counts is normal, estimate the percentage of children in the 1984 study who would fall below the 9.8 g/dl level (very low). (The actual percentage was about 1%.)

    (e) If one assumes that the distribution of hemoglobin counts is normal, estimate the percentage of children in the 1971 study who would fall below the 9.8 g/dl level. (The actual percentage was about 23%.)

    (f) Do your answers to (d) and (e) support or refute an assumption that the distributions are normal?

16. A study (P. L. Beyer et al., *Journal of the American Dietetic Association*, 1995) of 51 patients with Parkinson's disease yielded a mean weight change of −7.2 pounds (that is, a loss) with a sample standard deviation of 2.9 pounds, compared with a mean of 2.1 and sample standard deviation of 1.6 for a control group of 49 subjects.

    (a) Find the 95% confidence interval for the mean weight change in Parkinson's disease patients.

(b) Does this seem convincing evidence that Parkinson's disease is associated with weight loss? Explain.

(c) Find the 95% confidence interval for the mean weight change in the control group.

(d) Does this support or seem to contradict your conclusion in (b)? Explain.

Exercises 17–21 concern a study by K. K. Cook, N. R. Gregory, and C. M. Weaver published in 1990 in the *Journal of the American Dietetic Association*, which examined the sodium found in certain products as a percentage of the amount claimed on the product's label.

17. The study revealed the following percentages for desserts and snack foods:

> 130, 100, 103, 65, 80, 80, 86, 103, 124, 80, 92, 103, 111, 107, 65,
> 93, 55, 124, 80, 110, 66, 84, 80, 92, 79, 97, 72, 76, 118, 94.

Find the 95% confidence interval for the mean percentage.

18. The study revealed the following percentages for prepared vegetables:

> 85, 98, 20, 91, 111, 101, 104, 79, 174, 92, 122, 86, 60, 91, 98, 85, 110,
> 88, 122, 102, 44, 74, 74, 5, 96, 68, 101, 71, 98, 88, 44, 46, 73, 76.

Find the 95% confidence interval for the mean percentage.

19. The study concluded that package labeling was reasonably accurate for those trying to restrict sodium intake. Based on your answers to Exercises 17 and 18, do you agree with this conclusion? Explain.

20. Suppose you want to estimate the mean percentage of reported sodium actually found in prepared vegetables with a margin of error of at most 2 at the 95% confidence level. Based on the sample standard deviation found in Exercise 18, estimate the size of the sample you would need.

21. Suppose you want to estimate the mean percentage of reported sodium actually found in dessert and snack foods with a margin of error of at most 1 at the 95% confidence level. Based on the sample standard deviation found in Exercise 17, estimate the size of the sample you would need.

22. Names were chosen at random from the Fort Worth, Texas, 1998/1999 phone book and the number of letters in the last name counted. The frequency distribution of the name lengths is given in the table. (The 17-letter name was a hyphenated name.)

Length of Last Name	3	4	5	6	7	8	9	10	17
Frequency	2	3	16	11	7	2	4	4	1

Find the 99% confidence interval for the mean length of last names in the Fort Worth phone book.

23. Three of the authors' classes were surveyed to determine the number of siblings each student had. The findings are given in the table.

Number of Siblings	0	1	2	3	4	5	6	8
Frequency	7	33	30	10	5	2	1	1

Assuming that these students are a random sample of the student body, find the 92% confidence interval for the mean number of siblings in the student body.

24. The level of sulfite residues, in parts per million (ppm), in samples of maraschino cherries is given in the table.

Sulfite Level (ppm)	Frequency
<40	31
40–79.9	10
80–119	8
120–159	2
>160	2

Source: Julie A. Nordlee, Laura B. Martin, Steve L. Taylor, "Sulfite Residues in Maraschino Cherries." *Journal of Food Science* 50 (1985), 256–257.

Estimate the 85% confidence interval for the mean sulfite level. Note that you must choose a representative level for the >160 class.

Exercises 25–28 consider cases where the standard deviation, rather than the sample standard deviation, is given. In such cases, you must first find the sample standard deviation, s, by using the following conversion formula:

$$s = \sqrt{\frac{n}{n-1}} \cdot (\text{standard deviation of the sample}).$$

25. A sample of 151 young children living in an area near a lead smelter in Romania (M. M. Verberk et al., *Archives of Environmental Health*, 1996) found a mean lead concentration of 342 micrograms per liter of blood, with a *standard deviation* of 224 micrograms per liter. Find a 98% confidence interval for the mean level of lead in the entire population of young 8 children in that area.

26. A 1979 study by Friedrich V. Wong in the *Journal of Police Science and Administration* of death anxiety among 120 male police patrol officers consisted of a 15-question true-false questionnaire. The mean score was 7.23 with a *standard deviation* of 3.20. Find an 80% confidence interval for the mean anxiety score within the police force.

27. A test of shoulder joint flexibility measures mobility in degrees, with an accepted minimal threshold for adequate function of 120 degrees. A study (E. J. Bassey et al., *European Journal of Applied Physiology*, 1989) including 255 women between 65 and 74 years of age in the late 1980s revealed a mean of 124 degrees with a *standard deviation* of 19 degrees. Find a 98% confidence interval for the true mean of women in this age group.

28. Show that the given formula for converting from a standard deviation to a sample standard deviation is correct.

5.6 POLLS AND MARGIN OF ERROR

A poll with two possible answers, such as "yes" or "no," "agree" or "disagree," "approve" or "disapprove," may be thought of as a frequency distribution with values 1 for "yes" and 0 for "no." In this case, the mean of the frequency distribution is simply the proportion or percentage of the population in the 1 category. Here we will use π to denote the unknown proportion of the population falling into one of two categories and seek to estimate π by sampling. Once we know π, we know everything about this special frequency distribution, known as a **Bernoulli distribution**. We leave it as an exercise to show that the standard deviation of such a frequency distribution is $\sigma = \sqrt{\pi(1 - \pi)}$. We will use p to denote the proportion obtained in a random sample of size n. As in the general case of estimating a mean, p will serve as our estimate of π. However, we do not estimate the standard deviation in quite the same way in this special case, and we will take a more conservative approach. The graph in Figure 5.40 shows that the maximum value of $\sigma = \sqrt{\pi(1 - \pi)}$ for π between 0 and 1 is 0.5.

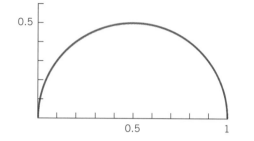

Figure 5.40
The graph of $\sqrt{\pi(1 - \pi)}$ for π between 0 and 1.

We will use 0.5 as the estimate for the standard deviation in our computation of confidence intervals, because the actual standard deviation cannot be any larger than this. Specializing our definition of confidence interval to this case, with this minor modification, we obtain the following definitions. We include assumptions on n and p that allow us to use the normal distribution.

Confidence Intervals and Margin of Error for Proportions

Consider a given confidence level $c\%$ and the value z from the table for the standard normal distribution that corresponds to a probability or area of $c\%/2$. Let p be the proportion of a sample of size n that falls into a certain category. Assume $n \geq 30$ with $np \geq 5$ and $n(1 - p) \geq 5$. Then a (two-tailed) $c\%$ **confidence interval** for the true proportion π is the interval from

$$p - \frac{z(0.5)}{\sqrt{n}} \text{ to } p + \frac{z(0.5)}{\sqrt{n}}.$$

The quantity $\dfrac{z(0.5)}{\sqrt{n}}$ is called the **margin of error** at the $c\%$ confidence level. The estimate p will be within the margin of error of the true population proportion π in about $c\%$ of all samples.

You may use either decimals or percentages to represent a proportion, whichever seems more appropriate, as long as you are consistent throughout the calculation. Recall that the confidence level does not address the possibility of errors due to bias in the sampling, confusing questions, or inaccurate responses.

Example 1 *Margin of Error in a Survey*

In a 1996 random survey of 150 people with a household income of at least $200,000 a year by the *Milwaukee Journal Sentinel*, 67% felt they were not saving enough for retirement. Find the margin of error at the 95% confidence level.

Solution: Looking up the z-score corresponding to an area of $0.95/2 = 0.475$,

"Yes. No. Sometimes. No. No. Yes.
Don't know. Sometimes. Yes. Yes. No."

we find $z = 1.96$. Thus, the margin of error for this survey is

$$1.96(0.5)/\sqrt{150} \approx 0.08002 \approx 8\%.$$

Note that we have rounded the 0.08002 to 8% so that the accuracy is consistent with the 67% found in the survey information.

∎

Technology Tip

Calculators and computers make it possible to compute exactly the confidence intervals for proportions as long as the sample size is not too large. This is better than the approximations we use based on z-scores. However, for the medium-sized samples we are considering, our approximations are usually fairly close to the exact confidence intervals.

Example 2   *A Gallup Poll*

When Gallup surveyed 1007 randomly chosen adults in 1993, 70% favored stricter gun laws. Find the 99% confidence interval for the percentage of the general, adult population favoring stricter gun laws.

Solution: For the 99% confidence level, we look up the z-score corresponding to an area of $0.99/2 = 0.495$, and use $z = 2.575$ because 0.495 is midway between 0.4949 and 0.4951. The margin of error is

$$2.575(0.5)/\sqrt{1007} \approx 0.04057 \approx 4\%.$$

The 99% confidence interval is from $70\% - 4\% = 66\%$ to $70\% + 4\% = 74\%$. With probability 99%, the proportion of the adult population favoring stricter gun laws is between 66 and 74%.

∎

BRANCHING OUT 5.4

ESTIMATING WILDLIFE POPULATIONS

Even when obtaining an actual count of an animal population in a region is impossible, it may be possible to obtain an estimate of the population by a very simple technique known as the *capture-recapture* or *mark-recapture method*. This method has been proposed as part of the statistical sampling techniques for the U.S. census in 2000. On the first of two counts, a number of animals are captured and tagged, or otherwise marked, and released. After a time that allows the captured animals to disperse within the region being surveyed, a number of animals are again captured or otherwise counted in a way that allows marked and unmarked animals to be distinguished. If n stands for the unknown animal population and m stands for the number of animals marked in the first count, then the basis of the population estimate is that the percentage p of marked animals "captured" in the second stage should be approximately the percentage of the entire population that has been marked, or $p \approx m/n$. Solving for n, we find that $n \approx m/p$. (This estimate is *statistically biased,* which means that its expected value is too large, as in this case, or too small. There is a modification of this formula giving an unbiased estimate.)

For instance, around 1919 K. Dahl marked 109 trout from certain Norwegian lakes and later captured 177 trout, of which 57 were marked. The rough estimate of the trout population is $109/(57/177) \approx 338$. (A correction for bias would lead to an estimate

of 337 trout. The difference would be more pronounced for smaller populations and samples.)

Applying this method has two requirements. Those animals that have been marked must have the same probability of being recaptured as the other animals have of being captured for the first time. Animals that have been captured may be wary of traps or, if traps are baited with especially appealing food, may become more attracted to traps. Often a different type of trap or bait is used. Furthermore, within the time period of the survey, the population in the region must be *closed*. In other words, there should not be significant changes in the population due to births, deaths, or migration. It is also possible for the mark to be lost, as often happens in the case of tagging fish. To allow for this, it is common to put two tags on the fish in order to get an estimate of the percentage of tags lost. More advanced methods attempt to account for such factors or to test the extent to which such potential problems occur.

In designing a poll, there are often times when a certain degree of accuracy is desired. Of course, you would prefer perfect accuracy, but you must balance this desire with the costs and time involved in gathering a sample. Once we have determined the confidence level and desired margin of error, we may use the formula for margin of error to find the required sample size.

Example 3   *A Political Poll*

Imagine yourself working for one of two mayoral candidates in your city. You want to determine the proportion of voters intending to vote for your candidate, with a margin of error of 5 percentage points at the 90% confidence level. How many random voters should be in your survey?

Solution: The area $0.90/2 = 0.45$ in Appendix A falls exactly midway between 0.4495 and 0.4505, with corresponding z-scores of 1.64 and 1.65. Thus, we will use 1.645 as our z-score. We want to find the value of n for which the margin of error is 5%. In other words, we want to solve the equation

$$\frac{1.645(0.5)}{\sqrt{n}} = 0.05.$$

Clearing the denominator yields

$$1.645(0.5) = 0.05\sqrt{n},$$
$$\sqrt{n} = \frac{1.645(0.5)}{0.05} = 16.45,$$
$$n = 16.45^2 = 270.6025.$$

Thus, our survey should include at least 271 voters to have a margin of error of at most 5%. Because nice, round numbers have some aesthetic appeal and perhaps even seem more credible to some, you may want to survey 275 or even 300 voters.

◼

There are also times when we know everything about a poll except the confidence level. We may use the formula for margin of error to find the confidence level whenever we know the margin of error and sample size. Because the margin of error is often rounded to the nearest percent, the confidence level we find may not be exactly that used for the poll. You should keep in mind that 95% is the most common confidence level.

Example 4   *Finding the Confidence Level*

A Gallup poll of 1002 adults early in 1996 found that 7% definitely intended to vote for Ross Perot in the coming fall election, with a margin of error of 3%. What is the most likely confidence level of the poll?

Solution: This time z is the unknown in the formula for the margin of error. We need to solve the equation

$$\frac{z(0.5)}{\sqrt{1002}} = 0.03$$

for z, and then translate this value into a confidence level. Solving for z, we find $z = 0.03 \cdot \sqrt{1002}/0.5 \approx 1.8993 \approx 1.90$. In Appendix A, we find the area 0.4713, which corresponds to a confidence level of $2(0.4713) = 0.9426$. Although the nearest integer percentage would be the 94% confidence level, the confidence level used in the poll was probably 95%.

To check whether 95% is reasonable, we could compute the margin of error for the 95% confidence level, using $z = 1.96$ (computed as in Example 1), obtaining $1.96(0.5)/\sqrt{1002} \approx 0.03096$, which would round to the given 3% margin of error.

◼

EXERCISES FOR SECTION 5.6

1. Fill in the margin of error for a proportion in the table for the given sample size n and confidence level. Give your answer as a percentage.

	Confidence Level			
n	60%	70%	80%	90%
50				
100				
500				
1000				

2. Fill in the margin of error for a proportion in the table for the given sample size n and confidence level. Give your answer as a decimal.

	Confidence Level			
n	96%	97%	98%	99%
100				
300				
900				
2500				

3. In a 1998 CBS News/*New York Times* poll of 1126 adults, 69% of those surveyed agreed with the statement "In many of our largest industries, one or two companies have too much control of the industry." Find the margin of error for the poll at the 95% confidence level.

4. A Gallup poll of 1010 American adults conducted in 1995 revealed that 77% favored President Clinton's proposal to raise the minimum wage from $4.25 to $5.15. Find the margin of error for the poll at the 99% confidence level. (The proposal passed the next year.)

5. In a 1996 *Washington Post* article, William Casey (of the Census Bureau, not the ex-director of the CIA) reported that he collected 11,902 coins lying on the ground over a 10-year period. Of the 1530 he found in 1995, 48.65% lay heads up.

   (a) If these represent a random sample of all coins on the ground, what is the margin of error in the 48.65% figure at the 96% confidence level?

   (b) Do you think there might be a reason that more coins were found lying tails up than heads up?

6. A Discovery News poll conducted by Chilton Research Services on May 17, 1997 found that 30% of 517 adults surveyed had tried marijuana at some point in their lives. Find the margin of error of this estimate at the 99.5% confidence level.

7. A 1997 poll commissioned by *Business Week* found that 259 of 1003 households sampled used the Internet or World Wide Web. Find a 97% confidence interval for the percentage of households using either the Internet or World Wide Web.

8. Out of a random 4770 healthy girls and women in Taiwan, 2934 tested positive for antibodies to rubella in a study published in 1993 (D.-B. Lin and C.-J. Chen, *Journal of Medical Virology*). Find the 92% confidence interval for the percentage of girls and women in Taiwan who had the antibodies at that time.

Exercises 9 and 10 are based on a 1979 study by Friedrich V. Wong of death anxiety among 120 male police patrol officers published in the *Journal of Police Science and Administration*. Assume that the 120 officers represent a random sample of male officers.

9. The study found that 62.5% of the officers in the study were married. Find a 91% confidence interval for the percentage of 1979 male police patrol officers who were married.

10. The study found that 43.5% had contact with death on their first year on the job. Find a 97% confidence interval for the percentage of 1979 police patrol officers who had contact with death on their first year on the job.

11. Suppose you are working in advertising and want to determine what percentage of adults know what product is being promoted in one of your extremely clever television ads. If you are satisfied with a 4% margin of error at the 99% confidence level, how many people must be included in your random survey?

12. Suppose a marketer in a pharmaceutical company wants to determine how many allergy sufferers are helped by the company's product, with a 3% margin of error at the 90% confidence level. How many subjects must be tested to determine the effectiveness of the product?

13. In 1996, opponents of a proposed sports complex for downtown Los Angeles released a poll including the result that only 77% of those surveyed had heard of the proposal. If the confidence level for the poll was 95% and its margin of error was 3.5%, estimate the number of respondents to the poll.

14. A U.S. Bureau of the Census report found that in 1994 13% of all 15- to 17-year-olds had been held back a year in school. If the confidence level for the poll was 95% and its margin of error was 1.5%, estimate the number included in the sample.

15. A 1996 *Congress Daily* poll of 500 young adults found that more than half believed the soap opera *General Hospital* would last longer than the Medicare program, with a margin of error of 4.4%. Estimate the confidence level of the poll.

16. A CNN/*USA Today*/Gallup poll of 1089 adults conducted just days before the House of Representatives voted to impeach President Bill Clinton on December 19, 1998, found that 62% of those polled wanted their Representative to vote against the impeachment of President Clinton. The poll had a margin of error of 3%. Estimate the confidence level used in this poll.

17. A Harris survey of 1004 adults in the United States conducted in 1996 found that 40% had guns in their homes, with a margin of error of 3%. Estimate the confidence level of the survey.

18. In a 1998 CBS News poll of 1118 adults, 74% of the respondents answered yes to the question "Do you think there was an official cover-up to keep the public from learning the truth about the Kennedy assassination?" For what confidence level would the margin of error be approximately 4%?

19. The *Arizona Republic* conducted a telephone poll of 800 adult residents of Arizona that had a 3.5% margin of error. Broken down to only the residents of Pima County, it had an 8% margin of error. Estimate the number of residents of Pima County included in the poll. (The confidence level was the same for all parts of the survey.)

20. A 1996 *Newsweek* poll found that 45% of registered voters were less likely to support

a candidate who favors gay rights, including marriage. The margin of error was 5%. The registered voters were a part of a survey of 779 adults, not all registered voters, that had a margin of error of 4%. Roughly how many registered voters were among the 779 adults? (The confidence level was the same for all parts of the survey.)

21. In a two-candidate race, explain how the margin of error in the proportion of those who favor one of the candidates is related to the margin of error for the difference in proportions between the two candidates.

22. Show that the standard deviation of a frequency distribution whose values 1 and 0 have relative frequencies π and $1 - \pi$, respectively, is $\sqrt{\pi(1 - \pi)}$.

5.7 GARBAGE IN, GARBAGE OUT: A LOOK AT MISLEADING USES OF STATISTICS AND AT SAMPLING TECHNIQUES

In the first six sections of this chapter, we saw many ways to depict data and to estimate characteristics of a population. A well-designed study identifies the population of interest and chooses members of the population to be sampled at random. It asks clear questions for which answers may be categorized and it asks the questions in such a way that the survey itself has limited influence on the subjects' answers. The report of the study's findings clearly identifies the group queried and the questions they were asked. In addition to means and proportions, the size of the sample and margin of error should be reported. Any subjective conclusions drawn from the data should be clearly recognizable as interpretations rather than hard facts. Unfortunately, a flaw in any of the steps leading up to our final result may negate its reliability. It is reasonable to be suspicious of numbers that do not seem right to you. They may be correct, but, if important, may merit further investigation. In this section we focus on surveys and polls because they involve so many ways to go astray. In an attempt to help the public draw educated conclusions from polls, the National Council on Public Polls has adopted the standard that polls prepared for public release should include information on the sponsor, dates and methods of interview, wording of the questions, size and composition of the sample, and percentages on which conclusions are based. The less such information provided, the more room there will be for a healthy suspicion.

We will examine three facets of a survey or poll. First is the source of the data. Where do the data come from? Who or what is surveyed? What is the size of the survey? The second facet we examine is the set of data itself. Are the questions and answers clear? Are there internal or external factors that influence the answers? Finally, we look at the conclusions drawn from the survey. Are the data pertinent to the conclusion? Is the evidence sufficiently strong to merit a conclusion?

The Source of the Data

Although no more misleading or malicious than a poorly designed or deceptive survey, the most flagrant abuse of statistics is a statistic with no evidence to back it up. For instance, in the early 1990s, several newspapers reported the claim that reports of domestic violence increased 40% on Super Bowl Sundays. It turns out that this figure was entirely anecdotal, with no supporting evidence. Of course,

Polls for the 1948 presidential election incorrectly predicted that Thomas Dewey would defeat President Harry Truman.

this does not mean that domestic violence is not a serious problem or even that the 40% figure is false. In fact, however, the evidence suggests only that there is an increase in domestic violence on many key days of the year, including Super Bowl Sunday.

A sample could be too small to yield meaningful conclusions. A sample mean given without a confidence level and margin of error means little. Moreover, a true population mean will fall outside the 95% confidence interval roughly 1 in 20 times. Similarly, when events occur randomly in time, there will be periods with many events and lulls with few events. Thus, reports of increases in crime, such as crimes against foreigners or arson or insurance fraud may (or may not) simply reflect a cluster of events that occasionally happens over time. Similarly, surveys with statements such as, "Seven out of 10 doctors recommend . . . ," may suffer from two different flaws. First, the wording is ambiguous. Does it mean that only 10 doctors were surveyed—not nearly enough—or that 70% of some unreported number of doctors recommended the product? Second, it is not unheard of for companies or advertisers to unethically repeat an experiment on a fairly small scale until the desired results are obtained.

The sample may not be representative of the target population due to a flawed sampling scheme. The standard on which error estimates are based is random sampling. If we were to use a math class using this text as a sample of college students, we would exclude students enrolled at other colleges and almost all students majoring in the sciences or engineering. Thus, our sample would not represent a random segment of all college students.

Some problems are more subtle. For instance, suppose we want to know what percentage of people in a city had tried a certain brand of soft drink. We construct our sample by selecting and telephoning numbers at random from a phone book. This method may not introduce the enormous errors we may encounter with our student sample, but there are numerous potential problems. Although it seems as if everyone has a telephone, roughly 5% of all households across the United States do not. That is, we will miss 5% of all households. (Do you think people in this 5% of households are more or less likely to have tried the soft drink?) We also miss those with unlisted phone numbers. Furthermore, our sample selects a random phone number, not a random person. Households with two phone numbers are twice as likely to be called as those with one. On the other hand, families with seven members have only the same chance of being called as a single-person household, but should be seven times more likely. The time of day we conduct the survey may influence who answers the phone. If we call during the day, we may be more likely to reach mothers with small children and those who work nights. If we want to ensure randomness of a telephone survey, we must make follow-up efforts to reach those phones that went unanswered. For instance, it has been found that Republicans are somewhat more likely than Democrats to go out on Friday nights.

We cite two clear-cut examples. Midway Airlines boasted in the mid-1980s that 84% of frequent business travelers from New York to Chicago preferred Midway to American, United, and TWA. With only 8% of the New York–Chicago traffic, Midway revealed in the advertisement's fine print that the survey was "conducted among Midway Metrolink passengers between New York and Chicago." Of the subjects in Kinsey's oft-quoted study of sexual habits and preferences, an abnormally high 1 in 4 had a prison record.

In these two examples, any margin of error due to randomness in sampling is rendered meaningless by the obvious sampling flaws. One method used to help ensure that a sample is representative is called *stratified sampling*. If a pollster has a breakdown of the population in question into groups that may be likely to have significantly different responses, then he or she may sample from each group randomly in proportion to its population. Differences in responses may be due to income, age, race, religion, and many other factors. By using stratified sampling, the pollster obtains a cross section that is more representative of the entire population than a single, unified sample might be. For instance, a study may attempt to follow up on those in the study but not interviewed, either because they could not be contacted or because they refused to cooperate. Results from these follow-ups, which could still be a disproportionately small number, could then be factored in with a weight equal to the proportion of the original sample that was missed.

We may decide that some of the factors that prevent our sample from being random are probably not relevant. For instance, if we want to determine the breakdown of colors of automobiles in a given neighborhood, it may be reasonable to simply survey the cars going through a particular intersection. The only way to detect problems in sampling methodology is to use a broader sample to determine if there are significant differences in results, and this can be expensive. Sometimes, one can get away with doing a smaller survey to check the validity of a survey method, and then to do a larger survey to get more precise results.

In general, it is helpful to know if the source of the study has an interest in its outcome. For instance, studies paid for by the manufacturer of a product that purport to show that product's safety should always be viewed with some skepticism.

The Data: Questions and Answers

We next turn our attention to problems that may affect the data obtained from a survey. First, there may be errors in the data. Clerical errors occur. A measurement may be inaccurate; a person's answer may be incorrect or fabricated. A wrong number may be typed when data are entered into a computer. The raw data could be correct, but means and margins of error may be computed incorrectly or recorded incorrectly. When these errors enter at the level of the indi-

BRANCHING OUT 5.5

DID MENDEL "PLANT" HIS DATA?

Suppose you flipped a coin 1000 times. You would probably be surprised if you got 800 heads, but you should also be somewhat surprised if you get exactly 500 heads, or even come within one or two of 500 heads. Gregor Mendel's famous experiments with peas looked at the percentage of offspring having dominant or recessive traits for several different characteristics. More details of the theory are in Section 4.8. Sir Ronald A. Fisher did some statistical computations in the 1930s which suggested that results as close to the theoretical values as those obtained by Mendel would occur by chance only 7 in 100,000 times. He conjectured that one of Mendel's assistants was guilty of falsifying the data. Articles by Alain Corcos and Floyd Monaghan and by Franz Welling in the 1980s have disputed Fisher's work, mostly based on the statistical methods he used, but also based somewhat on knowledge of genetics gained over the 50 years that elapsed between Fisher's work and their own work. Whatever the truth, there is no doubt that Mendel's intellectual achievement remains one of the greatest in genetics, and perhaps in science.

Gregor Mendel (1822–1884).

vidual respondent, they tend to have a minimal effect on the outcome of a survey. On the other hand, problems with the questions or the methodology of recording answers may invalidate the entire survey.

We begin with problems with the questions. A widely quoted Roper poll found that 22% of Americans found it possible that the Holocaust never happened and 12% did not know whether it had happened. However, the wording of the question, "Does it seem possible or does it seem impossible to you that the Nazi extermination of the Jews never happened?" was convoluted at best, so that recorded responses were not necessarily what those interviewed intended. In fact, a follow-up poll (of different individuals) found that 91% felt certain that the Holocaust really happened, with only 1% saying that it was possible it did not happen.

Even when the wording of a particular question is clear, there are many potential pitfalls. The order of choices in a multiple choice survey may influence the answer. It is well known among pollsters that the first choice in a written survey and the last choice in an oral survey tend to be selected more than they would be in another position. One way to address this problem is to vary the order of choices for different respondents. This carries over to elections, which are just (nonrandom) surveys. The first position on the ballot is a prized one. Some localities vary the order of positions among the ballots in order to negate this unfair advantage. In a German study done by the Institute für Demoskopie, when asked orally if traffic contributes more or less to air pollution than does industry, 45% said traffic contributes more and 32% said industry contributes more. When the order was reversed and interviewees were asked if industry contributes more or less to pollution than does traffic, only 24% said traffic contributes more, whereas 57% said industry contributes more. Two *Washington Post* polls in the early 1990s asked whether President Clinton's health plan was better or worse than the current system. In the first poll 52% said better, 34% said worse, and 14% volunteered that they did not know. In the second poll, interviewees were given the choices "better," "worse," or "don't know enough about the plan to say." Here, only 21% said better and 27% worse, whereas 52% chose the option that they did not know enough.

Surrounding questions may influence responses. In a Gallup survey, when asked whether they were worried about maintaining their standard of living over the next few years, 58% responded that they were worried. However, when the question was preceded with a question asking whether the interviewees were worried that they or their spouses would lose a job, only 48% were worried about maintaining their standard of living. Given the shifts in response from seemingly innocent shifts in a question, imagine what deliberately loaded questions could do.

The interviewers themselves may have great influence on the survey. When whites were asked in an ABC News/*Washington Post* poll whether or not they agreed with the statement "The problems faced by blacks were 'brought on by blacks themselves,'" 62% agreed when interviewed by whites, but only 46% agreed when interviewed by blacks. For similar reasons, tests of new drugs are often conducted by a double-blind study, where neither the patient nor the doctor knows whether a particular patient is taking the drug being tested as opposed to another drug or placebo.

We turn to the answers to surveys. Again there are many potential stumbling blocks. An answer to a survey question may be incorrect for a variety of reasons. When asked about events in the past, people do not always have a completely accurate recollection. When asked about sensitive subjects such as criminal history, drug use, sexual behavior, religion, or politics, responses may be deliberately inaccurate, even when confidentiality is guaranteed. People may answer questions without knowledge or conviction. Furthermore, to quote the 1995 article "Consulting the Oracle" from *U.S. News and World Report*, "Pollsters have long known that people tend to offer conventional opinions more than unconventional ones, socially accepted opinions rather than disreputable ones and any opinion instead of admitting ignorance. In one famous study, a third of people surveyed offered an opinion about a nonexistent 'Public Affairs Act.'"

There are examples that show how the process of being interviewed changes the future behavior of those interviewed. For instance, a voter who is polled about an upcoming election may be more likely to vote in the election as a result. Another instance when being part of a poll might change behavior is in the case of a person whose television-watching habits are monitored electronically to come up with ratings of television shows. When unmonitored, many of us will watch shows we would not admit to watching or think of turning on under the watchful eye of a monitor.

The Conclusions Drawn from the Data

We end with a very brief look at the conclusions drawn in a statistical study. For example, although it appears that women are more likely than men to believe in horoscopes and psychics, when one takes into account the amount of education in mathematics and science, the difference disappears.

We should always ask if the study really answers the intended question or another one. We sometimes hear advice not to buy lottery tickets or flight insurance because a person's chances of winning the lottery or dying in a plane crash are less than, say, the chance of getting hit by lightning. We can easily calculate the probability that a person buys a winning lottery ticket, but the probability of dying in a plane crash or getting hit by lightning may reasonably be calculated in several ways. Is the probability of dying in a plane crash based on those who fly or on the entire population? Is it based on one flight or all flights in a person's lifetime? What about the difference in risk between flying on personal aircraft and large commercial jets? Certainly the flying habits of the individual in question are critically important. When we turn to the probability of getting hit by lightning, those who work or play outdoors are at greatest risk. For instance, roofers and golfers get hit by lightning much more often than the general population. Our point is that although such comparisons of probabilities might seem reasonable at first glance, probabilities such as dying in a plane crash or getting hit by lightning allow so much latitude in interpretation that they are nearly meaningless without further details. Clearly, the answers to these questions determine the relevance of the information to you.

We conclude this section with a few words of advice: With so many ways to inadvertently or deliberately introduce errors, you should avoid a blind trust in statistical surveys and polls, especially those for which you do not know the reliability and motivation of the source.

EXERCISES FOR SECTION 5.7

1. The following appeared in the *New York Times:*

   > Under the redesigned survey, in the 12 months through August, the unemployment rate was 7.6%, a half-point above the 7.1% that the department had reported. The rate for women was 6.8%, rather than 6.0%; for men, the figure went to 6.9% from 6.7% . . . .

   Why should you suspect that something is awry?

2. An Eagleton Institute survey asked for agreement or disagreement with the statement "Abortion is a private matter that should be left to the women to decide without government intervention." The proportion of men agreeing with this statement increased from 70 to 77% and the proportion of women from 64 to 84% when asked by an interviewer of one particular sex rather than the other. Which sex of the interviewer should result in the higher percentage agreement? Explain your reasoning.

3. Some studies have concluded that the life expectancy of left-handers is as much as nine years less than that of right-handers. Pertinent to these studies is the fact that when surveys are done, the percentage of left-handers in older subjects in the study is lower than for younger subjects. Can you think of another explanation for this lower percentage other than a shorter life expectancy for left-handers?

4. *Time* magazine conducted a survey in the late 1950s of the incomes of the Yale class of 1924, based on a questionnaire mailed to those addresses available through Yale's alumni records. It found an average income of over $25,000, which translates to over $150,000 in today's dollars. Does it make sense to say that the average Yale graduate from 1924 had a mean income of over $25,000 at the time the survey was done? Explain.

5. A study of records of heart-attack victims revealed that men had a 21% higher rate of heart attacks on their birthdays, and women a 9% higher rate. Can you think of an explanation for the higher than normal heart-attack rates on birthdays?

6. A 1983 article in *U.S. News and World Report* found that from 1970 to 1982 the number of doctors in the United States increased from 334,000 to 480,000, whereas their average salary (in 1982 dollars) fell from $103,900 to $99,950. The article concluded that doctors were "growing in number but not in pay." Can you think of another, perhaps more plausible, explanation for the drop in average pay than that doctors' salaries were decreasing?

7. Would you expect the order of choices on a poll to matter more if the choices are long and involved or short and uncomplicated? Explain.

WRITING EXERCISES

1. A survey mentioned on National Public Radio found that men reported an average of 11 sexual partners per lifetime and women an average of 3. Average is usually taken as the mean. Is this possible? What is probably going on?

2. Which do you think is a better measure of the level of difficulty of an exam: the median score or the mean score? Give an argument to support your answer.

3. A 1995 Gallup poll asked the following question on Form A of a survey:

Which comes closer to your point of view:
(1) Congress should vote on a balanced budget amendment *only* after supporters identify the programs they would cut to achieve a balanced budget, or
(2) Congress should vote on a balanced budget amendment *whether or not* supporters have identified the programs they would cut to achieve a balanced budget.

On Form B of the survey the same questions were asked, but (1) and (2) were asked in reverse order. The results of the survey are shown in the table.

	Survey Responses		
	Form A	*Form B*	*Total*
Statement 1	69%	78%	74%
Statement 2	29%	17%	21%
No opinion	5%	5%	5%

Why do you think the survey results differed depending on the form that was being used?

4. A study published in the October 1997 issue of the journal *Pediatrics* found that 12% of children who were delayed in starting school displayed extreme behavior problems, compared with 7% of children whose ages were normal for their grade. Does this necessarily imply that delay in starting school causes extreme behavior problems? Explain your answer.

5. Teenage drivers are in a disproportionate share of automobile accidents. Some states have reacted by placing limitations on younger drivers, such as no driving past a certain hour at night or restricting the driver to be either alone or with an adult at night. Explain what statistics you would want to see before deciding whether to support such a policy.

PROJECTS

1. In surveys in the mid-1990s, more than half of the milk containers tested in U.S. schools had less than the amount on their labels. Milk suppliers reacted by slightly overfilling containers on average, but it was still possible for an individual container to be "short-filled." Consider a machine designed to fill 8-ounce milk cartons. Assume that the amount it dispenses to a random carton is normally distributed with standard deviation 0.14 ounce. The machine is set to dispense a mean of 8.1 ounces.

 (a) What percentage of milk cartons will have less than the 8 ounces on its label?
 (b) Consider a case of 24 cartons. (Assume that the amount in one carton of a case is independent of the amount in another.) Why can you assume that $\bar{x}$ is normally distributed even though $n < 30$?
 (c) What percentage of cases will average less than 8 ounces a carton?
 (d) What percentage of the time will a 500-case shipment average under 8 ounces a carton?
 (e) What percentage of the time will a 500-case shipment average under 8.05 ounces a carton?
 (f) If the standard deviation for one carton remains 0.14 ounce, what setting for the mean amount dispensed will yield only 5% of all cartons short-filled?
 (g) If a new machine can be designed to dispense a mean of 8.1 ounces with a smaller standard deviation, what value for the standard deviation will yield only 5% of all cartons short-filled?

2. Run a dice simulation to illustrate Theorem 5.1 by carrying out the following:

 (a) Plot two histograms of the theoretical probability distribution of one roll of a fair die. Start your first interval at 0.5 with width 1 for one histogram and 0.95 with width 0.1 for the second.
 (b) Roll a die twice. Compute the sample mean of the two rolls. Do this 100 times (more is even better) and plot two histograms of the relative frequencies. Start your first interval at 0.75 with width 0.5 for one histogram and at 0.95 with width 0.1 for the second.
 (c) Roll a die 10 times. Compute the sample mean of the 10 rolls. Do this 100 times (more is even better) and plot a histogram of the relative frequencies. Start your first interval at 0.95 and use widths of 0.1.
 (d) Keep track of all your individual rolls. Plot a histogram starting at 0.5 with width 1 and compare it to your theoretical distribution from part (a).
 (e) Comment on your empirical findings and how they relate to Theorem 5.1.

3. Suppose that 10% of a prospective jury population is black but that only 18 blacks are on a panel of 250 potential jurors for a high-profile trial. In looking for signs of discrimination in the jury process, one might start by computing the probability that 18 or fewer of 250 random people from the prospective jury population are black.

 (a) Calculate the probability using the normal approximation and the methods of Section 5.4. (The standard deviation of a frequency distribution whose values 1 and 0 have relative frequencies π and $1 - \pi$, respectively, is $\sqrt{\pi(1 - \pi)}$.)
 (b) Look up the binomial distribution in a probability or statistics book and calculate the probability exactly. Compare your answers.
 (c) Write a calculator program, spreadsheet, or computer program that adds up the individual probabilities for values within an interval for a random variable coming from a binomial distribution.

4. For various values of $k > 1$ (but not necessarily an integer), try to construct a list of numbers for which the fraction of data falling at least k standard deviations from the mean is as large as possible. For instance, the fraction can be at most 1/4 for $k = 2$. See if you can find this fraction as a function of k.

KEY TERMS

bar graph, 274
bell curve, 314
Bernoulli distribution, 343
bimodal, 291
categorical data, 272
Central Limit Theorem, 333
classes, 273
confidence interval, 336, 344
data, 272
deviation from the mean, 304
frequency distribution, 272

histogram, 277
margin of error, 336, 344
mean, 295, 296
median, 293
mode, 291
normal distribution, 314
normalized histogram, 314
numerical data, 272
pie chart, 279
population mean, 333
population standard deviation, 333

pth percentile, 324
random variable, 316
range, 304
relative frequency distribution, 272
sample mean, 333
sample standard deviation, 335
standard deviation, 305, 306
standard normal distribution, 317
two-tailed confidence interval, 336
unimodal, 291
z-score, 319

CHAPTER 5 REVIEW EXERCISES

1. The nations winning the World Cup soccer tournament are listed in the table.

1934	Italy	1974	Germany
1938	Italy	1978	Argentina
1950	Uruguay	1982	Italy
1954	Germany	1986	Argentina
1958	Brazil	1990	Germany
1962	Brazil	1994	Brazil
1966	England	1998	France
1970	Brazil		

(a) Compute the frequency distribution and relative frequency distribution of the data by country.

(b) Draw a bar graph representing the frequency distribution.

(c) Draw a pie chart labeled with the percentages for each country.

2. In 1995, there were about 509,000 men and women in the U.S. Army, 435,000 in the Navy, 175,000 in the Marine Corps, and 318,000 in the Air Force.

(a) Compute the relative frequency distribution for military personnel.

(b) Draw a bar graph of the relative frequency distribution.

(c) Draw a pie chart labeled with the percentages for each military branch.

3. The average game ticket prices at NFL stadiums for the 1997–98 season are listed in the table. Draw a histogram representing the relative frequency distribution with classes of width 10 and the first interval starting at 29.5.

Arizona	$40	Green Bay	$37	Oakland	$53
Atlanta	$32	Indianapolis	$34	Philadelphia	$38
Baltimore	$43	Jacksonville	$57	Pittsburgh	$36
Buffalo	$37	Kansas City	$42	San Diego	$54
Carolina	$55	Miami	$42	San Francisco	$50
Chicago	$38	Minnesota	$33	Seattle	$34
Cincinnati	$38	N.Y. Giants	$41	St. Louis	$34
Dallas	$43	N.Y. Jets	$36	Tampa Bay	$65
Denver	$36	New England	$39	Tennessee	$45
Detroit	$36	New Orleans	$34	Washington	$74

Source: © 1998 ESPN Internet Ventures.

4. The number of offshore tracts leased from the U.S. government in the years 1990 through 1995 were 825, 676, 204, 336, 560, 835, in order by year.

    (a) Find the median. (b) Find the mean.

    (c) Find the range. (d) Find the standard deviation.

5. Cuckoos are known to lay their eggs in the nests of other bird species. The lengths of eggs, in millimeters, for cuckoo eggs found in the nests of pied wagtails and wrens are given in the following table.

Lengths of Eggs (mm)

Pied Wagtail	21.05	21.85	21.85	21.85	22.05	22.45	22.65	23.05	23.05	23.25	23.45	24.05	24.05	24.05	24.85
Wren	19.85	20.05	20.25	20.85	20.85	20.85	21.05	21.05	21.05	21.25	21.45	22.05	22.05	22.05	22.25

Source: L. H. C. Tippett, *The Methods of Statistics.*[47]

    (a) Find the mean length of cuckoo eggs laid in nests of the pied wagtail.

    (b) Find the mean length of cuckoo eggs laid in nests of the wren.

    (c) Find the range and standard deviation in the lengths of cuckoo eggs laid in nests of the pied wagtail.

    (d) Find the range and standard deviation in the lengths of cuckoo eggs laid in nests of the wren.

6. The frequency distribution of the number of credit hours in which the students in a college class (taught by one of the authors) are currently enrolled is shown in the table.

Number of Credit Hours	Frequency
12.0	6
14.0	1
15.0	19
15.5	2
21.0	1

    (a) Find the median number of credit hours.

    (b) Find the mean number of credit hours.

    (c) Find the standard deviation of the credit hours.

7. The frequency distribution of the salaries of the 50 U.S. governors is shown in the table.

Salary (dollars)	Frequency
50,000–59,999	1
60,000–69,999	3
70,000–79,999	8
80,000–89,999	10
90,000–99,999	12
100,000–109,999	5
110,000–119,999	5
120,000–129,999	4
130,000–139,999	2

Source: 1998 Information Please®
Almanac.[48]

(a) Estimate the mean salary for the governors.

(b) Estimate the standard deviation of the salaries.

In Exercises 8–11, use the fact that U.S. Bureau of Labor Statistics data from 1994 reveal that the age of the civilian labor force was approximately normally distributed with mean 38.6 years old and a standard deviation of 13.1 years.

8. Estimate the percentage of the civilian labor force between the ages of 25 and 35.

9. Estimate the percentage of the civilian labor force over the age of 40.

10. Estimate the age that falls at the 65th percentile.

11. At what percentile does age 30 lie?

12. The mean uranium content of 153 soil samples from Illinois was 3.39 (milligrams per kilogram) with a sample standard deviation of 0.85 (R. L. Jones, *Soil Science Society of America Journal*, 1991).

(a) Find a 96% confidence interval for the mean uranium content in Illinois soil.

(b) The author of the study goes on to say that the data were approximately normally distributed. Assuming this, what percentage of soil samples should have more than 5 milligrams of uranium per kilogram of soil?

13. Dietary selenium has been associated with tumor protection in animal studies. Two regions in Brazil were studied by J. C. Chang et al. and published in *Chemosphere* in 1995.

(a) Values of selenium from 162 Brazil nuts from the Acre–Rondonia region ranged from 0.03 to 31.7 parts per million with a mean and sample standard deviation of 3.06 and 4.02, respectively. Find a 91% confidence interval for the mean selenium content in Brazil nuts from this region.

(b) Values of selenium from 162 Brazil nuts from the Manaus–Belem region ranged from 1.25 to 512.0 parts per million with a mean and sample standard deviation of 36.0 and 50.2, respectively. Find a 91% confidence interval for the mean selenium content in Brazil nuts from this region.

(c) Do the data support the conclusion that the selenium content in Brazil nuts depends heavily on the region? Explain.

14. A 1998 *Newsweek* poll of 752 adults found that 53% of those surveyed thought that the stories and reports of new organizations are often inaccurate. Find the margin of error for the proportion of the population that thinks the stories and reports of news organizations are often inaccurate at the 95% confidence level.

15. In a Gallup survey of 1000 adult Americans in 1995, 769 favored the death penalty. Find a 98% confidence interval for the percentage of adult Americans favoring the death penalty when the survey was conducted.

16. How many people must be included in a random survey to have a 0.5% margin of error at the 99% confidence level?

17. In 1994, Judge Jack B. Weinstein of the U.S. District Court for the Eastern district of New York used statistics from similar cases to assess the reasonableness of damages awarded an employee of Digital Equipment who suffered a repetitive stress injury. The plaintiff was awarded $100,000 for pain and suffering. In 21 similar cases the mean award for pain and suffering was $404,219 with a sample standard deviation of $476,984. Suppose we treat these cases as a random sample. Use the sample standard deviation as a basis for estimating the number of cases the judge would need to estimate the mean pain and suffering award to within $50,000 at the 95% confidence level. (Unfortunately, additional cases were probably unavailable.)

∫UGGESTED READINGS

Asher, Herbert. *Polling and the Public: What Every Citizen Should Know*, 3rd ed. Washington, D.C.: CQ Press, 1995. The nonmathematical issues that may lead to inaccurate polls.

Crossen, Cynthia. *Tainted Truth: The Manipulation of Fact in America*. New York: Simon & Schuster, 1994. Many examples of how the sponsor of a study may influence its conclusions.

Huff, Darrell. *How to Lie with Statistics*. New York: W. W. Norton, 1954. Brief, with an emphasis on ideas rather than mathematics.

Jaffe, A. J., and Herbert F. Spirer. *Misused Statistics: Straight Talk for Twisted Numbers*. New York: Marcel Dekker, 1987. Pitfalls in gathering and presenting data. Numerous real-life examples.

Maceli, John C. "How to Ask Sensitive Questions without Getting Punched in the Nose." In *Political and Related Models*, vol. 2, edited by Steven J. Brams, William F. Lucas, and Philip D. Straffin, Jr. New York: Springer-Verlag, 1983, 169–182. Ways to gather group data on sensitive issues without knowing individual responses.

Moore, David S. *Statistics: Concepts and Controversies*. New York: W. H. Freeman, 1991. Many examples with real data. Qualitative issues such as designing polls, data sources, misuse of significance tests.

Solomon, Herbert. "Confidence Intervals in Legal Settings." In *Statistics and the Law*, edited by Morris H. DeGroot, Stephen E. Fienberg, and Joseph B. Kadane. New York: John Wiley, 1986, 455–473. A series of case studies.

Creating communications networks, planning transportation routes, and designing manufacturing processes often involve thousands or even millions of pieces of information. The mathematical translation of the questions of optimal solutions to such problems brings us to graph theory, the focus of this chapter. Graph theory was launched by Leonhard Euler in 1736, when he analyzed the puzzle of the Königsberg bridges. Today, graph theory is routinely used in business and manufacturing. Computers, along with mathematical advances, have made it possible to solve some graph theory problems of enormous magnitude. The scope and complexity of problems in communications, transportation, and manufacturing would overwhelm the fastest foreseeable computers unless they are paired with a mathematical approach that reduces the scale of such problems. The rewards of applying this kind of combination of mathematics and computer science have included

savings of billions of dollars over the past 30 years.

Suppose we want to construct the least-expensive network linking a group of computers. Perhaps we need to construct the shortest route along city streets for mail delivery or garbage pickup. Suppose a salesman wants to visit a group of cities in the shortest possible round-trip. Each of these problems may be formulated in terms of graph theory. In this chapter, we examine these three problems. Collectively, they show some of the major ideas of applied graph theory. We will demonstrate a fast and simple solution to the computer network problem. Although there also exists a fast and effective method for solving the problem of mail delivery or garbage pickup, its complexity allows us only a glimpse of this solution. Finally, finding a fast method of determining the best route for the traveling salesman's trip is one of the most famous unsolved problems in mathematics. It is quite possible that no fast method exists. Currently we must be satisfied with finding solutions that are nearly optimal.

A view of a street network in Sun City, Arizona.

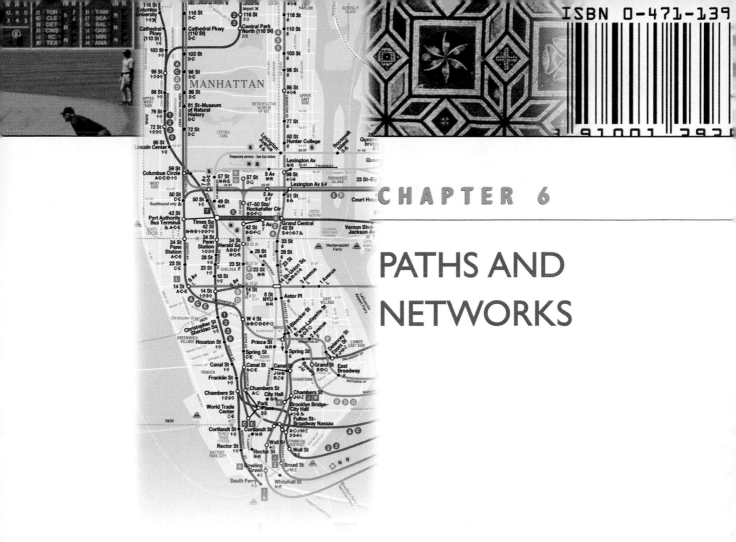

ISBN 0-471-139

CHAPTER 6

PATHS AND NETWORKS

6.1   EULERIAN PATHS AND CIRCUITS ON GRAPHS

In the eighteenth century, the Prussian town of Königsberg (now called Kaliningrad) was divided by the Pregel River, as shown in Figure 6.1. The town included the two islands in the middle of the river and land on each bank of the river. The islands and land were connected to one another by seven bridges.

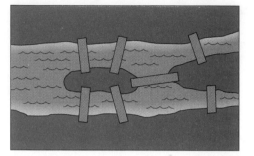

Figure 6.1
The Prussian town of Königsberg.

Leonhard Euler
(1707–1783).

Some of the townsfolk amused themselves by trying to find a path through the city that would cross each bridge exactly once. The path could start anywhere and need not end where it started. Try to trace out such a path yourself and you will see that it does not seem possible. This problem came to be known as the Königsberg bridge problem. The Swiss mathematician Leonhard Euler, then at St. Petersburg, considered the problem and published a paper in 1736 that not only proved that no such path existed but also laid the foundations for a new area of mathematics called graph theory. Euler's great contribution was to see how the complex picture of the map could be replaced by a simpler picture that captured the essence of the problem.

To solve the Königsberg bridge problem, Euler replaced each piece of land by a point and each bridge by a curve joining the points. In Figure 6.2, the four different pieces of land are labeled with points *A*, *B*, *C*, and *D* and curves are drawn between the points along each of the bridges.

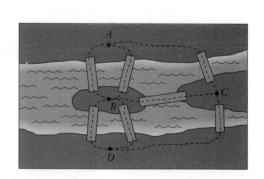

Figure 6.2

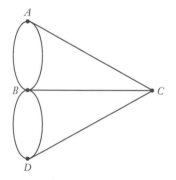

Figure 6.3
Königsberg bridge problem graph.

This map of land and bridges can be viewed more simply by erasing the underlying picture and smoothing out the curves. We are left with the simple figure in Figure 6.3.

Figure 6.3 is an example of a graph. In general, a **graph** consists of points called vertices (a single point is called a **vertex**) and lines or curves starting and ending at vertices called **edges**. The mathematical subject of **graph theory** is the study of graphs and their applications. Before looking at Euler's solution to the Königsberg bridge problem, we introduce a few general graph theory definitions.

An edge of a graph may connect two different vertices or it may start and end at the same vertex. If an edge starts and ends at the same vertex, it is called a **loop**. In Figure 6.4, the edge beginning and ending at vertex A is a loop.

Figure 6.4
A graph with a loop at vertex A.

Figure 6.5

A vertex with an odd number of edges attached to it (a loop at a vertex is counted twice) is called an **odd vertex**. A vertex with an even number of attached edges is called an **even vertex**. For instance, in the graph in Figure 6.5, the vertices B, C, and D are even: vertices B and D have four edges attached to them and vertex C has two edges attached to it. The vertices A and E are odd because A has one edge attached to it and E has five edges attached to it, where the loop at E has been counted twice.

When drawing graphs it is sometimes convenient or necessary to have edges that cross over one another, as they do in the graph shown in Figure 6.6. Note here that you can tell that the point at which the edge between A and D and the edge between B and C cross is not a vertex because we will always indicate vertices by dots. We will not consider these two edges to be intersecting, but instead think of one of the edges as passing above the other.

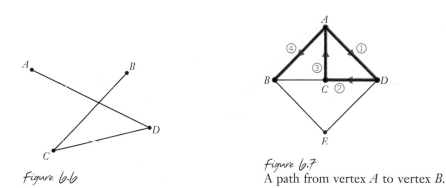

Figure 6.6

Figure 6.7
A path from vertex A to vertex B.

A **path** on a graph is a route along the edges that starts at a vertex and ends at a vertex. In Figure 6.7, a path from vertex A to vertex B is marked in blue. The direction of the path is indicated by the arrows and the edges have been numbered in the order in which they are traveled. A path that begins and ends on the same vertex is called a **circuit**. A graph is **connected** if for any two of its vertices there is at least one path connecting them. Thus, a graph is connected if it consists of one piece. If a graph is not connected, it is **disconnected**. The graph in Figure 6.8(a) is connected, whereas the graph in Figure 6.8(b) is disconnected.

BRANCHING OUT 6.1

GRAPH THEORY AND THE SOLUTION OF THE FOUR COLOR PROBLEM

How many different colors would be needed to color any map in such a way that no two regions, for instance countries or states, with a common border are colored with the same color? (Countries that only touch at single points are not considered to have a common border.) To color the map of South America, which includes Brazil, Bolivia, Paraguay, and Argentina, we see that we would need four colors because each of these four countries has a common border with each of the other three.

The question of whether or not four colors would suffice for coloring any map was first formally asked in 1852 by Francis Guthrie, a graduate student at University College in London, and the problem eventually became known as the Four Color Problem. Even though no map could be found that required more than four colors, a proof that four colors would work for *every* map remained elusive in spite of the efforts of many mathematicians. The fact that four colors appeared to be enough did not ensure that they were, in fact, enough. Perhaps there was some strange map that no one had thought of that would require more than four colors.

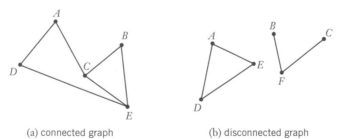

Figure 6.8   (a) connected graph   (b) disconnected graph

In the Königsberg bridge problem, we are trying to find a path that traverses each edge of the graph exactly once. We have the following important definitions:

A path that includes each edge of a graph exactly once is called an **Eulerian path**, and an Eulerian path that is also a circuit is called an **Eulerian circuit**.

The Four Color Problem can be viewed as a problem in graph theory by translating the map into a graph. This is done by representing each country by a vertex, and joining two vertices with an edge if and only if the corresponding countries have a common border. For instance, we see here how the map of five states is translated to a graph.

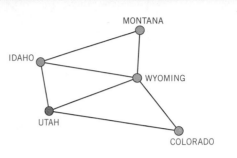

The problem of coloring the map can be seen as the problem of coloring the vertices of the corresponding graph in such a way that no two vertices connected by an edge have the same color. The vertices of the graph corresponding to the map of the five states are colored in this way. Using this graph-theoretical approach, American mathematicians Kenneth Appel and Wolfgang Haken proved in 1976 that four colors would suffice to color any map. Their proof enlisted the help of a computer program that checked approximately 1500 special cases, requiring about 1200 hours of computer time. To date, no one has found a proof that does not make use of computer calculations.

The path from *D* to *C* in Figure 6.9 is an Eulerian path because it includes each edge of the graph exactly once. See if you can find some of the other Eulerian paths in the graph.

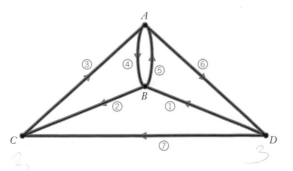

Figure 6.9

The path shown in Figure 6.10 is an Eulerian circuit because it includes each edge of the graph exactly once and it begins and ends at the same vertex, *A*. You should try to find some of the other Eulerian circuits.

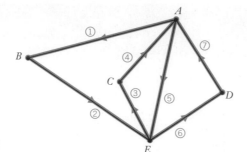

Figure 6.10

We now see that the Königsberg bridge problem is the problem of finding an Eulerian path on the graph in Figure 6.3.

Eulerian paths and circuits have many practical applications. In garbage collection, the workers must complete a route that covers every street in an area. For maximum efficiency, it is desirable to cover each street exactly once. In mail delivery or reading parking meters, a worker may need to cover both sides of some streets, but these routes can also be modeled by graphs for which we want to find Eulerian paths or circuits. We look at some applications of this type later in this section and in the exercises.

To understand how Euler analyzed the Königsberg bridge problem, we consider the graph in Figure 6.11. We will see that this graph has an Eulerian path but not an Eulerian circuit. The path in Figure 6.12 shows an Eulerian path. So far, so good.

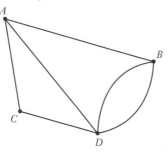

Figure 6.11

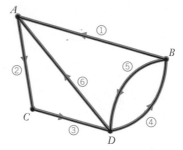

Figure 6.12

What happens if you try to trace out an Eulerian circuit on the graph? After many tries, you might guess that there is none. Let's look at the Eulerian path we discovered for clues to what is happening. Edges meeting at a vertex, except for the vertices at the beginning and end of the path, are covered two at a time: one on the way into the vertex and one on the way out. This is a general property holding for every Eulerian path or circuit. Therefore, the vertices in a graph for which an Eulerian path or circuit exists must all be even, with the possible exception of the initial and final vertices. In an Eulerian circuit, the path begins and ends at the same vertex, so that this vertex will also be even. In an Eulerian path with different initial and final vertices, both of these vertices will be odd. Therefore, there cannot be an Eulerian circuit on a graph unless all vertices are even and there cannot be an Eulerian path on a graph unless all the vertices are even or only the initial and final vertices are odd. Euler discovered this and stated that these conditions are enough to guarantee the existence of an Eulerian circuit or path, although it took over 100 years for a formal proof of the existence to be published by mathematician Carl Hierholzer. This result came to be known as **Euler's Theorem**, which we will now formally state.

> ## Euler's Theorem
>
> **(a)** A graph has an Eulerian circuit if and only if it is connected and all of its vertices are even.
>
> **(b)** A graph has an Eulerian path if and only if it is connected and has either no odd vertices or exactly two odd vertices. If two of the vertices are odd, then any Eulerian path must begin at one of the odd vertices and end at the other.

Note that if a graph has an Eulerian circuit, then it automatically has an Eulerian path, because the Eulerian circuit is just a special kind of Eulerian path.

We now return to the problem we began with—namely, whether or not it is possible to find a path through the city of Königsberg that crosses each bridge exactly once. As we have seen, this is equivalent to the question of whether or not there exists an Eulerian path on the graph in Figure 6.13.

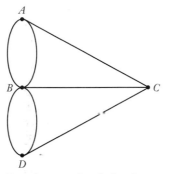

Figure 6.13
Königsberg bridge
problem graph.

Because each of the four vertices in the Königsberg bridge graph in Figure 6.13 is odd, Euler's Theorem tells us that there is no Eulerian path on this graph, and therefore it is impossible to walk through the city of Königsberg along a path that crosses each bridge exactly once.

In Example 1, we look for Eulerian paths and circuits. If we are trying to find an Eulerian circuit, then we can start on any vertex. However, if we are trying to find an Eulerian path on a graph with two odd vertices, we must start the path at one of the odd vertices.

Example 1   *Eulerian Paths and Circuits*

For each graph in Figures 6.14, 6.15, and 6.16, determine whether the graph has an Eulerian circuit, and if it does, give one. If the graph does not have an Eulerian circuit, determine whether it has an Eulerian path, and if it does, give one.

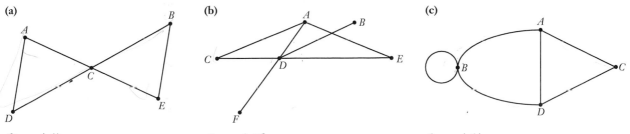

Figure 6.14   **Figure 6.15**   **Figure 6.16**

Solution:

(a) The vertices of this graph are all even, so we know it has an Eulerian circuit. One circuit is shown in Figure 6.17.

(b) The four vertices A, B, D, and F are all odd, so by Euler's Theorem we know that the graph has neither an Eulerian circuit nor an Eulerian path.

(c) Vertices B and C are even, whereas vertices A and D are odd. The graph does not have an Eulerian circuit, but it does have an Eulerian path starting and ending at the odd vertices. One such Eulerian path is shown in Figure 6.18.

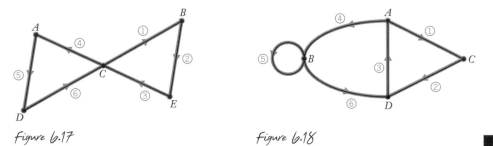

figure 6.17   *figure 6.18*

Note that if a graph has one Eulerian path or circuit, it may have several different ones. For small graphs, using trial and error to find Eulerian paths or circuits works fine. However, with a larger graph a somewhat more systematic approach might be helpful. One such method simply requires that, at each step in tracing out a path or circuit, we should never cover an edge that will cause the uncovered parts of the graph to become disconnected. This method is illustrated in the next example.

Example 2   *Finding an Eulerian Path*

Find an Eulerian path in the graph in Figure 6.19.

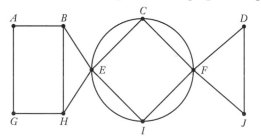

Figure 6.19

Solution: Because the graph has exactly two odd vertices, B and H, we know that it has an Eulerian path but not an Eulerian circuit. We start at vertex B (the other odd vertex, H, would be okay too). Beginning with the path marked in Figure 6.20, we have chosen edges that have kept the uncovered part connected up to this stage.

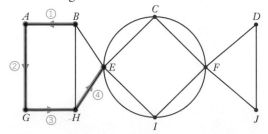

 Figure 6.20

At this point, we need to make a careful choice about what to do next. If we now choose to move back to vertex B, the uncovered parts will be disconnected. Therefore, we should choose any of the other edges going out from vertex E. We will choose to go to vertex I next. Continuing in this way, and being careful not to choose edges that will cause the uncovered part to become disconnected, one possible path is shown in Figure 6.21.

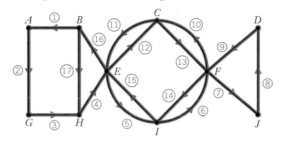

Figure 6.21

You should trace this out yourself to see that it works. You might also try to find some of the other Eulerian paths in the graph.

We will now apply graph theory to a practical problem. The first step is to find a graph that models the problem we are trying to solve. This idea is illustrated in the next example.

Example 3   *Postal Worker Route*

Suppose a postal worker must deliver mail to the four-block neighborhood shown in Figure 6.22. She plans to park her truck at one of the street intersections and deliver mail to each of the houses. The streets on the outside of the neighbor-

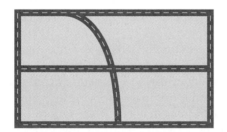

Figure 6.22
Street layout of the neighborhood.

hood have houses on only one side, and the interior streets have houses on both sides. She must walk down each of the interior streets twice (once for each side) and each of the outside streets once. Draw a graph corresponding to the walking route she needs to cover. Determine whether or not the graph has an Eulerian circuit, and if so, find one.

Solution: In the graph corresponding to this problem, each street intersection in the neighborhood is a vertex. The streets with houses on only one side must be covered only once, so they are represented by one edge. The streets with houses on both sides must be covered twice, so they are represented by two edges. We ignore the lengths of the streets and how they are curved. The graph corresponding to the route the postal worker needs to cover is shown in Figure 6.23. Because each of the vertices in this graph is even, we know that it has an Eulerian circuit. There are many possibilities; find a few yourself. One possible Eulerian circuit is shown in Figure 6.24.

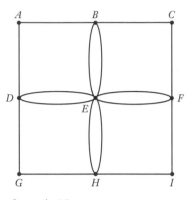

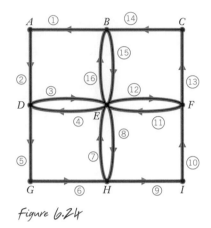

figure 6.23
Graph model of the walking route.

figure 6.24

Eulerization

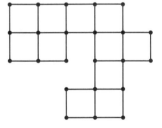

figure 6.25
Street layout of the neighborhood.

Suppose a newspaper delivery person must deliver papers in a neighborhood with streets laid out as in the graph in Figure 6.25. Because he delivers papers to houses on both sides of the street at once, he would like to travel down each street exactly once. He wants to begin and end his route at his home, which lies on a corner in the neighborhood. The question of whether or not such a route exists is equivalent to the question of whether or not the graph has an Eulerian circuit. Because the graph has several odd vertices, we know by Euler's Theorem that such a route is not possible.

To cover all of the streets of the neighborhood, the newspaper delivery person will have to go down some streets more than once. But how can he do this efficiently?

Our newspaper route example leads to the more general question of how to cover all of the edges in a connected graph having some odd vertices in such a way that the number of edges that are reused is kept small. We can visualize the reused edges by altering the original graph, adding new edges corresponding to the edges we will reuse. These new edges are colored green to signify that they were not part of the original graph and will always represent duplicating an existing edge. To get a new graph that has an Eulerian circuit, we want to add new edges in such a way that the altered graph has only even vertices. This process of adding edges in order to guarantee an Eulerian circuit in the new graph is called **eulerization**. *It is essential to note that in eulerization, we are only allowed to add new edges between vertices that have an edge between them in the original graph.*

To see how eulerization works, consider the simple graph in Figure 6.26(a). Because this graph has two odd vertices, it cannot have an Eulerian circuit. However, if an edge is added between the odd vertices *A* and *C*, then the resulting graph, shown in Figure 6.26(b), has all even vertices and therefore has an Eulerian circuit.

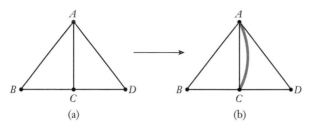

(a)   (b)

Note that traveling along the added edge on the altered graph actually represents reusing the edge in the original graph. Another way to eulerize the graph would be to add edges between vertices A and B and between vertices B and C, but this would require adding two new edges.

Ideally, in eulerizing a graph, we want to minimize the total length of new edges, for this would minimize the length of the complete circuit. For instance, in the case of the newspaper route, we want to minimize the length of the entire route. The problem of finding such an eulerization is called the *Chinese postman problem*, and it was first studied by the Chinese mathematician Meigu Guan. However, for simplicity, we will not worry about the lengths of the edges and instead will try to keep the total number of new edges to a minimum. This problem is sometimes called the *simplified Chinese postman problem*. Note, however, that given any graph, if we break down the edges into smaller edges, so that all of the edges of the graph have approximately equal length, then the problem of minimizing the total number of new edges is essentially equivalent to the problem of minimizing the total length of the new edges.

There is an algorithm for eulerizing a graph while keeping the number of new edges to a minimum, but it is complicated in the general case, so we will simply use the following guidelines in finding eulerizations:

Guidelines for Eulerizing a Graph

1. Circle all of the odd vertices.
2. Pair each odd vertex off with another odd vertex that is close to it in the graph.
3. For each pair of odd vertices, find the path with the fewest edges connecting them in the original graph, and then duplicate the edges along this path.

A bit of exploration will convince you that there are always an even number of odd vertices, so it is possible to pair them off as required in the first step. These guidelines for eulerizing may not give the best eulerization because it is not always easy to see the best way to pair off the odd vertices. However, the guidelines work reasonably well for finding a good eulerization.

Example 4   *Eulerization*

Find an eulerization of the graph in Figure 6.27.

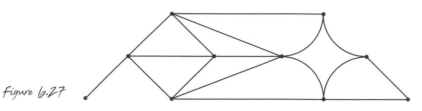

Figure 6.27

Solution: We first circle all of the odd vertices as in Figure 6.28.

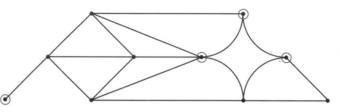

Figure 6.28

Pairing the two rightmost odd vertices, pairing the two odd vertices to the left of them, and duplicating edges along a shortest path between each of these pairs, we get the eulerization of the graph shown in Figure 6.29.

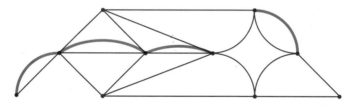

Figure 6.29

There are two other paths with just three edges between the two odd vertices on the left (find them), and therefore two other ways to eulerize the graph by duplicating only four edges. To see that this is the minimal number of edges we need to add, first note that the odd vertex farthest to the left is at least three edges away from any of the other three odd vertices, and connecting the remaining two odd vertices will require adding at least one more edge.

■

Let's now return to our newspaper route question and use eulerization to find a reasonable route.

Example 5   *Newspaper Route*

Find an eulerization of the graph corresponding to the neighborhood shown in Figure 6.25 that will allow an Eulerian circuit that would be a reasonably efficient newspaper route.

Solution: First, we circle the odd vertices of the graph corresponding to the neighborhood as in Figure 6.30. One possible eulerization is shown in Figure 6.31.

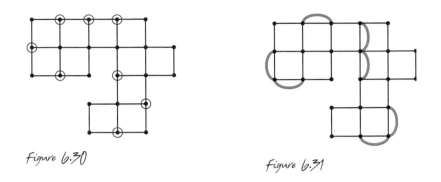

Figure 6.30   *Figure 6.31*

The eulerized graph in Figure 6.31 has an Eulerian circuit that corresponds to an efficient newspaper route.

■

In applications involving Eulerian paths or circuits, there may be additional complications we have not considered. For instance, it may very well be that the route that a truck must cover includes some one-way streets. Another complication that arises with services, such as street sweeping or snow removal, is that the vehicles must travel down only one side of the street at a time in the direction of traffic. These issues and others must be dealt with when they arise in practice.

*E*XERCISES FOR SECTION 6.1

In Exercises 1–4, list all of the odd vertices of the graph.

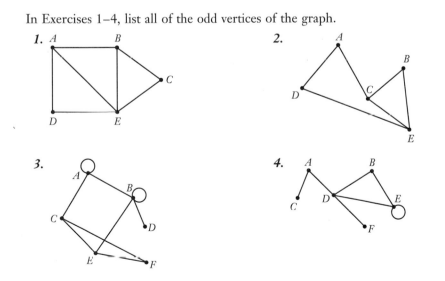

In Exercises 5–14, determine whether the graph has an Eulerian circuit, and if it does, give one. If it does not, explain why not. If the graph does not have an Eulerian circuit, determine whether it has an Eulerian path, and if it does, give one. If it does not, explain why not.

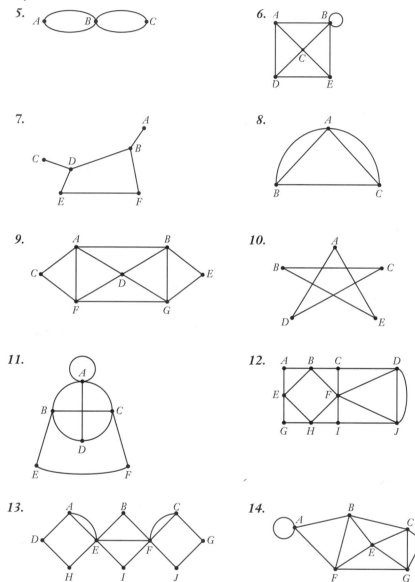

5.

6.

7.

8.

9.

10.

11.

12.

13.

14.

15. Today the city of Königsberg is called Kaliningrad, and in Kaliningrad there are two more bridges in addition to those in the original Königsberg bridge problem. A picture of the current situation is shown here. With these additional bridges, we can again ask if it is possible to start somewhere in the town and walk about the town crossing each bridge exactly once. The walk does not have to start and end at the same place. Draw the corresponding graph with the pieces of land as vertices and the bridges as edges, and answer this question.

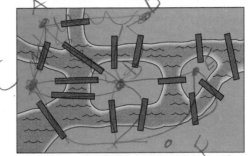

16. In his article on the Königsberg bridge problem, Euler also considered the same question for the hypothetical city illustrated here.

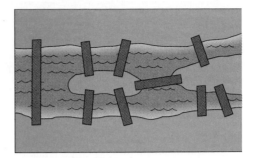

(a) Draw a graph with the pieces of land as vertices and the bridges as edges.

(b) Is it possible to start somewhere in the town and walk about the town crossing each bridge exactly once? The walk does not have to start and end at the same place. If it is possible, find such a path. If not, explain why not.

17. A meter reader wants to read the meter at each house in the three-block neighborhood pictured. He plans to park his truck at one of the street intersections and walk to each of the houses. All of the streets have houses on both sides, and he needs to travel down each street twice (once for each side). Draw a graph corresponding to the walking route he needs to cover. Determine whether or not the graph has an Eulerian circuit, and if so, give one. If it does not, explain why not.

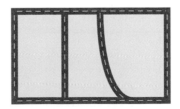

18. The sidewalks in a park are as illustrated in the figure. A city worker must sweep all of the sidewalks, and would like to cover each sidewalk exactly once. Draw a graph corresponding to this situation, and determine whether or not the graph has an Eulerian circuit, and if so, give one. If it does not, then explain why not. If the graph does not have an Eulerian circuit, determine whether it has an Eulerian path, and if it does, give one. If it does not, then explain why not.

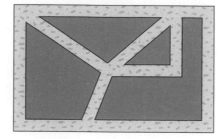

In Exercises 19 and 20, the parking meters along the streets of a city are indicated by dots in the given picture. A worker has the job of checking each of these meters and would like to find a route for which she will walk past each of the meters exactly once and not walk down streets without meters. Note that for the streets with meters on both sides, she will need to travel down the street twice. Draw a graph corresponding to the walking route she needs to cover. Determine whether or not the graph has an Eulerian circuit, and if so, give one. If it does not, then explain why not. If the graph does not have an Eulerian circuit, determine whether it has an Eulerian path, and if it does, give one. If it does not, then explain why not.

19.   **20.**

In Exercises 21 and 22, the floor plan for a house is shown.

(a) Draw a graph representing this floor plan by letting the rooms and the area outside the house be vertices and drawing an edge between any two vertices that have a door connecting them.

(b) Use the graph from part (a) to determine whether or not it is possible to start outside, go through each door of the house exactly once and end up outside again. If it is possible, give such a path. If it is not possible, explain why not.

21. 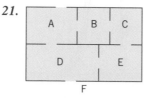   **22.**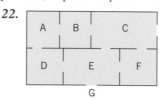

In Exercises 23–28, find an eulerization of the graph.

23. **24.**

25. 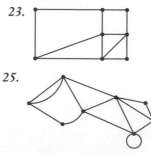   **26.**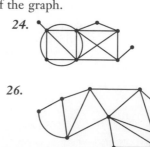

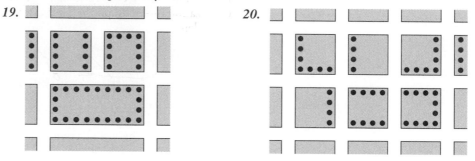

27. 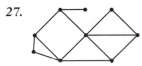   28.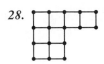

29. Suppose a garbage truck must pick up garbage in a neighborhood with streets laid out as in the graph. The workers pick up from both sides of the street at the same time, so ideally they would like to travel down each street as few times as possible. Find an eulerization of the graph that will allow the garbage truck to make a reasonably efficient Eulerian circuit through the neighborhood.

30. The layout of the garden at the Bonnefont Cloister in New York City is shown in the figure. Suppose a visitor to the garden wants to walk down each interior path at least once. Draw a graph corresponding to the walking route the visitor wants to cover. Find an eulerization of the graph that will allow the visitor to make an Eulerian circuit of the garden.

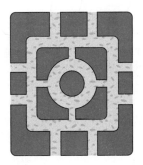

31. The sidewalks in a park are as shown in the figure. Suppose a woman wants to walk down each of the sidewalks at least once.

   (a) Draw a graph corresponding to the walking route the woman wants to cover.
   (b) Find an eulerization of the graph that will allow the woman to make an Eulerian circuit of the park.
   (c) Assuming the lengths of the sidewalks in the picture are drawn to scale, does the eulerization you found in part (b) give the shortest tour of the park? If not, find one that gives a shorter tour.

32. Explain why every graph has an even number of odd vertices.

Sir William Rowan
Hamilton (1805–
1865).

In the last section, we were concerned with finding paths or circuits that cover every edge of a graph. We now look at the problem of visiting all of the vertices of a graph, without concern for whether or not all edges have been covered. This is a somewhat different way of thinking about graphs that comes up naturally in many real-life problems. For instance, a delivery person may be required to deliver packages to several locations. If we view the delivery locations as the vertices on a graph and the roads between the locations as the edges, then we see that the delivery person is only concerned with finding a path that touches each vertex and is not concerned with what edges are covered.

The problem of finding paths that visit vertices was studied by the Irish mathematician and astronomer Sir William Rowan Hamilton. Hamilton considered the problem of finding a circuit that goes through every vertex exactly once, returning to the starting vertex. These kinds of circuits are now named after him, and we have the following definition.

> A circuit that begins at some vertex, goes through every other vertex exactly once, and returns to the starting vertex is called a **Hamiltonian circuit**.

A path that is not a circuit and goes through each vertex exactly once is called a *Hamiltonian path*. However, we will be concerned primarily with Hamiltonian circuits in this section.

Example 1   *Hamiltonian Circuits*

Find a Hamiltonian circuit, if possible, on the graphs in Figures 6.32 and 6.33.

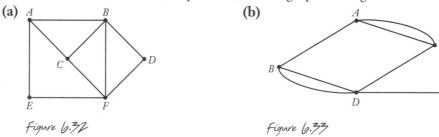

(a)   (b)

figure 6.32   figure 6.33

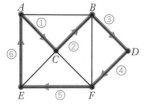

figure 6.34

Solution:
(a) The circuit in Figure 6.34 is one possible answer. See if you can find other Hamiltonian circuits starting either at *A* or at some other vertex.
(b) There are no Hamiltonian circuits on this graph because any circuit that goes through vertex *E* must go through vertex *D* at least twice.
∎

It would be nice to have a quick method for determining whether a graph has a Hamiltonian path or circuit, just as we are able to use Euler's Theorem to test whether a graph has Eulerian paths or circuits. Unfortunately, there is no known quick method for doing this. However, there are results that give us some information about particular graphs. One such result says that if a graph has at least three vertices and each vertex is connected to more than half of the other vertices in the graph, then the graph must have a Hamiltonian circuit.

Rather than studying the general problem of finding Hamiltonian circuits, we will consider a related problem, known as the traveling salesman problem, which has many practical applications. Before stating this problem precisely, we consider a specific instance of the traveling salesman problem.

Suppose a salesperson, starting in Philadelphia, must visit the cities of Cleveland, New York, and Pittsburgh on one car trip, and then return home to Philadelphia. If each city must be visited exactly once, what is the route that will minimize the distance of the entire trip? A map of the cities is shown in Figure 6.35, and the road mileage between these cities is listed in Table 6.1.

Figure 6.35

Table 6.1	Road Mileage			
	Cleveland	*New York*	*Philadelphia*	*Pittsburgh*
Cleveland	—	473	413	129
New York	473	—	101	368
Philadelphia	413	101	—	288
Pittsburgh	129	368	288	—

We can use this information to construct a new kind of graph, a weighted graph. A **weighted graph** is a graph for which a number, called a **weight**, is assigned to each edge of the graph. In the graph modeling our current situation, each city will be represented by a vertex. There will be an edge between any two vertices and the edge will be assigned a weight equal to the road mileage between the two corresponding cities. The resulting weighted graph is shown in Figure 6.36.

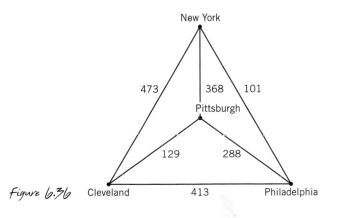

Figure 6.36

Notice that the edges of the graph are not drawn to scale, nor are the cities laid out as they are on the map. We could draw this weighted graph as it appears on a map, but it is not necessary. Our salesman's route question can now be stated in terms of graph theory as follows: Find a Hamiltonian circuit that begins in Philadelphia for which the sum of the lengths of the edges along the circuit is a minimum. In this example, it is fairly easy to answer this question by simply checking all of the possible Hamiltonian circuits beginning in Philadelphia (there are only six in this case), and finding the one whose edges have the minimum total length. If you try this, you should find that the best route for the salesman is to follow the circuit shown in Figure 6.37, going first from Philadelphia to New York, next to Cleveland, then Pittsburgh, and finally back to Philadelphia, or the reverse route that goes first from Philadelphia to Pittsburgh, then to Cleveland, next to New York, and then back to Philadelphia.

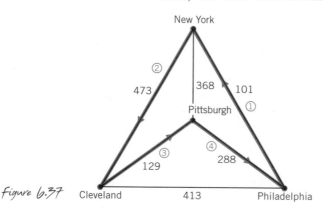

Figure 6.37

The total mileage for either of these trips would be

$$101 + 473 + 129 + 288 = 991 \text{ miles.}$$

The problem we just looked at is only one example of the traveling salesman problem. The general problem applies to an entire class of graphs, known as complete weighted graphs. A graph is called **complete** if there is exactly one edge between every pair of vertices in the graph. A complete graph with six vertices is shown in Figure 6.38.

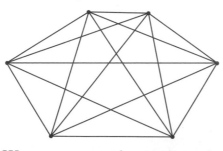

Figure 6.38
A complete graph with six vertices.

We can now state the traveling salesman problem.

> The **traveling salesman problem** is the problem of finding a Hamiltonian circuit in a complete weighted graph for which the sum of the weights of the edges is a minimum.

We refer to the sum of the weights of the edges of a circuit as the **weight of the circuit**. Thus, the solution to the traveling salesman problem is the circuit of minimum weight. It is important to note that in searching for a circuit of minimum weight, we can consider any particular vertex in the graph as the starting point for all of the circuits because every circuit can be considered as starting from any particular vertex lying on it. In most applications we will have a particular vertex in mind as the starting vertex.

The example of a salesman traveling to several cities that we looked at earlier is one instance of the traveling salesman problem. However, the traveling salesman problem also arises in other situations. A few such situations are as follows:

1. A water meter reader must find a walking route through a neighborhood for visiting each of the water meters he must read. In this case the vertices of the graph are the water meters, and the weighted edges are the routes between the meters with weight equal to the walking distance between each meter.

2. A technician must determine the order of a series of jobs on a machine so that the total amount of time required to go through all of the jobs is minimized. If any order of the jobs is possible, then we can think of the jobs as the vertices of a complete graph, and the time required to reset the machine from performing one job to performing the other as the weight of the edge joining these two jobs. For this situation, we are trying to minimize the total time rather than the total distance.

3. A district supervisor of several offices needs to find the most "efficient" route to visit each of the offices and return to her home office. In this case the vertices of the graphs are the offices. The weighted edges are the routes between the offices with weight equal to the distance or the driving time between the offices, depending on whether the supervisor is more interested in minimizing time or distance. Note that due to speed limits and traffic, minimizing time and distance may be very different problems.

We take this opportunity to emphasize that a weighted graph is an abstraction of the relevant features of a problem, and may not depict other features of the problem's setting. For instance, on the map in Figure 6.39(a), we see that Memphis lies between Little Rock and Nashville on Interstate 40. A complete weighted graph, based on mileage, for these three cities is shown in Figure 6.39(b). In traveling *directly* from Little Rock to Nashville along Interstate 40, you would have to pass through Memphis, but this would not be indicated when you travel along the edge of the corresponding graph from Little Rock to Nashville. However, because a direct trip would not include a stop in Memphis, it is irrelevant that it is along the route and nothing is really lost in the translation to the graph.

Figure 6.39
Comparing a map to a graph.

(a)   (b)

How can we solve the traveling salesman problem for a given complete weighted graph? In other words, how do we find the Hamiltonian circuit of minimum weight? One method that is sure to work is to find the weight of every Hamiltonian circuit in the graph. Then the solution is the circuit or circuits (if there is a tie) whose weight is a minimum. This is how we found the solution for our salesman who was traveling from Philadelphia to Cleveland, New York, Pittsburgh, and back to Philadelphia. However, in that case the graph had only four vertices and there were only six possible circuits to check. To see why there were only six possibilities, note that after leaving Philadelphia, the salesman has three choices for the first city to visit. After making that choice, he has only two choices for the next city to visit, because he must go to one of the two remaining cities he has not visited. After making the choice for the second city to visit, the salesman has only one choice for the third city, because it is the only city left to visit before returning to Philadelphia. Therefore, the total number of Hamiltonian circuits beginning in Philadelphia is

$$3 \cdot 2 \cdot 1 = 6.$$

For a graph with n vertices the same argument applies, and we see that the total number of different Hamiltonian circuits starting at a particular vertex on the graph is given by

$$(n - 1) \cdot (n - 2) \cdots 3 \cdot 2 \cdot 1.$$

For any positive integer k, the number $k \cdot (k - 1) \cdots 3 \cdot 2 \cdot 1$ is called k *factorial* and is denoted by $k!$. Therefore, for a complete graph with n vertices, there are $(n - 1)!$ Hamiltonian circuits starting at a particular vertex. As the number of vertices, n, gets larger, the number $(n - 1)!$ gets large quickly, as we see in Table 6.2.

Table 6.2	Growth of $(n - 1)!$
n	$(n - 1)!$
4	6
8	5,040
12	39,916,800

With 12 vertices, it would still be possible to check all of the 39,916,800 possible Hamiltonian circuits fairly quickly on a computer. However, suppose we wanted to solve the traveling salesman problem on a complete weighted graph with 30 vertices. Then the number of Hamiltonian circuits to check is $29! \approx 10^{31}$. Even if we could check these circuits at a rate of one billion per second on a computer, it would still take over 280 trillion years to check them all. For comparison, the estimated age of the universe is less than 20 billion years.

There are better methods than trying every possibility. For instance, we would no longer continue trying the different possibilities for a partial path that was doomed to exceed a value we have already found. Also, many problems have characteristics that allow us to rule out large numbers of paths as contenders for an optimal solution. In these kinds of problems, a traveling salesman problem with 30 vertices can be transformed into an easy computer exercise. Reducing the number of cases by deductive reasoning is a very important mathematical problem-solving technique that sometimes allows solutions to problems that would otherwise be overwhelming. An algorithm that takes a "reasonable" amount of time given the scale of the problem is called an *efficient algorithm*. Advances in computer hardware alone cannot keep pace with the number of vertices arising in practical applications. The attempt to find an efficient algorithm for solving the traveling salesman problem has taken on increased importance.

Although there is no known efficient algorithm that will solve the traveling salesman problem for every graph, there are methods that enable us to find a Hamiltonian circuit for which the weight is often fairly close to the minimum possible weight. These methods are called *approximate algorithms*. We will look at two relatively simple approximate algorithms that may yield fairly good answers for the traveling salesman problem. There are more advanced and better algorithms that, although rather difficult to carry out by hand, generally give excellent approximations when run on a computer.

The Nearest Neighbor Algorithm

The first approximate algorithm we will consider is called the **nearest neighbor algorithm**, and it is implemented as follows:

The Nearest Neighbor Algorithm for the Traveling Salesman Problem

1. Start at the vertex on which the circuit is supposed to begin and end.
2. At each vertex in the circuit, the next vertex in the circuit should be chosen to be the nearest vertex that has not yet been visited—in other words, the one connected to the current vertex by the edge of least weight. If there is a tie, choose any one of the nearest vertices. Mark the edges covered as you go along the path.
3. When all of the vertices have been visited, return to the starting vertex.

Example 2 *Nearest Neighbor Algorithm*

Use the nearest neighbor algorithm to find an approximate solution to the traveling salesman problem for a circuit starting at vertex *C* on the graph in Figure 6.40.

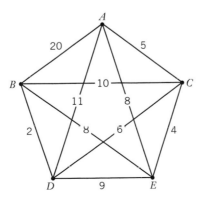

Figure 6.40

Solution: Starting at vertex *C*, the nearest neighbor is vertex *E*, so it should be the next vertex in the circuit, and we mark off our path as in Figure 6.41.

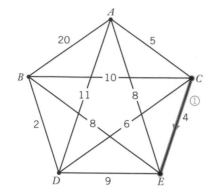

Figure 6.41

The nearest remaining vertices to *E* are vertices *A* and *B*. We may choose either one, so we choose vertex *A* and mark the edge we travel along as in Figure 6.42.

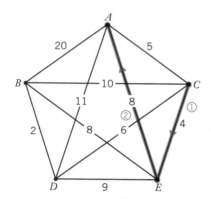

Figure 6.42

At this stage, we notice that vertex *C* is the nearest vertex to *A*, but we cannot go to vertex *C* because it has already been visited. Instead, we go to vertex *D*, which is the nearest vertex to *A* that has not yet been visited, and we mark the edge between *A* and *D* as in Figure 6.43.

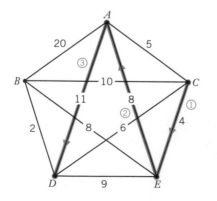

Figure 6.43

Continuing the algorithm, we go to vertex B and then return to C. We get the circuit in Figure 6.44. The weight of this circuit is $4 + 8 + 11 + 2 + 10 = 35$.

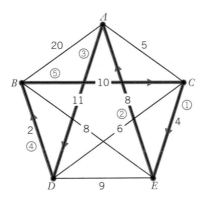

Figure 6.44

In Example 2, the nearest neighbor algorithm did not give the minimum weight circuit. In fact, by checking all circuits starting at C, we would find that one of the two circuits of minimum weight is the one shown in Figure 6.45, and the other circuit of minimum weight is the one given by traveling this circuit in the opposite direction.

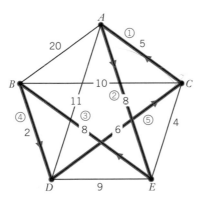

Figure 6.45

These circuits both have weight 29. However, the circuit we found using the nearest neighbor algorithm was of much smaller weight than the maximum weight circuit starting at C, whose edges sum to 54. (See if you can find such a circuit.)

The Greedy Algorithm

We now look at another approximate algorithm for the traveling salesman problem, which we will call the **greedy algorithm**. It is sometimes called the *sorted edges algorithm*. It is similar to the nearest neighbor algorithm because we will choose shortest edges. However, instead of choosing the edge attached to the nearest vertex to our current vertex, we can choose the lowest weight edge available with a couple of restrictions that ensure we end up with a Hamiltonian circuit. More precisely, this algorithm can be stated as follows:

The Greedy Algorithm for the Traveling Salesman Problem

1. Begin by choosing the edge of least weight and marking the edge. If there is a tie between edges, choose any one.

2. At each stage, the next edge marked should be the unmarked edge of least weight unless it creates a circuit that does not visit every vertex or unless it results in three marked edges coming out of the same vertex in the graph. If there is a tie between edges, choose any one.

3. When the marked edges form a Hamiltonian circuit, the algorithm has been completed. The approximate solution is the marked circuit that begins at the starting vertex, travels along the marked edges in *either direction*, and returns to the starting vertex.

Example 3   *Greedy Algorithm*

We will now use the greedy algorithm to find an approximate solution to the traveling salesman problem we considered in Example 2: finding a circuit starting at vertex C on the graph in Figure 6.40.

Solution: Starting with the graph in Figure 6.40, we see that the edge of least weight is the edge of weight 2 between vertices B and D. We mark this edge as in Figure 6.46.

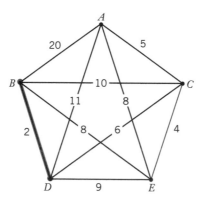

Figure 6.46

The unmarked edge of least weight at this stage is the edge between vertices C and E of weight 4. We mark it next as in Figure 6.47.

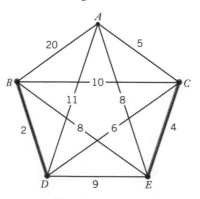

Figure 6.47

Next, we mark the edge of weight 5 between vertices A and C as in Figure 6.48.

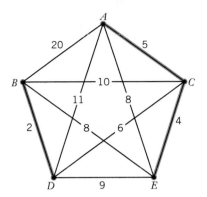

Figure 6.48

At this point, the unmarked edge of the least weight is the edge between vertices C and D of weight 6. However, we cannot mark this edge, because doing so would result in three marked edges coming out of vertex C. Therefore, we move on to consider the two edges of weight 8. In the case of a tie, we would normally be able to choose either one. However, in this case, marking the edge between vertices A and E would create a circuit. Therefore, we mark the edge between vertices B and E as in Figure 6.49.

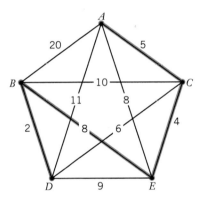

Figure 6.49

Continuing the greedy algorithm, the next edge that we will mark is the edge of weight 11 between vertices A and D. The last edge we add using the greedy algorithm is always the only unmarked edge left that will close the circuit, as shown in Figure 6.50.

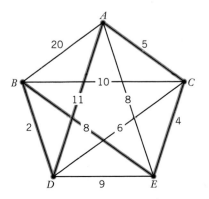

Figure 6.50

We are looking for a circuit starting and ending at vertex C. Starting at vertex C and following the edges marked in Figure 6.50, we get either the circuit shown in Figure 6.51 or the same circuit in reverse. Each of these circuits is an approximate solution given by the greedy algorithm, and each of them has weight $4 + 8 + 2 + 11 + 5 = 30$.

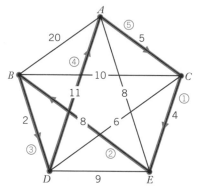

Figure 6.51

In Examples 2 and 3 we found approximate solutions to the same traveling salesman problem using two different algorithms. The nearest neighbor algorithm gave a circuit of weight 35 and the greedy algorithm gave a circuit of weight 30. Although neither of these gave the optimal circuit, which has weight 29, they both gave solutions much better than the worst possible circuit, which has weight 54. For some graphs, the nearest neighbor algorithm gives a better approximation than the greedy algorithm, and sometimes they give the same approximation. In fact, in the problem we looked at in Examples 2 and 3, the nearest neighbor algorithm would have given the same circuit as the greedy algorithm if we had made a different choice in the case of the tied edges. With some luck, either of these approximate algorithms may give the optimal circuit and, therefore, the solution to the traveling salesman problem.

Example 4   *Delivery Truck Route*

A copying service must make five deliveries downtown at the locations labeled on the map in Figure 6.52 with the letters A, B, C, D, and E. The copying shop is located at the spot labeled S. All of the blocks are of equal size. The deliveries must be made along a route that begins and ends at the copying shop and includes visits to each delivery location exactly once.

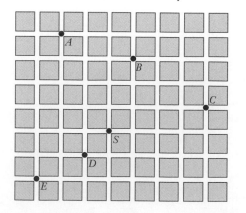

Figure 6.52
Map of the town.

(a) Draw a complete weighted graph corresponding to the problem.

(b) Find a route and its length using the nearest neighbor algorithm.

(c) Find a route and its length using the greedy algorithm.

Solution:

(a) The vertices of the complete weighted graph corresponding to this situation are the copying shop and the five delivery locations. The weight of an edge between two vertices is the distance between the corresponding locations on the map, measured in blocks that must be traveled along the roads. We will always take the shortest distance possible. For instance, the weight of the edge between vertex E and S is 5, because we must travel three blocks east and two blocks north to get from E to S. Finding all of the weights in this way, we get the graph in Figure 6.53.

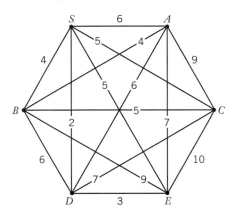

Figure 6.53

Notice that all of the relevant geometry from the original problem has been incorporated in the weights for this complete graph.

(b) When we apply the nearest neighbor algorithm for a circuit starting at the copying shop S, we do not run into any step where we have a tie, and the resulting circuit is drawn in Figure 6.54. The length of the entire route is $2 + 3 + 7 + 4 + 5 + 5 = 26$ blocks.

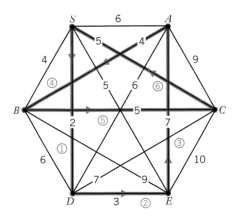

Figure 6.54

A driving route corresponding to the circuit we found with the nearest neighbor algorithm is drawn in Figure 6.55. Note, however, that this is just one of many routes going through the points in the same order and covering the same distance. For instance, in going from point S to point D, the route could have gone one block west and then one block south, rather than as it was drawn going south first and then west.

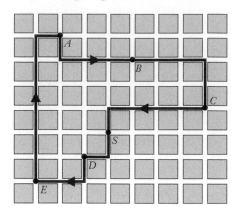

Figure 6.55

(c) With the greedy algorithm, we begin by marking the edge of length 2 and then the edge of length 3, as in Figure 6.56.

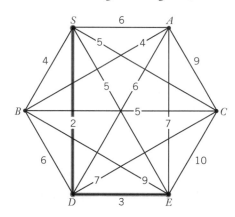

Figure 6.56

Now we must choose one of the edges of length 4. We can choose either, and in this graph, we then choose the other one next, as in Figure 6.57.

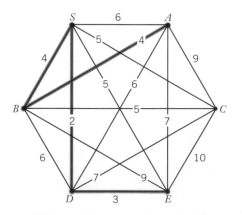

Figure 6.57

We see that we cannot mark any of the edges of length 5, 6, or 7, because each one will create a circuit or result in three marked edges coming out of the same vertex. We also cannot mark the edge of weight 9 between vertices *B* and *E*, so we mark the edge of weight 9 between vertices *A* and *C*, as in

Figure 6.58.

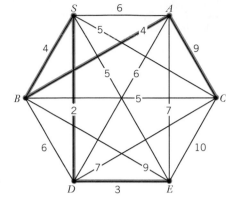

figure 6.58

We now complete the algorithm by marking the edge between vertices *C* and *E*, and we get the circuit in Figure 6.59.

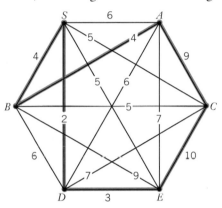

figure 6.59

Our circuit must start and end at vertex *S*, so the greedy algorithm has yielded the circuit in Figure 6.60 or the reverse circuit.

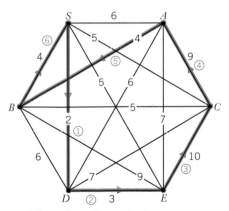

figure 6.60

The lengths of the corresponding routes are both $2 + 3 + 10 + 9 + 4 + 4 = 32$ blocks. We see that, in this case, the nearest neighbor algorithm gave a better approximation to the traveling salesman problem than did the greedy algorithm. In fact, the nearest neighbor algorithm gave the optimal route.

BRANCHING OUT 6.2

WHY EFFICIENT ALGORITHMS ARE IMPORTANT

Unlike questions in other areas, questions about specific graphs, such as finding the Hamiltonian circuit of minimal weight on a complete weighted graph, are always theoretically solvable. Because the graph is finite, we need only check all possibilities, for instance all possible Hamiltonian circuits. The only issue is the practical one of time. Thirty vertices is fairly small for a typical application. We live in a world where computers can execute a billion instructions per second. Yet even if computers were a thousand times faster than this, all of the computers in the world working together could not carry out a brute force check of the weights of all of the Hamiltonian circuits in a complete weighted graph with 30 vertices within our lifetimes.

A great algorithm would process the input data, such as the weights of the edges of a complete weighted graph, and determine the answer, such as the Hamiltonian circuit of minimal weight, in a time proportional to the size of the input. Thus, doubling the amount of data would approximately double the time taken to determine the answer, tripling the input would triple the time, and so forth. Such an algorithm is said to work in *linear time*. Translating passages from one language to another approximates the idea of a linear time algorithm. A more mathematical example would be adding two numbers, where the size of the input is the number of digits of the numbers. For more complex problems, it is too much to hope for a linear time algorithm. For instance, the length of time required to multiply two numbers goes up by a factor of four when the numbers double in length, by a factor of nine when the numbers triple in length, and so forth. Standard multiplication is thus an example of a *quadratic time* algorithm, with computing times growing in a manner similar to that of the square of the size of the input, or n^2, where n is the size of the input. More generally, we could speak of a *polynomial time* algorithm, with computing times growing in a manner similar to n^k, where n is the size of the input and k is a positive integer. Whether there exists a polynomial time algorithm for solving

the traveling salesman problem is one of the most famous unsolved problems in all of mathematics.

Whether an algorithm is linear, quadratic, polynomial, or worse is a statement about the relative long-term efficiency of the algorithm. However, this is only half of the story. If an algorithm takes a million years to solve even a small-scale problem, it does not matter much whether the algorithm is linear. Thus, a practical measure of an algorithm is how long it takes to solve a problem. The speed of the computer or computers is as much a factor of whether an algorithm is practical or not as the actual steps of the algorithm. Thus, an algorithm that is currently viewed as impractical may become practical as technology improves. Another measure of the practicality of an algorithm is how large a problem it can solve in a "reasonable" amount of time. One of *Discover* magazine's highlights for 1992 was the solution of a 3038-city traveling salesman problem by David Applegate (AT&T Bell Labs), Robert Bixby (Rice University), Vasek Chvátal (Rutgers University), and William Cook (Bellcore). The solution to this problem is shown here. It took 50 computer workstations a combined one and a half years of CPU time to arrive at the solution. This particular problem arose from circuit board design. The same researchers solved a problem with 7397 cities in 1994. Traveling salesman problems with 100 vertices now usually take only a few minutes of computer time and those with 1000 vertices typically take only a few days. Unfortunately, many real-world applications have far more than 1000 vertices.

In practical settings, any improvement of the current practice translates into real monetary savings. Thus, finding approximate solutions to problems such as the traveling salesman problem has immense practical importance. In many, if not most, practical problems, the weight of the edge between two vertices cannot exceed the combined weight obtained by passing through one or more intermediate vertices. In such instances, a simple variation of Prim's algorithm for finding the minimal weight network, treated in the next section, easily finds a Hamiltonian circuit that is

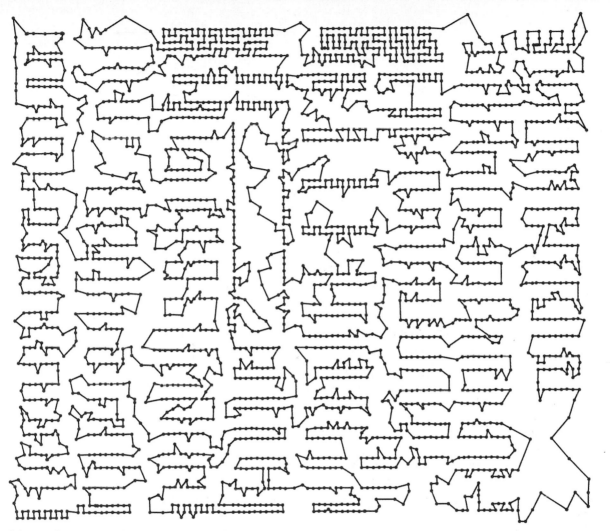

Solution of a traveling salesman problem with 3038 cities.

no more than twice the weight of the minimal weight circuit. A modification of this idea can decrease this factor from 2 to 1.5. More sophisticated methods, such as the Lin–Kernighan method, yield circuits that are usually within a few percent of the optimal solution. Such techniques take only a few minutes when applied to a problem with 1000 vertices. Even a problem with a million vertices can usually be solved to within a few percent of the optimum in several hours on a fast computer. In the context of a massive network such as an oil pipeline or the circuit layout used in manufacturing millions of computer chips, saving even a few percent can represent millions of dollars. It is far from unusual for applications of graph theory to yield savings of 25% or more. Thus, the search for even slight improvements in existing algorithms is an important and active area of research in mathematics and computer science.

EXERCISES FOR SECTION 6.2

In Exercises 1–6, find a Hamiltonian circuit, if possible, in the given graph. If it is not possible to find a Hamiltonian circuit, explain why not.

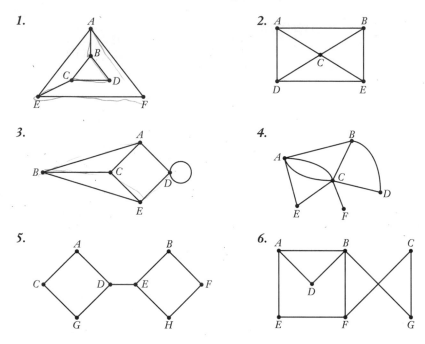

1.

2.

3.

4.

5.

6.

In Exercises 7–14, find an approximate solution to the traveling salesman problem for a circuit starting at vertex A on the given weighted graph and find the weight of the circuit using **(a)** the nearest neighbor algorithm,   **(b)** the greedy algorithm.

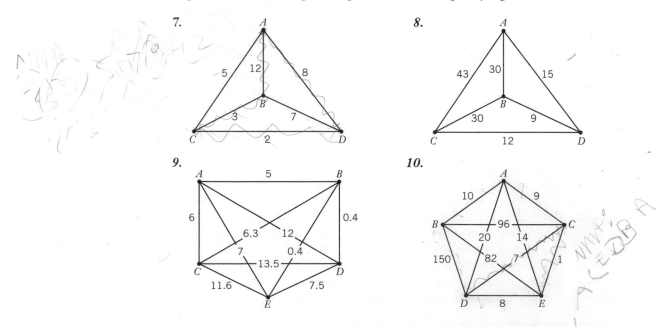

7.

8.

9.

10.

NAA→ACEBP+A
ACEBP+A
18+4+11+4+7=138
92

11.

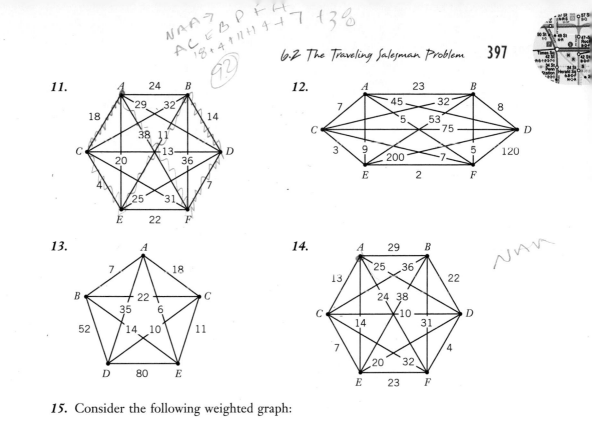

12.

13.

14.

15. Consider the following weighted graph:

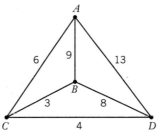

(a) Use the nearest neighbor algorithm to find an approximate solution to the traveling salesman problem for a circuit starting at vertex *C*, and find the weight of this circuit.

(b) Use the greedy algorithm to find an approximate solution to the traveling salesman problem for a circuit starting at vertex *C*, and find the weight of this circuit.

(c) By checking all possible circuits, find an exact solution of the traveling salesman problem for a circuit starting at vertex *C*.

16. Consider the following weighted graph:

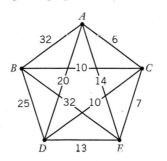

(a) Use the nearest neighbor algorithm to find an approximate solution to the traveling salesman problem for a circuit starting at vertex D, and find the weight of this circuit.

(b) Use the greedy algorithm to find an approximate solution to the traveling salesman problem for a circuit starting at vertex D, and find the weight of this circuit.

(c) It turns out that neither the nearest neighbor algorithm nor the greedy algorithm gives a circuit of minimum total weight starting at vertex D. Find some circuit starting at D of total weight smaller than those of the circuits you found in parts (a) and (b).

17. A person starting in Wichita must visit Kansas City, Omaha, and St. Louis, and then return home to Wichita in one car trip. A map including these cities and the road mileage between them is shown.

	Kansas City	*Omaha*	*St. Louis*	*Wichita*
Kansas City	—	201	256	190
Omaha	201	—	449	298
St. Louis	256	449	—	447
Wichita	190	298	447	—

(a) Draw a weighted graph corresponding to this situation.

(b) Use the nearest neighbor algorithm to find an approximate solution to the traveling salesman problem for a circuit starting at Wichita, and find the length of this circuit.

(c) Use the greedy algorithm to find an approximate solution to the traveling salesman problem for a circuit starting at Wichita, and find the length of this circuit.

(d) By checking all possible routes, find the route that will minimize the distance of the entire trip, and find the total length of this route. How does the length of this route compare to the lengths of the routes you found in parts (b) and (c)?

18. In August 1997, the one-way airline fares on American Airlines were those listed in the following table. Suppose a company president, who lived in Nashville, needed to travel to offices in Los Angeles, New York, Oklahoma City, and Washington, D.C., and then return to Nashville in one trip.

	Los Angeles	*Nashville*	*New York City*	*Oklahoma City*	*Washington, D.C.*
Los Angeles	—	$299	$495	$626	$854
Nashville	$299	—	$497	$399	$520
New York City	$495	$497	—	$672	$ 95
Oklahoma City	$626	$399	$672	—	$614
Washington, D.C.	$854	$520	$ 95	$614	—

(a) Draw a weighted graph corresponding to this situation.

(b) Use the nearest neighbor algorithm to find a travel route and find the total cost of traveling this route.

New York City
subway map.

(c) Use the greedy algorithm to find a travel route and find the total cost of traveling this route.

19. A tourist visiting New York City wants to visit the Empire State Building, the Statue of Liberty, the Brooklyn Bridge, and Times Square. She will begin and end her tour at Grand Central Station and travel to each of her stops by subway. The estimated travel time, in minutes, on the New York subway between each of the stops is given in the table.

	Grand Central Station	Empire State Building	Statue of Liberty	Brooklyn Bridge	Times Square
Grand Central Station	—	15	30	5	5
Empire State Building	15	—	30	20	5
Statue of Liberty	30	30	—	30	20
Brooklyn Bridge	5	20	30	—	15
Times Square	5	5	20	15	—

(a) Draw a weighted graph corresponding to this situation.

(b) Use the nearest neighbor algorithm to find a route, and find the estimated time required to travel this route.

(c) Use the greedy algorithm to find a route, and find the estimated time required to travel this route.

20. A pizza delivery man must deliver pizzas to the four locations on the following map labeled with the letters A, B, C, and D. The pizza shop is labeled with the letter P. All of the blocks are of equal size. The delivery man needs to travel a route that begins and ends at the pizza shop and includes a visit to each delivery location exactly once.

(a) Draw a complete weighted graph corresponding to the problem.

(b) Find a route and its length using the nearest neighbor algorithm.

(c) Find a route and its length using the greedy algorithm.

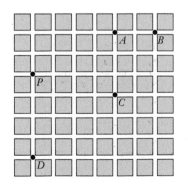

21. A truck driver must deliver newspapers to each of the five newsstands sitting on street corners in a small town. The truck's route will begin and end at the newspaper's main office. The map shows the location of the main office labeled with the letter M and the locations of the newsstands labeled with the letters A, B, C, D, and E. All blocks are of equal size. The truck needs to travel a route that begins and ends at the main office and includes a visit to each newspaper stand exactly once.

(a) Draw a complete weighted graph corresponding to the problem.

(b) Find a route and its length using the nearest neighbor algorithm.

(c) Find a route and its length using the greedy algorithm.

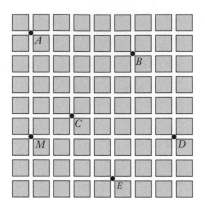

For Exercises 22 and 23, consider the following: Obtaining viewing times on the large telescopes at the major observatories is often a highly competitive process. Because such opportunities are rationed out, it is desirable to minimize the time spent moving the telescope into position. Furthermore, minimizing the telescope's movement may prolong its life. Thus, it may be reasonable to minimize the total angular change as the telescope is adjusted to view different objects.

22. Suppose an astronomer in New Mexico wants to view five stars around 3:30 A.M. on January 14, 2001. The angles (in degrees) between the stars and between the stars and the base position of the telescope are given in the table.

	Base	*Polaris*	*Betelgeuse*	*Rigel*	*Vega*	*Alphard*
Base	—	8.0	80.8	96.6	54.7	91.7
Polaris	8.0	—	82.1	96.8	51.6	98.8
Betelgeuse	80.8	82.1	—	17.2	132.8	55.3
Rigel	96.6	96.8	17.2	—	144.1	61.3
Vega	54.7	51.6	132.8	144.1	—	131.4
Alphard	91.7	98.8	55.3	61.3	131.4	—

(a) Draw a weighted graph corresponding to this situation.

(b) Use the nearest neighbor algorithm to find an approximate solution to the traveling salesman problem for a circuit starting at the base position, and find the total angular change when traveling this circuit.

(c) Use the greedy algorithm to find an approximate solution to the traveling salesman problem for a circuit starting at the base position, and find the total angular change when traveling this circuit.

23. Suppose an astronomer in New Mexico wants to view seven comets around midnight on June 11, 2003. The angles (in degrees) between the comets and between the comets and the base position of the telescope are given in the table. *Without drawing the corresponding graph*, use the nearest neighbor algorithm to find an approximate solution to the traveling salesman problem for a circuit starting at the base position, and find the total angular change when traveling this circuit.

	Base	Barnard 3	du Toit–Hartley	Gunn	Harrington–Wilson	Neujmin 2	Shoemaker–Levy 7	Swift
Base	—	46.4	53.7	85.0	71.1	76.9	61.7	119.1
Barnard 3	46.4	—	59.4	39.2	93.4	55.8	20.0	72.8
du Toit–Hartley	53.7	59.4	—	73.0	34.8	114.8	78.2	102.4
Gunn	85.0	39.2	73.0	—	104.3	61.9	32.3	35.8
Harrington–Wilson	71.1	93.4	34.8	104.3	—	146.3	112.8	123.4
Neujmin 2	76.9	55.8	114.8	61.9	146.3	—	39.1	64.7
Shoemaker–Levy 7	61.7	20.0	78.2	32.3	112.8	39.1	—	58.5
Swift	119.1	72.8	102.4	35.8	123.4	64.7	58.5	—

24. A rabid Chicago Cubs fan decides to spend his summer vacation visiting each of the stadiums of the National Baseball League. The mileage between each of the National League cities is given in the chart. He plans to begin and end his trip in Chicago and pass through each city exactly once. *Without drawing the corresponding graph,* use the nearest neighbor algorithm to find a route.

	Atlanta	Chicago	Cincinnati	Denver	Houston	Los Angeles	Miami	Milwaukee	Montreal	New York	Philadelphia	Pittsburgh	Phoenix	St. Louis	San Diego	San Francisco
Atlanta	—	674	440	1398	789	2182	655	761	1181	841	741	687	1793	541	2126	2496
Chicago	674	—	287	996	1067	2054	1329	90	828	802	738	452	1713	289	2064	2142
Cincinnati	440	287	—	1173	1029	2179	1095	374	805	647	567	295	1804	340	2155	2362
Denver	1398	996	1173	—	1019	1059	2037	1029	1815	1771	1691	1411	792	857	1108	1235
Houston	789	1067	1029	1019	—	1538	1190	1142	1827	1608	1508	1313	1149	839	1482	1912
Los Angeles	2182	2054	2179	1059	1538	—	2687	2087	2873	2786	2706	2426	389	1845	124	379
Miami	655	1329	1095	2037	1190	2687	—	1416	1686	1308	1208	1200	2390	1196	2672	3053
Milwaukee	761	90	374	1029	1142	2087	1416	—	915	889	825	539	1751	363	2102	2175
Montreal	1181	828	805	1815	1827	2873	1686	915	—	378	449	583	2519	1101	2923	2961
New York	841	802	647	1771	1608	2786	1308	889	378	—	101	368	2411	948	2762	2934
Philadelphia	741	738	567	1691	1508	2706	1208	825	449	101	—	288	2331	868	2682	2866
Pittsburgh	687	452	295	1411	1313	2426	1200	539	583	368	288	—	2051	588	2402	2578
Phoenix	1793	1713	1804	792	1149	389	2390	1751	2519	2411	2331	2051	—	1470	358	763
St. Louis	541	289	340	857	839	1845	1196	363	1101	948	868	588	1470	—	1821	2089
San Diego	2126	2064	2155	1108	1482	124	2672	2102	2923	2762	2682	2402	358	1821	—	504
San Francisco	2496	2142	2362	1235	1912	379	3053	2175	2961	2934	2866	2578	763	2089	504	—

Exercises 25 and 26 illustrate the variety of possible applications for the following: When the vertices of a graph are located in a plane and the weight of an edge is simply the distance between the two vertices, we can use the distance formula to find the length of the edges. The distance between points with coordinates (x_1, y_1) and (x_2, y_2) is

$$\sqrt{(x_2 - x_1)^2 + (y_2 - y_1)^2}.$$

25. A drill must drill holes at four points on a circuit board. The points have coordinates $(1, 0)$, $(4, 1)$, $(0, 2)$, and $(5, 4)$, measured in centimeters. The drill begins at the point $(1, 0)$ and must complete a round-trip through each of the other three locations. (The round-trip is so it can drill the next board in the production line.)

(a) Draw a weighted graph corresponding to this situation.

(b) Use the nearest neighbor algorithm to find an order in which to drill the holes, and find the length the drill must be moved as it follows this order.

(c) Use the greedy algorithm to find an order in which to drill the holes, and find the length the drill must be moved as it follows this order.

26. A golfer hits three balls from the point $(0, 0)$ that come to rest at the points $(97, -15)$ $(89, 2)$, and $(95, 11)$, measured in yards. The golfer wishes to retrieve the balls and go back to the starting point to hit them again.

   (a) Draw a weighted graph corresponding to this situation.

   (b) Use the nearest neighbor algorithm to find an order in which the golfer should retrieve the balls and the length of the corresponding walking route.

   (c) Use the greedy algorithm to find an order in which the golfer should retrieve the balls and the length of the corresponding walking route.

27. Sir William Rowan Hamilton, after whom Hamiltonian circuits were named, marketed a game called Icosian consisting of a wooden dodecahedron (the regular solid with 12 pentagonal faces as shown in Figure 7.47) with pegs attached to each corner and a string. The corners of the dodecahedron were labeled with the names of different cities of the world. The object of the game was to travel from city to city along the edges without going through any city twice and then return to the starting point. The circuit was marked off using the string. When a dodecahedron is flattened out it looks like the graph shown, where the edges and vertices of the graph correspond to the edges and corners of the dodecahedron. Therefore, finding a Hamiltonian circuit on this graph is equivalent to the object of Hamilton's game. Find a Hamiltonian circuit on this graph starting at vertex D.

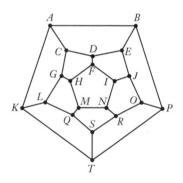

Exercises 28 and 29 consider another method for finding an approximate solution to the traveling salesman problem. With this method the nearest neighbor algorithm is applied repeatedly, using each of the vertices of the graph as the starting vertex in the algorithm. From these circuits, the one with the shortest length can be rewritten so that the starting vertex is the one specified by the particular problem, giving an approximate solution. Apply this method to find an approximate solution to the traveling salesman problem starting at vertex B, and give the weight of this circuit.

28.   29.

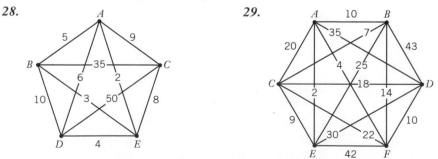

30. Draw an example of a graph that has an Eulerian circuit but does not have a Hamiltonian circuit.

31. Draw an example of a graph that has a Hamiltonian circuit but does not have an Eulerian circuit.

32. Draw an example of a complete weighted graph with at least four vertices for which the nearest neighbor algorithm starting at some vertex gives the worst possible Hamiltonian circuit. In other words, all other Hamiltonian circuits, except for the same circuit in reverse order, should have smaller weight.

33. Draw an example of a complete weighted graph with at least four vertices for which the greedy algorithm gives the worst possible Hamiltonian circuit. In other words, all other Hamiltonian circuits, except for the same circuit in reverse order, should have smaller weight.

34. Suppose you want to find the length of the longest Hamiltonian circuit in a complete weighted graph. How can the idea of the nearest neighbor algorithm be adapted to give an algorithm for finding an approximate solution to this problem?

6.3   EFFICIENT NETWORKING: MINIMAL SPANNING TREES

There are many instances when we want to connect several points, but when a direct connection between each pair of points is not necessary. Consider, for instance, eight computers in a building that need to be networked—that is, connected to one another by cable. Suppose it is most convenient to install the cable alongside the existing wiring in the building, and the eight computers and current wiring are laid out as shown in Figure 6.61, where the boxes indicate the computers, the edges indicate the existing wiring, and the numbers along each edge indicate the length of the edge in yards.

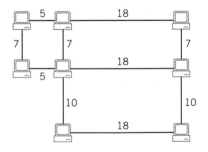

Figure 6.61
Computers and existing wiring.

One way to network all of the computers is shown in Figure 6.62, where the lines marked in blue indicate where the new cable will be installed.

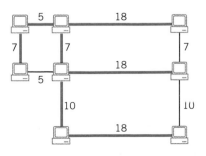

Figure 6.62

Notice that in Figure 6.62, each computer is connected to every other computer by cable, although some of the connections are not direct because they go through one or more other computers. There is no need for new cabling along any of the unused edges in Figure 6.62 because all computers are now networked. However, we might wonder if we networked the computers using the least amount of cable possible. Adding up all of the edges with cable in Figure 6.62, we see that we would need $5 + 18 + 7 + 7 + 18 + 10 + 18 = 83$ yards of cable. It turns out that we could have done better than this if we had installed the cable as in Figure 6.63.

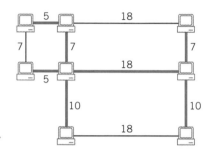

Figure 6.63

The total length of the cable in Figure 6.63 is $5 + 7 + 7 + 5 + 18 + 10 + 10 = 62$ yards, a significant improvement over the 83 yards used in Figure 6.62. Using a method we will learn later in this section, we can show that this is the least amount of cable that can be used to form the network.

In terms of graph theory, a networking problem asks us to efficiently connect all of the vertices of a given graph. The solution will always be a graph within the original graph that is connected, contains all of the vertices, and contains no circuits. We will usually be interested in minimizing the weight of such a network. Some terms from graph theory are needed to help us to describe networking problems.

A graph lying within the original graph is called a subgraph. A **subgraph** is any collection of edges and vertices from the original graph that themselves form a graph. The graph colored blue within the graph in Figure 6.64 is an example of a subgraph.

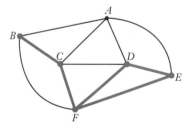

Figure 6.64
A subgraph.

We have the following important definitions:

> Any graph that is connected and contains no circuits is called a **tree**. A subgraph that is a tree containing all of the vertices of the graph is called a **spanning tree** for the graph.

In Figure 6.65(a), we see a subgraph that is a spanning tree. The subgraph in Figure 6.65(b) is not a tree because it contains a circuit and therefore it is not a spanning tree. The subgraph in Figure 6.65(c) is not a tree because it is not connected, and therefore it cannot be a spanning tree. The subgraph in Figure 6.65(d) is a tree because it is connected and contains no circuits, but it is not a spanning tree because it does not contain the vertex F.

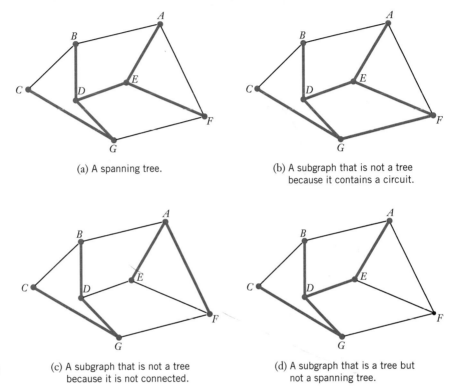

(a) A spanning tree.

(b) A subgraph that is not a tree because it contains a circuit.

(c) A subgraph that is not a tree because it is not connected.

(d) A subgraph that is a tree but not a spanning tree.

Figure 6.65

Because spanning trees give a connection between every pair of vertices in the graph, they provide a network connecting all of the vertices of the graph. Because there are no circuits in a spanning tree, there are no unnecessary edges. In many applications, the graph will be weighted, and we will be interested in finding a minimum spanning tree, which is defined as follows:

> A **minimum spanning tree** for a weighted graph is a spanning tree for which the sum of the weights of the edges is a minimum.

Analogous to what we did with circuits, we call the sum of the weights of the edges of a spanning tree the **weight of the tree**.

Prim's Algorithm

The problem of finding a minimum spanning tree for a weighted graph is quite similar to solving the traveling salesman problem, where we are looking for the minimum Hamiltonian circuit. Surprisingly, even though no efficient algorithm is known for solving the traveling salesman problem, there are simple and efficient algorithms for finding a minimum spanning tree. One such algorithm was presented by Robert Prim in 1957 and is known as **Prim's Algorithm**. It is similar to the nearest neighbor algorithm for the traveling salesman problem.

Prim's Algorithm for Finding a Minimum Spanning Tree

1. Start at any vertex in the graph. This vertex will be the starting tree for the algorithm.

2. Find the nearest vertex to the current tree that is not already contained in the current tree. In other words, from among those vertices not in the current tree, find the one that may be connected to a vertex in the current tree by an edge of the least weight. Connect this vertex to the current tree by the edge of least weight connecting them. If there is a tie between nearest vertices, choose any one of them.

3. Continue the procedure in step 2 until all of the vertices are contained in the tree.

Example 1   *Prim's Algorithm*

Using Prim's algorithm, find a minimum spanning tree for the weighted graph in Figure 6.66 and give the weight of the tree.

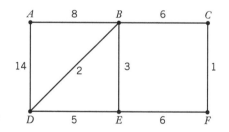

Figure 6.66

Solution: We can start at any vertex we wish, so we start with vertex B. This is the start of our minimum spanning tree. The nearest vertex to B is vertex D, so we choose the edge between B and D and mark it as in Figure 6.67.

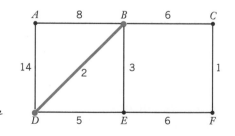

Figure 6.67

The nearest vertex to the current tree is vertex E. We connect it to vertex B and get the tree in Figure 6.68.

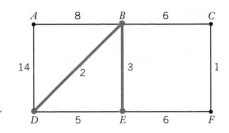

Figure 6.68

The nearest vertices to the current tree that are not already contained in it are vertices *C* and *F*. Because it is a tie, we can connect either one. Connecting vertices *C* and *B* results in the tree in Figure 6.69.

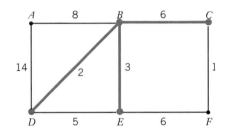

Figure 6.69

The nearest vertex to this tree is vertex *F*, and connecting it to vertex *C* we get the tree in Figure 6.70.

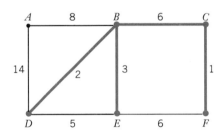

Figure 6.70

Finally, we connect the last vertex *A* to vertex *B* and we have found the minimum spanning tree shown in Figure 6.71.

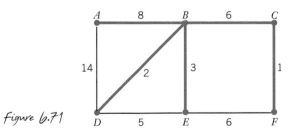

Figure 6.71

The weight of this spanning tree is the sum of the weights of the edges: $8 + 6 + 2 + 3 + 1 = 20$.

BRANCHING OUT 6.3

STEINER POINTS: IMPROVING ON THE MINIMAL SPANNING TREE

In planning communication networks, several points, such as cities, must be connected. It is natural to think of these points as the vertices in a graph and of the straight lines between the points as the edges. However, when trying to find a minimal network connecting these points, it is often desirable to add other vertices and edges. This may reduce the length of the network. For instance, suppose we want to connect three cities lying at the corners, *A*, *B*, and *C*, of an equilateral triangle. If the sides of the triangle have length 1, then one minimal spanning tree of the graph with vertices *A*, *B*, and *C*, and edges formed by the triangle is shown in figure (a). We see that the minimal spanning tree has length 2. However, if we augment the graph by adding a vertex at the center of the triangle and joining each of the original vertices to this new vertex by an edge, then the minimal spanning tree of this new graph, shown in figure (b), has length $\sqrt{3} \approx 1.732$. This is the best we can do for connecting vertices *A*, *B*, and *C*. When our vertices are points in the plane, added vertices are called *Steiner points*. Allowing the addition of Steiner points gives many more possibilities for a minimal network.

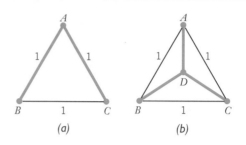

(a)   (b)

It lets us reduce the total length, but makes the problem of being sure we have the best solution more difficult. Using the algorithm of Ernest J. Cockayne and Denton E. Hewgill of the University of Victoria, the minimal network connecting 29 cities in North America and allowing Steiner points was found and the total length is about 7400 miles. It is shown in figure (c).

By comparison, the minimal spanning tree of the complete graph formed by the 29 cities and the straight lines connecting them to one another, which is shown in figure (d), has a total length of about 7600 miles.

(c) *Minimal network allowing Steiner points.*

(d) Minimal spanning tree using only the original vertices.

Although many properties of Steiner points are known, such as that three edges meet at 120° angles at any Steiner point in a minimal configuration, there is no known algorithm in the spirit of Prim's algorithm for finding Steiner points. Mathematicians Frank Hwang and Ding-Zhu Du proved in 1990 that in going from the minimal spanning tree to the minimal network with Steiner points allowed, the weight of the network is reduced by at most $(2 - \sqrt{3})/2 \approx 13.4\%$. This was the savings obtained in the example of figure (b). Although the minimal spanning tree is not a bad approximation to the minimal length network including Steiner points, you should remember that with enormous networks, a few percent savings may translate to millions of dollars.

An interesting alternative to computational algorithms that sometimes improves on the minimal spanning tree is to place pegs on a surface to mark the locations of vertices and to stretch soap bubbles across these pegs. Soap

bubbles are known to form in ways that tend to minimize area, which in this case is equivalent to the length of the network. A soap bubble solution for the 29-cities problem we looked at appears in figure (e). Notice that this soap bubble network has Steiner points. Although this solution is a short network, it is not the minimal one allowing Steiner points in this case.

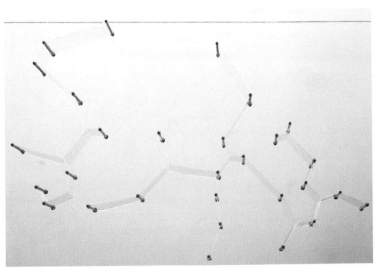

(e) Soap bubble network.

A graph may have only one minimal spanning tree or several minimum spanning trees. Which spanning tree we find using Prim's algorithm can depend on the choices we make when we have to decide between the two or more nearest vertices in the case of a tie. For example, another minimum spanning tree for the graph in Example 1 is shown in Figure 6.72.

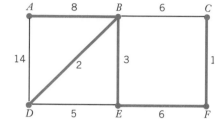

figure 6.72

However, when the edges of a weighted graph have distinct weights, the graph has only one minimum spanning tree.

Earlier in this section, we discussed the problem of networking the computers in a building, where the cable must be installed alongside the already existing wiring in the building. The eight computers and current wiring were laid out as shown in Figure 6.61. This situation can be easily represented by a weighted graph, where the vertices represent the computers and the edges represent the possible places that the cable can be installed. If we use Prim's algorithm to find a minimum spanning tree, one of the possible resulting minimum spanning trees corresponds to the network shown in Figure 6.63. (You should check this yourself.) Therefore, the minimum amount of cable that can be used to network the computers is the weight, 62 yards, of this spanning tree.

In the next example, we see another application of networking.

Example 2   *Bike Trails*

Suppose five towns that are connected by roads would like to build bike trails on some of the roads in such a way that a biker could travel from any of the five towns to another along bike trails, though possibly by an indirect route. If the towns and the current roads and their lengths in miles are as shown in Figure 6.73, along which roads should the bike trails be built in order to keep the total length of the trails to a minimum?

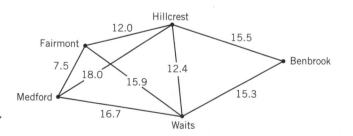
figure 6.73

Solution: The towns can be considered vertices and the roads edges of a weighted graph. To keep the total length of the trails to a minimum, they should be built along the roads forming a minimum spanning tree in this graph. If we use Prim's algorithm starting at the town of Fairmont, then we first connect Fairmont and Medford. Next, we connect Hillcrest to Fairmont, then Waits to Hillcrest. Last of all, we connect Benbrook to Waits. The resulting minimum spanning tree is as shown in Figure 6.74.

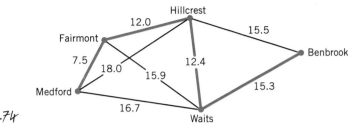

Figure 6.74

EXERCISES FOR SECTION 6.3

In Exercises 1–6, determine whether or not the subgraph colored blue is a tree. If it is not a tree, explain why not. If it is a tree, determine whether or not it is a spanning tree.

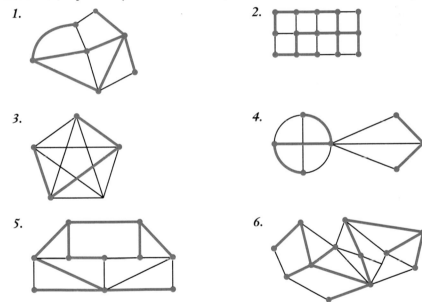

In Exercises 7 and 8, find three different spanning trees of the graph.

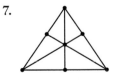

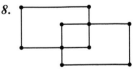

In Exercises 9 and 10, find all possible spanning trees of the graph.

In Exercises 11–18, use Prim's algorithm to find a minimum spanning tree for the given weighted graph and find the weight of this tree.

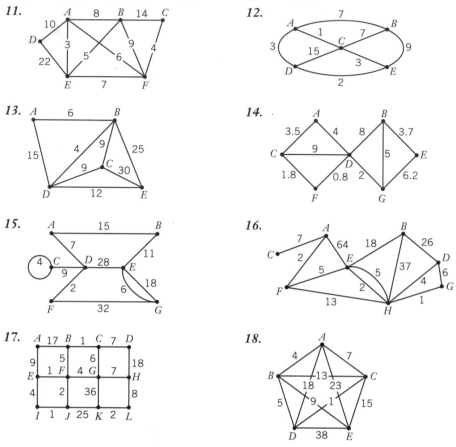

11.

12.

13.

14.

15.

16.

17.

18.

19. Six small towns are currently connected by dirt roads as illustrated in the figure, where the numbers indicate the lengths of the roads in miles. Which of these roads should be paved so it would be possible to travel from any town to another along only paved roads and so the total length of the roads paved is kept to a minimum? What is the minimum length?

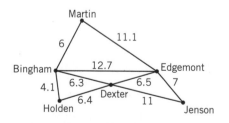

20. Seven buildings on a high school campus are connected by the sidewalks shown in the figure, where the numbers indicate the length of each sidewalk in yards. A heavy snow has fallen, and the groundskeeper must shovel the sidewalks quickly. The groundskeeper therefore decides to shovel as little as possible and still ensure that a person walking from any of the buildings to another will be able to do so along only cleared sidewalks and through buildings. Draw a graph corresponding to this situation, and indicate those sidewalks the worker should shovel. Also, give the total length of the sidewalks the groundskeeper should shovel.

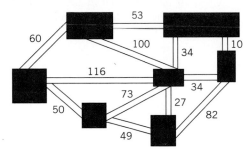

21. Suppose a small airline in the northwestern United States currently has flights between some of the cities in the figure, where the mileage between the cities with flights between them is indicated on the edge between the cities. The airline wants to reduce the total number of different flights it offers but wants to be able to provide travel between any two cities, where in some cases it may be necessary to take more than one flight. Which of the current flights should be retained in order to do this in such a way that the total length of the remaining flights is as short as possible?

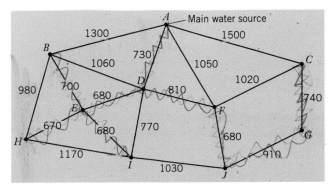

22. An irrigation system is to be installed. The main water source is located at vertex A in the graph and the other vertices represent the proposed sprinkler head locations. The edges indicate all possible choices for installing pipes. For the irrigation system to work, each sprinkler head must be connected to the main water source through one or more pipes. The costs of installing the pipes, in dollars, are given as the weights of the corresponding edges. Find the least expensive way of installing the irrigation system, and give the cost.

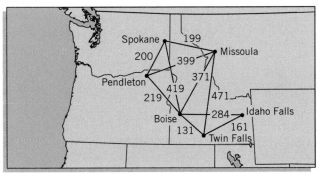

23. The highway distances, in miles, between the cities of Boston, Cincinnati, Cleveland, Detroit, New York, and Philadelphia are given as the weights of the edges between each pair of cities in the graph. If a high-speed fiber optic cable is to be installed along these highways, where should the cable be installed so the minimal amount of cable is used to connect the cities? What is the total length of cable needed?

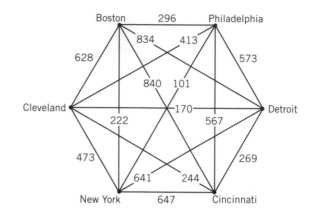

24. Five computers in an office, which we will label *A*, *B*, *C*, *D*, and *E*, must be networked with cable. The cost of connecting each pair of computers by cable is given in the table. Draw a graph corresponding to this situation. Between which computers should the cable be installed so as to keep the total cost of networking to a minimum? What is the total cost of this least expensive network?

	A	*B*	*C*	*D*	*E*
A	—	$120	$250	$180	$225
B	$120	—	$ 85	$130	$ 85
C	$250	$ 85	—	$200	$175
D	$180	$130	$200	—	$220
E	$225	$ 85	$175	$220	—

25. A company needs to provide a computer network between its four main offices in Baltimore, Charlotte, Chicago, and Los Angeles. Suppose the monthly cost for leasing a computer line between the four cities is as shown in the table. Draw a graph corresponding to this situation, and find the minimum cost network. What is the total monthly cost of this network?

	Baltimore	*Charlotte*	*Chicago*	*Los Angeles*
Baltimore	—	$ 820	$3500	$2000
Charlotte	$ 820	—	$1100	$1750
Chicago	$3500	$1100	—	$1300
Los Angeles	$2000	$1750	$1300	—

WRITING EXERCISES

1. Explain why the traveling salesman problem is an inappropriate model for a round-trip tour of all the planets in the solar system.

2. Explain why, in building some networks, we might want more connections than just those of a minimal spanning tree.

3. Give an example of a weighted graph modeling a real situation where for any two vertices in the graph connected by an edge, the weight of the edge between the two vertices is at most the sum of the weights of the edges of any other path connecting the two vertices. Would you expect this to be true for all weighted graphs modeling real situations? Explain why or why not.

4. In Exercise 19 of Section 6.3, we considered the problem of paving the roads between six towns that are currently connected by dirt roads in such a way that it would be possible to travel from any town to another along only paved roads and so the total length of the roads paved is kept to a minimum. Suppose you are to present the proposed solution of paving the minimum spanning tree, shown in the figure, to the assembled councils of the towns. What sorts of comments or questions would you anticipate, and how would you respond to them? How might you propose that the cost of paving be allocated, and how would you justify your proposal?

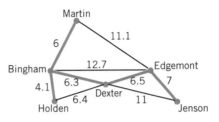

5. The version of the traveling salesman problem we have considered is called the *symmetric traveling salesman problem* because the weight of the path from vertex A to vertex B is the same as that from B to A.

   (a) Can you think of some applications for which the weights would not be equal?

   (b) Can the nearest neighbor algorithm and the greedy algorithm be applied in this asymmetric case? If so, what adaptations must be made? If not, why not?

PROJECTS

1. Invent an approximate algorithm for solving the traveling salesman problem that is different from the nearest neighbor and greedy algorithms. Test your algorithm on several graphs and compare it to the nearest neighbor and greedy algorithms.

2. The basic idea of an algorithm may be applied in several different settings. For instance, the nearest neighbor algorithm and Prim's algorithm are essentially the same algorithm applied in the settings of the traveling salesman problem and of finding the minimum spanning tree.

   (a) Describe how to adapt the greedy algorithm from the context of the traveling salesman problem to the context of finding the minimum spanning tree. (This algorithm should be different from Prim's algorithm.)

   (b) By trying a variety of examples, see if you think your algorithm always gives the minimum spanning tree.

3. Many calculators have a random number generator. Use one that does, or the random number generator on a computer, to generate random weights for each of the six edges of a complete graph on vertices A, B, C, and D. (Often values will be between 0 and 1, but any range of positive numbers will do.) For each of 20 trials (more if you are part of a group), compute the weight of the minimal circuit, the weight of the circuit obtained by the nearest neighbor algorithm starting at vertex A, and the weight of the circuit obtained by the greedy algorithm. Find the average of each of these three values over the 20 trials. What percentage of the time does each algorithm give a circuit

of minimal total weight? What does this suggest to you about choosing between these methods for solving problems arising in practice?

4. **(a)** What is the least number of edges needed on a graph with n vertices to be sure that all vertices are connected even if one edge is destroyed (say, by a natural disaster)?

   (b) What is the least number of edges needed on a graph with n vertices to be sure that all vertices are connected even if two edges are destroyed?

5. Consider either the nearest neighbor algorithm or Prim's algorithm. In writing a computer program to implement this algorithm, what quantities need to be stored in memory? Describe the steps that would have to be programmed. Write a program implementing your choice of algorithm.

KEY TERMS

CHAPTER 6 REVIEW EXERCISES

In Exercises 1–4, determine whether or not the graph has an Eulerian circuit, and if it does, give one. If it does not, explain why not. If the graph does not have an Eulerian circuit, determine if it has an Eulerian path, and if it does, give one. If it does not, explain why not.

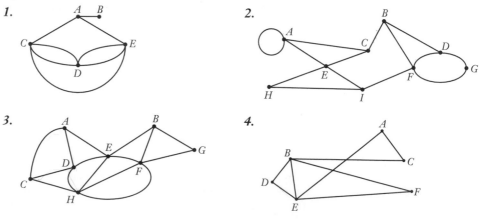

1.

2.

3.

4.

In Exercises 5 and 6, find an eulerization of the graph.

5.   **6.**

In Exercises 7 and 8, find a Hamiltonian circuit, if possible, in the given graph. If it is not possible to find a Hamiltonian circuit, explain why not.

7.   **8.**

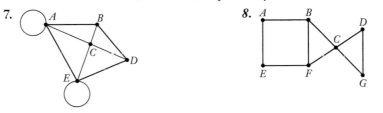

In Exercises 9 and 10, find an approximate solution to the traveling salesman problem for a circuit starting at vertex A on the given weighted graph and find the weight of the circuit using

(a) the nearest neighbor algorithm,   **(b)** the greedy algorithm.

9.   **10.**

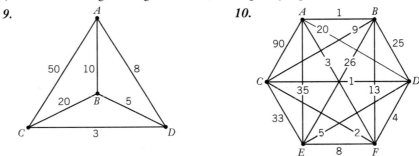

In Exercises 11 and 12, use Prim's algorithm to find a minimum spanning tree for the given weighted graph, and find the weight of this tree.

11.   **12.**

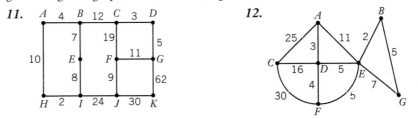

13. A political campaign volunteer wants to deliver flyers to each of the homes in the neighborhood pictured. The streets on the outside of the neighborhood have houses on only one side, and the interior streets have houses on both sides. The volunteer plans to park her car at one of the street intersections and walk down each of the interior streets twice (once for each side) and each of the outside streets once. Draw a graph corresponding to this situation. Determine whether or not the graph has an Eulerian circuit, and if so, give one. If it does not, explain why not.

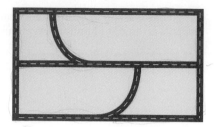

14. A burglar, when apprehended, claimed that he entered a room in the house shown in the figure by forcing a window open at the location marked **w**. He then went through the house, passing through each door exactly once, and left through the back door at **x**. Why must the burglar be lying?

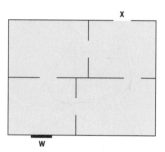

15. A trucker starting in Denver must travel to Cheyenne, Chicago, Des Moines, and Minneapolis, and then return to Denver in one trip. The road mileage between these cities is listed in the following table:

	Cheyenne	*Chicago*	*Denver*	*Des Moines*	*Minneapolis*
Cheyenne	—	954	100	627	870
Chicago	954	—	996	327	405
Denver	100	996	—	669	920
Des Moines	627	327	669	—	252
Minneapolis	870	405	920	252	—

(a) Draw a weighted graph corresponding to this situation.

(b) Use the nearest neighbor algorithm to find a travel route, and find the length of this route.

(c) Use the greedy algorithm to find a travel route, and find the length of this route.

16. The homes in a rural community, indicated by vertices, and the possible choices for installing a gas pipeline system with the corresponding costs, in dollars, are shown in the figure. Find the cheapest way of installing the pipeline so that all of the homes are connected to it, and give the cost.

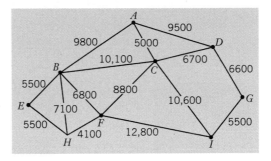

17. For what values of *n* does a complete graph with *n* vertices have an Eulerian circuit?

SUGGESTED READINGS

Bern, Marshall W., and Ronald L. Graham. "The Shortest-Network Problem." *Scientific American* 260:1 (January 1989), 84–89. Introduction to the problem of finding minimal networks that allow the inclusion of Steiner points.

Euler, Leonhard. "The Koenigsberg Bridges," trans. James Newman. *Scientific American* 189:1 (July 1953), 66–70. English translation of Euler's original paper on the Königsberg bridge problem.

Lewis, Harry R., and Christos Papadimitriou. "The Efficiency of Algorithms." *Scientific American* 238:1 (January 1978), 96–109. Discussion of efficient algorithms in the context of graph theory.

Rosen, Kenneth H. *Discrete Mathematics and Its Applications*. New York: McGraw-Hill, 1995. One of many discrete mathematics textbooks that include a somewhat more technical, but still accessible, treatment of graph theory.

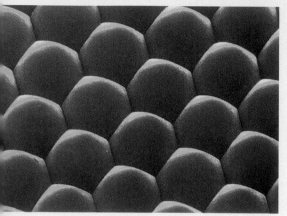

Geometrical patterns and symmetry can be seen in many objects in nature, for instance in the compound eyes of a fly and in a group of soap bubbles. Because of the beauty and usefulness of such patterns, we also find them in man-made objects.

A magnified view of the compound eye of a fly.

Consider, for example, the patterns that appear on the Dome of the Rock in Jerusalem, a beautiful masterpiece of Islamic art and architecture built in 692 A.D.

In this chapter we will see the geometry that lies behind such patterns and shapes. We begin with a review of polygons and angle measure. We then apply these ideas to explore some of the kinds of patterns that can arise in tiling a flat surface with polygons. We conclude the chapter by looking at regular and semiregular polyhedra. Throughout the chapter we will see that the ways that shapes can be fitted together are governed by simple geometric relations.

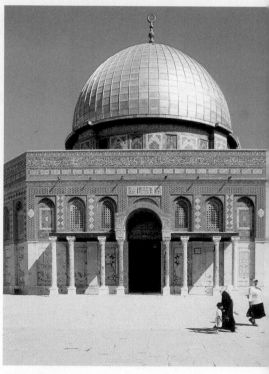

The Dome of the Rock in Jerusalem (known in Arabic as the Qubbat as-Sakhra).

Soap bubbles.

ISBN 0-471-13935-3

CHAPTER 7

TILINGS AND POLYHEDRA

7.1   POLYGONS

Many patterns in nature and art are composed of one or more geometrical shapes fitted together in a pattern. The Roman mosaic in Figure 7.1 is an example of such a pattern. In this tiling, as in many patterns, the basic geometric shapes are polygons.

Figure 7.1
A Roman mosaic.

We use the word **polygon** to refer to a *simple* polygon, a connected figure in the plane consisting of a finite number of line segments each of whose endpoints intersects an endpoint of exactly one of the other line segments and such that there are no other points where two of the line segments intersect. The shapes in Figures 7.2(a) and 7.2(b) are both polygons.

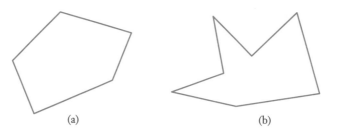

Figure 7.2
Polygons.

(a)   (b)

The shape in Figure 7.3(a) is not a polygon because two of the line segments have endpoints that do not intersect the endpoint of another line segment, and the shape in Figure 7.3(b) is not a polygon by our definition because it has sides that intersect in a place other than their endpoints. We will also use the word polygon to refer to a polygon along with its interior because we will have no need to distinguish between whether or not the interior is included.

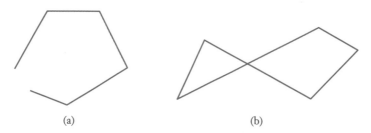

Figure 7.3
These figures are not polygons.

(a)   (b)

Each of the line segments forming a polygon is called a **side**, and each point at which two sides touch is called a **vertex**. (The plural of vertex is vertices.) The name of a polygon is based on how many sides or angles it has. The names for several different polygons are given in Table 7.1.

Table 7.1	The Names of Polygons	
Number of Sides	**Name of the Polygon**	**Example**
3	Triangle	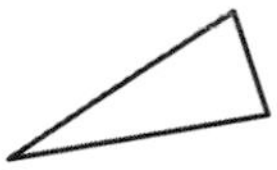
4	Quadrilateral	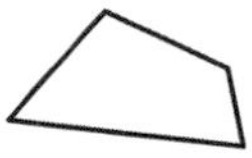
5	Pentagon	
6	Hexagon	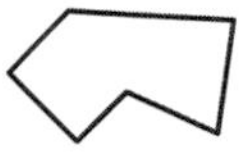
7	Heptagon	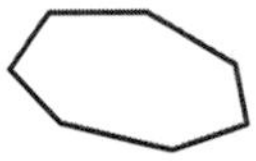
8	Octagon	
9	Nonagon	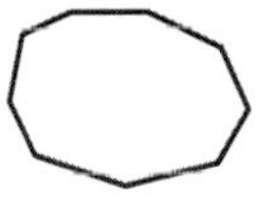
10	Decagon	
12	Dodecagon	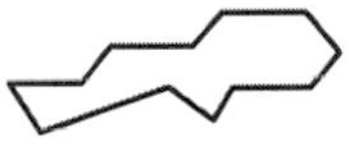

A polygon with *n* sides, especially when it is not one of the ones included in Table 7.1, is called an **n-gon**. For instance, a polygon with 17 sides is called a 17-gon.

Polygons can be classified as convex or concave. A polygon is called **convex** if every line segment drawn between two points in the polygon lies completely inside the polygon. A polygon that is not convex is called **concave**. The polygon in Figure 7.4 is convex. To convince yourself of this, try to connect any two points within the polygon, and you will see that the connecting line segment stays inside the polygon.

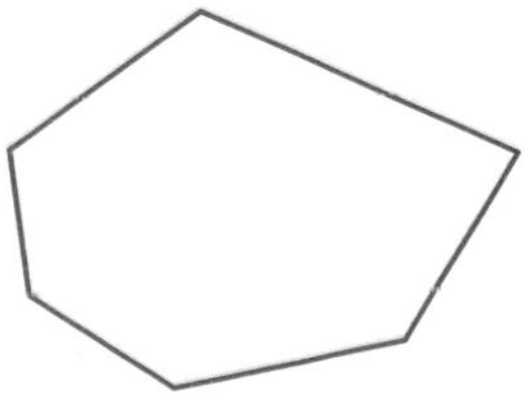

When we try to connect the two points *A* and *B* inside the polygon in Figure 7.5, the line segment connecting them falls outside the polygon. Therefore, the polygon is concave.

Figure 7.4
A convex polygon.

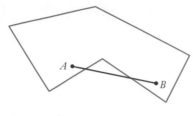

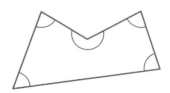

Figure 7.5
A concave polygon.

Figure 7.6
The interior angles of a polygon.

The inside angle at which two sides of a polygon meet is called an *interior angle*. In Figure 7.6, the interior angles of the polygon are marked. For brevity, we call an interior angle simply an **angle**, and it is to be understood that we are referring to an interior angle. The angles of a convex polygon all measure at most 180°, whereas a concave polygon has at least one angle measuring more than 180°.

We are particularly interested in looking at the most symmetric of polygons, called the regular polygons. They are defined as follows:

A **regular polygon** is a polygon with sides of equal length and angles of equal measure.

The first six regular polygons are shown in Figure 7.7.

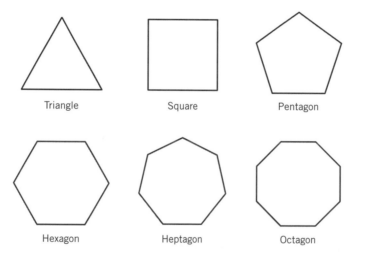

Figure 7.7
Regular polygons.

A regular triangle is more commonly called an equilateral triangle, and a regular quadrilateral is a square. Notice that a regular *n*-gon has rotational symmetry in the sense that if we spin it one *n*th of a full turn about its center, it appears exactly as it did before the turn. It is important to note that for a polygon to be regular, it is not enough to require that the edges be of equal length. For instance, the hexagon in Figure 7.8(a) is not regular for, although the sides are of equal length, the angles are not of equal measure. Notice that this hexagon does not have the kind of rotational symmetry that a regular hexagon has.

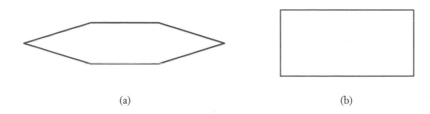

Figure 7.8
Nonregular polygons.   (a)   (b)

It is also not enough to require that angles be of equal measure. For instance, the rectangle in Figure 7.8(b) is a quadrilateral with angles of equal measure, but it is not regular. It is not the most symmetrical of quadrilaterals.

In the next section, we will construct patterns involving regular polygons, and we will need to know the measure of an angle of a regular polygon. To find a formula for this, we first use the fact that the sum of the angles of any triangle is 180° to find a formula for the sum of the angles of any n-gon, not necessarily regular. Let's first look at a convex quadrilateral. If we draw the line segment from any vertex to its opposite vertex, then the quadrilateral is divided into two triangles as shown in Figure 7.9(a). From this picture, we see that the sum of the angles of the quadrilateral is equal to the sum of the angles of the two triangles. Thus, it follows that the sum of the angles of a quadrilateral is $2 \cdot 180° = 360°$. Similarly, for any convex pentagon, we can draw two straight lines from any vertex to the vertices opposite it as in Figure 7.9(b), and this will divide the pentagon into three triangles. The sum of the angles of these three triangles is equal to the sum of the angles of the pentagon, so we see that the sum of the angles of a pentagon is given by $3 \cdot 180° = 540°$. The same idea works for any convex hexagon, and because in this case the hexagon can be divided into four triangles as in Figure 7.9(c), we find that the sum of the angles of a hexagon is $4 \cdot 180° = 720°$.

Figure 7.9
Breaking convex
polygons into triangles.

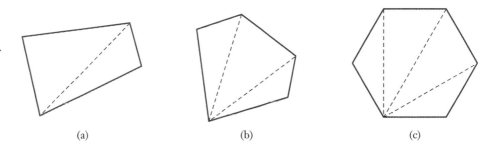

            (a)   (b)   (c)

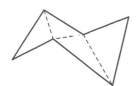

Figure 7.10
Breaking a concave
hexagon into triangles.

In a similar way, any convex n-gon can be broken into $n - 2$ triangles. By looking at a few cases, you should be able to convince yourself that it is also possible to break any concave n-gon into $n - 2$ triangles. For instance, the concave hexagon in Figure 7.10 has been broken into 4 triangles. Therefore, to find the sum of the angles of any n-gon, a similar argument can be used to arrive at the following result:

The sum of the angles of any n-gon is $(n - 2)180°$.

Example 1   *Sum of the Angles of an n-gon*

Find the sum of the angles of a dodecagon.

Solution: A dodecagon has 12 sides, so using the formula for the sum of the angles of a 12-gon, we see that the angle sum is given by

$$(12 - 2)180° = 10 \cdot 180° = 1800°.$$

■

Because the n angles of a regular n-gon are all equal, we can find the measure of each angle of a regular n-gon by dividing the sum of the angles by n to get

$$\frac{(n - 2)180°}{n} = \frac{180°n - 360°}{n} = 180° - \frac{360°}{n}.$$

Therefore, we have the following formula:

Angle of a Regular n-gon

The measure of an angle of a regular n-gon is

$$180° - \frac{360°}{n}.$$

Example 2   *Measure of the Angle of a Regular n-gon*

Find the measure of the angle of a regular

(a) pentagon,
(b) 14-gon.

Solution:
(a) Using the formula for the measure of the angle of a regular n-gon with $n = 5$, we see that the measure of the angle of a regular pentagon is

$$180° - \frac{360°}{5} = 180° - 72° = 108°.$$

(b) The measure of the angle of a regular 14-gon is

$$180° - \frac{360°}{14} \approx 180° - 25.7143° = 154.2857°.$$

■

EXERCISES FOR SECTION 7.1

In Exercises 1–8, determine whether the figure is a polygon. If it is not, explain why not.

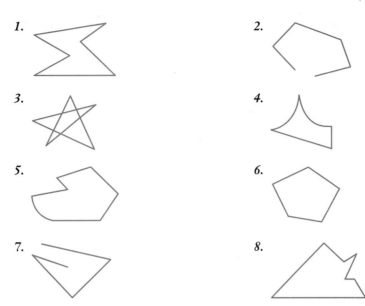

1.

2.

3.

4.

5.

6.

7.

8.

In Exercises 9–16, determine whether the given polygon is convex or concave, whether or not it appears regular, and state the polygon's name.

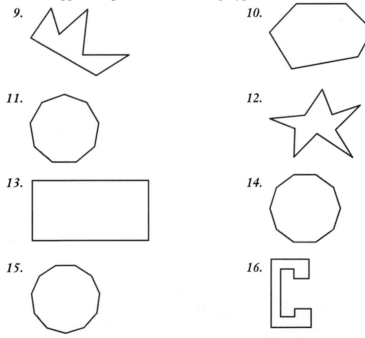

9.

10.

11.

12.

13.

14.

15.

16.

In Exercises 17–22, find the sum of the angles of the given polygon.

17. decagon   18. octagon

19. 17-gon   20. heptagon

21. nonagon   22. 16-gon

In Exercises 23–28, find the measure of an angle of the given polygon.

23. regular octagon   24. regular hexagon

25. regular heptagon

26. regular 11-gon

27. regular 30-gon

28. regular dodecagon

29. Three of the four angles of the following polygon are given. Find the measure of the remaining angle.

30. Four of the five angles of the following polygon are given. Find the measure of the remaining angle.

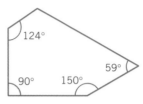

31. Three of the angles of a pentagon are 35°, 42°, and 93°. The other two angles are equal. What is the measure of each of the other angles?

32. Two of the angles of a quadrilateral are 37° and 161°. One of the two remaining angles is twice as large as the other remaining angle. What are the measures of the two remaining angles?

7.2   TILINGS

Certain sorts of polygons and other basic shapes can be placed side-by-side to form geometric patterns that completely cover a flat surface. These kinds of patterns can be found in the tiling of floors, in fabric and wallpaper designs, in nature, and in art. The design of the quilt we see in Figure 7.11 is based on a pattern of regular hexagons, whereas in Figure 7.12 we see a floor tiling pattern made up of squares and hexagons that are not regular. The patterns in both of these pictures can be extended in every direction to cover as large an area as we wish.

Figure 7.11

Figure 7.12

We think of the basic shapes that are used in forming these patterns as tiles. A pattern in which one or more repeated tiles are used to cover a plane without leaving gaps or overlapping and that can be extended forever in every direction is called a **tiling** or **tessellation**.

Tilings that use only one size and shape of tile, where the tile may be used face up or face down, are called **monohedral** tilings. We will look at both monohedral tilings and tilings using more than one kind of tile. However, we will restrict our attention to tilings for which each edge of a tile coincides exactly with the edge of a bordering tile. Tilings of this kind are called **edge-to-edge**. In Figure 7.13(a), we see a tiling that is edge-to-edge, and in Figure 7.13(b) we see one that is not edge-to-edge.

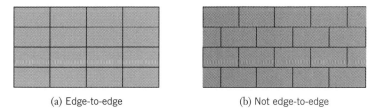

Figure 7.13

        (a) Edge-to-edge   (b) Not edge-to-edge

The **vertices** of an edge-to-edge tiling are the vertices of the tiles, each of which belongs to more than one tile. In Figure 7.14, we see an edge-to-edge tiling in which each vertex belongs to three, four, or five tiles.

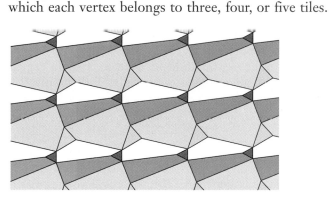

Figure 7.14
An edge-to-edge tiling by triangles, quadrilaterals, and pentagons.

Regular Tilings

We begin by asking which regular polygons can be used to tile the plane by themselves. You have almost certainly seen tilings formed using equilateral triangles, squares, or regular hexagons, as shown in Figure 7.15.

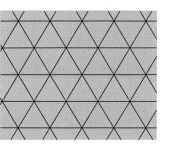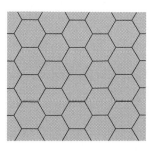

Figure 7.15
Tilings by equilateral triangles, squares, and regular hexagons.

Notice that in each case the tiles fit together exactly and the pattern can be extended in every direction. These three tilings are regular tilings, which are defined as follows:

A **regular tiling** is an edge-to-edge monohedral tiling for which the tiles are regular polygons.

Figure 7.16

Are the three regular tilings in Figure 7.15 the only possible regular tilings? To answer this question, it is helpful to understand why these three regular tilings work. Let's first look at the tiling with equilateral triangles. Because the triangles must fit about each vertex without overlapping or leaving any gaps as shown in Figure 7.16, we know that the angle of an equilateral triangle must divide 360°, the total angle about a vertex, evenly. In fact, the angle of an equilateral triangle is 60°, and 360°/60° = 6, the number of triangles about each vertex.

Similarly, the angle of a square is 90°, and 90° divides 360° evenly, giving 360°/90° = 4, the number of squares about each vertex as seen in Figure 7.17.

Figure 7.17

For the case of the regular hexagons, we first use the formula for the angle of a regular n-gon to find that the angle of a regular hexagon is

$$180° - \frac{360°}{6} = 180° - 60° = 120°.$$

The angle 120° divides 360° evenly, giving 360°/120° = 3, which is the number of hexagons about each vertex as we see in Figure 7.18.

If any other n-gon could be used to make a regular tiling, then it must be true that the angle of this n-gon divides 360° evenly. The angle of a regular pentagon is given by

Figure 7.18

$$180° - \frac{360°}{5} = 180° - 72° = 108°,$$

which does not divide 360° evenly. The angle of any regular polygon with more than six sides is larger than 120°, the angle of a regular hexagon, but smaller than 180°. Because 360°/180° = 2 and 360°/120° = 3, we see that when 360° is divided by any number between 180° and 120°, the result is a number between 2 and 3, and thus not an integer. Therefore, we have the following fact:

The only regular tilings are those with equilateral triangles, with squares, or with regular hexagons.

Semiregular Tilings

Having found all of the regular tilings, we will now look for edge-to-edge tilings that involve more than one kind of regular polygon. As with regular tilings, the sum of the angles of the polygons around each vertex must add up to exactly 360°. The arrangement of polygons around a vertex determines the vertex type.

BRANCHING OUT 7.1

THE HEXAGON PATTERN OF HONEYBEE COMBS

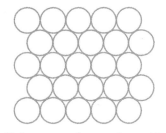

the same size and the bees arrange themselves in a triangular array with the cylinders just touching as shown in the figure.

One of the places in nature where the hexagon regular tiling pattern appears is in the familiar honeybee comb. It seems quite remarkable that the bees construct such a symmetric and beautiful pattern. However, the pattern emerges very simply as a result of the manner in which the bees build the comb. A comb is constructed by a group of bees, each one building a cylinder of wax about itself. The cylinders are about

This is an efficient use of space in packing the cylinders. That the bees arrange themselves in this pattern is not the result of careful planning but instead follows as a result of the bees exerting approximately equal pressures in trying to make their own circles as large as possible. Because the wax is soft, the cylinders then flow together to form the regular hexagonal pattern.

Two vertices are said to be of the same **vertex type** if the same number and kind of polygons are arranged about them in the same order, except that one arrangement may be clockwise whereas the other may be counterclockwise.

In Figure 7.19(a) and (b), the indicated vertices are not of the same vertex type because, even though the number and types of polygons are the same, the arrangements of the polygons about the vertices are different.

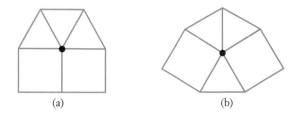

figure 7.19
Vertices of different types.

(a)   (b)

The indicated vertices in Figure 7.20(a), (b), and (c) are all of the same vertex type, even though the clockwise order of the polygons about the vertices in Figure 7.20(a) and (b) is the counterclockwise order in Figure 7.20(c).

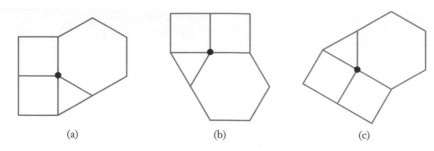

Figure 7.20
Vertices of the same
type.

     (a)   (b)   (c)

We now have the following definition:

> A **semiregular tiling** is an edge-to-edge tiling by at least two different regular polygons such that all of the vertices are of the same type.

There are only eight semiregular tilings. As with regular tilings, these can be found first by looking for sets of regular polygons that will fit around a single vertex and then by checking whether these patches of tiles can be extended to cover the plane. We will find one of these semiregular tilings in the following example and will look for others in the exercises.

Example 1   *Finding a Semiregular Tiling*

Find a semiregular tiling that has three tiles about each vertex, at least one of which is an octagon.

Solution: The angles of the three regular polygons about each vertex must add up to 360°, and the angle of a regular octagon is 180° − 360°/8 = 135°. Therefore, the angles of the remaining two polygons about each vertex must add up to 360° − 135° = 225°. Because 225°/2 = 112.5°, one of the two remaining tiles about each vertex has to have an angle of at most 112.5°. Recall that a regular hexagon has angles measuring 120°, and regular n-gons with $n > 6$ have even larger angles. Therefore, we can conclude that one of the two remaining polygons must be a regular pentagon with 108° angles, a square with 90° angles, or an equilateral triangle with 60° angles. If one of the polygons is a regular pentagon, the third polygon must have angles measuring 225° − 108° = 117°. But there is no regular polygon with an angle of this size. If one of the polygons is a square, the third polygon must have angles measuring 225° − 90° = 135°, the measure of the angle of a regular octagon. There is only one way to arrange two octagons and a square about a vertex and it is shown in Figure 7.21. By forming a tiling where every vertex is of this type, we get the familiar tiling in Figure 7.22.

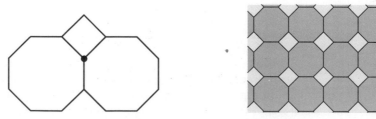

Figure 7.21   *Figure 7.22*

Technology Tip

When trying to construct tilings such as in Example 1, it is helpful to use a computer drawing program to construct the tiles. These tiles can then be easily duplicated and moved around. If a drawing program is not available to you, the same idea can be carried out with paper cutouts of tiles.

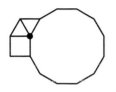

Figure 7.23

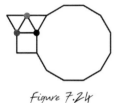

Figure 7.24

It is important to note here that even when a set of regular polygons fits about a vertex, we cannot be sure that they will give a semiregular tiling. For instance, two equilateral triangles, one square, and one regular 12-gon fit about a vertex as illustrated in Figure 7.23. However, a semiregular tiling with vertices of this type is not possible. To see why, consider what happens when we try to continue this tiling so that all the vertices are of the same type. In Figure 7.24, we see that another triangle must be placed at the blue vertex so that the blue vertex is of the same type as the original black vertex. However, this forces the green vertex to be of a different type because it includes more than two triangles.

There are 21 different vertex types consisting of regular polygons arranged about a vertex. However, only 11 of these give either regular or semiregular tilings. We will explore these further in the exercises. Note that the order in which the polygons are arranged about a vertex may be important. For instance, you are asked to show in an exercise that it is not possible to make a semiregular tiling where the vertices are of the type shown in Figure 7.20, but a different arrangement of the same polygons about a vertex will give a semiregular tiling.

Although we have focused our discussion of tilings by regular polygons on the regular and semiregular tilings, these 11 tilings are certainly not the only possibilities for edge-to-edge tilings of the plane with regular polygons. When we take away the restriction of having each vertex be of the same type, the possibilities are limitless. Figure 7.25, which is a tiling with regular polygons where three different vertex types appear, is just one of the infinite number of possibilities.

Figure 7.25
An edge-to-edge tiling by regular polygons that is not a regular or semiregular tiling. The three different vertex types are marked by the dots.

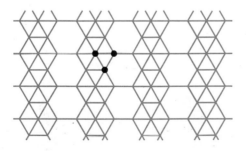

Tilings with Nonregular Polygons

We know that the only regular polygons that can be used to form an edge-to-edge monohedral tiling are equilateral triangles, squares, and regular hexagons. We now look at the more general idea of monohedral tilings where the tile used is in the shape of a polygon that is not regular.

It is possible to form an edge-to-edge tiling of the plane with any kind of triangle. One way to do this is to attach two copies of the triangle together to form a parallelogram—that is, a quadrilateral with opposite sides parallel to one another, as shown in Figure 7.26(a). These parallelograms can then be used to tile the plane as shown in Figure 7.26(b).

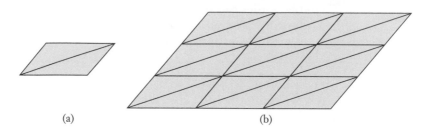

Figure 7.26
Tiling with a triangle.

(a) (b)

Next we might ask whether it is possible to form an edge-to-edge tiling of the plane with any kind of quadrilateral. As with triangles, the answer is yes. A tiling with any quadrilateral can be formed by first attaching two copies of the quadrilateral in such a way that a hexagon with opposite sides parallel is formed as shown in Figure 7.27(a). These hexagons can then be used to form an edge-to-edge tiling as shown in Figure 7.27(b).

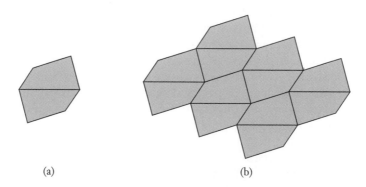

Figure 7.27
Tiling with a convex
quadrilateral.

(a) (b)

This method of tiling with a quadrilateral also works for concave quadrilaterals, as illustrated in Figures 7.28(a) and 7.28(b).

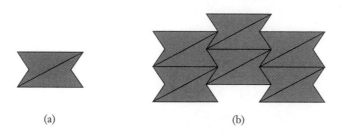

Figure 7.28
Tiling with a concave
quadrilateral.

(a) (b)

Because it is possible to tile with regular hexagons, it is natural to wonder whether we can tile with any hexagon. It turns out that some but not all hexagons can be used to tile the plane. In 1918, mathematician K. Reinhardt showed that only three different types or families of convex hexagons can tile the plane. They are shown in Figure 7.29. Notice, however, that tiling with a hexagon of Type 2 requires that some of the tiles be flipped over so that they are laying face down. Recall that this is allowed in the definition of a monohedral tiling.

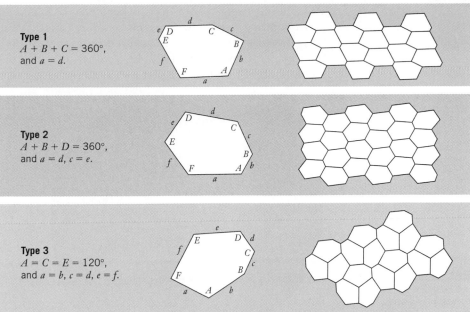

Figure 7.29
The three families of
convex hexagons that
can tile the plane.

Type 1
$A + B + C = 360°$,
and $a = d$.

Type 2
$A + B + D = 360°$,
and $a = d$, $c = e$.

Type 3
$A = C = E = 120°$,
and $a = b$, $c = d$, $e = f$.

We know that the regular pentagon cannot tile the plane, but are there other pentagons that can? The situation is similar to that for hexagons in that some but not all pentagons can be used in a monohedral tiling. For instance, any pentagon with two sides parallel to one another can be used to tile the plane. An example of such a tiling is shown in Figure 7.30.

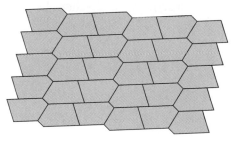

Figure 7.30
A tiling by a pentagon
with two sides parallel.

For convex pentagons, it is now known that at least 14 different families can tile the plane, but not all of these families give rise to edge-to-edge tilings. It is not known whether there are any others. Interestingly, four of the families of pentagons that give tilings were discovered in the 1970s by amateur mathematician Marjorie Rice, who had no formal mathematics education beyond high school. One of the tilings she discovered is shown in Figure 7.31.

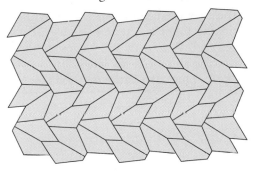

Figure 7.31
A tiling by pentagons
discovered by Marjorie
Rice in 1976. (This
tiling is not edge-to-
edge.)

What about tiling with *n*-gons, where $n \geq 7$? It is possible to tile the plane with some concave *n*-gons with seven or more sides. A beautiful spiral monohedral tiling by concave 9-gons, discovered by Heinz Voderberg in 1936, is shown in Figure 7.32. However, in 1927, K. Reinhardt proved that it is not possible to form a monohedral tiling of the plane with a convex *n*-gon for $n \geq 7$.

Figure 7.32
A monohedral tiling by concave 9-gons discovered by Heinz Voderberg.

Tilings with Other Shapes

Some of the most beautiful tilings involve tiles that are not in the shape of polygons. In the work of Dutch artist M. C. Escher (1898–1972), we find tilings with animals and other creatures. Figures 7.33 and 7.34 are two examples of his many tilings.

Figure 7.33   *Figure 7.34*

Escher designed the tiles used in his work by starting with a simple polygon or polygons and modifying them in a systematic way. Escher was not the first to use recognizable figures such as animals and leaves in tilings. Examples from the end of the nineteenth century precede his work, but it was Escher who popularized the style. We will look at just two of the methods for modifying polygons to form Escher-like monohedral tilings.

BRANCHING OUT 7.2

NAPOLEON AND TILINGS

Napoleon Bonaparte (1769–1821).

Napoleon Bonaparte, general and emperor of France, was quite interested in geometry. A result in geometry—attributed to him and known as Napoleon's Theorem—says that given any triangle, if we attach equilateral triangles along each side, then the centers of these attached triangles form an equilateral triangle, as shown in figure (a).

(a)

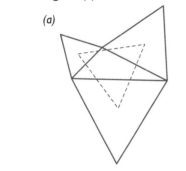

Napoleon's Theorem can be proved by using the fact that the plane can be tiled using the given triangle and the attached equilateral triangles as shown in figure (b).

(b)

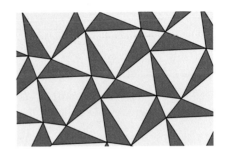

The theorem is proved by showing that the centers of the equilateral triangles in this tiling form the vertices of a regular tiling with triangles.

There is some doubt as to whether Napoleon actually proved the theorem named after him. According to a story, sometime before he became the emperor of France, Napoleon had a discussion with the French mathematicians Lagrange and Laplace. In the course of the conversation, Laplace told him "The last thing we want from you, general, is a lesson in geometry." Apparently Napoleon took the comment well, because Laplace later became Napoleon's chief military engineer.

In the first method, we start with a rectangle as in Figure 7.35(a). One side—in our case the top side—of the rectangle is modified as in Figure 7.35(b), and this modification is copied on the opposite side, as illustrated in Figure 7.35(c). Thus the tiles fit into one another like a jigsaw puzzle. We can then modify one of the two remaining sides (or leave them unmodified) and copy that change on the opposite side as illustrated in Figure 7.35(d). The resulting tile can now be used to tile the plane as shown in Figure 7.35(e).

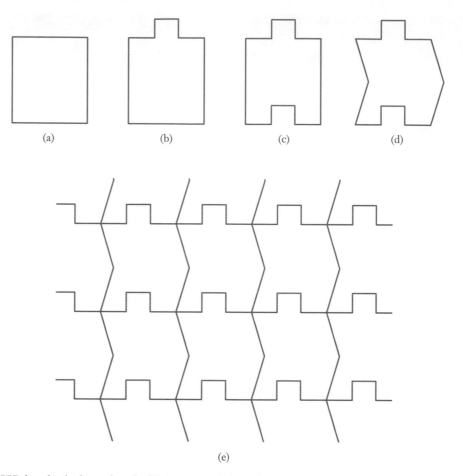

Figure 7.35
Tiling with a modified
rectangle.

With a little bit of embellishment of the tile, we get Figures 7.36(a) and 7.36(b).

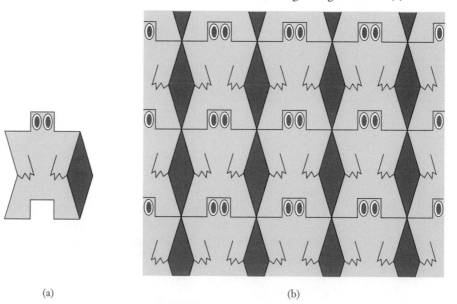

Figure 7.36
Embellished tiling by a
modified rectangle.

Another method for forming a tiling is to start with any triangle as shown in Figure 7.37(a). We first modify one half of one side of the triangle, in our case the upper right side as in Figure 7.37(b), and then carry this modification over

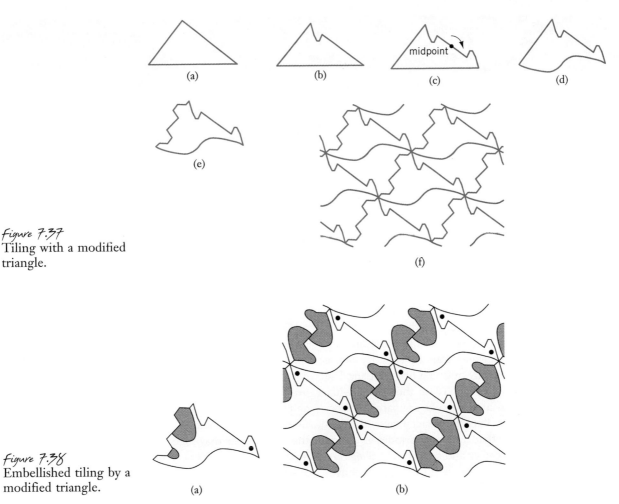

Figure 7.37
Tiling with a modified triangle.

Figure 7.38
Embellished tiling by a modified triangle.

to the other half of the side by rotating the side about its center 180° as shown in Figure 7.37(c). A similar modification can be made to each of the other two sides, if desired, as shown in Figures 7.37(d) and 7.37(e). A tiling can now be made from the resulting tile as shown in Figure 7.37(f). Notice that this tiling is based on the kind of tiling by a triangle shown in Figure 7.26. Again, we add a creative touch as in Figure 7.38(a) and the tiles can now be fitted together to arrive at the tiling in Figure 7.38(b).

These are just two of the methods that can be used to make tilings. We have only touched on the possibilities.

Technology Tip

When constructing Escher-like monohedral tilings, a computer drawing program is quite helpful. There are also commercial software packages available that partially automate some of the steps needed to construct the tile and then draw the tiling with a single command. These packages include many different methods for constructing tiles.

BRANCHING OUT 7.3

PENROSE TILINGS

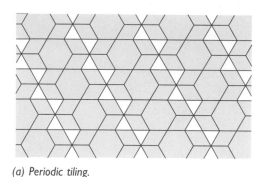

(a) Periodic tiling.

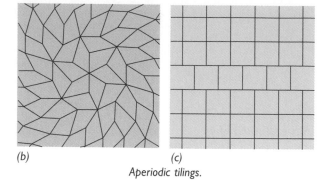

(b)   *(c)*

Aperiodic tilings.

Most of the tilings we consider in this text are *periodic* tilings. To explain what we mean by periodic, consider any tiling and imagine a transparent copy of the same tiling lined up exactly over the original tiling. The tiling is periodic if we can translate the copy—without any rotation—in at least two nonparallel directions so that it again lines up exactly over the original tiling. A tiling that is not periodic is called *aperiodic*. All of the regular and semiregular tilings are periodic as is the tiling in figure (a). The tilings shown in figures (b) and (c) are both aperiodic. In figure (b), the pattern cannot be translated back onto itself in any direction (although it can be rotated back onto itself), whereas in figure (c) the pattern can be translated horizontally, but not in a second direction.

It was at one time believed that if a set of one or more tiles could be used to construct an aperiodic tiling, then the same set of tiles could be used to construct a periodic tiling. However, in 1964, the logician Hao Wang discovered a set of over 20,000 tiles for which this assertion failed. Later, smaller sets of tiles with the same property were found, and finally in 1975 Roger Penrose, a mathematical physicist

at Oxford, found a set containing only two tiles that could tile the plane aperiodically but not periodically. The two tiles Penrose used, called kites and darts, are cut out from a special quadrilateral as shown in figure (d). The angles of each of the tiles are labeled with an H for head or T for tail, and the rule for fitting the two Penrose tiles together is that only angles with the same letter may meet. If this rule were not in effect, then the tiles could be fitted together as they are in the quadrilateral and the plane could then be periodically tiled with this quadrilateral.

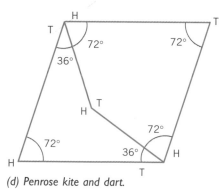

(d) Penrose kite and dart.

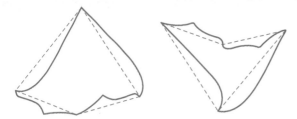

(e) Modifications of the Penrose kite and dart tiles that avoid the necessity of the head and tail rule for matching.
From: TILINGS AND PATTERNS by Grunbaum and Shephard ©1987 by W. H. Freeman and Company. Used with permission.

To avoid the necessity of imposing the head and tail rule for matching, the kites and darts can be modified as in figure (e) so that the head and tail rule is forced by the shapes of the tiles. Now we have two tiles that can only tile the plane aperiodically no mat-ter how we try to fit them together. However, the Penrose tiling is usually presented in the kites and darts version with the head and tail rule in force, because it is more convenient to work with them and they are attractive. This Penrose tiling is shown in figure (f). In recognition of his wide-ranging service to science, Roger Penrose was knighted in 1994.

In a rather peculiar turn of events, Sir Roger Penrose noticed one day that the pattern on a roll of Kleenex quilted toilet paper, shown in figure (g), strongly resembled one of his own nonperiodic patterns. In 1997, he and Pentaplex Ltd., which markets his patterns, filed suit against Kimberly-Clark Ltd. for copyright infringement.

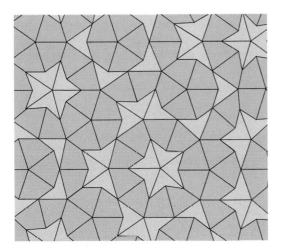

(f) Penrose tiling.

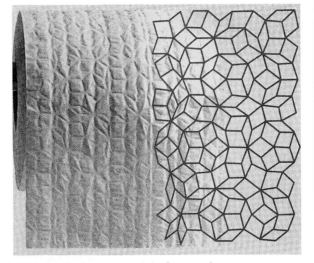

(g) Kleenex toilet paper and the Penrose tiling.

EXERCISES FOR SECTION 7.2

In Exercises 1 and 2, explain why the given tiling is not a regular tiling.

1.

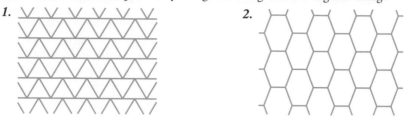

2.

In Exercises 3–6, give at least one reason why the given tiling is not a semiregular tiling.

3.

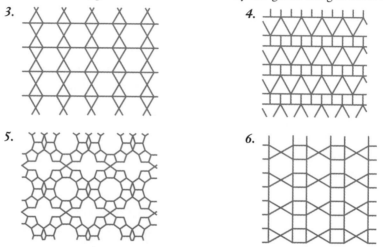

4.

5.

6.

In Exercises 7–10, state the number of different vertex types in the given tiling and draw a sketch of each.

7.

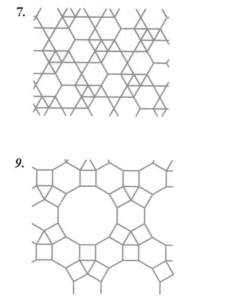

8.

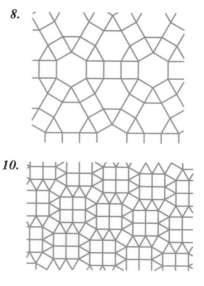

9.

10.

11. Find two different semiregular tilings that have two squares and three triangles about each vertex.

12. Find a semiregular tiling that does not include squares and that has four tiles about each vertex.

13. Find a semiregular tiling that does not include squares and that has five tiles about each vertex.

14. **(a)** Explain why it is not possible to construct a semiregular tiling where each vertex is of the type shown here.

   (b) Find a semiregular tiling with two squares, an equilateral triangle, and a regular hexagon at each vertex. (The vertices have to be of a type different from the one shown in part (a).)

15. Explain why it is not possible to construct a semiregular tiling where each vertex is of the type shown here.

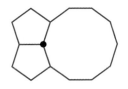

16. There is a vertex type that includes a regular 18-gon and two other regular polygons such that a semiregular tiling with the vertices of this type is not possible. Find the other two regular polygons in this vertex type.

17. There is a vertex type that includes a regular 24-gon and two other regular polygons such that a semiregular tiling with the vertices of this type is not possible. Find the other two regular polygons in this vertex type.

In Exercises 18–23, tile a sufficiently large region of the plane to illustrate how the given polygon can be used to tile the plane.

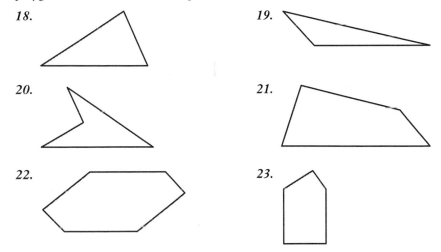

18.

19.

20.

21.

22.

23.

24. Suppose a rectangle has two sides modified as shown in the figure.

    (a) Modify the remaining two sides by the method illustrated in Figure 7.35.
    (b) Use the modified tile from part (a) to form a tiling.

25. Suppose a triangle has one half of one of its sides modified as shown in the figure.

    (a) Modify the other half of the side by the method illustrated in Figure 7.37, leaving the other sides unchanged.
    (b) Use the modified tile from part (a) to form a tiling.

26. Start with any rectangle and modify it by the method illustrated in Figure 7.35. Use this modified tile to form a tiling.

27. Start with any triangle and modify it by the method illustrated in Figure 7.37. Use this modified tile to form a tiling.

28. Using only regular polygon tiles, draw a tiling, which is not regular or semiregular, that has a pattern that can be extended to the entire plane.

In Exercises 29–32, tile a sufficiently large region of the plane to illustrate how the given polygon can be used to tile the plane. We have not looked at a systematic approach for doing this, so you will have to use trial and error.

33. Consider the tiles shown in the figures. By flipping over the white tile, we can get the tile that is shaded blue. The definition of monohedral tiling allows for this kind of flipping of the tiles. Find at least two different tilings of the plane by this triangular shape, with some of the tiles face up and some flipped over. Tile a sufficiently large region to illustrate how the tiling can be extended to the entire plane.

34. Explain why there cannot be more than five regular polygons around a vertex in a semiregular tiling.

7.3   POLYHEDRA

Having considered the kinds of patterns that arise when tiling the plane with polygons, we will now look at the kinds of patterns that occur when polygons are used to form the surface of a solid figure. Such shapes and patterns can be seen in nature and in man-made objects; for instance, in the beautiful garnet crystals in Figure 7.39, in the Great Pyramid of ancient Egypt in Figure 7.40, and in the modern architecture in Figure 7.41.

Figure 7.39
Garnet crystals.

Figure 7.40
The Great Pyramid.

Figure 7.41
Engineer Janos Baracs and sculptor P. Granche designed this sculpture for a subway station in Montreal.

A solid figure, whose surface does not intersect itself, with polygonal sides, called **faces**, that are arranged so that every side of each face coincides entirely with a side of a bordering face is called a **polyhedron**. (The plural of polyhedron is polyhedra.) We further restrict our definition of polyhedron to prohibit cases of two faces sharing more than one side. A line forming the side of two faces of a polyhedron is called an **edge**. The **vertices** of a polyhedron are the vertices of the faces, each of which belongs to more than one face. A polyhedron is called **convex** if every line segment drawn between two points inside the polyhedron lies completely inside the polyhedron. A polyhedron that is not convex is called **concave**. The polyhedron in Figure 7.42(a) is convex, whereas the ones in Figures 7.42(b) and 7.42(c) are concave.

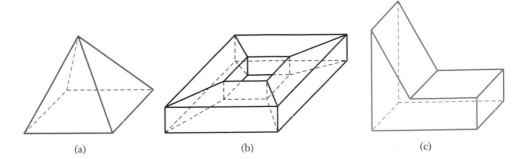

Figure 7.42
Polyhedra.

(a)   (b)   (c)

There is a relationship between the number of vertices, edges, and faces of a convex polyhedron. To discover this relationship ourselves, let's look at the few simple cases listed in Table 7.2. We use the letters V, E, and F to denote the number of vertices, edges, and faces, respectively. The first case is a cube, and it has $V = 8$, $E = 12$, and $F = 6$. You should verify that the number of vertices, edges, and faces for the other three convex polyhedra in Table 7.2 are correct.

Table 7.2	Vertices, Edges, and Faces of Some Polyhedra		
Polyhedron	**Vertices: V**	**Edges: E**	**Faces: F**
	8	12	6
	6	9	5
	10	15	7
	8	13	7

By looking at the numbers in Table 7.2, perhaps you already see a relationship between the number of vertices, edges, and faces. If not, try some more polyhedra on your own. After looking at many cases and searching for a relationship between V, E, and F, you will most likely happen upon the same formula discovered by the great mathematician Leonhard Euler (1707–1783):

Euler's Formula for Convex Polyhedra

If V is the number of vertices, E the number of edges, and F the number of faces of a convex polyhedron, then

$$V - E + F = 2.$$

Euler's formula is valid for all polyhedra, convex or concave, as long as they do not have any "holes" going through them. Polyhedra without holes are called *simple polyhedra*. The polyhedra in Figures 7.42(a) and 7.42(c) are simple, whereas the polyhedron in Figure 7.42(b) has a hole through it and therefore is not simple.

Example 1   *Verifying Euler's Formula*

Verify Euler's formula for the polyhedron in Figure 7.43.

Solution: The polyhedron shown has 7 vertices, 11 edges, and 6 faces, so we have $V = 7$, $E = 11$, and $F = 6$. Therefore,

$$V - E + F = 7 - 11 + 6 = 2,$$

and Euler's formula is satisfied.

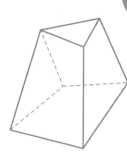

Figure 7.43

Example 2   *Using Euler's Formula*

If a convex polyhedron has 14 vertices and 21 edges, how many faces does it have?

Solution: Substituting $V = 14$ and $E = 21$ into Euler's formula $V - E + F = 2$, we get

$$
\begin{aligned}
14 - 21 + F &= 2 \\
-7 + F &= 2 \\
F &= 9,
\end{aligned}
$$

so the polyhedron has 9 faces.

Example 3   *Finding an Unknown Polyhedron*

Suppose a convex polyhedron has 7 faces and 6 vertices. How many triangles, quadrilaterals, pentagons, and so on make up the faces of such a polyhedron? Sketch one polyhedron of this type.

Solution: Substituting $F = 7$ and $V = 6$ into Euler's formula, we get

$$
\begin{aligned}
6 - E + 7 &= 2 \\
13 - E &= 2 \\
E &= 11,
\end{aligned}
$$

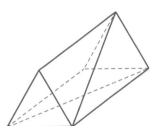

Figure 7.44

so the polyhedron has 11 edges. Because each of the 11 edges belongs to exactly 2 faces, the polygonal faces have a total of $11 \cdot 2 = 22$ sides altogether. Because each of the 7 polygonal faces must have at least 3 sides, we see that the only way the total number of sides can add up to 22 is to have 6 triangular faces and 1 quadrilateral face. A sketch of one convex polyhedron with 6 triangular faces, 1 quadrilateral face, 11 edges, and 6 vertices is shown in Figure 7.44.

Example 4   *Finding an Unknown Polyhedron*

Suppose a convex polyhedron has 9 edges and 5 faces. Determine the number of sides each face must have and sketch one polyhedron of this type.

Solution: Letting $E = 9$ and $F = 5$ in Euler's formula, we see that

$$V - 9 + 5 = 2$$
$$V - 4 = 2$$
$$V = 6,$$

so the polyhedron has 6 vertices. Because the polyhedron has 9 edges, the polygonal faces have $9 \cdot 2 = 18$ sides altogether. Each of the 5 faces must have at least 3 sides, but this leaves three more sides. There are three possible combinations of 5 faces for which the total number of sides is 18:

> **1.** 4 triangles and 1 hexagon,
>
> **2.** 3 triangles, 1 quadrilateral, and 1 pentagon,
>
> **3.** 2 triangles and 3 quadrilaterals.

Will 4 triangles and 1 hexagon work? Let's see. The hexagonal face would have 6 vertices and attaching one triangular face would have to contribute at least one more vertex, bringing the total number of vertices to be at least 7. But there are only supposed to be 6 vertices, so 4 triangles and 1 hexagon will not work. What about the second possibility of 3 triangles, 1 quadrilateral, and 1 pentagon? The pentagonal face would have 5 vertices and attaching the quadrilateral face would have to contribute at least two more vertices, bringing the total number of vertices to at least 7, so this will not work either. In the third case, with 2 triangle faces and 3 quadrilateral faces, we do not run into this kind of problem. One polyhedron of this type is shown in Figure 7.45.

Figure 7.45

Regular Polyhedra

Figure 7.46

Which polyhedra are the most symmetric? A cube is very symmetric. As we see in Figure 7.46, all of the faces of a cube are regular polygons of the same size and shape and three edges meet at every vertex. The view from any face of a cube is the same as the view from any other face. In general, the most symmetric kind of polyhedra are the regular polyhedra, defined as follows:

> A **regular polyhedron** is a convex polyhedron with all of its faces regular polygons of the same size and shape and with the same number of edges meeting at each vertex.

A cube is an example of a regular polyhedron, but there are others. The regular polyhedra were studied by the ancient Greeks, and the Greek mathematician Theaetetus (414–369 B.C.) proved that there are only five possible regular polyhedra. This proof is included as the final result in Euclid's *Elements* (circa 350 B.C.). A list of these five solids along with the shape of each face and the number of faces is given in Table 7.3, and the regular solids are shown in Figure 7.47.

Table 7.3	The Five Regular Polyhedra	
Name	*Shape of the Faces*	*Number of Faces*
Tetrahedron	Equilateral triangles	4
Cube	Squares	6
Octahedron	Equilateral triangles	8
Dodecahedron	Regular pentagons	12
Icosahedron	Equilateral triangles	20

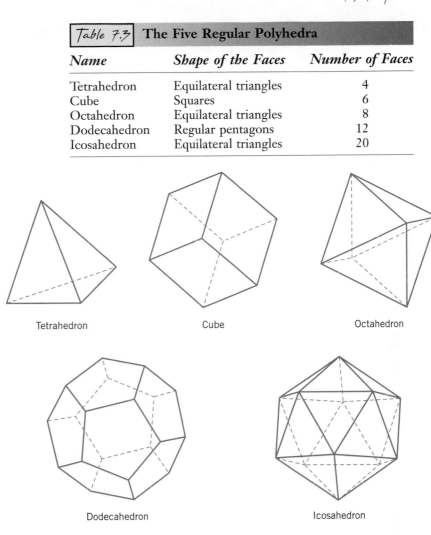

Tetrahedron   Cube   Octahedron

Dodecahedron   Icosahedron

Figure 7.47
The five regular
polyhedra.

Notice that each of the regular polyhedra in Figure 7.47 is very symmetrical. One way to visualize this symmetry is to notice that if the polyhedron is rotated about its center so that one face is moved into the previous position of any other face, then the polyhedron would look exactly the same.

The five polyhedra in Figure 7.47 are the only regular polyhedra. The fact that no others exist can be proven by using Euler's formula. To see this, suppose we have a regular polyhedron whose faces are regular n-gons, and such that r edges meet at each vertex. Let V, E, and F be the number of vertices, edges, and faces of the polyhedron. Because each face has n edges and each edge belongs to two faces, we see that $nF = 2E$. Solving for F, we have

$$F = \frac{2E}{n}.$$

Similarly, because each vertex has r edges meeting it and each edge belongs to two vertices, we have $rV = 2E$, and therefore

$$V = \frac{2E}{r}.$$

Now, substituting these equations for F and V into Euler's formula $V - E + F = 2$, we get

$$\frac{2E}{r} - E + \frac{2E}{n} = 2.$$

Solving for E in this equation, we have

$$\left(\frac{2}{r} - 1 + \frac{2}{n}\right)E = 2,$$

$$E = \frac{2}{\left(\dfrac{2}{r} - 1 + \dfrac{2}{n}\right)}.$$

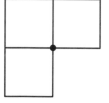

Figure 7.48

We now have an equation for the total number of edges, E, in terms of the number of edges, r, meeting at each vertex and the number of edges on each face, n. Consider the faces around a particular vertex and imagine taking these faces and flattening them out, as illustrated for a vertex of a cube in Figure 7.48. In general, the sum of the angles of the r faces of the regular n-gons at a vertex must be less than 360°. Because r is at least 3, the angle of each n-gon face must be less than $360/3 = 120°$, the measure of the angle of a regular hexagon. Therefore, the n-gon faces have at most 5 sides, and we can conclude that $n = 3, 4,$ or 5. Furthermore, because the angle of any regular n-gon is at least 60°, we know that there are fewer than $360°/60° = 6$ n-gons meeting at each vertex, and therefore $r = 3, 4,$ or 5. Trying the nine possible combinations of $n = 3, 4,$ or 5 and $r = 3, 4,$ or 5, the formula

$$E = \frac{2}{\left(\dfrac{2}{r} - 1 + \dfrac{2}{n}\right)}$$

gives a positive integer value for E only in the five cases: $n = 3$, $r = 3$; $n = 3$, $r = 4$; $n = 3$, $r = 5$; $n = 4$, $r = 3$; and $n = 5$, $r = 3$. (We will leave the verification of this as an exercise.) But these are precisely the cases of the tetrahedron, octahedron, icosahedron, cube, and dodecahedron, respectively. Therefore, the five regular solids we have seen are the only ones possible.

The five regular polyhedra are sometimes called the **Platonic solids** because Plato (429–348 B.C.) attempted to relate the four basic elements of fire, earth, air, and water, plus the universe as a whole, to regular polyhedra. Plato was not the last to attempt to find deep significance in the fact that there are only five regular polyhedra. Johannes Kepler (1571–1630), who first correctly described the elliptical orbits of the planets, originally proposed a rather peculiar theory of the solar system involving the regular solids. According to this theory, by circumscribing a dodecahedron about the sphere containing the orbit of Earth, and then circumscribing a sphere about the dodecahedron, the sphere containing the orbit of Mars can be found. Next, if a tetrahedron is circumscribed about the orbit of Mars, then the sphere surrounding this tetrahedron contains the orbit of Jupiter. Similarly, the sphere of the orbit of Venus lies inside the icosahedron inscribed in the sphere containing the orbit of Earth, and finally, the sphere orbit of Mercury lies inside the octahedron inscribed in the sphere containing the orbit of Venus. Kepler's own drawing illustrating this theory is shown in Figure 7.49.

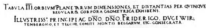

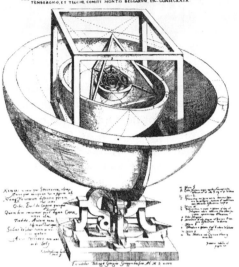

Figure 7.49
Kepler's solar system
model.

These ideas of Plato and Kepler were a bit off the mark, but regular polyhedra
do arise in nature. Some crystals grow in the shape of regular polyhedra because
of the arrangement of atoms within them—for instance, the octahedral fluorite
crystal in Figure 7.50. Some of the tiny ocean animals called radiolaria, two of
which are shown in Figure 7.51, have appendages that lie at the vertices of a regular polyhedron.

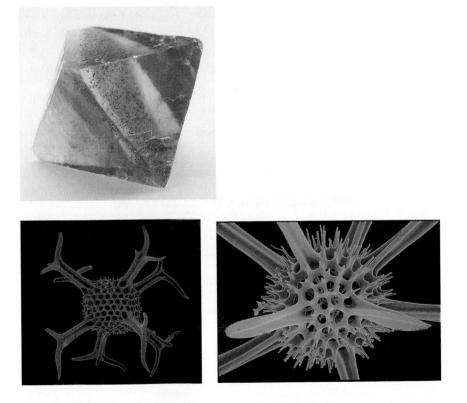

Figure 7.50
Fluorite crystal.

Figure 7.51
Skeletons of radiolaria
in the shape of a cube
and an icosahedron.

BRANCHING OUT 7.4

HOW MANY SIDES CAN A FAIR DIE HAVE?

Common dice are cubes. Because of its symmetry, a cube makes a fair die. When rolled, all six faces are equally likely to come out on the top. Similarly, fair dice can also be constructed in the shapes of the other four regular polyhedra. However, because a tetrahedron will land with a vertex up rather than a face up, we consider the face that is on the bottom to be the face that was rolled. These kinds of unusual dice can be found in some games. The game of Scattergories™ uses the icosahedron die shown in figure (a). The 20 faces are labeled with the letters of the alphabet (excluding Q, U, V, X, Y, and Z).

(a) Die from the Scattergories™ game.

(b) 10-sided fair die.

What about fair dice of other shapes? In figure (b) we see a fair die with 10 sides. This and the regular polyhedra are the most common dice in adventure games such as Dungeons and Dragons™ and Star Trek. This 10-sided die is symmetric in that all its faces are of the same size and shape and have identical surroundings. If the die were unlabeled, each of the faces would be indistinguishable from the others. For what number of faces is it possible to construct such a symmetrical die? It is easy to construct one when the number of faces desired is an even number that is at least six. To build such a die with $2n$ faces, where $n \geq 3$, simply begin with a regular n-gon and build a pyramid with identically shaped, triangular faces above and below it. The die in figure (b) is constructed slightly differently, but is in the same spirit. It turns out that constructing a symmetrical die with an odd number of faces is impossible.

Is it possible to construct fair dice that are not so symmetrical? With a die whose faces have distinct shapes, an analysis of the physics of the die and surface and of the manner in which it is rolled would allow computation of the probability of landing on each of the faces, but such an analysis is not practical to carry out. Of course, a very large number of rolls of any particular die would likely approximate the true probabilities quite well. However, it is very unlikely that by good fortune one could happen upon a non-symmetrical die all of whose faces are equally likely to be rolled.

Semiregular Polyhedra

With tilings, we saw that there were only three regular tilings and eight semiregular tilings of the plane. Because there are only five regular polyhedra, it is natural to ask how many semiregular polyhedra are possible. A semiregular polyhedron is defined as follows:

A **semiregular polyhedron** is a polyhedron with all of its faces regular polygons, all of its vertices of the same type, with at least two different kinds of regular polygons as faces, and an additional symmetry condition that, up to mirror image, the polyhedron looks exactly the same from every vertex.

As in the case of tilings, we consider two vertices to be of the same **vertex type** if the same number and kind of polygons are arranged about them in the same order, except that one arrangement is possibly clockwise whereas the other is counterclockwise. Two infinite families of semiregular polyhedra—the prisms and antiprisms—are illustrated in Figure 7.52. **Prisms** have two regular n-gons for top and bottom, aligned with one directly above the other, and n square sides. **Antiprisms** again have two regular n-gons for top and bottom, with the top twisted by $180/n$ degrees, so that $2n$ equilateral triangles make up the sides.

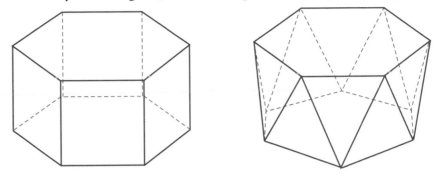

Figure 7.52   (a) A prism with $n = 6$ (b) An antiprism with $n = 6$

There are only 13 other possible convex semiregular polyhedra, which are sometimes called the **Archimedean solids** after the Greek scholar Archimedes (287–212 B.C.). (Two of the Archimedean solids have mirror images that are different from themselves. Each of these mirror image pairs has been counted as a single Archimedean solid.) Four of the Archimedean solids are illustrated in Figure 7.53.

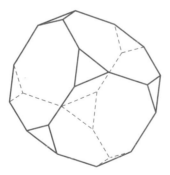

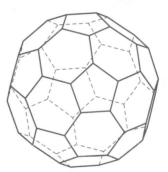

(a) Truncated cube   (b) Truncated icosahedron

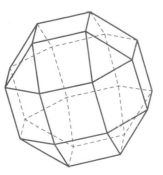

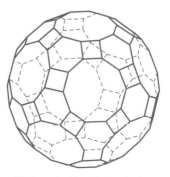

Figure 7.53
Four of the 13
Archimedean solids.   (c) Rhombicuboctahedron   (b) Truncated icosidodecahedron

As required, for each of the semiregular polyhedra shown in Figures 7.52 and 7.53, all vertices of the polyhedron are of the same type. For instance, the vertices of the rhombicuboctahedron in Figure 7.53(c) all have three squares and one equilateral triangle surrounding them. Notice that the angles of these four polygons add up to $90° + 90° + 90° + 60° = 330°$ and not $360°$, as would the angles at a vertex on a flat surface.

Surprisingly, all of the Archimedean solids can be constructed by starting with a regular polyhedron and altering it by slicing it and sometimes distorting it. For instance, the truncated cube in Figure 7.53(a) is just a cube whose corners have been sliced off as shown in Figure 7.54 so that the resulting faces are regular octagons and equilateral triangles.

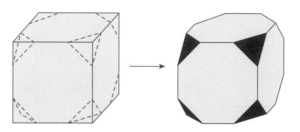

Figure 7.54
Constructing a truncated cube.

There is one more polyhedron whose faces are regular polygons and whose vertices are of the same type, but it does not satisfy the symmetry condition that would classify it as a semiregular polyhedron. This polyhedron is called the pseudo rhombicuboctahedron, and it may be formed by twisting the bottom third of the rhombicuboctahedron in Figure 7.53(c) by $45°$.

If the truncated icosahedron in Figure 7.53(b) looks familiar, that is probably because the pattern in this solid is the same pattern used on a regulation soccer ball, although the ball is inflated into a spherical shape, as shown in Figure 7.55.

The truncated icosahedron also appears in chemistry. The molecular structure of C_{60}, a form of carbon molecule, is in the shape of a truncated icosahedron. In these molecules, which look like tiny soccer balls, 60 carbon atoms lie at the 60 vertices of the truncated icosahedron. A model of C_{60} is shown in Figure 7.56, where the edges between the atoms represent the bonds between the atoms. The molecule C_{60} was discovered by chemists Richard Smalley and Robert Curl in 1985, and in 1996 they were awarded the Nobel Prize in chemistry for this work. The C_{60} molecule is called a *buckminsterfullerene* because it resembles the geodesic domes found in the designs of the architect R. Buckminster Fuller (1895–1983). These molecules are commonly called *buckyballs*. The buckyball is one of a family of carbon molecules called *fullerenes*. In a fullerene, the carbon atoms and their bonds form a polyhedron with hexagonal and pentagonal faces. Each carbon atom is joined to three others. Using these facts, combined with Euler's formula, it can be shown that any fullerene must have exactly 12 pentagonal faces. We leave the verification of this as an exercise. A buckyball has 12 pentagonal faces and 20 hexagonal faces. Since the discovery of the buckyball, other fullerenes have been found. One of them, C_{70}, has an oblong shape, and it has 12 pentagonal and 25 hexagonal faces. Another, C_{44}, is a smaller fullerene made up of 12 pentagons and 12 hexagons. There are also very large fullerenes such as C_{540}. This certainly invites the question of what fullerenes are geometri-

Figure 7.55
A regulation soccer ball.

cally possible given the limitation of having only pentagonal and hexagonal faces and three edges at each vertex. This question was answered by mathematicians Branko Grünbaum and Theodore S. Motzkin in 1963, more than 30 years before the discovery of fullerenes. Their results showed that there cannot be a fullerene with exactly one hexagonal face, but any other number of hexagonal faces, including possibly none, is geometrically possible.

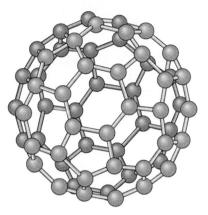

Figure 7.56
The buckyball molecule, C_{60}.

BRANCHING OUT 7.5

TEXAS POLITICIANS THINK VERY SMALL

New York has its official state muffin, the apple muffin, Arizona has the bolo tie as its official state neckwear, and Ohio has honored tomato juice by naming it the official state beverage. But only Texas can lay claim to an official state molecule. After Rice University professors Richard Smalley and Robert Curl won the Nobel Prize in chemistry for their work on C_{60}, the buckyball, state Representative Scott Hochberg, a 1975 Rice alumnus, proposed legislation that would name the buckyball the state molecule of Texas. In spite of some competition from a molecule called Texaphyrin, which was engineered by chemists at the University of Texas, the buckyball proposal passed in 1997. At the same time, the Texas sweet onion became the state vegetable and picante sauce was named the state sauce. Texas has more state symbols than any other state. As explained by University of Texas Professor Janice May, "It's the second-biggest state, but first in braggarts." Nevertheless, Texas has yet to name an official state muffin.

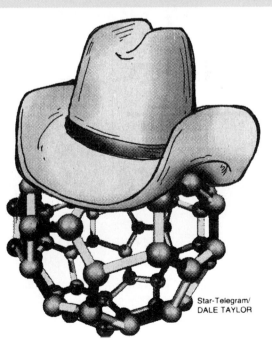

Star-Telegram/
DALE TAYLOR

EXERCISES FOR SECTION 7.3

In Exercises 1–4, determine whether the polyhedron is convex or concave.

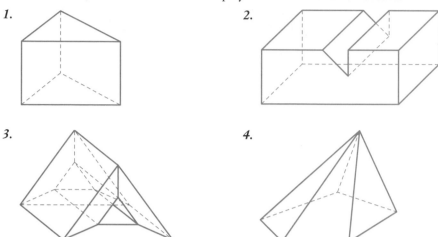

1.

2.

3.

4.

In Exercises 5–8, verify Euler's formula for the polyhedron.

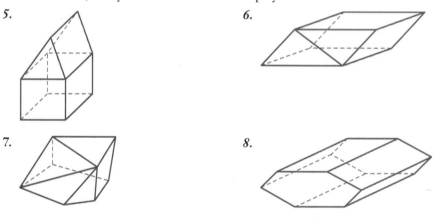

5.

6.

7.

8.

In Exercises 9–13, find the number of vertices, edges, and faces and verify Euler's formula for the given regular polyhedron.

   9. tetrahedron   10. cube
   11. octahedron   12. dodecahedron
   13. icosahedron

14. If a convex polyhedron has 25 edges and 12 faces, how many vertices does it have?

15. If a convex polyhedron has 20 edges and 10 vertices, how many faces does it have?

16. If a convex polyhedron has 33 faces and 33 vertices, how many edges does it have?

17. Suppose a convex polyhedron has six faces and five vertices. How many triangles, quadrilaterals, pentagons, and so on make up the faces of such a polyhedron? Sketch one polyhedron of this type.

18. Suppose a convex polyhedron has 13 edges and 8 vertices and all its faces are triangles and quadrilaterals. Find the number of triangles and quadrilaterals making up the faces of such a polyhedron and sketch one polyhedron of this type.

19. Suppose a convex polyhedron has 10 edges and 6 vertices. How many triangles, quadrilaterals, pentagons, and so on might make up the faces of such a polyhedron? (There are two possible answers.) Sketch an example of a polyhedron of each of these two types.

20. Suppose a convex polyhedron has 7 faces and 10 vertices and none of its faces are triangles. How many quadrilaterals, pentagons, hexagons, and so on make up the faces of such a polyhedron? Sketch one polyhedron of this type.

21. Suppose a convex polyhedron has eight faces, of which four are triangles and four are quadrilaterals.

   (a) How many edges does the polyhedron have?

   (b) How many vertices does the polyhedron have?

   (c) Sketch one polyhedron of this type.

22. Suppose a convex polyhedron has 12 faces, of which 4 are triangles, 7 are quadrilaterals, and 1 is a hexagon.

   (a) How many edges does the polyhedron have?

   (b) How many vertices does the polyhedron have?

Euler's formula holds for all polyhedra, convex or concave, as long as they do not have any "holes" through them as in the polyhedron in Figure 7.42(b). Polyhedra without holes are called simple polyhedra. In Exercises 23 and 24, verify that Euler's formula holds for the simple concave polyhedron shown.

23.   **24.**

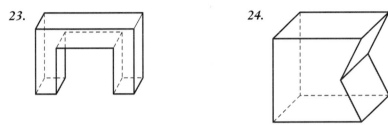

In Exercises 25–28, find the number of vertices, edges, and faces of the polyhedron.

25.

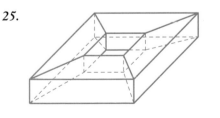

26.

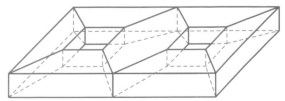

27.

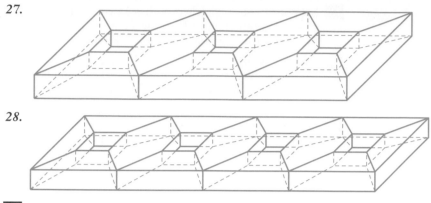

28.

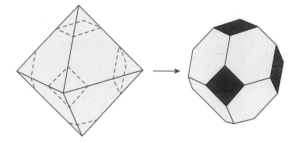

29. There is a formula similar to Euler's formula that relates the number of vertices, edges, and faces of a polyhedron with *n* holes. Based on the answers to Exercises 25–28, try to find this formula.

30. A cuboctahedron is a semiregular solid that can be constructed by slicing off the corners of a cube through the midpoints of the edges as shown here.

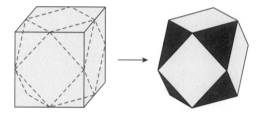

(a) How many triangular faces does a cuboctahedron have?

(b) How many square faces does a cuboctahedron have?

31. A truncated octahedron is a semiregular solid that can be constructed by slicing off the corners of an octahedron so that the resulting faces are regular hexagons and squares as shown in the figure.

(a) How many hexagonal faces does a truncated octahedron have?

(b) How many square faces does a truncated octahedron have?

32. The cuboctahedron, constructed in Exercise 30 by slicing off the corners of a cube, can also be constructed by slicing off the corners of another regular solid. Which regular solid is this?

33. A truncated tetrahedron is a semiregular solid formed by slicing off the corners of a tetrahedron. What types of regular polygons appear as faces of a truncated tetrahedron, and how many faces are there of each type?

If a cube is drawn and the centers of each face of the cube are connected by lines, an octahedron lying within the cube is formed as shown in the figure.

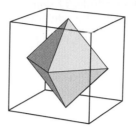

We say that the octahedron is the dual of the cube. In general, the dual of any regular polyhedron is another regular polyhedron formed by connecting the centers of the faces of the original polyhedron. In Exercises 34–37, find the dual of the given regular polyhedron.

34. tetrahedron   **35.** octahedron

36. dodecahedron   **37.** icosahedron

38. In the text we saw that if a regular polyhedron has E edges, faces that are regular n-gons, and r edges meeting at each vertex, then the following equation is satisfied:

$$E = \frac{2}{\left(\dfrac{2}{r} - 1 + \dfrac{2}{n}\right)}.$$

We also saw that $n = 3, 4,$ or 5 and $r = 3, 4,$ or 5.

(a) In the following table, use this formula for E to fill in the value of E for each of the nine possible pairs of values for n and r.

n	3	3	3	4	4	4	5	5	5
r	3	4	5	3	4	5	3	4	5
E									

(b) Show how the pairs of values for n and r that gave positive integer values for E in part (a) correspond to the five regular polyhedra: the tetrahedron, cube, octahedron, dodecahedron, and icosahedron.

39. Suppose a fullerene has p pentagonal faces, h hexagonal faces, and a carbon atom at each vertex. Each carbon atom is joined by three bonds to three other carbon atoms, and these bonds form the edges of a polyhedron.

(a) Explain why the fullerene has $\frac{1}{3}(5p + 6h)$ vertices.

(b) Explain why the fullerene has $\frac{1}{2}(5p + 6h)$ edges.

(c) Use the results of parts (a) and (b) and Euler's formula to show that any fullerene must have exactly 12 pentagonal faces.

40. Use the results of Exercise 39 to find the number of hexagonal faces in the fullerene C_{540}.

41. Show that for any polyhedron with V vertices, E edges, and F faces,

(a) $E \geq \frac{3}{2}F$,

(b) $E \geq \frac{3}{2}V$.

42. Using Euler's formula and the results of Exercise 41, find all of the pairs of values of E and V for which there may be a convex polyhedron with seven faces.

43. Using Euler's formula and the results of Exercise 41, find all of the pairs of values of E and V for which there may be a convex polyhedron with eight faces.

44. Show that for any polyhedron, the number of faces that have an odd number of sides is even.

WRITING EXERCISES

1. Explain some of the practical advantages of using an edge-to-edge tiling when tiling a floor and the possible disadvantages of laying a brick wall in an edge-to-edge pattern.

2. Discuss some of the practical advantages of using square tiles instead of tiles of some other shape.

3. Explain why you think cubical dice seem to be the most common randomizing device for board games, rather than a die in the shape of some other regular polyhedron or a different type of randomizing device such as a spinner or drawing chips with numbers.

4. A die in the shape of a pyramid with a square base and four triangular faces with a common vertex above the center of the base may be tall and thin or short and wide. Explain why adjusting these relative proportions should allow one to construct a fair die.

PROJECTS

1. We saw that a quadrilateral of any shape can be used in a monohedral tiling of the plane. In our construction, every vertex contains all four angles of the quadrilateral. For certain special quadrilaterals, another way of arranging the tiles yields a different monohedral tiling. Find some examples using rhombuses with particular angles. (A *rhombus* is a quadrilateral with all four sides of equal length. It is necessarily a parallelogram, with opposite sides parallel.)

2. A pentomino is a polygon built up from five unit squares. They must be connected through shared edges (not partial edges or just corners). Two such pentominoes are illustrated here.

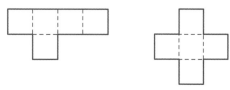

   (a) Show that either of these pentominoes can be used in a monohedral tiling of the plane.

   (b) There are twelve different pentominoes (if you are allowed to flip them over), any of which tile the plane in a monohedral tiling. Find as many as you can and show that they tile the plane.

3. In addition to the translational symmetry exhibited by periodic tilings as described in Branching Out 7.3, many tilings also exhibit rotational symmetry. A tiling has rotational symmetry if it is possible to rotate the entire tiling about some point in the tiling so that the rotated tiling lines up exactly with the original tiling.

(a) Which regular and semiregular tilings exhibit rotational symmetry? What are the points about which you can rotate and the angles through which you can rotate?

(b) Can you find other tilings with rotational symmetry? Try to produce some different sets of angles through which you can rotate.

4. Find examples of tessellations and polyhedra that occur in nature. Comment on any structural or evolutionary advantages conferred by such patterns.

5. We can alter a hexagon by adding or subtracting identical small, equilateral triangles from certain edges, as in the three examples shown here.

(a) Which of these shapes can be used in a monohedral tiling of the plane? Consider many different ways tiles could fit together. Use deductive reasoning to keep the number of cases manageable. Explain how you can eliminate certain cases.

(b) Design some tiles built up in a similar way from hexagons, and determine whether they can be used to form a monohedral tiling of the plane.

6. The *angle defect* of a vertex of a polyhedron is defined to be 360° minus the sum of the face angles at the vertex. (The angle defect may be negative.) The *total angle defect* of a polyhedron is the sum of the angle defects over all vertices of a polyhedron.

(a) Find the total angle defect for each of the regular polyhedra. What do you get?

(b) Find the total angle defect for some other convex polyhedra. Make a conjecture about what can be said about the total angle defect for convex polyhedra.

(c) Suppose a convex polyhedron has n_3 triangular faces, n_4 quadrilateral faces, n_5 pentagonal faces, and so forth. Express V, E, F in terms of the total angle defect and $n_3, n_4, n_5, \ldots$. Try to use Euler's formula to prove your conjecture from part (b).

(d) What happens in concave polyhedra or in polyhedra with holes such as those in Exercises 25–28 of Section 7.3?

KEY TERMS

CHAPTER 7 REVIEW EXERCISES

1. Sketch a concave hexagon.

2. Sketch a convex 11-gon.

3. Find the measure of an angle of a regular 15-gon.

4. Five of the six angles of the following polygon are given. Find the remaining angle.

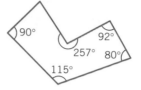

5. Tile a sufficiently large region of the plane to illustrate how the triangle shown here can be used to tile the plane.

6. Tile a sufficiently large region of the plane to illustrate how the quadrilateral shown here can be used to tile the plane.

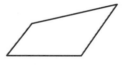

7. State the number of different vertex types in the given tiling and draw a sketch of each one.

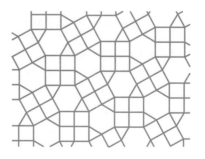

8. Find a semiregular tiling that does not include octagons and that has three tiles about each vertex. (There are two possible answers.)

9. There is a vertex type that includes a regular 42-gon and two other regular polygons such that a semiregular tiling with vertices of this type is not possible. Find the other two regular polygons in this vertex type.

10. Explain why it is not possible to construct a semiregular tiling where each vertex is of the type shown in the figure.

11. The tiling shown here from the coloring book *Altair Design, Book 1*, appears, at first glance, to be a tiling with regular polygon tiles. Explain why it is impossible for all of these polygons to be regular.

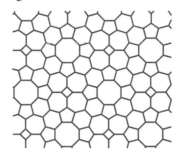

12. Verify Euler's formula for the following polyhedron.

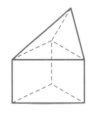

13. If a convex polyhedron has 14 faces and 13 vertices, how many edges does it have?

14. If a convex polyhedron has 44 edges and 22 faces, how many vertices does it have?

15. Suppose a convex polyhedron has eight edges and five faces. How many triangles, quadrilaterals, pentagons, and so on make up the faces of such a polyhedron? Sketch a picture of one polyhedron of this type.

16. A truncated dodecahedron is a semiregular solid that can be constructed by slicing off the corners of a dodecahedron so that the resulting faces are regular triangles and decagons as shown in the figure.

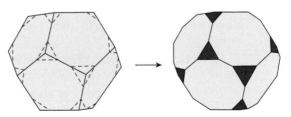

(a) How many triangular faces does a truncated dodecahedron have?

(b) How many decagonal faces does a truncated dodecahedron have?

(c) Find the number of vertices, faces, and edges and verify Euler's formula for the truncated dodecahedron.

Suggested Readings

Danzer, Ludwig, Branko Grünbaum, and G. C. Shephard. "Equitransitive Tilings, or How to Discover New Mathematics." *Mathematics Magazine* 60:2 (April 1987), 67–88. An investigation of a class of tilings more general than semiregular tilings, written in a way that illustrates the process of mathematical research.

Gardner, Martin. "Extraordinary Nonperiodic Tiling That Enriches the Theory of Tiles." *Scientific American* 236:1 (January 1977), 110–121. Reprinted with additional updates in Martin Gardner, *Penrose Tiles to Trapdoor Ciphers*. New York: Freeman, 1989, 1–29. An introduction to Penrose tilings.

Gardner, Martin. "On Tessellating the Plane with Convex Polygon Tiles." *Scientific American* 233:1 (July 1975), 112–117, 132. Reprinted with an addendum in Martin Gardner, *Time Travel and Other Mathematical Bewilderments*. New York: Freeman, 1988, 163–176. The original article listed the known results about monohedral tilings in the plane with convex polygons, including R. B. Kershner's incorrect claim to have found all convex pentagon tilings. The addendum to the article presented new pentagon tilings found by Richard E. James III and Marjorie Rice.

Grünbaum, Branko, and G. C. Shephard. "Some Problems on Plane Tilings." In *The Mathematical Gardner*, edited by David A. Klarner. Boston: Prindle, Weber & Schmidt, 1981, 167–196. On testing a tile to see whether it can be used in a monohedral tiling. Heesch's problem.

Grünbaum, Branko, and G. C. Shephard. *Tilings and Patterns*. New York: Freeman, 1987. A comprehensive treatment of the subject. The standard text in the field.

Kepes, Gyorgy. *Module, Proportion, Symmetry, Rhythm*. New York: George Braziller, 1966. Patterns in nature, art, and architecture.

O'Daffer, Phares G., and Stanley R. Clemens. *Geometry: An Investigative Approach*. Reading, Massachusetts: Addison-Wesley, 1992. An introductory geometry textbook emphasizing exploration.

Rigby, J. F. "Napoleon, Escher, and Tessellations." *Mathematics Magazine* 64:4 (October 1991), 242–246. Proofs of Napoleon's Theorem and Escher's Theorem based on tessellations.

Schattschneider, Doris. "In Praise of Amateurs." In *The Mathematical Gardner*, edited by David A. Klarner. Boston: Prindle, Weber & Schmidt, 1981, 140–166. Monohedral tilings with pentagons, emphasizing the discoveries of amateur mathematician Marjorie Rice.

Schattschneider, Doris. "The Plane Symmetry Groups: Their Recognition and Notation." *The American Mathematical Monthly* 85:6 (June–July 1978), 439–450. Possible symmetries of tilings.

Schattschneider, Doris. "Tiling the Plane with Congruent Pentagons." *Mathematics Magazine* 51:1 (January 1978), 29–44. Monohedral tilings with pentagons.

Schattschneider, Doris. *Visions of Symmetry*. New York: Freeman, 1990. An extensive treatment of the work of M. C. Escher, including mathematical explanations, pages from his notebooks, and over 350 illustrations.

Seymour, Dale, and Jill Britton. *Introduction to Tessellations*. Palo Alto, Calif.: Dale Seymour, 1989. A nontechnical introduction to tiling. The chapter on creating Escher-like tilings shows some other techniques that are not included in this text.

Stevens, Peter S. *Patterns in Nature*. Boston: Little, Brown, 1974. Includes tiling patterns and polyhedra that occur in nature.

Thompson, D'Arcy. *On Growth and Form*. London: Cambridge University Press, 1961. Covers honeycombs of bees and the skeletons of radiolaria.

Washburn, Dorothy K., and Donald W. Crowe, *Symmetries of Culture*. Seattle: University of Washington Press, 1988. Includes a chapter on tiling patterns found in various cultures.

Number theory is the mathematics of numbers, especially the integers, ..., −3, −2, −1, 0, 1, 2, 3, ... In number theory, we look at divisibility of numbers, patterns in the integers, and solutions to equations. Until recently, number theory was almost exclusively considered to be "pure" mathematics. By this we mean mathematics that is studied for its own intrinsic beauty rather than for its applications. However, today, number theory has found its way into many practical applications. For instance, with modern electronic communications, encoding messages has become an area of vital importance in which number theory plays a central role. In this chapter, we begin by looking at some of the basic ideas of number theory and some fascinating unsolved problems. We then go on to study a special kind of arithmetic used in number theory, called modular arithmetic, and some of its applications.

0-471-13

CHAPTER 8

NUMBER THEORY

8·1   DIVISIBILITY AND PRIMES

The first numbers we encounter are those used in counting: 1, 2, 3, ... However, when we subtract these numbers from one another, we encounter negative numbers and zero. For this reason, we will consider the entire set of **integers**, ..., $-3, -2, -1, 0, 1, 2, 3, \ldots$. When given a particular integer such as the number 4, the first thing we might try to do with it is break it down into other positive integers. This can be done through addition—for instance, with 4 as

$$4 = 3 + 1 = 2 + 2 = 2 + 1 + 1 = 1 + 1 + 1 + 1.$$

This leads us to the major area of number theory called *additive number theory*. However, we will focus on writing integers as products of other positive or negative integers, or *multiplicative number theory*. For example, the number 4 can be written as

$$4 = 4 \cdot 1 = 2 \cdot 2 = (-4)(-1) = (-2)(-2).$$

In this case, we say that 1, -1, 2, -2, 4, and -4 are all factors of 4, and that 4 is divisible by each of these numbers. More generally, we have the following definitions:

> If a and b are integers and there is an integer c such that $a = b \cdot c$, then we say that b **divides** a or a is **divisible** by b, and write $b \mid a$. In this case, we say that b is a **factor** or **divisor** of a.

For example, because $18 = 6 \cdot 3$, we see that $6 \mid 18$ and $3 \mid 18$. Similarly, $18 = (-9)(-2)$, so $(-9) \mid 18$ and $(-2) \mid 18$. Because $18/5 = 3.6$ is not an integer, we see that 5 does not divide 18. Starting from 1 and -1, we can list all of the divisors of 18: 1, -1, 2, -2, 3, -3, 6, -6, 9, -9, 18, and -18. We note here that the number 0 is divisible by any integer b because $0 = b \cdot 0$.

We just saw that the number 18 has six positive divisors. There are some positive integers—for example, the number 7—with only two positive divisors. We have a special name for such integers.

> A **prime number** is an integer greater than 1 whose only positive divisors are 1 and itself. An integer greater than 1 that is not prime can be written as a product of integers greater than 1 and is called a **composite number**.

The first six prime numbers are 2, 3, 5, 7, 11, and 13. Prime numbers are the basic multiplicative building blocks of all integers. This fact, called the **Fundamental Theorem of Arithmetic**, was known to the ancient Greeks. It can be stated formally as follows:

> ## The Fundamental Theorem of Arithmetic
>
> Every integer greater than 1 can be expressed as a product of primes, and this representation is unique up to the order in which the factors occur.

To find the prime factorization of a given composite integer, first factor the integer into any two positive factors. If the factors are primes, then we are done. If any of the factors are not prime, continue factoring any composite factors into two factors until all the numbers in the factorization are prime. This technique is illustrated in the following example:

Example 1   *Prime Factorization*

Find the prime factorization of the number 3150.

Solution: Integers ending in a 0 are divisible by 10, so first we can factor 3150 as

$$3150 = 10 \cdot 315.$$

Now factoring the 10 into $2 \cdot 5$ and the 315 into $5 \cdot 63$, we have

$$3150 = 2 \cdot 5 \cdot 5 \cdot 63.$$

The factor 63 is the product of 7 and 9, and therefore is not prime. We can factor it to get

$$3150 = 2 \cdot 5 \cdot 5 \cdot 7 \cdot 9.$$

Finally, factoring the only remaining composite factor, 9, into $3 \cdot 3$ gives the prime factorization

$$3150 = 2 \cdot 5 \cdot 5 \cdot 7 \cdot 3 \cdot 3.$$

It is standard to write the prime factors from the smallest to the largest and to combine any powers of the same prime factor. Doing this for our factorization of 3150, we get

$$3150 = 2 \cdot 3^2 \cdot 5^2 \cdot 7.$$

■

Prime Numbers

Prime numbers have always fascinated professional and amateur mathematicians. There are many questions we might ask about prime numbers—some of these have been answered, whereas others remain unresolved. A reasonable question to begin with is "How many prime numbers are there?" This question was answered by the Greek mathematician Euclid (circa 350 B.C.) in his famous work, the *Elements*, in which he gives an elegant proof that there are an infinite number of primes.

A CLOSER LOOK 8.1

How Do We Know There Are an Infinite Number of Primes?

The proof that there are an infinite number of primes dates back at least to the time of Euclid, and most likely was known even earlier. The very simple and convincing proof found in the *Elements* is based on the following idea: Suppose we take the first four prime numbers 2, 3, 5, and 7. Consider the number

$$2 \cdot 3 \cdot 5 \cdot 7 + 1 = 211.$$

We know that 211 is not divisible by 2, 3, 5, or 7, because when dividing by any of these numbers there would be a remainder of 1. Therefore, 211 must be either prime or divisible by some prime other than 2, 3, 5, or 7. In either case, we know of the existence of some prime other than the four we started with.

We can use this idea to show that there are an infinite number of primes. Suppose we multiply the first k prime numbers, 2, 3, 5, ..., p_k, together and add 1 to get

$$N = 2 \cdot 3 \cdot 5 \cdots p_k + 1.$$

We know that N is not divisible by any of the first k prime numbers, 2, 3, 5, ..., p_k. Therefore, N is either prime itself or it is the product of some other primes that are not in our current list. In either case, we have found a new prime. Therefore, we see that if we list the first k primes, we can always find another bigger prime. From this we can conclude that there must be an infinite number of primes.

Another question one might ask is how to go about finding the prime numbers. The Greek scholar Eratosthenes of Cyrene (276–194 B.C.) developed a technique known as the **Sieve of Eratosthenes** for finding all of the prime numbers less than or equal to some given number. To see how the sieve works, suppose we want to find all prime numbers less than or equal to 50. We first list all integers between 2 and 50. The first number, 2, is prime, so we draw a box around it to indicate this. Next, we cross out any of the remaining numbers that are divisible by 2, in other words all of the remaining even numbers 4, 6, 8, ... The next number after 2 that has not been crossed out is 3, which must be prime because it is not divisible by any smaller prime number. We now draw a box around 3 to indicate that it is prime, and then we proceed to cross out any of the other numbers that are divisible by 3—in other words, every third number 6, 9, 12, ... Some of these numbers already have been crossed out at this point. Continuing in this way, we see that after crossing out the multiples of 7 nothing more is crossed out. We arrive at the list in Figure 8.1 with all of the primes boxed and all of the composite numbers crossed out.

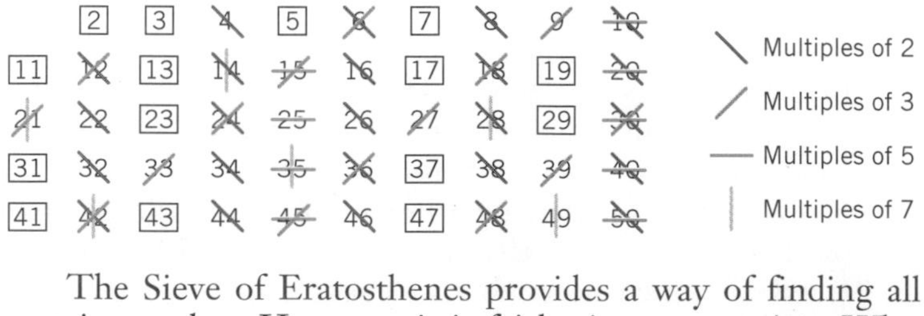

Figure 8.1
The Sieve of Eratosthenes.

The Sieve of Eratosthenes provides a way of finding all primes less than a certain number. However, it is fairly time consuming. What if we just want to know whether or not one particular positive integer n is prime? One way to tell would be to divide n by every smaller positive integer. However, this is more than we have to do, for if n is composite, then it can be written as $n = a \cdot b$ with a and

b both integers greater than 1. It cannot be true that both of these numbers are larger than $\sqrt{n}$, for if a and b were both greater than $\sqrt{n}$, we would have

$$n = a \cdot b > \sqrt{n} \cdot \sqrt{n} = n.$$

Because this is not possible, we know that either a or b is no larger than $\sqrt{n}$, and this factor must have a prime factor no larger than $\sqrt{n}$. We have deduced the following test for primality:

> ### Test for Primality
>
> An integer n greater than 1 is prime if it is not divisible by any prime number in the range from 2 to $\sqrt{n}$, inclusive.

Of course, if you do not have a list of all prime numbers up to $\sqrt{n}$, then you can just check to see if n is divisible by any integer in the range from 2 and $\sqrt{n}$, inclusive. Note that from this test for primality it follows that, when finding all primes less than or equal to n with the Sieve of Eratosthenes, we need only cross out multiples of all primes up to $\sqrt{n}$.

Example 2   *Primality Testing*

Determine whether or not the number 617 is prime.

Solution: Because $\sqrt{617} \approx 24.84$, we only have to see if 617 is divisible by primes in the range from 2 to 24.84. Looking at the primes from our Sieve of Eratosthenes in Figure 8.1, we see that we only have to check for divisibility by the primes 2, 3, 5, 7, 11, 13, 17, 19, and 23. Because 617 is odd, 2 does not divide 617. Because

$$617/3 \approx 205.67,$$

3 does not divide 617. Similarly,

$$617/5 = 123.4,$$

so 617 is not divisible by 5. (Or we could have seen that 617 is not divisible by 5 because 617 does not have last digit 5 or 0.) Similarly, we can show that 617 is not divisible by 7, 11, 13, 17, 19, or 23. Therefore, we conclude that 617 is prime. ∎

Example 3   *Primality Testing*

Determine whether or not the number 259 is prime.

Solution: Because $\sqrt{259} \approx 16.09$, we only have to see if 259 is divisible by any of the primes 2, 3, 5, 7, 11, or 13. Because 259 is odd, it is not divisible by 2. Computing $259/3 \approx 86.33$ and $259/5 = 51.8$, we see that 259 is not divisible by 3 or 5. Next we compute

$$259/7 = 37,$$

and we see that $259 = 7 \cdot 37$, so 259 is not prime. ∎

Much more sophisticated tests for primality are available today. These tests can determine whether an integer with 200 digits is prime in a few seconds of computer time. These tests are not able to factor a number, but merely determine whether or not it is prime. You might ask why we should want to find large primes. As we will see in Sections 8.6 and 8.7 of this chapter, large primes play a key role in certain secret code systems, so there is now a great practical interest in finding primes that are several hundred digits long. However, the problem of finding the *largest* known prime, which as of June 1999 was the number $2^{6,972,593} - 1$, a number with 2,098,960 digits (almost 500 pages worth), is a challenge that will continue to be of interest even though there are no immediate applications for primes this large.

Throughout mathematical history, many have attempted to find quick and easy formulas that will generate prime numbers as far out as one would care to go. All of these attempts failed. One formula that gives primes for many values of n is

$$n^2 - n + 41.$$

For instance,

when $n = 0$, $0^2 - 0 + 41 = 41$ is prime,

when $n = 1$, $1^2 - 1 + 41 = 41$ is prime,

when $n = 2$, $2^2 - 2 + 41 = 43$ is prime, and

when $n = 3$, $3^2 - 3 + 41 = 47$ is prime.

The formula $n^2 - n + 41$ continues to generate primes until we get to $n = 41$, in which case we get

$$41^2 - 41 + 41 = 1681 = 41 \cdot 41.$$

Another formula that showed some promise of giving only primes was proposed by the famous French mathematician Pierre de Fermat (1601–1665). Fermat noted that for $n = 0, 1, 2, 3,$ and 4, the formula

$$F_n = 2^{2^n} + 1$$

gives only primes. For instance, when $n = 0$,

$$F_0 = 2^{2^0} + 1 = 2^1 + 1 = 3$$

is prime, and when $n = 1$,

$$F_1 = 2^{2^1} + 1 = 2^2 + 1 = 5$$

is prime. Fermat believed that the numbers F_n, which are now called **Fermat numbers**, were prime for all integers $n \geq 0$. Unfortunately, in 1732 he was proved wrong by another great mathematician, Leonhard Euler (1707–1783), who showed that

$$F_5 = 2^{2^5} + 1 = 4,294,967,297 = 641 \cdot 6,700,417.$$

Many other Fermat numbers have now been shown to be composite. So far, only the Fermat numbers F_n with $n = 0, 1, 2, 3,$ and 4 are known to be prime, and no one knows if there are any more prime Fermat numbers. It is known that F_n is composite for $n = 5$ through 30 with the possible exception of F_{24}. Although no prime factors for F_{14}, F_{20}, and F_{22} are known, primality tests have shown them to be composite. The problem with testing larger Fermat numbers for primality is that they get large so quickly that even very efficient primality tests take too much computation time. For instance, F_{24} has over five million digits! Because F_{25} through F_{30} have moderately small prime factors, they were shown to be composite without undergoing a more extensive primality test.

Another question concerning primes is whether or not there are infinitely many pairs of consecutive odd numbers that are primes. These pairs of primes are called **twin primes**. For example, 5 and 7 are twin primes, as are the pair 11 and 13. Although the twin prime pairs become more scarce as we look at larger and larger numbers, no one knows if they continue forever. As of early 1999, the largest known twin prime pair was

$$835,335 \cdot 2^{39,014} - 1 \quad \text{and} \quad 835,335 \cdot 2^{39,014} + 1.$$

These numbers have 11,751 digits. It is interesting to note that both the largest prime and the largest twin prime pair were discovered by members of large, worldwide groups searching for primes on individual personal computers.

Christian Goldbach (1690–1764) raised another question about primes in a 1742 letter to Euler in which he made a conjecture that can be generalized slightly to the statement, "Every even integer greater than 4 can be written as a sum of two odd prime numbers." This conjecture came to be known as **Goldbach's conjecture**. It is easy to show that Goldbach's conjecture is true for the first several even integers greater than 4:

$$6 = 3 + 3,$$
$$8 = 3 + 5,$$
$$10 = 3 + 7 = 5 + 5,$$
$$12 = 5 + 7,$$
$$14 = 3 + 11 = 7 + 7.$$

Notice that in some cases there is more than one way to write the integer as the sum of two odd prime numbers. Goldbach's conjecture is easy to verify on a computer for a particular even integer greater than 4, and it has been shown to hold for all even integers less than 400 trillion. However, it is not known if it holds for all even integers greater than 4, so it remains one of the most famous unsolved problems in number theory.

There are numerous other unanswered questions about primes. It is these mysteries that make prime numbers so fascinating.

The Division Algorithm

Returning to the idea of divisors, we now examine one basic theorem that will serve as the foundation of much of modular arithmetic, which will be introduced in the next section and used in applications in the remaining sections of this chapter. This theorem is called the **division algorithm**, and it is really nothing more than the idea that when an integer is divided by a positive integer we get a quotient and a remainder. This idea is introduced in elementary school when division is first learned. We use the traditional name for this theorem, even though the division algorithm is not an algorithm. The precise statement of it is given here:

The Division Algorithm

If a is any integer and b is a positive integer, then there exist unique integers q and r such that

$$a = q \cdot b + r \qquad \text{with } 0 \leq r < b.$$

The number q is called the **quotient** and r is called the **remainder** in the division of a by b.

BRANCHING OUT 8.1

FERMAT'S LAST THEOREM: A MATH PROBLEM THAT TOOK OVER 350 YEARS TO SOLVE

Pierre de Fermat (1601–1665) was a great mathematician who pursued mathematics only as a hobby. He trained as a lawyer and worked as a magistrate. He had no formal training in higher mathematics. As was the custom at that time, he published very little of his mathematical work, and instead recorded his discoveries in letters of correspondence and in unpublished notes. Fermat is best known for a problem he suggested that turned out to be one of the most notorious mathematical problems of all time. This problem is known as Fermat's Last Theorem, and it arises naturally from the fact that there are solutions to the equation $x^2 + y^2 = z^2$, where x, y, and z are positive integers. One solution is given by the right triangle with sides of length 3, 4, and 5 (shown in the figure), where we know from the Pythagorean Theorem that $3^2 + 4^2 = 5^2$.

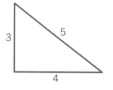

In fact, there are an infinite number of solutions to the equation $x^2 + y^2 = z^2$. Because of this, we might expect that equations like $x^3 + y^3 = z^3$ or $x^4 + y^4 = z^4$ also have some solutions in the positive integers. However, if you try to find such solutions, you are doomed to failure, for there are none! This is what Fermat asserted: For any positive integer $n \geq 3$, the equation $x^n + y^n = z^n$ has no solutions in the positive integers. Fermat made this claim in the margin of his copy of Diophantus' *Arithmetica*, presumably in about the year 1637. However, instead of proving this claim, he wrote in the margin, "I have assuredly found an admirable proof of this, but the margin is too narrow to contain it." Fermat's claim came to be known as Fermat's Last Theorem, though he had not given a proof of it.

Fermat's Last Theorem has a long and colorful history. Fermat himself provided a proof in the case where $n = 4$ by showing that the equation $x^4 + y^4 = z^4$ has no solutions in the positive integers, but progress on the problem beyond that point was very slow. In the eighteenth century, the great mathematician Euler provided a substantially correct proof for the case $n = 3$. The cases of $n = 5$, $n = 14$, and

Pierre de Fermat (1601–1665).

Example 4   *The Division Algorithm*

Find the quotient and remainder when the following divisions are performed according to the division algorithm:

(a) 51 divided by 8,

(b) 117 divided by 13.

$n = 7$ were proven in the early nineteenth century. In 1847, Ernst Kummer showed that Fermat's Last Theorem was true for an entire class of exponents, but he still had not shown that it was true in general. In 1908, 100,000 marks was bequeathed by Paul Wolfskehl to the Academy of Science in Göttingen, to be awarded to anyone who could give a complete proof of Fermat's Last Theorem. This resulted in an onslaught of incorrect attempts by amateur mathematicians. Because only printed solutions were eligible for the prize, the problem probably has the distinction of being the mathematical problem for which the largest number of false proofs have been published. After runaway inflation in Germany in the 1920s nearly wiped out the Wolfskehl prize, it rebounded to a value of about $50,000, when it was awarded to Andrew Wiles.

Even without the draw of a large monetary reward, professional mathematicians continued to attack the now very famous problem. The armor of Fermat's Last Theorem was gradually chipped away by advances in mathematics, until finally on June 23, 1993, Professor Andrew Wiles of Princeton University announced that he had found a proof. News of his proof made the front page of the *New York Times* and a full-page article in *Time,* and it was even covered in the national evening newscasts—quite unusual for a mathematical result. *People* magazine declared Professor Wiles one of "The 25 Most Intriguing People of the Year" in 1993, putting him on the eclectic list that included, among others, Yasir Arafat, Oprah Winfrey,

Howard Stern, and Michael Jordan. Professor Wiles' proof turned out to have a flaw, but after over a year of effort he corrected the proof on September 19, 1994, and it was published in 1995.

Dr. Wiles first came across Fermat's Last Theorem at the age of 10 in a book. When he finally solved the problem at age 40, Dr. Wiles confessed his bittersweet feelings about the toppling of one of mathematics' great unsolved problems. "For many of us, his problem drew us in and we always considered it something you dream about but never actually do." Now, he said, "There is a sense of loss, actually."

The question of whether or not Fermat had a proof of his theorem as he had claimed in the marginal note will never be answered. However, given the very complex nature of the mathematical ideas that Wiles used in constructing his proof, few believe that Fermat could have had a short, simple proof that somehow was not rediscovered despite serious efforts by great mathematicians over a period of 350 years.

Professor Andrew Wiles (1953–   ).

Solution:

(a) First we compute $51/8 = 6.375$, and rounding this down to the nearest integer we see that $q = 6$. To find the remainder r, we write the division algorithm formula $a = q \cdot b + r$ with $a = 51$, $b = 8$, and $q = 6$ and solve for r to get

$$51 = 6 \cdot 8 + r$$
$$51 = 48 + r$$
$$r = 51 - 48 = 3.$$

Therefore, $51 = 6 \cdot 8 + 3$, and we have $q = 6$ and $r = 3$. Alternatively, we could find the quotient and remainder by performing long division.

(b) Dividing 117 by 13, we have $117/13 = 9$. In this case 13 divides 117 evenly and the remainder is 0. We have $117 = 9 \cdot 13 + 0$, so $q = 9$ and $r = 0$. ∎

Notice that in the division algorithm, the number a may be a negative number. In this case we can still find a quotient q, which will be negative, such that the remainder r satisfies $0 \le r < b$. This is illustrated in the next example.

Example 5   *The Division Algorithm*

Find the quotient and remainder when the following divisions are performed according to the division algorithm:

(a) -57 divided by 4,
(b) -1456 divided by 16.

Solution:

(a) First we compute $-57/4 = -14.25$. In order for the remainder r to be positive, the quotient q must be found by rounding -14.25 down to the negative integer just below it in value. So we have $q = -15$ in this case. To find r, we write out the division algorithm formula $a = q \cdot b + r$ with $a = -57$, $b = 4$, and $q = -15$ and solve for r to get

$$-57 = (-15) \cdot 4 + r$$
$$-57 = -60 + r$$
$$r = -57 + 60 = 3.$$

Therefore, $-57 = (-15) \cdot 4 + 3$, and we have $q = -15$ and $r = 3$.

(b) Because $-1456/16 = -91$ exactly, no rounding down is needed, and we have $-1456 = (-91) \cdot 16 + 0$, so $q = -91$ and $r = 0$. ∎

The Greatest Common Divisor

Another idea of much importance in arithmetic and number theory is that of the greatest common divisor of two numbers, which is defined as follows:

If a and b are integers with at least one of them different from 0, then the **greatest common divisor** of a and b is the largest positive integer that divides both a and b. It is denoted by

$$\gcd(a,b).$$

We first run into the idea of greatest common divisor when we learn to reduce fractions in elementary school. To reduce the fraction a/b, we divide both the numerator and denominator by $\gcd(a,b)$. For instance, to reduce the fraction $4/6$, we divide both the numerator and the denominator by $\gcd(4,6) = 2$ to get the reduced fraction $2/3$. The idea of the greatest common divisor is important in some of the applications of number theory we will look at later in this chapter.

Example 6 *The Greatest Common Divisor*

Find the following:

(a) $\gcd(8,12)$,

(b) $\gcd(15,0)$,

(c) $\gcd(-21,35)$.

Solution:

(a) The positive divisors of 8 are 1, 2, 4, and 8, and the positive divisors of 12 are 1, 2, 3, 4, 6, and 12. Therefore,

$$\gcd(8,12) = 4.$$

(b) Because 0 is divisible by any integer, we know that it is divisible by every divisor of 15. The largest divisor of 15 is itself, and therefore

$$\gcd(15,0) = 15.$$

(c) The positive divisors of -21 are 1, 3, 7, and 21, and the positive divisors of 35 are 1, 5, 7, and 35. Therefore,

$$\gcd(-21,35) = 7.$$

Technology Tip

Mathematical software, spreadsheets, and even some calculators may have the greatest common divisor function built in.

Finding the greatest common divisor by listing the positive divisors of the two numbers and choosing the largest divisor they have in common, as we did in Example 6, is not a very efficient method when working with larger numbers. Another way to find $\gcd(a,b)$, where a and b are both integers greater than 1, is to first find the prime factorizations of a and b. Then $\gcd(a,b)$ is the product of all of the prime power factors common to both a and b, unless a and b have no factors in common, in which case $\gcd(a,b) = 1$. Furthermore, because

$$\gcd(a,b) = \gcd(-a,b) = \gcd(a,-b) = \gcd(-a,-b),$$

this method can also be used to find greatest common divisors involving negative numbers. This idea is illustrated in the next example.

Example 7   *The Greatest Common Divisor*

Find the following:

(a) $\gcd(660, 1400)$,
(b) $\gcd(272, -825)$.

Solution:
(a) The prime factorizations of 660 and 1400 are given by

$$660 = 2^2 \cdot 3 \cdot 5 \cdot 11,$$
$$1400 = 2^3 \cdot 5^2 \cdot 7.$$

A CLOSER LOOK 8.2

THE GREATEST COMMON DIVISOR AND THE EUCLIDEAN ALGORITHM

Even using the latest techniques in mathematics, it is impossible to factor most extremely large numbers quickly. It is for this reason that the RSA cryptosystem (discussed later in this chapter) is secure. You might expect that finding the greatest common divisor of two large numbers would be equally difficult. However, a method known as the *Euclidean algorithm* allows us to find the greatest common divisor of two large numbers fairly rapidly by avoiding factorization. When we apply the division algorithm to divide a by b, we obtain quotient q and remainder r. The key to the Euclidean algorithm is that the greatest common divisor of a and b is exactly the same as the greatest common divisor of b and r. We repeat the division algorithm, replacing a and b with b and r until the remainder is 0. At this stage, the number we have just divided by is the greatest common divisor of the original a and b.

Let's use the method to find the greatest common divisor of 25,443 and 80,441. We let $a = 80,441$,

the larger of the two numbers, and $b = 25,443$. Then

$$80,441 = 3 \cdot 25,443 + 4112.$$

Now we apply the division algorithm with $a = 25,443$ and $b = 4112$, obtaining

$$25,443 = 6 \cdot 4112 + 771.$$

We need two more applications to get remainder 0:

$$4112 = 5 \cdot 771 + 257,$$
$$771 = 3 \cdot 257 + 0.$$

Thus, the greatest common divisor of 25,443 and 80,441 is 257.

This method allows us to compute the greatest common divisor of two 100-digit numbers in at most 478 steps. A slight refinement of the method that permits negative remainders cuts this to at most 262 steps.

The highest power of 2 that appears in both of these prime factorizations is 2^2 and the highest power of 5 that appears in both is $5^1 = 5$. The numbers 660 and 1400 have no other prime factors in common. Therefore,

$$\gcd(660, 1400) = 2^2 \cdot 5 = 20.$$

(b) The prime factorizations of 272 and 825 are

$$272 = 2^4 \cdot 17,$$
$$825 = 3 \cdot 5^2 \cdot 11.$$

Because these prime factorizations have no factors in common, we see that

$$\gcd(272, 825) = 1,$$

and therefore

$$\gcd(272, -825) = 1.$$

Two integers a and b with $\gcd(a,b) = 1$, like 272 and -825 in Example 7, are called **relatively prime**.

Now that we have a basic foundation in the ideas of primes and divisibility, we will go on to use these ideas in the next section to develop a new kind of arithmetic, called modular arithmetic, in which divisibility plays a key role.

EXERCISES FOR SECTION 8.1

In Exercises 1–12, determine whether the statement is true or false.

1. $7|42$

2. $4|18$

3. $12|40$

4. $18|0$

5. $0|28$

6. $32|160$

7. $6|-34$

8. $15|-55$

9. $4|0$

10. $0|24$

11. $-10|240$

12. $-5|20$

In Exercises 13–18, find the prime factorization of the given number.

13. 132

14. 112

15. 1625

16. 1755

17. 6800

18. 8100

19. Use the Sieve of Eratosthenes to find all primes less than or equal to 100.

20. Use the Sieve of Eratosthenes to find all primes less than or equal to 150.

In Exercises 21–26, determine whether or not the given number is prime.

21. 751

22. 979

23. 1817

24. 1247

25. 2579

26. 2131

27. Show that the Fermat number $F_3 = 2^{2^3} + 1$ is prime.

28. Show that $n^2 - n + 41$ is prime for $n = 22$.

Goldbach's conjecture states that every even integer greater than 4 can be written as a sum of two odd prime numbers. In Exercises 29–34, write the given number as the sum of two odd prime numbers. (There may be more than one possible answer.)

29. 16   **30.** 18

31. 46   **32.** 38

33. 98   **34.** 54

In 1949, H. E. Richert proved that every integer $n > 6$ can be written as a sum of one or more distinct primes. In Exercises 35–38, write the given number in this way. (There may be more than one possible answer.)

35. 86   **36.** 73

37. 232   **38.** 147

In Exercises 39–48, find the quotient and remainder when the division is performed according to the division algorithm.

39. 68 divided by 7   **40.** 92 divided by 8

41. 204 divided by 12   **42.** 192 divided by 11

43. -36 divided by 5   **44.** -18 divided by 4

45. -123 divided by 9   **46.** -270 divided by 15

47. -87 divided by 10   **48.** -55 divided by 6

In Exercises 49–60, evaluate the greatest common divisor.

49. $\gcd(9,15)$ **50.** $\gcd(18,30)$

51. $\gcd(0,24)$ **52.** $\gcd(-25,-12)$

53. $\gcd(-16,40)$ **54.** $\gcd(6,0)$

55. $\gcd(77,130)$ **56.** $\gcd(168,700)$

57. $\gcd(650,475)$ **58.** $\gcd(170,-81)$

59. $\gcd(-525,-231)$ **60.** $\gcd(770,900)$

For Exercises 61–63, consider the following: A positive integer is called a *perfect number* if it is equal to the sum of all of its positive divisors except itself. For instance, the number 6 is perfect because $6 = 1 + 2 + 3$. It is not known whether there are infinitely many perfect numbers or whether there are any odd perfect numbers.

61. Show that the number 10 is not perfect.

62. Find the next perfect number after 6.

63. The third perfect number is 496. Verify that it is perfect.

64. Suppose a positive integer n has at least three prime factors. Let p be the smallest prime factor of n. In terms of n, what is the largest that p can be? Explain your answer.

65. (a) For what positive integers n is $n(n + 1)/2$ an integer? Explain your answer.

(b) For what positive integers n is $n(n + 1)/2$ an even integer? Explain your answer.

66. Find all positive integers n for which $n^2 - 1$ is a prime number. Explain how you know you have found them all.

8.2 MODULAR ARITHMETIC

Many applications of number theory use a special kind of arithmetic known as **modular arithmetic**. In modular arithmetic, rather than thinking of integers in the usual way, we instead consider only their remainders after division by a given integer n. In this way, modular arithmetic can be thought of as an arithmetic of remainders. The idea of modular arithmetic was first introduced by the German mathematician Carl Friedrich Gauss (1777–1855) in his work *Disquisitiones Arithmeticae*. Gauss is considered by many to be the greatest mathematician who ever lived, and *Disquisitiones Arithmeticae* was published when he was only 24 years old.

BRANCHING OUT 8.2

CARL FRIEDRICH GAUSS: THE PRINCE OF MATHEMATICIANS

It is difficult, if not impossible, to bestow the title of the "best mathematician who ever lived" to any individual. Because the work that mathematicians do depends so much on the previous work of mathematicians, it is impossible to judge just who made the most remarkable discoveries. Nevertheless, many argue that Carl Friedrich Gauss is the greatest mathematician of all time.

Unlike most other distinguished scholars of his time, Gauss did not come from a privileged family. He was born in Germany in 1777 to a poor and uneducated family, but as a young child his intellectual gifts were apparent. His mathematical prowess was so great that when Gauss reached the age of 10, his schoolmasters claimed that they had nothing else to teach him. Fortunately, his gifts caught the attention of Duke Ferdinand of Brunswick, who saw to it that

Gauss received a proper education. While studying at the University of Göttingen and when he was only 18 years old, Gauss proved that a regular 17-gon could be constructed with only an unmarked ruler and compass. This was an amazing discovery, for at the time no new constructions of regular polygons had been discovered since the time of Euclid and no one knew that such a construction was even possible. This discovery was just the first of many in Gauss' long and distinguished career. Although Gauss made major contributions to many areas of mathematics, he was particularly fond of number theory. He is credited with saying "Mathematics is the Queen of the sciences, and the theory of numbers is the Queen of mathematics." His publication, *Disquisitiones Arithmeticae,* laid the foundation for the field of modern number theory.

Gauss on a German 10-mark note.

It may at first seem strange to be concerned only with the remainder of an integer after division by a particular integer, but such situations arise fairly naturally. For instance, if a clock now reads 7:00, and you want to know what time the clock will read 38 hours from now, you need only know the remainder when $7 + 38 = 45$ is divided by 12. By the division algorithm, $45 = 3 \cdot 12 + 9$. Therefore, because 9 is the remainder, the clock will read 9:00. In the remaining sections of this chapter, we will see many practical applications that are made easier through the use of modular arithmetic.

In modular arithmetic, we have the special notation for remainders given next.

If n is a positive integer and a is any integer, then

$$a \bmod n$$

is the remainder obtained when a is divided by n according to the division algorithm. Note that by the definition of the remainder, we know that $0 \leq a \bmod n < n$.

Example 1   *Computing a mod n*

Evaluate each of the following quantities:

(a) 11 mod 4,
(b) 253 mod 35,
(c) 18 mod 6,
(d) −43 mod 5,
(e) −1272 mod 24.

Solution:
(a) By the division algorithm, $11 = 2 \cdot 4 + 3$, so the remainder of 11 when divided by 4 is 3. Therefore, 11 mod 4 = 3.
(b) Because $253 = 7 \cdot 35 + 8$, we have 253 mod 35 = 8.
(c) We have $18 = 3 \cdot 6 + 0$, so 18 mod 6 = 0.
(d) We have $-43 = (-9) \cdot 5 + 2$, and therefore −43 mod 5 = 2.
(e) Because $-1272 = (-53) \cdot 24 + 0$, it follows that −1272 mod 24 = 0.

Note that parts (c) and (e) of Example 1 illustrate the fact that $n|a$ is equivalent to $a \bmod n = 0$. ■

Technology Tip

Most number theory or general mathematics software, spreadsheets, and some calculators can compute $a \bmod n$ automatically.

In modular arithmetic, we work with one fixed **modulus** n and call the arithmetic we are doing **arithmetic modulo n**. To define what we mean by arithmetic modulo n, first we must define what it means for two numbers to be "equal" modulo n. Rather than using the word equal to signify that two things are equivalent in modular arithmetic, we use the word congruent. The precise definition for two numbers to be congruent modulo n is as follows:

Congruence Modulo n

Let n be a positive integer. Two numbers a and b are said to be **congruent modulo n** if they both have the same remainder when divided by n—in other words, if

$$a \bmod n = b \bmod n.$$

We use the notation

$$a \equiv b \pmod{n}$$

to indicate that a is congruent to b modulo n. If a is not congruent to b modulo n, we write

$$a \not\equiv b \pmod{n}.$$

A commonly used equivalent definition is that a and b are congruent modulo n if $n \mid (a - b)$.

Example 2   *Congruence Modulo n*

(a) Determine whether 35 is congruent to 67 modulo 8.
(b) Determine whether -26 is congruent to 15 modulo 12.

Solution:
(a) Because 35 leaves remainder 3 when divided by 8 and 67 leaves remainder 3 when divided by 8, we see that $35 \bmod 8 = 67 \bmod 8 = 3$, and therefore $35 \equiv 67 \pmod 8$.
(b) By the division algorithm, $-26 = (-3) \cdot 12 + 10$, and therefore $-26 \bmod 12 = 10$. Because $15 \bmod 12 = 3$, it follows that $-26 \not\equiv 15 \pmod{12}$.

We are now ready to introduce arithmetic modulo n. The basic rules of addition, subtraction, and multiplication are similar to those we use in regular arithmetic. The main difference is that two numbers or calculations are equal if they are congruent modulo n. Three major properties of modular arithmetic are listed next.

Properties of Modular Arithmetic

If $a \equiv b \pmod{n}$ and $c \equiv d \pmod{n}$, then

1. $a + c \equiv b + d \pmod{n}$,
2. $a - c \equiv b - d \pmod{n}$,
3. $ac \equiv bd \pmod{n}$.

For example, because $14 \equiv 4 \pmod 5$ and $48 \equiv 3 \pmod 5$, Property 1 tells us that

$$14 + 48 \equiv 4 + 3 \pmod 5.$$

Because $14 + 48 = 62$ and $4 + 3 = 7$, we see that

$$62 \equiv 7 \pmod 5.$$

Similarly, Property 2 tells us that

$$14 - 48 \equiv 4 - 3 \pmod 5,$$

or

$$-34 \equiv 1 \pmod 5.$$

It follows from Property 3 that

$$14 \cdot 48 \equiv 4 \cdot 3 \pmod 5,$$

or

$$672 \equiv 12 \pmod 5.$$

When we are working in arithmetic modulo n, we usually want the numbers we are working with to be small. The properties of modular arithmetic allow us to substitute a small number b in place of a larger number a in calculations involving addition, subtraction, and multiplication modulo n as long as $a \equiv b \pmod n$. In applications, we usually want our final answer to lie between 0 and $n - 1$, so we want to give the answer mod n. These ideas are illustrated in the following example.

Example 3   *Modular Arithmetic*

Perform the following calculations. In each case, the final answer must be a number between 0 and $n - 1$, where n is the modulus:

(a) $(17 + 86) \bmod 4$,
(b) $(23 \cdot 32 \cdot 2) \bmod 20$,
(c) $(157(13 - 16)) \bmod 17$,
(d) $(37^5 - 3 + 42) \bmod 7$.

Solution:
(a) Using the fact that $17 \equiv 1 \pmod 4$ and $86 \equiv 2 \pmod 4$, we see that

$$17 + 86 \equiv 1 + 2 \equiv 3 \pmod 4.$$

Because $0 \leq 3 < 4$, we have

$$(17 + 86) \bmod 4 = 3.$$

(b) In this case we use the fact that $23 \equiv 3 \pmod{20}$ and $32 \equiv 12 \pmod{20}$ to find that

$$23 \cdot 32 \cdot 2 \equiv 3 \cdot 12 \cdot 2 \equiv 72 \equiv 12 \pmod{20},$$

and therefore, because $0 \leq 12 < 20$,

$$(23 \cdot 32 \cdot 2) \bmod 20 = 12.$$

(c) Using the fact that $157 \equiv 4 \pmod{17}$, we have

$$157(13 - 16) \equiv 4(-3) \equiv -12 \equiv 5 \pmod{17},$$

so, noting that $0 \le 5 < 17$,

$$(157(13 - 16)) \bmod 17 = 5.$$

(d) We know $37 \equiv 2 \pmod 7$ and $42 \equiv 0 \pmod 7$, so

$$37^5 - 3 + 42 \equiv 2^5 - 3 + 0 \equiv 32 - 3 \equiv 29 \equiv 1 \pmod 7.$$

Because $0 \le 1 < 7$, we have

$$(37^5 - 3 + 42) \bmod 7 = 1.$$

There are instances where it is helpful to substitute a negative number for a positive number in a modular arithmetic calculation, as in the following example.

Example 4   *Modular Arithmetic*

Find $32^{50} \bmod 33$.

Solution: The number 32^{50} is quite large. However, if we use the fact that $32 \equiv -1 \pmod{33}$, we have

$$32^{50} \equiv (-1)^{50} \equiv ((-1)^2)^{25} \equiv 1^{25} \equiv 1 \pmod{33}.$$

Therefore, because $0 \le 1 < 33$,

$$32^{50} \bmod 33 = 1.$$

The properties of modular arithmetic provide a simple way to find a negative number mod n, as illustrated in the following example.

Example 5   *Finding a Negative Number mod n*

Evaluate $-136 \bmod 7$.

Solution: Because $7 \equiv 0 \pmod 7$, adding multiples of 7 does not change the value of a quantity in arithmetic modulo 7. To evaluate $-136 \bmod 7$, we add just enough multiples of 7 to change the value -136 to a number between 0 and 6 in arithmetic modulo 7. In this case, because $136/7 \approx 19.43$, we add $20 \cdot 7$ to get

$$-136 \equiv -136 + 20 \cdot 7 \equiv -136 + 140 \equiv 4 \pmod 7,$$

so

$$-136 \bmod 7 = 4.$$

You may have noticed that we have added, subtracted, and multiplied in modular arithmetic, but we have not divided. In regular arithmetic, we can divide by any number except 0. In arithmetic modulo n, it is only possible to divide by a number k for which $\gcd(k, n) = 1$. We will address this idea later in this chapter when it comes up in some of our applications.

Exercises for Section 8.2

In Exercises 1–20, evaluate the given quantity.

1. 48 mod 9

2. 8 mod 3

3. −38 mod 12

4. −64 mod 8

5. 396 mod 7

6. 67 mod 8

7. 14 mod 2

8. −2115 mod 4

9. −342 mod 6

10. 135 mod 24

11. −54 mod 4

12. −73 mod 10

13. 45,237 mod 10

14. −136 mod 7

15. −252 mod 11

16. 461 mod 12

17. 2295 mod 16

18. −3 mod 9

19. −30,881 mod 50

20. 4312 mod 5

In Exercises 21–40, perform the given calculation. In each case, the answer must be a number between 0 and $n - 1$, where n is the modulus.

21. $(9 + 14) \bmod 5$

22. $(46 + 27) \bmod 6$

23. $(33 \cdot 96 \cdot 11) \bmod 10$

24. $(32 - 58) \bmod 2$

25. $(15^3 - 25) \bmod 4$

26. $(44 \cdot 38 \cdot 9 \cdot 72) \bmod 7$

27. $(47(42 - 18)) \bmod 9$

28. $12^{99} \bmod 3$

29. $(8 \cdot 9 \cdot 10 \cdot 11 \cdot 12 \cdot 13) \bmod 7$

30. $(42(32 - 45)) \bmod 12$

31. $(2^{310} + 1) \bmod 3$

32. $(14^{387} \cdot 16^{563}) \bmod 15$

33. $(83 + 30 + 15 + 3) \bmod 9$

34. $(31 \cdot 34) \bmod 30$

35. $(47 \cdot 50) \bmod 44$

36. $((15 + 18)(61 - 90)) \bmod 2$

37. $(42^5 + 17^{28}) \bmod 8$

38. $(456 + 321 + 9872 + 400) \bmod 10$

39. $(823 + 620) \bmod 100$

40. $(23 \cdot 38) \bmod 7$

If a positive integer is divided by 10, then the remainder is its last digit. Therefore, a positive integer is congruent to its last digit mod 10. In Exercises 41–44, use this fact and modular arithmetic to find the last digit of the given number.

41. 9^{2001}

42. 3^{1000}

43. 7453^{56}

44. 123456789^{999}

45. To pass the time, a general lines up his soldiers in 100 equal rows, with three soldiers left over. He then lines them up in 99 equal rows, this time with two soldiers left over. The general has under 10,000 soldiers. Exactly how many soldiers does he have? Explain how you arrived at your answer.

46. Two pirates were marooned on an island with a monkey and a "treasure" consisting of a pile of coconuts. Not trusting the other pirate, the first pirate rises in the night, divides the coconuts evenly into two piles, with one coconut left over, which he gives to the monkey. He then hides his pile of coconuts, leaves the remaining coconuts, and goes back to sleep. Later, the second pirate rises, divides the remaining coconuts evenly into two piles, with one coconut left over, which he also gives to the monkey. After hiding his pile of coconuts, he leaves the remaining coconuts and then goes back to sleep. Morning comes. The pirates divide their "treasure" into halves, with one coconut left over for the monkey.

(a) Find the number of coconuts mod 8 in the original pile.

(b) What is the smallest number of coconuts the original pile could have contained?

8.3 DIVISIBILITY TESTS

A **divisibility test** is a method for determining whether a given integer is divisible by another given integer without actually performing the division. There are several simple divisibility tests that make use of modular arithmetic. These tests can be used not only to test divisibility, but, as we will see later, also to check calculations.

The divisibility tests we will look at make use of the fact that our numbers are written in base 10 notation. For example, the number 357 stands for

$$357 = 3 \cdot 100 + 5 \cdot 10 + 7 = 3 \cdot 10^2 + 5 \cdot 10 + 7,$$

and the expression $3 \cdot 10^2 + 5 \cdot 10 + 7$ is called the base 10 expansion of 357. Similarly, the base 10 expansion of the number 2,508,623 is given by

$$2{,}508{,}623 = 2 \cdot 10^6 + 5 \cdot 10^5 + 0 \cdot 10^4 + 8 \cdot 10^3 + 6 \cdot 10^2 + 2 \cdot 10 + 3.$$

In general, the base 10 expansion of a positive integer $N = a_m a_{m-1} \cdots a_1 a_0$, where the number a_k is the $(k + 1)$st digit of N from the right, is given by

$$N = a_m \cdot 10^m + a_{m-1} \cdot 10^{m-1} + \cdots + a_1 \cdot 10 + a_0.$$

The first divisibility test we will look at is one for testing divisibility by 9. This test is based on the fact that

$$10 \equiv 1 \ (\mathrm{mod}\ 9).$$

Therefore, in arithmetic modulo 9, we can substitute 1 for 10 in the base 10 expansion for $N = a_m a_{m-1} \cdots a_1 a_0$ to get

$$\begin{aligned} N &\equiv a_m \cdot 10^m + a_{m-1} \cdot 10^{m-1} + \cdots + a_1 \cdot 10 + a_0 \\ &\equiv a_m \cdot 1^m + a_{m-1} \cdot 1^{m-1} + \cdots + a_1 \cdot 1 + a_0 \\ &\equiv a_m + a_{m-1} + \cdots + a_1 + a_0 \ (\mathrm{mod}\ 9). \end{aligned}$$

In other words, N is congruent to the sum of its digits modulo 9. Because $N \bmod 9$ is the remainder we obtain when dividing N by 9, it follows that $(a_m + a_{m-1} + \cdots + a_1 + a_0) \bmod 9$ is the remainder we obtain when dividing N by 9. When adding the digits modulo 9, we can ignore any 9's or any combination of digits that add up to 9 because they are congruent to 0 modulo 9. This process for finding the remainder obtained after dividing by 9 is called **casting**

out nines. Because a number is divisible by 9 if and only if the remainder obtained after dividing by 9 is 0, we have the following divisibility test:

> ### A Test for Divisibility by 9 by Casting Out Nines
> A positive integer is divisible by 9 if and only if the sum of its digits mod 9 is 0.

We illustrate this divisibility test in the following example.

Example 1 *Casting Out Nines*

Use casting out nines to determine whether or not the following numbers are divisible by 9:

(a) 37,940,531,
(b) 6,206,309,719,290.

Solution:
(a) Summing the digits of 37,940,531 modulo 9, casting out nines, and finding combinations of digits that add up to 9, we get

$$3 + 7 + 9 + 4 + 0 + 5 + 3 + 1 \equiv 3 + 7 + 4 + (5 + 3 + 1)\,(\text{mod } 9).$$

Now casting out the combination of digits that adds up to 9, we have

$$3 + 7 + 9 + 4 + 0 + 5 + 3 + 1 \equiv 3 + 7 + 4 \equiv 14 \equiv 5\,(\text{mod } 9).$$

Therefore, the sum of the digits mod 9 is 5, so 37,940,531 is not divisible by 9 (and, in fact, it leaves remainder 5 when divided by 9).

(b) Adding up the digits of 6,206,309,719,290 modulo 9 and casting out nines and combinations of digits that add up to 9, we have

$$6 + 2 + 0 + 6 + 3 + 0 + 9 + 7 + 1 + 9 + 2 + 9 + 0$$
$$\equiv 6 + 2 + (6 + 3) + 1 + (7 + 2)$$
$$\equiv (6 + 2 + 1) \equiv 0\,(\text{mod } 9).$$

Therefore, 6,206,309,719,290 is divisible by 9. ∎

Using the same ideas as we did in arriving at casting out nines, we can find a test for divisibility by 11. We start with the fact that

$$10 \equiv -1\,(\text{mod } 11).$$

If $N = a_m a_{m-1} \cdots a_1 a_0$, then in arithmetic modulo 11 we can substitute -1 for 10 in the decimal expansion of N to get

$$N \equiv a_m \cdot 10^m + a_{m-1} \cdot 10^{m-1} + \cdots + a_1 \cdot 10 + a_0$$
$$\equiv a_m \cdot (-1)^m + a_{m-1} \cdot (-1)^{m-1} + \cdots + a_1 \cdot (-1) + a_0.$$

Writing the sum in reverse order, we see that

$$N \equiv a_0 - a_1 + a_2 - a_3 + \cdots + (-1)^m \cdot a_m\,(\text{mod } 11).$$

Thus N is congruent to the "alternating sum" of its digits modulo 11, where by an alternating sum of its digits we mean the first digit on the right minus the second digit from the right plus the third digit, and so on. Because $N \bmod 11$ is the remainder we obtain after dividing N by 11, it follows that $(a_0 - a_1 + a_2 - a_3 + \cdots + (-1)^m \cdot a_m) \bmod 11$ is the remainder we obtain after dividing N by 11. When calculating the alternating sum mod 11, many of the numbers will cancel out with regular addition. We can also throw out any combination of digits that add up to 11 or -11, so this method of finding $N \bmod 11$ is sometimes called **casting out elevens**. We have the following test for divisibility by 11:

A Test for Divisibility by 11 by Casting Out Elevens

A positive integer $N = a_m a_{m-1} \cdots a_1 a_0$ is divisible by 11 if and only if the alternating sum of its digits mod 11 is 0—in other words, if and only if

$$(a_0 - a_1 + a_2 - a_3 + \cdots + (-1)^m \cdot a_m) \bmod 11 = 0.$$

Example 2   *Casting Out Elevens*

Use casting out elevens to determine whether or not the following numbers are divisible by 11:

(a) 4,506,073,
(b) 771,003,239.

Solution:

(a) The alternating sum of the digits of 4,506,073 mod 11 is given by

$$3 - 7 + 0 - 6 + 0 - 5 + 4 \equiv 3 - 7 - (6 + 5) + 4$$
$$\equiv 3 - 7 + 4 \equiv 0 \,(\bmod\ 11),$$

and therefore 4,506,073 is divisible by 11.

(b) For 771,003,239, the alternating sum of the digits mod 11 is

$$9 - 3 + 2 - 3 + 0 - 0 + 1 - 7 + 7 \equiv (9 + 2) - 3 - 3 + 1$$
$$\equiv -5 \equiv -5 + 11 \equiv 6 \,(\bmod\ 11),$$

and therefore 771,003,239 is not divisible by 11. ∎

We derive one more divisibility test, that for divisibility by 4. As you might guess by now, the idea for this comes from looking at the decimal expansion of the number $N = a_m a_{m-1} \cdots a_1 a_0$ modulo 4. First we note that $4 | 10^2$, so $10^2 \equiv 0 \,(\bmod\ 4)$. In general, we can see that because $4 | 10^k$ for $k \geq 2$, we have

$$10^k \equiv 0 \,(\bmod\ 4) \qquad \text{for } k \geq 2.$$

Therefore, in arithmetic modulo 4, whenever $k \geq 2$ we can substitute 0 for 10^k in the base 10 expansion for $N = a_m a_{m-1} \cdots a_1 a_0$ to get

$$N \equiv a_m \cdot 10^m + a_{m-1} \cdot 10^{m-1} + \cdots + a_2 \cdot 10^2 + a_1 \cdot 10 + a_0$$
$$\equiv a_m \cdot 0 + a_{m-1} \cdot 0 + \cdots + a_2 \cdot 0 + a_1 \cdot 10 + a_0$$
$$\equiv a_1 \cdot 10 + a_0 \,(\text{mod } 4).$$

However, $a_1 \cdot 10 + a_0$ is just the two-digit number $a_1 a_0$ given by the last two digits of N. It follows that N and the number given by its last two digits are congruent modulo 4 and therefore leave the same remainder when divided by 4. Thus, we have the following test for divisibility by 4:

Test for Divisibility by 4

A positive integer $N = a_m a_{m-1} \cdots a_1 a_0$ is divisible by 4 if and only if the two-digit number $a_1 a_0$ given by its last two digits is divisible by 4.

Example 3   *Divisibility by 4*

Use the test for divisibility by 4 to determine whether or not the following numbers are divisible by 4:

(a) 9,903,562,748,

(b) 7,903,603.

Solution:

(a) The number formed by the last two digits of 9,903,562,748 is 48, and $48 = 4 \cdot 12$ is divisible by 4. Therefore, 9,903,562,748 is also divisible by 4.

(b) The number formed by the last two digits of 7,903,603 is 03 = 3. Because 3 is not divisible by 4, we see that 7,903,603 is not divisible by 4.

There are tests for divisibility by other numbers that are based on ideas similar to those used in the tests for divisibility by 9, 11, and 4. A few of these tests are summarized here.

Other Divisibility Tests

A positive integer is divisible by

- 2 if and only if its last digit is even,
- 3 if and only if the sum of its digits mod 3 is 0,
- 5 if and only if its last digit is 0 or 5,
- 6 if and only if it is divisible by both 2 and 3,
- 7 if and only if when its last digit is doubled and then subtracted from the number formed by the remaining digits, the resulting number is divisible by 7,
- 8 if and only if the number formed by its last three digits is divisible by 8,
- 10 if and only if its last digit is 0.

We will look at all of these tests in the exercises at the end of this section. In the next example, we see how the test for divisibility by 7 works.

Example 4   *Divisibility by 7*

Use the test for divisibility by 7 to determine whether or not 4183 is divisible by 7.

Solution: When the last digit, 3, of 4183 is doubled and subtracted from the remaining digits, we have

$$
\begin{array}{r}
418 \\
-\ 6 \\
\hline
412
\end{array}
$$

Now we see that 4183 is divisible by 7 if and only if 412 is divisible by 7. To see whether 412 is divisible by 7, we apply the test for divisibility by 7 to it. Doubling the last digit of 412 and subtracting it from the remaining digits gives

$$
\begin{array}{r}
41 \\
-4 \\
\hline
37
\end{array}
$$

We know that 37 is not divisible by 7. Therefore, we can conclude that 412 is not divisible by 7, and hence, neither is the number we started with, 4183.   ■

Sometimes divisibility tests can supply missing information about a number, as in the next example.

Example 5   *Finding the Missing Digit*

Suppose you purchased 22 identically priced drinks with a check. The entry in the checkbook is difficult to read, but you know the total price was \$14.#6, where # is an illegible digit. Assuming there was no sales tax, what is the illegible digit?

Solution: Because $22 = 2 \cdot 11$ divides into the number 14#6, we know that $11|14\#6$. Therefore, the alternating sum of the digits of 14#6 must be congruent to 0 mod 11. This gives us the equation

$$
6 - \# + 4 - 1 \equiv 9 - \# \equiv 0 \,(\text{mod } 11).
$$

Therefore, the digit # must be a 9.   ■

Now we will see how divisibility tests can be used to provide a partial check on an arithmetic calculation. In particular, we will look at how casting out nines can be used to check multiplication. Suppose we want to multiply two positive integers N and M. If, for instance, $N \bmod 9 = 5$ and $M \bmod 9 = 3$, then by the properties of modular arithmetic we know that $N \cdot M \equiv 5 \cdot 3 \equiv 15 \equiv 6 \,(\text{mod } 9)$. Therefore, after we have multiplied N and M in regular arithmetic, we can partially check the calculation by seeing if the answer is 6 modulo 9. If it is not, then we know the calculation is incorrect. This idea is illustrated in the next example.

Example 6   *Checking a Calculation*

Suppose you multiply the two numbers 4569 and 3428 and arrive at the answer 15,662,542. Use casting out nines to check the answer.

Solution: We know that every number is congruent to the sum of its digits modulo 9, so we have

$$4569 \equiv 4 + 5 + 6 + 9 \equiv (4 + 5) + 6 \equiv 6 \,(\text{mod } 9)$$

and

$$3428 \equiv 3 + 4 + 2 + 8 \equiv (3 + 4 + 2) + 8 \equiv 8 \,(\text{mod } 9).$$

Therefore, we know from the properties of modular arithmetic that

$$4569 \cdot 3428 \equiv 6 \cdot 8 \equiv 48 \equiv 3 \,(\text{mod } 9).$$

Now finding what the answer 15,662,542 is modulo 9 by summing its digits modulo 9, we have

$$
\begin{aligned}
15{,}662{,}542 &\equiv 1 + 5 + 6 + 6 + 2 + 5 + 4 + 2 \\
&\equiv 1 + 5 + 6 + 6 + 2 + (5 + 4) + 2 \\
&\equiv 1 + 5 + 6 + 6 + 2 + 2 \equiv 22 \equiv 4 \,(\text{mod } 9).
\end{aligned}
$$

Because this does not agree with the fact that $(4569 \cdot 3428) \bmod 9 = 3$, we know that the calculation $4569 \cdot 3428 = 15{,}662{,}542$ must be wrong. (In fact, the correct calculation is $4569 \cdot 3428 = 15{,}662{,}532$.)

■

It is important to note that casting out nines provides only a partial check that a calculation has been done correctly. It is quite possible to arrive at an incorrect answer that just happens to be congruent to the correct answer modulo 9. For example, suppose we multiply 29 and 84 and arrive at the incorrect answer 2976. (The correct answer is 2436.) We have $29 \equiv 2 + 9 \equiv 2 \,(\text{mod } 9)$ and $84 \equiv 8 + 4 \equiv 12 \equiv 3 \,(\text{mod } 9)$, and therefore

$$29 \cdot 84 \equiv 2 \cdot 3 \equiv 6 \,(\text{mod } 9).$$

Because

$$2976 \equiv 2 + 9 + 7 + 6 \equiv (2 + 7) + 6 \equiv 6 \,(\text{mod } 9),$$

our check does not tell us that the calculation $29 \cdot 84 = 2976$ is incorrect. However, even with this flaw, casting out nines does catch most errors, so it provides a nice partial check. We can also use casting out elevens or other divisibility tests to check calculations.

BRANCHING OUT 8.3

CALCULATING A BILLION DIGITS OF PI

We know that a circle of radius r has area equal to πr^2 and perimeter equal to $2\pi r$, where the constant π is a number approximately equal to 3.14. But what is the exact value of π? The Greek scholar Archimedes, in his work *On the Measurement of the Circle,* approximated π by inscribing and circumscribing polygons in a circle as in the figure and computing the perimeters of those polygons.

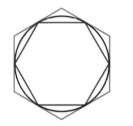

Using this method with regular polygons with 96 sides, Archimedes showed that $3\frac{10}{71} < \pi < 3\frac{1}{7}$, or in decimal notation $3.140845\ldots < \pi < 3.142857\ldots$.

In 1771, Johann Lambert (1728–1777) proved that π is irrational, and therefore that its decimal expansion is not finite nor does it repeat in a pattern. Before Lambert's proof, π had been calculated to 112 decimal places:

$\pi \approx 3.1415926535897932384626433832795028841$
$9716939937510582097494459230781640628620899862803482534211706798214808651 3.$

The search for even more precise calculations of π has been carried out by both amateur and professional mathematicians as a kind of sport.

In 1989, brothers Gregory V. and David V. Chudnovsky of Columbia University calculated over a billion digits of π. Using the formula

$$\pi = \left(\frac{1}{426,880\sqrt{10,005}} \sum_{n=0}^{\infty} \left[\frac{(6n)!(545,140,134n + 13,591,409)}{n!^3(3n)!(-640,320)^{3n}} \right] \right)^{-1}$$

and the help of high-powered computers, they computed π to 1,011,196,691 digits. You might think it is unlikely for a computer to make a calculation error if it has been programmed correctly, but in the context of calculations of this magnitude, errors are to be expected. Because of the possibility of calculation errors on the computer, the Chudnovsky brothers used modular arithmetic to check the calculations as they went along. Using these modular arithmetic checks along with other checks, the probability of an undetected error at any step was less than 1 in 10^{290}.

As of 1997, the record for the number of digits of π was 51,539,600,000 digits, calculated by a Japanese team led by Yasumasa Kanada. This staggering number of digits would fill about 26 million pages.

EXERCISES FOR SECTION 8.3

In Exercises 1–6, use casting out nines to determine whether or not the given number is divisible by 9.

1. 7893

2. 327

3. 1,957,195,819,861,989

4. 123,456,789,123,456,789

5. 5,356,990,218,031,912,008

6. 80,907,704,561,223,513,069

In Exercises 7–12, use casting out elevens to determine whether or not the given number is divisible by 11.

7. 9123

8. 51,457

9. 741,050,902

10. 45,477,432

11. 123,456,789,987,654,321

12. 43,991,704,651,030,209

In Exercises 13–16, use the test for divisibility by 4 to determine whether or not the given number is divisible by 4.

13. 78,406

14. 816

15. 82,168,003,234,084

16. 780,811,234

In Exercises 17–20, use the divisibility tests to determine whether or not the given number is divisible by each of the following:

(a) 2   **(b)** 3   **(c)** 4   **(d)** 5   **(e)** 6
(f) 7   **(g)** 8   **(h)** 9   **(i)** 10   **(j)** 11

In each case, show how you used the divisibility test to arrive at your answer.

17. 62,952

18. 857,304

19. 2,756,768,175

20. 531,468,300

In Exercises 21–24, use arguments similar to the ones used in the text to arrive at the tests for divisibility by 9, 11, and 4.

21. Show that a positive integer is divisible by 8 if and only if the number formed by its last three digits is divisible by 8.

22. Show that a positive integer is divisible by 5 if and only if its last digit is 0 or 5.

23. Show that a positive integer is divisible by 3 if and only if the sum of its digits mod 3 is 0.

24. Come up with a test for divisibility by 16 (it is similar to those for divisibility by 4 and 8) and show why the test works.

25. A child bought nine identical candy bars for a total of \$4.#1, where the missing digit was smeared with chocolate. Assuming there was no sales tax, how much did each bar cost?

26. A storekeeper purchased 144 identically priced hammers from a wholesaler. When setting her selling price for the hammers, she checked the shop's record for the wholesale price. Her accountant had written

$$\$1\#25.4\#,$$

where the #'s represent illegible digits, which may or may not be different. Determine the wholesale price of a single hammer.

27. A *palindrome* is a number that reads the same backwards as forwards. For example, the numbers 151 and 12,033,021 are palindromes. Explain why any palindrome with an even number of digits is divisible by 11.

28. Use casting out nines to determine whether the following calculation *could* be correct:

$$89,034,690,257 \cdot 739,005,139,645 = 65,797,093,306,623,605,938,765.$$

29. Use casting out nines to determine whether the following calculation *could* be correct:

$$730{,}963{,}709 \cdot 28{,}100{,}923{,}561{,}398 = 20{,}450{,}755{,}312{,}764{,}971{,}305{,}182.$$

30. Use casting out elevens to determine whether the calculation in Exercise 28 *could* be correct.

31. Use casting out elevens to determine whether the calculation in Exercise 29 *could* be correct.

8.4 CHECK DIGITS

Identification numbers are widespread in society today. We find them on credit cards, driver's licenses, retail items, bank checks, and many other places. For many of these kinds of numbers, the last digit is an extra digit that is mathematically related to all of the other digits in the number. We call this type of special digit a **check digit**. As we will see, the check digit provides a way to detect errors that may occur when the number is recorded. The check digit also makes it somewhat more difficult to come up with fraudulent identification numbers for items like credit cards and bank checks because the person making up the identification numbers would have to know what method is normally used to generate the check digit. There are many different methods used to generate check digits, and modular arithmetic plays a role in most of these methods. We begin by looking at some of the simpler methods, and then go on to look at some more complicated methods.

Throughout our discussion of identification numbers, we will refer to the part of an identification number that does not include the check digit as the **main part** of the number.

The method of generating check digits used in the identification numbers on airline tickets is quite simple. The check digit in these cases is just the main part mod 7. For instance, the airline ticket in Figure 8.2 has stock control number 1915240876, where 191524087 is the main part and the last digit 6 is the check digit. The check digit is 6 in this case because

$$191524087 \bmod 7 = 6.$$

Figure 8.2
American Airlines ticket with stock control number 1915240876 and serial number 21939933142.

Similarly, the serial number for the ticket is 21939933142. The main part of the serial number is 2193993314 and the check digit is 2 because

$$2193993314 \bmod 7 = 2.$$

Example 1   *Airline Ticket*

Suppose an airline ticket has an 11-digit stock control number with a 10-digit main part 5182007843. If the check digit is given by the main part mod 7, what is the entire number?

Solution: Because $5182007843 \bmod 7 = 5$, the check digit is 5 and the entire stock control number is 51820078435.

■

A check digit that is generated by taking the main part mod 7 will not detect every possible error. For instance, if a 0 is typed in place of the 7 in the stock control number 1915240876, the check digit 6 would still be correct. More generally, the errors of substituting a 0 for a 7, a 1 for an 8, or a 2 for a 9, or vice versa, in the main part will not be detected by the check digit, because the original check digit would still be equal to the main part mod 7 with these substitutions. This kind of error—where one digit is replaced by a different digit in either the main part or check digit—is called a **single-digit error**.

The check digit method used for the serial numbers on U.S. Postal Service money orders is also quite simple. The serial numbers consist of eleven digits, the first ten of which are the main part, and the eleventh digit is a check digit that is equal to the main part mod 9. For example, the money order in Figure 8.3 has identification number 64865294458. Recalling that a number is congruent to the sum of its digits modulo 9, we have

$$
\begin{aligned}
6486529445 &\equiv 6 + 4 + 8 + 6 + 5 + 2 + 9 + 4 + 4 + 5 \\
&\equiv 6 + 4 + 8 + 6 + 2 + (5 + 4) + (4 + 5) \equiv 26 \equiv 8 \ (\mathrm{mod}\ 9),
\end{aligned}
$$

so the check digit 8 is equal to the main part of the serial number mod 9 as we would expect.

Figure 8.3
U.S. Postal Service money order with serial number 64865294458.

Example 2 *U.S. Postal Service Money Order*

Suppose a document that appears to be a U.S. Postal Service money order has serial number 56347894283. Could it be an authentic money order?

Solution: The check digit for this serial number should be

$$5634789428 \equiv 5 + 6 + 3 + 4 + 7 + 8 + 9 + 4 + 2 + 8$$
$$\equiv (5 + 4) + (6 + 3) + (7 + 2) + 8 + 4 + 8 \equiv 20 \equiv 2 \pmod 9.$$

The check digit on the presumed money order is 3, and therefore it cannot be authentic.

■

The check digit scheme for U.S. Postal Service money orders will not detect the error of replacing a 0 by a 9 or vice versa in the main part of the serial number. However, it will detect any other single-digit error. Unfortunately, other kinds of errors would go undetected. For instance, suppose the serial number for the money order in Figure 8.3 is typed in as 64856294458, with the 65 typed in incorrectly as 56. This kind of error—where two digits are interchanged—is called a **transposition error**. Because every number is congruent to the sum of its digits modulo 9, the only transposition errors that would be detected are those involving the check digit itself. Furthermore, any rearrangement of the digits in the main part of the serial number would still result in the same check digit, and therefore would not be detected.

For an 11-digit identification number (including the check digit), the mod 9 method detects about 98.0% of the possible single-digit errors but only about 10.0% of possible errors consisting of the transposition of adjacent digits. By contrast, the mod 7 method detects approximately 93.9% of possible single-digit errors and 94.0% of possible errors consisting of the transposition of adjacent digits. One study of the frequency of certain types of errors committed by humans when working with identification numbers found that 79.1% of the errors that occur are single-digit errors, whereas only 10.2% of the errors are transpositions of adjacent digits. It would be fair to assume that when identification numbers are read by a scanner rather than a human, single-digit errors account for an even higher percentage of the errors because a machine is unlikely to transpose or rearrange digits. It would therefore be nice to have a check digit scheme that would detect all single-digit errors. The next three methods we will look at all do this.

The check digit method employed by the U.S. banking system in assigning identification numbers to banks detects all single-digit errors and about 88.9% of errors consisting of the transposition of adjacent digits. Every U.S. bank has a nine-digit identification number, which we write as

$$a_8 a_7 a_6 a_5 a_4 a_3 a_2 a_1 a_0,$$

where $a_8 a_7 a_6 a_5 a_4 a_3 a_2 a_1$ is the main part and a_0 is the check digit. In this case the check digit a_0 is given by the following formula:

Bank Identification Number Check Digit Formula

$$a_0 = (7a_8 + 3a_7 + 9a_6 + 7a_5 + 3a_4 + 9a_3 + 7a_2 + 3a_1) \bmod 10$$

The bank identification number on personal checks is the first number on the bottom left. For example, the bank identification number for the BayBank check in Figure 8.4 is 011300605. In this case, the main part of the identification number is 01130060 and the number 5 is the check digit. To see that it is indeed given by our check digit formula, let's compute the check digit ourselves. In doing calculations modulo 10, it is helpful to use the fact that because the remainder of a positive integer when divided by 10 is just its last digit, *a positive integer is congruent to its last digit mod 10*. For the number 011300605, we have

$$a_0 \equiv 7 \cdot 0 + 3 \cdot 1 + 9 \cdot 1 + 7 \cdot 3 + 3 \cdot 0 + 9 \cdot 0 + 7 \cdot 6 + 3 \cdot 0$$
$$\equiv 3 + 9 + 21 + 42 \equiv 3 + 9 + 1 + 2 \equiv 15 \equiv 5 \,(\text{mod } 10),$$

so the check digit should be 5.

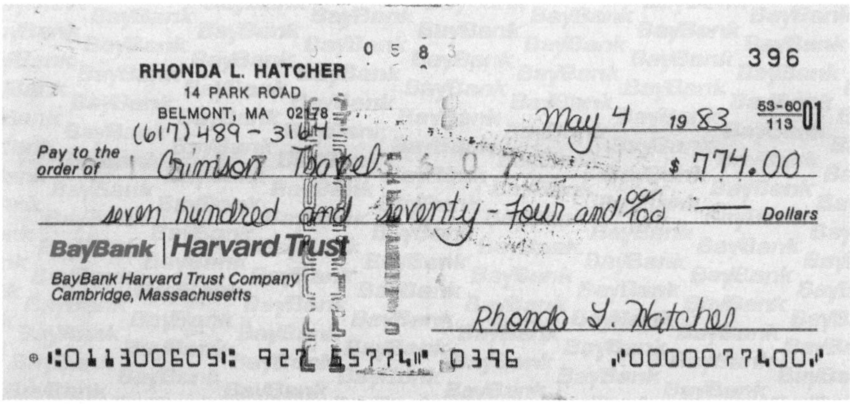

Figure 8.4
Bank check with bank identification number 011300605.

The following example gives an instance of a transposition error that is not caught by the U.S banking system check digit.

Example 3   *Bank Check*

The bank identification number of BancFirst-Austin is 114912819. Suppose it is accidentally typed in as 119412819. Will the transposition error of replacing the 49 in the correct number by 94 be detected by the check digit?

Solution: The check digit for the incorrect number would be

$$a_0 \equiv 7 \cdot 1 + 3 \cdot 1 + 9 \cdot 9 + 7 \cdot 4 + 3 \cdot 1 + 9 \cdot 2 + 7 \cdot 8 + 3 \cdot 1$$
$$\equiv 7 + 3 + 81 + 28 + 3 + 18 + 56 + 3$$
$$\equiv 7 + 3 + 1 + 8 + 3 + 8 + 6 + 3 \equiv 39 \equiv 9 \,(\text{mod } 10),$$

so the check digit of 9 is correct for the mistyped number. Therefore, the error will not be detected.

■

Another check digit method we look at here is the method used in the UPC bar codes found on retail items. The UPC (Universal Product Code) bar code first came into use on grocery store items in 1973 and can now be found on most retail items. In Figure 8.5 we have an example of this very familiar bar code.

Figure 8.5
UPC bar code for
Kellogg's® Crispix®.

Most UPC numbers consist of 12 digits, which are usually written as numbers as well as translated into bars that can be read by a laser scanner. There is also a version of the UPC number consisting of only seven digits for use on small items like candy bars or on some round containers like jelly jars. We consider only the full 12-digit version.

The last digit, a_0, of a 12-digit UPC number, $a_{11}a_{10}a_9a_8a_7a_6a_5a_4a_3a_2a_1a_0$, is a check digit given by the following formula:

UPC Number Check Digit Formula

$$a_0 = -(3a_{11} + a_{10} + 3a_9 + a_8 + 3a_7 + a_6 + 3a_5 + a_4 + 3a_3 + a_2 + 3a_1) \bmod 10.$$

The idea behind this formula is that the check digit a_0 is chosen so that

$$(3a_{11} + a_{10} + 3a_9 + a_8 + 3a_7 + a_6 + 3a_5 + a_4 + 3a_3 + a_2 + 3a_1 + a_0) \bmod 10 = 0.$$

This check digit method detects all single-digit errors and about 88.9% of errors consisting of the transposition of adjacent digits.

Consider the UPC number for a 31.4-ounce box of Crispix® cereal shown in Figure 8.5. Using the UPC number check digit formula, we see that the check digit 9 was given by the calculation

$$\begin{aligned}
a_0 &\equiv -(3 \cdot 0 + 3 + 3 \cdot 8 + 0 + 3 \cdot 0 + 0 + 3 \cdot 9 + 9 + 3 \cdot 2 + 2 + 3 \cdot 0) \\
&\equiv -(3 + 24 + 27 + 9 + 6 + 2) \equiv -(3 + 4 + 7 + 9 + 6 + 2) \\
&\equiv -31 \equiv -31 + 4 \cdot 10 \equiv 9 \pmod{10}.
\end{aligned}$$

Example 4   UPC Number

The UPC number for a $10\frac{3}{4}$-ounce can of Campbell's® Cream of Mushroom Soup is 051000012616 as shown in Figure 8.6. If the scanner accidentally reads the digit 2 as a 3, will the check digit detect the error?

Figure 8.6
UPC bar code for
Campbell's® Cream of
Mushroom Soup.

Solution: If the main part of the UPC number is read as 05100001361, then the scanner will expect to see the check digit 3 because the check digit formula gives

$$a_0 \equiv -(3 \cdot 0 + 5 + 3 \cdot 1 + 0 + 3 \cdot 0 + 0 + 3 \cdot 0 + 1 + 3 \cdot 3 + 6 + 3 \cdot 1)$$
$$\equiv -(5 + 3 + 1 + 9 + 6 + 3) \equiv -27 \equiv -27 + 3 \cdot 10 \equiv 3 \,(\text{mod } 10).$$

However, the correct check digit is 6, so this error would be detected.

■

Of the four check digit methods we have looked at, none of them detect all single-digit errors and transposition errors, but there are methods available that do so. One such method is the one used in the International Standard Book Number (ISBN number). The check digit for ISBN numbers comes from a formula similar to the ones for the UPC number and the U.S. banking system, except that the arithmetic is done modulo 11. Unfortunately, this allows the two-digit number 10 as a possible check digit. However, this problem is easily remedied by using an X to represent the check digit 10. The ISBN number is a 10-digit number $a_9 a_8 a_7 a_6 a_5 a_4 a_3 a_2 a_1 a_0$, where the last digit a_0 is a check digit computed using the following formula:

ISBN Check Digit Formula

$$a_0 = (a_9 + 2a_8 + 3a_7 + 4a_6 + 5a_5 + 6a_4 + 7a_3 + 8a_2 + 9a_1) \,\text{mod } 11$$

For example, the ISBN number of the paperback version of John Grisham's novel *The Rainmaker* is 0-440-22165-X. The check digit is X because for this ISBN number the check digit formula gives

$$a_0 \equiv 0 + 2 \cdot 4 + 3 \cdot 4 + 4 \cdot 0 + 5 \cdot 2 + 6 \cdot 2 + 7 \cdot 1 + 8 \cdot 6 + 9 \cdot 5$$
$$\equiv 8 + 12 + 10 + 12 + 7 + 48 + 45 \equiv 142 \equiv 10 \,(\text{mod } 11).$$

Example 5   *ISBN Number*

The main part of the ISBN number for *Green Eggs and Ham* by Dr. Seuss is 0-394-80016. What is the check digit?

Solution: The check digit is given by

$$a_0 \equiv 0 + 2 \cdot 3 + 3 \cdot 9 + 4 \cdot 4 + 5 \cdot 8 + 6 \cdot 0 + 7 \cdot 0 + 8 \cdot 1 + 9 \cdot 6$$
$$\equiv 6 + 27 + 16 + 40 + 8 + 54 \equiv 151 \equiv 8 \,(\text{mod } 11).$$

Therefore, the check digit is 8.

■

You might wonder whether it is possible to find a check digit method that detects all single-digit and transposition errors and only generates 0 through 9 as check digits. In 1969 a method satisfying these criteria was discovered by

BRANCHING OUT 8.4

THE UPC BAR CODE

The 12-digit UPC number of a product is made up of four parts as shown in the UPC number for a box of Q-tips® at right.

The first digit represents the type of product. For instance, a 3 is used for drugs and some other health products. The next five digits are the manufacturer number. In the case of the Q-tips®, the manufacturer is Chesebrough-Pond's USA Co. The five digits following the manufacturer's number are the product number. It is assigned by the manufacturer and identifies the product precisely, including such details as size or weight. The last digit, as we know, is the check digit. To translate the digits of a UPC number into a bar code, the digits are first encoded as seven-digit numbers consisting of only 0's and 1's. This kind of code is called a binary code. The code used for the first digit and the manufacturer's number is the same, whereas the code used for the product number and the check digit can be obtained from the code for the manufacturer's numbers by replacing each 0 by a 1 and each 1 by 0 as shown in the table.

Q-tips® UPC bar code.

	UPC Coding	
Digit	**Manufacturer's Number and First Digit**	**Product Number and Check Digit**
0	0001101	1110010
1	0011001	1100110
2	0010011	1101100
3	0111101	1000010
4	0100011	1011100
5	0110001	1001110
6	0101111	1010000
7	0111011	1000100
8	0110111	1001000
9	0001011	1110100

Using the coded representation, each digit is represented in the bar code by seven spaces, called modules, of equal width corresponding to the seven digits of the binary code. The spaces corresponding to 1's are dark and those corresponding to 0's are light as shown in the figure at right.

An entire UPC bar code starts with a left-hand guard bar, three modules wide, corresponding to the binary digits 101. Therefore, it is a light module surrounded by two dark modules. Similarly, it ends with a right-hand guard bar of the same type. These guard bar patterns define the thickness of a module for the scanner. The bars representing the manufacturer's number and the product number are separated by a center bar pattern, five modules wide, corresponding to the binary digits 01010. In addition, each bar code must have margins at least 11 modules wide on the left and the right. Because the binary codes used for the manufacturer numbers have an odd number of 1's, whereas the codes used for the product numbers have an even number of 1's, the scanner will be able to detect whether the code is being scanned from left to right or vice versa, and therefore scanning can be done in either direction.

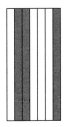

UPC bar coding of 0110001.

J. Verhoeff. This method uses ideas from an area of mathematics known as group theory.

There are many other methods available for computing check digits. In the exercises, you will look at two other methods commonly used today.

Exercises for Section 8.4

1. A particular United Airlines ticket has a 13-digit serial number with main part 161311619298 followed by a check digit given by the main part mod 7. What is the check digit?

2. A particular United Airlines ticket has an 11-digit stock control number 66131949785, where the last digit is a check digit given by the main part mod 7. If the number is accidentally entered as 66131249785, will the check digit detect the error?

3. Suppose the stock control number for an airline ticket is the 10-digit number 7481035881, where the last digit is a check digit given by the main part mod 7. If the number is accidentally recorded as 7981035881, will the check digit detect the error?

4. A U.S. Postal Service money order has serial number with main part 6486745669. What is the entire 11-digit serial number?

5. Suppose the serial number for a U.S. Postal Service money order is 57923518026. If the number is accidentally recorded as 57923158026, will the check digit detect the error?

6. Suppose the serial number on a U.S. Postal Service money order is 89012791172. Could it be authentic?

7. The main part of the nine-digit bank identification number of NationsBank of Texas is 11100002. What is the entire bank identification number?

8. The bank identification number of the First National Bank of Northfield, Minnesota (once robbed by the Jesse James gang), is 091901477. Suppose it is accidentally typed in as 091907417. Will the error be detected by the check digit?

9. Suppose the bank identification number on a check is 100646023. Could it be an authentic check from a U.S. bank?

10. Suppose the bank identification number on a valid U.S. bank check is 025500#16, where # is a digit that is illegible. What must the digit # be?

11. The UPC number for the 30-ounce box of Nestlé® Quik is 028000242800. If the scanner accidentally reads the digit 4 as a 9 instead, what check digit will the scanner expect to see?

12. The UPC number for a $10\frac{1}{2}$-ounce bag of Kraft® Miniature Marshmallows is 021000660728. If the scanner accidentally reads the digit 7 as a 9, what check digit will the scanner expect to see?

13. The UPC number for a pack of two 9-volt Energizer® batteries is 039800036780. If the number is accidentally entered as 039800036870, will the check digit detect the error?

14. The UPC number for a box of Contac® cold medicine is 345800238209. If the number is accidentally entered as 345800283209, will the check digit detect the error?

15. Suppose you know that the UPC number of a certain product is 3008108#0249, where # is a digit that is illegible. What must the digit # be?

16. The main part of the ISBN number of *The Wizard of Oz* by L. Frank Baum is 0-688-06944. Find the check digit.

17. The main part of the ISBN number for *Webster's II New Riverside Dictionary* is 0-395-33957. Find the check digit.

18. Suppose you know that the ISBN number of a certain book is 0-590-5#880-9, where # is a digit that is illegible. What must the digit # be?

For Exercises 19–22, use the fact that the serial numbers on American Express travelers cheques are 10-digit numbers $a_9a_8a_7a_6a_5a_4a_3a_2a_1a_0$, where $a_9a_8a_7a_6a_5a_4a_3a_2a_1$ is the main part and a_0 is a check digit given by the formula

$$a_0 = -(a_9 + a_8 + a_7 + a_6 + a_5 + a_4 + a_3 + a_2 + a_1) \bmod 9.$$

19. If the main part of the serial number of an American Express travelers cheque is 417948919, find the entire 10-digit serial number.

20. Suppose the serial number of a document that appears to be an American Express travelers cheque is 2893459113. Could it be authentic?

21. The serial number of an American Express travelers cheque is 2570113098.

    (a) If the number is accidentally recorded as 2670113098, will the check digit detect the error?

    (b) Is there any kind of single-digit error that would not be detected by the check digit?

22. The serial number on the American Express travelers cheque shown here is 4370633811.

    (a) If the number is accidentally recorded as 4730633811, will the check digit detect the error?

    (b) Is there any kind of transposition error that will be detected by the check digit?

For Exercises 23–26, consider the following: Some identification codes use both digits and letters. One such type of identification number, known as Code 39, is used by the U.S. Department of Defense, automotive companies, and the health industry. Code 39 identification codes are composed of digits and the 26 uppercase letters A through Z. The identification codes have 15 characters, $a_{14}a_{13}a_{12}a_{11}a_{10}a_9a_8a_7a_6a_5a_4a_3a_2a_1a_0$, where the last character a_0 is a check "character" computed as follows: Any of the main part characters a_k that are letters A through Z are converted to the numbers 10 through 35, respectively. In other words, any A's are converted to 10's, B's are converted to 11's, and so on. The check character is then given by the formula

$$a_0 = -(15a_{14} + 14a_{13} + 13a_{12} + 12a_{11} + 11a_{10} + 10a_9 + 9a_8$$
$$+ 8a_7 + 7a_6 + 6a_5 + 5a_4 + 4a_3 + 3a_2 + 2a_1) \bmod 36,$$

and if a_0 is between 10 and 36, it is converted to the equivalent letter.

23. If the main part of a Code 39 identification code is 34T1CR9244YD3W, find the check character.

24. If the main part of a Code 39 identification code is V97MKU357H8PQB, find the check character.

25. If the Code 39 identification code T313HK21R7732Y5 is typed in as T313HK21R7723Y5, will the check digit detect the error?

26. If the Code 39 identification code KPT26853NT993NR is typed in as JPT26853NT993NR, will the check digit detect the error?

27. Explain why the bank identification number check digit formula may be written alternatively as

$$a_0 = -(3a_8 + 7a_7 + a_6 + 3a_5 + 7a_4 + a_3 + 3a_2 + 7a_1) \bmod 10.$$

8.5 TOURNAMENT SCHEDULING

Modular arithmetic can be used to set up a schedule for a round-robin tournament. In a **round-robin tournament**, each team must play every other team exactly once in a series of rounds. For instance, in the preliminary round of the Olympic Hockey tournament, groups of four teams are scheduled to play round-robin tournaments. In the preliminary round of the 1998 Olympics, Bulgaria, France, Germany, and Japan played in one group as shown in Table 8.1.

Table 8.1	**Round-Robin Schedule in Olympic Hockey**		
	First Round, February 7	*Second Round, February 9*	*Third Round, February 10*
Bulgaria	France	Germany	Japan
France	Bulgaria	Japan	Germany
Germany	Japan	Bulgaria	France
Japan	Germany	France	Bulgaria

France vs. Japan in the 1998 Winter Olympic Games.

If there are an even number of teams in a round-robin tournament, then every team should be playing in every round. So, for instance, with six teams, it would take five rounds to complete the tournament. If there are an odd number of teams, then in every round one team would have to be scheduled to sit out with a **bye**, and the remaining teams would all play. Therefore, for example, with seven teams in the tournament, there would have to be seven rounds.

Setting up a round-robin tournament for a fairly small number of teams can be done by trial and error because there are relatively few choices to make. However, for a large number of teams it is quite difficult to avoid all of the possible conflicts. (If you are not convinced of this, try to schedule a round-robin tournament even with just 10 teams.) In order to discover a general method that works for any number of teams, we will try to create a schedule with a regular pattern. An irregular construction is unlikely to always work, and if it does, we would have trouble describing it precisely. It turns out to be easier to discover a pattern when we start with an odd number of teams, so we will start with five teams. We give each of the teams a number 1 through 5 as a name. To complete the tournament schedule, we need to schedule five rounds, with one team not playing in each round. The table entry will be the team's opponent for that round. To begin, notice that we can rearrange the rounds of any schedule so that Team 1 plays Teams 2, 3, 4, 5 in order and ends with its bye. We have started such a schedule in Table

8.2, also filling in the schedule for those teams scheduled to play Team 1. This idea illustrates a useful problem-solving technique of reducing the scope of the problem in a way that allows all cases to be constructed from the restricted cases. In our situation, we could find more general schedules by reordering the rounds.

Table 8.2	**Start of a Round-Robin Schedule for Five Teams**				
			Round		
	First	*Second*	*Third*	*Fourth*	*Fifth*
Team 1	2	3	4	5	Bye
Team 2	1				
Team 3		1			
Team 4			1		
Team 5				1	

We must now decide how to proceed; there are many choices. One of the nicest patterns would be if we could, roughly speaking, increase the number of the opponent by one each round. Let's try this for Team 2 and see how it goes, which we have done in Table 8.3.

Table 8.3	**Part of a Round-Robin Schedule for Five Teams**				
			Round		
	First	*Second*	*Third*	*Fourth*	*Fifth*
Team 1	2	3	4	5	Bye
Team 2	1	2?	3	4	5
Team 3		1	2		
Team 4			1	2	
Team 5				1	2

Notice that a nice pattern appears to be forming, but we have the problem that Team 2 is scheduled to play itself in round 2. However, the obvious solution is to give Team 2 a bye here. With the optimism spurred by our initial success, we fill in the rest of the schedule by extending this pattern. Whenever a team is scheduled to play itself, it gets a bye. This works out to give the complete round-robin tournament schedule shown in Table 8.4.

Table 8.4	**Round-Robin Schedule for Five Teams**				
			Round		
	First	*Second*	*Third*	*Fourth*	*Fifth*
Team 1	2	3	4	5	Bye
Team 2	1	Bye	3	4	5
Team 3	5	1	2	Bye	4
Team 4	Bye	5	1	2	3
Team 5	3	4	Bye	1	2

Careful examination of Table 8.4 may remind you somewhat of an addition table. The way it "wraps around" (see the rows for Teams 3 and 5 in particular) suggests modular arithmetic, possibly arithmetic mod 5 because only five possible values exist. We let $T_{m,r}$ be the team assigned to play Team m in round r. Both m and r can take on any value from 1 through 5. Then in round r, the entry in Table 8.4 is given by $T_{m,r} \equiv r + 2 - m \pmod 5$. We give a bye to the team scheduled to play itself. It is easy to add a sixth team; simply schedule it to play the team with a bye, as in Table 8.5.

Table 8.5	Round-Robin Schedule for Six Teams				
	Round				
	First	*Second*	*Third*	*Fourth*	*Fifth*
Team 1	2	3	4	5	6
Team 2	1	6	3	4	5
Team 3	5	1	2	6	4
Team 4	6	5	1	2	3
Team 5	3	4	6	1	2
Team 6	4	2	5	3	1

Does such a procedure work in general? It is not obvious that it does because there are several potential pitfalls. In describing a general scheduling procedure, we will simplify slightly by replacing $r + 2$ with r in our formula for $T_{m,r}$. The only effect of this change is to reorder the rounds. Suppose there are N teams in a round-robin tournament, with N odd. We need to set up N rounds. In round r, we begin by letting $T_{m,r}$ be the number between 1 and N such that

$$T_{m,r} \equiv r - m \pmod N.$$

For example, if $N = 7$, then in the third round Team 1 will be assigned to play Team 2, because

$$T_{2,3} \equiv 3 - 2 \equiv 1 \pmod 7.$$

Similarly, in the third round we have

$$T_{3,3} \equiv 3 - 3 \equiv 0 \equiv 7 \pmod 7,$$

so Team 3 will play Team 7. Notice that we could not let $T_{3,3} = 0$ because all of the $T_{m,r}$ must lie between 1 and 7. That is why we adjusted the number 0 by adding 7 in the last step in finding $T_{3,3}$.

The formula we have for $T_{m,r}$ does give some sort of team assignments, but at this point we do not know if the assignments it gives correspond to a round-robin tournament.

One thing that will need to be true for the team assignments given by $T_{m,r}$ is that if Team B is assigned to play Team A in a particular round, then Team A is also assigned to play Team B in that round. To check that this is the case, suppose that Team B is assigned to play Team A. Then it must be true that

$$T_{A,r} \equiv r - A \equiv B \pmod N.$$

If we add A and subtract B from both sides of the congruence equation $r - A \equiv B \pmod N$, we get

$$r - A + A - B \equiv B + A - B \pmod N,$$

and therefore

$$r - B \equiv A \pmod{N}.$$

Thus, $T_{B,r} \equiv r - B \equiv A \pmod{N}$, so Team A is assigned to play Team B. Similar computations using modular arithmetic show that in a given round r, different values m in $T_{m,r} \equiv r - m \pmod{N}$ will give different values of $T_{m,r}$, so two different teams are not assigned to the same opponent in the same round. And in a like way you can show that no team is assigned to play the same team in two different rounds. These points are covered in exercises.

Although it looks like this scheme might work, we have another potential problem. With an odd number of teams, there must be one and only one team assigned to play itself. Because we have paired off N teams in each round and N is odd, it must be that in each round we have assigned at least one team to play itself. Could more than one team be assigned a bye in a given round? We leave it as an exercise to show that only one team has been assigned to play itself. With an even number of teams, we schedule the first $N - 1$ teams in this way and let the team that was assigned to play itself play Team N instead.

We can summarize our method for setting up a round-robin tournament schedule as follows:

Round-Robin Scheduling Method

Suppose the number of teams in the tournament is N, with N odd. Then for $m = 1, 2, \ldots, N$, in round r Team $T_{m,r}$ is assigned to play Team m, where $T_{m,r}$ is the number between 1 and N such that

$$T_{m,r} \equiv r - m \pmod{N},$$

unless $T_{m,r} = m$, in which case Team m is assigned a bye. When the number of teams is N, with N even, we schedule teams 1 through $N - 1$ as for a tournament with $N - 1$ teams, except that a team that would be assigned a bye is instead assigned to play Team N.

Example 1 *Round-Robin Tournament for Seven Teams*

Using our round-robin scheduling method, set up a round-robin tournament schedule for seven teams.

Solution: We have $N = 7$, which is odd. In the first round, Team $T_{1,1}$ is assigned to play Team 1, where

$$T_{1,1} \equiv 1 - 1 \equiv 0 \equiv 7 \pmod{7}.$$

So Team 7 is assigned to play Team 1, and we know, without computing $T_{7,1}$, that Team 1 will be assigned to play Team 7. Next, we see that Team $T_{2,1}$ is assigned to play Team 2, where

$$T_{2,1} \equiv 1 - 2 \equiv -1 \equiv -1 + 7 \equiv 6 \pmod{7},$$

so Teams 2 and 6 play each other. Similarly, we have

$$T_{3,1} \equiv 1 - 3 \equiv -2 \equiv -2 + 7 \equiv 5 \pmod{7},$$

and therefore Teams 3 and 5 play each other. Next we see that

$$T_{4,1} \equiv 1 - 4 \equiv -3 \equiv -3 + 7 \equiv 4 \,(\text{mod } 7),$$

so Team 4 has been assigned to play itself and will get a bye in round 1. We also should have known that Team 4 would get a bye even before computing $T_{4,1}$ in the first round because it was the only team left to be assigned. We can now complete the schedule for the first round as in Table 8.6.

Table 8.6	**Beginning of a Round-Robin Schedule for Seven Teams**						
				Round			
	First	*Second*	*Third*	*Fourth*	*Fifth*	*Sixth*	*Seventh*
Team 1	7						
Team 2	6						
Team 3	5						
Team 4	Bye						
Team 5	3						
Team 6	2						
Team 7	1						

Moving on to the second round, we see that

$$T_{1,2} \equiv 2 - 1 \equiv 1 \,(\text{mod } 7),$$

so Team 1 will have a bye this round. We next find that

$$T_{2,2} \equiv 2 - 2 \equiv 0 \equiv 7 \,(\text{mod } 7),$$

and therefore Team 2 and Team 7 will play each other. Similarly, because

$$T_{3,2} \equiv 2 - 3 \equiv -1 \equiv 6 \,(\text{mod } 7),$$

we see that Team 3 plays Team 6. This leaves out only Teams 4 and 5, so they must play each other. We have now completed the schedule for the second round. The remaining rounds are computed in a similar manner, and we arrive at the complete schedule in Table 8.7.

Table 8.7	**Round-Robin Schedule for Seven Teams**						
				Round			
	First	*Second*	*Third*	*Fourth*	*Fifth*	*Sixth*	*Seventh*
Team 1	7	Bye	2	3	4	5	6
Team 2	6	7	1	Bye	3	4	5
Team 3	5	6	7	1	2	Bye	4
Team 4	Bye	5	6	7	1	2	3
Team 5	3	4	Bye	6	7	1	2
Team 6	2	3	4	5	Bye	7	1
Team 7	1	2	3	4	5	6	Bye

Note that after we have assigned all of the teams for the first six rounds, we can figure out the team assignments for the seventh round simply by assigning to each team the team they have not yet been assigned. However, because the calculations are fairly simple, you may want to go ahead and do them to double-check your work.

Example 2   *Round-Robin Tournament for Eight Teams*

Using our round-robin scheduling method, set up a round-robin tournament schedule for eight teams.

Solution: We need to make a schedule as for seven teams and let the team assigned a bye in a given round play the Team 8 instead. For example, when we found

$$T_{4,1} \equiv 1 - 4 \equiv -3 \equiv -3 + 7 \equiv 4 \, (\text{mod } 7)$$

for the first round in Example 1, we would schedule Team 4 to play Team 8 in round 1. For the second round,

$$T_{1,2} \equiv 2 - 1 \equiv 1 \, (\text{mod } 7),$$

so Team 1 will play Team 8 in that round. Otherwise proceeding just as we did for seven teams in Example 1, we create a round-robin schedule for eight teams, given in Table 8.8.

Table 8.8   **Round-Robin Schedule for Eight Teams**

				Round			
	First	*Second*	*Third*	*Fourth*	*Fifth*	*Sixth*	*Seventh*
Team 1	7	8	2	3	4	5	6
Team 2	6	7	1	8	3	4	5
Team 3	5	6	7	1	2	8	4
Team 4	8	5	6	7	1	2	3
Team 5	3	4	8	6	7	1	2
Team 6	2	3	4	5	8	7	1
Team 7	1	2	3	4	5	6	8
Team 8	4	1	5	2	6	3	7

Another issue that may arise in setting up a round-robin tournament is in designating which team will be the home team and which will be the away team for each game. To do this fairly, you would like to assign the teams a roughly equal number of home games. Two possible schemes for doing this are explored in the exercises.

EXERCISES FOR SECTION 8.5

In Exercises 1–8, using our round-robin scheduling method, set up a round-robin tournament for the given number of teams.

1. 9 teams

2. 11 teams

3. 10 teams

4. 12 teams

5. 15 teams

6. 13 teams

7. 16 teams

8. 14 teams

9. For a tournament with 20 teams, use our round-robin scheduling method to find the team that will be assigned to play Team 9 in round 7.

10. For a tournament with 24 teams, use our round-robin scheduling method to find the team that will be assigned to play Team 14 in round 5.

11. For a tournament with 31 teams, use our round-robin scheduling method to find the team that will be assigned to play Team 22 in round 12.

12. For a tournament with 17 teams, use our round-robin scheduling method to find the team that will be assigned to play Team 9 in round 3.

13. For a tournament with 22 teams, use our round-robin scheduling method to determine the round in which Team 7 will play Team 10.

14. For a tournament with 19 teams, use our round-robin scheduling method to determine the round in which Team 8 will have a bye.

15. For a tournament with 17 teams, use our round-robin scheduling method to determine the round in which Team 12 will have a bye.

16. For a tournament with 28 teams, use our round-robin scheduling method to determine the round in which Team 15 will play Team 26.

In many tournaments, each game must have a designated home team and away team. One method for doing this in a round-robin tournament is as follows: In any round, if Team i plays Team j, then

i.   if $i + j$ is even, the team with smallest number i or j is the home team,

ii. if $i + j$ is odd, the team with largest number i or j is the home team.

In Exercises 17–20, use this method to set up a round-robin tournament with home and away teams designated for the given number of teams.

17. 10 teams

18. 12 teams

19. 15 teams

20. 13 teams

21. Suppose the number of teams, N, in a tournament is odd and the home and away assignments are made using the method of Exercises 17–20. Explain why each team will be assigned $(N - 1)/2$ home games.

22. Suppose the number of teams, N, in the tournament is even and the home and away assignments are made using the method of Exercises 17–20. Explain why each team will be assigned either $N/2$ or $(N - 2)/2$ home games.

Another method for designating home and away teams is as follows: In any round, if Team i plays Team j, where $j > i$, then

    i.   if $j - i \leq N/2$, Team i is the home team,

   ii.   if $j - i > N/2$, Team j is the home team.

In Exercises 23–26, use this method to set up a round-robin tournament with home and away teams designated for the given number of teams.

23. 10 teams   **24.** 12 teams

25. 15 teams   **26.** 13 teams

27. Suppose the number of teams, N, in a tournament is odd and the home and away assignments are made using the method of Exercises 23–26. Explain why each team will be assigned $(N - 1)/2$ home games.

28. Suppose the number of teams, N, in the tournament is even and the home and away assignments are made using the method of Exercises 23–26. Explain why each team will be assigned either $N/2$ or $(N - 2)/2$ home games.

29. Let N be odd. Suppose that the method for scheduling a round robin tournament given in this section is altered so that in round r Team $T_{m,r}$ is assigned to play Team m, where $T_{m,r}$ is the number between 1 and N such that

$$T_{m,r} \equiv 2r - m \pmod{N}$$

unless $T_{m,r} = m$, in which case Team m is assigned a bye. Set up a round-robin tournament schedule for seven teams using this method. How does the schedule differ from the one we arrived at in Example 1?

30. Let N be odd. Suppose that the method for scheduling a round-robin tournament given in this section is altered so that in round r Team $T_{m,r}$ is assigned to play Team m, where $T_{m,r}$ is the number between 1 and N such that

$$T_{m,r} \equiv r + 2 - m \pmod{N}$$

unless $T_{m,r} = m$, in which case Team m is assigned a bye. Set up a round-robin tournament schedule for eight teams on the basis of this method. How does the schedule differ from the one we arrived at in Example 2?

31. Let N be odd. When you use the relation $T_{m,r} \equiv r - m \pmod{N}$ to assign teams, explain why the same opponent cannot be assigned to two different teams in the same round.

32. Let N be odd. When you use the relation $T_{m,r} \equiv r - m \pmod{N}$ to assign teams, explain why no team is assigned to play the same team in two different rounds.

33. Let N be odd. When you use the relation $T_{m,r} \equiv r - m \pmod{N}$ to assign teams, explain why exactly one team is assigned to play itself.

34. Show that there is exactly one way to complete the schedule shown in the table to create a round-robin tournament schedule for five teams.

	\multicolumn{5}{c}{*Round*}
	First	*Second*	*Third*	*Fourth*	*Fifth*
Team 1	2	3	4	5	Bye
Team 2	1	Bye	3	4	5
Team 3		1	2		
Team 4			1	2	
Team 5				1	2

8.6 INTRODUCTION TO CRYPTOLOGY

From the time that people first began to communicate through writing to the present day, there has been an interest in sending written messages in secret codes. Before the advent of postal systems, messages were often sent by private messenger. However, because the messenger could be captured or could be disloyal, it was in the interest of the sender to have the message in secret code. With the widespread use of electronic communication today and the ability of others to intercept these communications, the need for secret codes has become an even more important issue. Number theory plays a key role in many kinds of secret codes.

Before we look at specific secret codes, we will introduce some terminology. The entire discipline of encoding and decoding secret messages is known as **cryptology**. A method for encoding messages is a **cipher**. We call a message that is to be encoded **plaintext**. After the message has been encoded in secret code, we will refer to it as **ciphertext**. The process of encoding a message is sometimes called **enciphering** or **encryption**, and decoding a message is called **deciphering** or **decryption**.

We will limit our discussion of cryptology to the case of encoding and decoding text consisting only of the 26 letters of the alphabet. All of the methods we will consider can be easily adapted to larger character sets including, for instance, numbers, punctuation, or blank spaces.

The Caesar Cipher

One of the earliest known ciphers was used by Julius Caesar and is now called the **Caesar cipher**. With the Caesar cipher, each letter of the message is replaced by the letter three places beyond it in the alphabet, with the last three letters of the alphabet shifted to the first three letters of the alphabet. Caesar used the Roman alphabet, but we will adapt this cipher to the English alphabet. To illustrate the encoding of a message with the Caesar cipher, suppose we start with the plaintext message

I AM IN THE YARD.

Because L is three letters beyond I in the alphabet, we replace I by L. We replace A by D because D is three letters beyond A. Similarly, we replace the remaining letters, and we get the following ciphertext message:

L DP LQ WKH BDUG.

Notice that the Y in the plaintext message was replaced by a B, because in going three spaces forward, A and B follow Z. One problem with this method of enciphering jumps to mind. Because the letter L in the ciphertext stands for a one-letter word, a person trying to decipher the message would already know that the L must stand for A or I because they are the only one-letter words in the English language. Similarly, if the message were longer, the word THE might appear many times in the plaintext, and therefore the word WKH would appear often in the ciphertext. This would give a person trying to decipher the message a strong hint that the ciphertext word WKH is a fairly common word in the English language. Other hints based on particular words that appear in the ciphertext might provide enough information to make the message easy to decipher. Therefore, it is wise to eliminate this problem by breaking the original plaintext message into blocks of letters of the same length. For example, we could break our plaintext message above into blocks of length five and get

IAMIN THEYA RD.

Notice, of course, that the last word was not of length five because we ran out of letters. Enciphering this plaintext message with the Caesar cipher will now give

LDPLQ WKHBD UG.

We have still not eliminated the problem that a very common sequence of letters such as THE in the English language will also appear as a common sequence of letters in the enciphered message (possibly across two blocks). We will see ways to deal with this problem when we look at more complex methods of enciphering in the next section.

We can describe the Caesar cipher and many other ciphers in terms of modular arithmetic. In each of these ciphers, after breaking the letters into blocks, we translate the plaintext message into the integers between 0 and 25 as in Table 8.9.

Table 8.9	Numerical Equivalents of Letters

A	B	C	D	E	F	G	H	I	J	K	L	M	N	O	P	Q	R	S	T	U	V	W	X	Y	Z
0	1	2	3	4	5	6	7	8	9	10	11	12	13	14	15	16	17	18	19	20	21	22	23	24	25

With the Caesar cipher, after the plaintext is converted to numbers, we replace each number in the plaintext, P, by a number in ciphertext, C, given by the following formula:

Enciphering Formula for the Caesar Cipher

$$C = (P + 3) \bmod 26$$

This formula gives a value for C between 0 and 25. After converting each number in plaintext to a number in ciphertext with this formula, we then convert the enciphered message back to letters. We see how this whole process works in the following example.

Example 1   *Encoding with the Caesar Cipher*

Encipher the following message by breaking it into five-letter blocks and using the Caesar cipher:

MY OFFER IS FIRM.

Solution: First we break it into blocks and get

MYOFF ERISF IRM.

Now, converting the letters to their numerical equivalents we get

12 24 14 5 5   4 17 8 18 5   8 17 12.

We now use the Caesar cipher formula $C = (P + 3) \bmod 26$ to transform each number. For instance, for the first number, 12, we get

$$C \equiv 12 + 3 \equiv 15 \ (\text{mod } 26),$$

so $C = 15$. Similarly, the second number, 24, is transformed to 1 because

$$C \equiv 24 + 3 \equiv 27 \equiv 1 \ (\text{mod } 26).$$

Transforming all of the numbers in the plaintext message in this way, we get the following ciphertext message:

$$15 \ 1 \ 17 \ 8 \ 8 \qquad 7 \ 20 \ 11 \ 21 \ 8 \qquad 11 \ 20 \ 15.$$

Now converting the ciphertext numbers to their letter equivalents, we get the enciphered message

PBRII HULVI LUP,

and this is the message that is sent.

■

When the enciphered message reaches the recipient, it must be deciphered. After converting the enciphered message into numbers, the recipient must transform the ciphertext numbers into plaintext numbers. To decipher a message enciphered with the Caesar cipher, we use the fact that the formula for encoding was $C = (P + 3) \bmod 26$. Starting with this equation and subtracting 3 from both sides, we get

$$C \equiv P + 3 \ (\text{mod } 26),$$
$$C - 3 \equiv P \ (\text{mod } 26),$$

and rearranging we have the following formula for P:

Deciphering Formula for the Caesar Cipher

$$P = (C - 3) \bmod 26$$

Example 2   *Decoding with the Caesar Cipher*

Decipher the following message, which was enciphered with the Caesar cipher:

BRXFD QVWDU WQRZ.

Solution: We begin by converting the letters to their numerical equivalents, and we get

$$1 \ 17 \ 23 \ 5 \ 3 \qquad 16 \ 21 \ 22 \ 3 \ 20 \qquad 22 \ 16 \ 17 \ 25.$$

Next we use the deciphering formula for the Caesar cipher to convert each ciphertext number to a plaintext number. Starting with the first number, 1, we use the decoding formula $P = (C - 3) \bmod 26$ to find that

$$P \equiv 1 - 3 \equiv -2 \equiv 24 \ (\text{mod } 26),$$

so the 1 will be converted to 24. The second number, 17, will be converted to 14 because

$$P \equiv 17 - 3 \equiv 14 \,(\text{mod } 26).$$

Similarly, we convert the rest of the ciphertext numbers into plaintext numbers, obtaining

$$24\ 14\ 20\ 2\ 0 \qquad 13\ 18\ 19\ 0\ 17 \qquad 19\ 13\ 14\ 22.$$

Now, converting these numbers into their letter equivalents we have

YOUCA NSTAR TNOW,

and breaking the letters into words, we see that the original message was

YOU CAN START NOW.

You may now be thinking that we have gone to a great deal of trouble with our formulas for the Caesar cipher. After all, when encoding and decoding with this cipher, we are just shifting the letters back and forth by 3. This can easily be accomplished without introducing modular arithmetic. However, as we will see next, the ideas we have just used can be generalized to give us a large number of different ciphers.

Affine Ciphers

We could generalize the idea of the Caesar cipher by trying a transformation of the form

$$C = (P + b) \,\text{mod } 26,$$

where b is some integer between 0 and 25. However, this is still just a shift of the letters. To mix things up a bit more, let's instead try the formula

$$C = (aP + b) \,\text{mod } 26,$$

where a and b are both integers between 0 and 25. In order for this idea to be of use, we would have to be able to decipher a message that had been encoded with this formula. This means we would have to be able to solve for P. Starting with the enciphering formula and trying to solve for P using the same ideas we use in solving algebraic equations in regular arithmetic, we have

$$C \equiv aP + b \,(\text{mod } 26),$$
$$C - b \equiv aP \,(\text{mod } 26),$$
$$aP \equiv C - b \,(\text{mod } 26).$$

To finish solving for P in this equation in regular arithmetic, we would multiply both sides by a^{-1}. But what is a^{-1} in arithmetic modulo 26? Let's look at this next.

For any number a, if there exists a number x with $0 \leq x < n$ such that

$$ax \equiv 1 \,(\text{mod } n),$$

then we call x the **inverse** of a mod n and denote it by $\bar{a}$ mod n or more briefly

For example, because

$$3 \cdot 9 \equiv 27 \equiv 1 \pmod{26},$$

we see that $\overline{3} \bmod 26 = 9$. It turns out that not all numbers will have an inverse mod n. For instance, if 13 had an inverse, x, mod 26, then we would have

$$13x \equiv 1 \pmod{26}.$$

Multiplying both sides of this equation by 2, we get

$$26x \equiv 2 \pmod{26},$$

and because $26 \equiv 0 \bmod 26$, we would have

$$0 \equiv 2 \pmod{26}.$$

This last statement is not true, so 13 must not have an inverse mod 26. In general, we have the following fact:

> The integer a has an inverse mod n if and only if
> $$\gcd(a,n) = 1.$$

There is a systematic method based on the Euclidean algorithm for finding the inverse of an integer mod n that works well even for large values of n. However, because we are now interested only in finding the inverses of integers mod 26, we will simply find them through trial and error. We first note that we need only look for inverses of those integers a between 0 and 25 for which $\gcd(a,26) = 1$—in other words, only the numbers that are not divisible by 2 or 13. We have already noted that $\overline{3} \bmod 26 = 9$, and it follows that $\overline{9} \bmod 26 = 3$. We also see that 1 is its own inverse mod 26. Next, let's try to find $\overline{5} \bmod 26$. We try the numbers between 0 and 25 that are not divisible by 2 or 13, except 1, 3, and 9 for which we already know the inverses, until we find that

$$5 \cdot 21 \equiv 105 \equiv 1 \pmod{26}.$$

Table 8.10	Inverses mod 26
a	$\overline{a} \bmod 26$
1	1
3	9
5	21
7	15
9	3
11	19
15	7
17	23
19	11
21	5
23	17
25	25

So 5 and 21 are inverses of each other mod 26. Now, to find $\overline{7} \bmod 26$, we try the numbers between 0 and 25 that are not divisible by 2 or 13, except 1, 3, 5, 9, and 21, until we find that $\overline{7} \bmod 26 = 15$. Continuing in this way, we find all of the inverses mod 26 as given in Table 8.10.

Looking back now at the idea of using the formula

$$C = (aP + b) \bmod 26$$

to encode a message, recall that in order to decode the message we needed to solve the equation

$$aP \equiv C - b \pmod{26}.$$

Multiplying both sides of this equation by $\overline{a} \bmod 26$ gives

$$\overline{a}aP \equiv \overline{a}(C - b) \pmod{26},$$
$$P \equiv \overline{a}(C - b) \pmod{26}.$$

This formula for P works whenever a has an inverse mod 26. Therefore, we have a way of encoding that works. Codes of the form $C = (aP + b) \bmod 26$, with a and b both integers between 0 and 25, and $\gcd(a, 26) = 1$, are called **affine ciphers**. Next, we summarize what we have discovered about affine ciphers.

Affine Ciphers

If a and b are both integers between 0 and 25, and $\gcd(a, 26) = 1$, then an affine cipher is given by the enciphering formula

$$C = (aP + b) \bmod 26.$$

It can be deciphered using the formula

$$P = (\overline{a}(C - b)) \bmod 26.$$

Example 3   *Encoding with an Affine Cipher*

Encipher the following message by breaking it into five-letter blocks and using the affine cipher $C = (5P + 12) \bmod 26$:

THE TIME IS NOW.

Solution: Breaking it into five-letter blocks, we have

THETI MEISN OW.

Changing the letters to their numerical equivalents gives

19 7 4 19 8   12 4 8 18 13   14 22.

Now we are ready to use the transformation $C = (5P + 12) \bmod 26$. Transforming the first number, 19, into ciphertext, we get

$$C \equiv 5 \cdot 19 + 12 \equiv 107 \equiv 3 \ (\bmod\ 26),$$

so $C = 3$. The second letter, 7, is transformed to 21 because

$$C \equiv 5 \cdot 7 + 12 \equiv 47 \equiv 21 \ (\bmod\ 26).$$

Similarly, we transform the remaining numbers and arrive at the ciphertext numbers

3 21 6 3 0   20 6 0 24 25   4 18,

and transforming these into their letter equivalents gives the ciphertext message

DVGDA UGAYZ ES.

Example 4   *Decoding with an Affine Cipher*

Decipher the following message, which was enciphered using the affine cipher $C = (17P + 4) \bmod 26$:

PTUMI RPHEM PKYIS EW.

BRANCHING OUT 8.5

MECHANICAL CIPHER DEVICES: FROM ANCIENT GREECE TO ENIGMA TO SECRET DECODER RINGS

(a) A scytale.

Mechanical devices for encrypting messages go back 2500 years to Sparta and the scytale shown in figure (a). The scytale consists of a rod of a specific diameter. To encode a message, a ribbon is wrapped around the scytale and a message is written horizontally along the rod, usually along a single row. Other rows of nonsense letters are also written. Once the message and other letters are written, the ribbon is removed and delivered. To decode the message, the recipient rewraps the message around a rod of the same diameter as the one used for encoding. Such a system is not very secure because of the regular appearance of meaningful letters, for instance every sixth letter. Also, the code can be broken easily by anyone who obtains the correct rod, which is the complete encoding and decoding device. With more advanced and secure mechanical devices, codes cannot be broken immediately by someone who obtains the device because an additional key, which corresponds to the way the device is set up for a particular message, is needed.

A device that is used for encryption and decryption of simple ciphers such as the Caesar cipher is the cipher disk, an example of which is shown in figure (b). The cipher disk was invented by the fifteenth-century Italian Leon Battista Alberti. To use the disk, the inside wheel is set to a specific rotation. The plaintext letters are then on the outside with the corresponding ciphertext letters inside. The Captain Midnight secret decoder ring from the mid-twentieth century was the children's version of this cipher disk. These devices make it easy to encode and decode quickly and minimize errors.

(c) A replica of the Jefferson cipher wheel.

In the 1790s, Thomas Jefferson invented a wheel cipher shown in figure (c). This cipher consisted of several different rotating disks, each containing the 26 letters of the alphabet in some scrambled order. To

(b) A British cipher disk (circa 1880).

send a message the sender simply lines up the message along any one row and writes down any of the other 25 rows. This process is repeated until the full message is encoded. To decode the message, a recipient, who has an identical wheel cipher with disks placed in the same order as the sender, just lines up the encoded message along some row. Then the encoded message will appear on some other row. This scheme provided a cipher that was much more difficult to break than the simple substitution schemes. This system was rediscovered and used by the U.S. Army in World War I and later by the U.S. Navy.

In 1817, Colonel Decius Wadsworth invented a geared cipher disk, which was manually rotated as each letter was enciphered, thus changing the key with each successive letter. In 1867, Charles Wheatstone invented a very similar, though slightly less secure, device. These manual devices were the forerunners of machines with rotors that automatically changed the key with each letter. The patent application for a rotor-based machine filed in the Netherlands by Hugo Alexander Koch in 1919 led to the development of the infamous German Enigma machine from World War II shown in figure (d). The interactions between three rotational rotors, a reflecting rotor, and a plugboard that permuted the text before it was enciphered, along with the enormous number of possible keys, made the Enigma largely unassailable to attack by hand. However, during the war, with the help of the Colossus, according to some the world's first electronic computer, and a model of the Enigma built by Polish mathematicians, the British developed methods for discovering Enigma's keys from encoded messages.

Between spies, traitors, and losses on the battlefield, it is not unusual for mechanical devices to fall into the hands of the enemy. An exception was the Japanese PURPLE machine from World War II. Although American cryptanalysts were eventually able to break the code and deduce the workings of the machine, an intact PURPLE machine was never captured. Ironically, just before the bombing of Pearl Harbor, when Japanese security precautions were at their highest and when U. S. codebreakers were able to partially decipher some Japanese messages only through painstaking work, the Japanese embassy in Washington allowed their American custodian to dust and clean the room containing several PURPLE machines.

(e) A modern secret decoder ring.

We conclude this excursion by noting that the secret decoder rings of the authors' youths have evolved into reality. Offered for sale by Dallas Semiconductor and Jostens Inc., the ring pictured in figure (e) is to be used along with a touch-memory probe mounted on a computer to decode encrypted data files without having to type in a lengthy password or key.

(d) The German Enigma machine.

Solution: Converting the ciphertext letters to their numerical equivalents, we have

$$15\ 19\ 20\ 12\ 8 \qquad 17\ 15\ 7\ 4\ 12 \qquad 15\ 10\ 24\ 8\ 18 \qquad 4\ 22.$$

We know that the deciphering formula is $P = \left(\overline{17}(C - 4)\right) \bmod 26$. Because $\overline{17} \bmod 26 = 23$, we can write the deciphering formula in the form

$$P = (23(C - 4)) \bmod 26.$$

The first ciphertext number, 15, is transformed to 19 because

$$P \equiv 23(15 - 4) \equiv 23 \cdot 11 \equiv 253 \equiv 19 \ (\bmod\ 26).$$

The second ciphertext number, 19, is transformed to 7 because

$$P \equiv 23(19 - 4) \equiv 23 \cdot 15 \equiv 345 \equiv 7 \ (\bmod\ 26).$$

Similarly, we transform the remaining ciphertext numbers, and we find that the entire message becomes

$$19\ 7\ 4\ 2\ 14 \qquad 13\ 19\ 17\ 0\ 2 \qquad 19\ 8\ 18\ 14\ 10 \qquad 0\ 24.$$

Converting to the letter equivalents, we get

THECO NTRAC TISOK AY,

so the original message was

THE CONTRACT IS OKAY.

∎

The affine ciphers we have looked at are not the only possible ciphers that arise from rearranging the letters of the alphabet. In fact, because only 12 of the numbers between 0 and 25 have inverses mod 26, there are only $26 \cdot 12 - 1 = 311$ such ciphers (not counting the "cipher" $C = P \bmod 26$). However, there are $26! \approx 4 \cdot 10^{26}$ ways to permute the letters of the alphabet. Even with this very large number of ways of rearranging the letters, ciphers that are based on this simple idea are fairly easy to decipher if the message is reasonably long. The reason for this is that the approximate frequency with which each letter occurs in ordinary English text is well known. One such list of frequencies is given in Table 8.11.

Table 8.11	**Frequency (in percent) of Occurrence of Letters in English Text**																								
A	B	C	D	E	F	G	H	I	J	K	L	M	N	O	P	Q	R	S	T	U	V	W	X	Y	Z
8	2	3	4	13	2	2	6	7	<1	1	4	2	7	8	2	<1	6	6	9	3	1	2	<1	2	<1

Source: H. Becker and F. Piper, *Cipher Systems: The Protection of Communications.* New York: Wiley, 1982.

This information can be used to attempt to decipher a message that was encoded by a rearrangement of the letters of the alphabet. For instance, if the most frequent letter occurring in a ciphertext message of at least moderate length is R, then there is a good possibility that the corresponding plaintext letter is E. Similarly, letters that occur very infrequently in the ciphertext probably correspond to one of the uncommon letters in Table 8.11. In addition, certain letter combinations can be helpful in decoding. For instance, if the letter T in ciphertext is always followed by an N (possibly across two blocks), then it may be that the corresponding plaintext letters are Q and U. Similarly, if the three-letter sequence YCP occurs often in a ciphertext message, then the corresponding plaintext letters may very well be THE or AND or some other common three-letter sequence. Using these kinds of clues along with the letter frequency data makes codes based on rearranging the alphabet too easy to break if the message is long. In the next section, we will look at ciphers that encode blocks of letters. These kinds of codes can be difficult or nearly impossible to break.

EXERCISES FOR SECTION 8.6

1. Encipher the message

   ANY TIME IS OKAY

   by breaking it into five-letter blocks and using the Caesar cipher.

2. Encipher the message

   THIS IS THE WAY TO GO

   by breaking it into five-letter blocks and using the Caesar cipher.

3. Decipher the following message, which was enciphered with the Caesar cipher:

   BRXJR WLWUL JKW.

4. Decipher the following message, which was enciphered with the Caesar cipher:

   DZDBZ HJR.

In Exercises 5–8, encode the given message by breaking it into five-letter blocks and using the given affine cipher.

5. YOU SHOULD STAY, $C = (5P + 14) \bmod 26$

6. IT IS THE RIGHT SIZE, $C = (11P + 6) \bmod 26$

7. HERE IS THE SECRET, $C = (19P + 7) \bmod 26$

8. NOW WE KNOW, $C = (9P + 10) \bmod 26$

In Exercises 9–12, decode the given message, which was encoded using the given affine cipher.

9. RPQCU TQGQU D, $C = (7P + 16) \bmod 26$

10. FMZMI GCMOD ZGSTN M, $C = (15P + 4) \bmod 26$

11. ATWEE OLWFE VLDX, $C = (23P + 9) \bmod 26$

12. VLPXV FULF, $C = (3P + 11) \bmod 26$

13. Suppose that the formula $C = (6P + 5) \bmod 26$ is used for encoding.

    (a) Show that the letter G is encoded as P.

    (b) Show that the letter T is also encoded as P.

    (c) Why will the formula $C = (6P + 5) \bmod 26$ not work as an affine cipher?

14. Suppose you need a cipher for encoding numbers rather than words (for instance encoding Social Security numbers). Explain how you could adapt the affine cipher we used for letters to get a cipher for encoding numbers.

15. Suppose you need a cipher for encoding messages that include numbers, words, and blank spaces. Explain how you could adapt the affine cipher we used for letters to get a cipher for doing this.

8.7   ADVANCED ENCRYPTION METHODS

We have seen that ciphers that only rearrange the letters of the alphabet can be broken by use of letter frequency or other clues. To make letter frequency data unusable, we need a code that does more than rearrange individual letters. We will now look at ciphers that encode blocks of letters or blocks of digits coming from letters.

In the first cipher we will look at in this section, the Hill cipher, blocks of letters of a certain length in plaintext are replaced by blocks of letters of the same length in ciphertext. For instance, for a particular code the three-letter block THE in plaintext may be replaced by YWP in ciphertext, whereas the block TIO may be replaced by NCJ. Notice that although the T was encoded as Y in the first case, it was encoded as N in the second case. In block codes of this type, each block is always encoded the same way, but each individual letter may be encoded various ways. This prohibits a person who is trying to decode the message from using single-letter frequency data as an aid.

With the second kind of cipher we will look at in this section, the RSA public key system, instead of encoding blocks of letters of a particular length, the code works on blocks of digits of the same length. However, the basic idea is the same as with blocks of letters, and again the code is not simply a rearrangement of the alphabet. As we will see later, the RSA public key system also has other features that make it a particularly attractive method.

The Hill Cipher

The **Hill cipher** was presented by Lester S. Hill in a paper published in 1931. It works for blocks of letters of any size, but for simplicity we will first look at how it works in the case of two-letter blocks.

With the Hill cipher that we will discuss, each block of two plaintext letters P_1P_2 is encoded as a block of two ciphertext letters C_1C_2, which are given by two equations of the form

$$C_1 = (aP_1 + bP_2) \bmod 26,$$
$$C_2 = (cP_1 + dP_2) \bmod 26,$$

where a, b, c, and d are integers between 0 and 25. We will see later that in order for this type of code to work, we must also require that $\gcd(ad - bc, 26) = 1$. Before seeing why this is the case and finding the deciphering formula, let's look at an example of encoding using a Hill cipher for two-letter blocks.

Example 1   *Encoding with the Hill Cipher*

Use the Hill cipher for two-letter blocks given by

$$C_1 = (12P_1 + 3P_2) \bmod 26,$$
$$C_2 = (5P_1 + 6P_2) \bmod 26$$

to encipher the message

CALL TODAY.

Solution: First we break the message into the two-letter blocks and add a "dummy" letter at the end so that the last block is two letters long. Using the dummy letter T (any letter will do) we have

CA LL TO DA YT.

Next, the letters are translated into their numerical equivalents and we get

2 0   11 11   19 14   3 0   24 19.

Now, we use the Hill cipher formula to encode the first block 2 0 with $P_1 = 2$ and $P_2 = 0$ and we get

$$C_1 \equiv 12 \cdot 2 + 3 \cdot 0 \equiv 24 \ (\mathrm{mod}\ 26),$$
$$C_2 \equiv 5 \cdot 2 + 6 \cdot 0 \equiv 10 \ (\mathrm{mod}\ 26).$$

Therefore, the block 2 0 will be encoded as the block 24 10. Similarly, for the second block 11 11, we set $P_1 = 11$ and $P_2 = 11$, and we have

$$C_1 \equiv 12 \cdot 11 + 3 \cdot 11 \equiv 165 \equiv 9 \ (\mathrm{mod}\ 26),$$
$$C_2 \equiv 5 \cdot 11 + 6 \cdot 11 \equiv 121 \equiv 17 \ (\mathrm{mod}\ 26).$$

Thus, the block 11 11 is encoded as 9 17. Continuing in this way, we can encode the remaining three blocks in the plaintext message, and we get the ciphertext message in numerical form:

24 10   9 17   10 23   10 15   7 0.

Translating back to letters, we get the ciphertext message

YK JR KX KP HA.

In an affine cipher, the letter A is always encoded as the same letter. Notice that, under the Hill cipher we just considered, the block CA was encoded as YK, whereas the block DA was encoded as KP. Similarly, the block LL was encoded as JR. This is to be expected because we are not just rearranging the alphabet. Any particular letter may be replaced by any other letter. However, any particular two-letter block will always be encoded the same way.

To see how to decode a message that was encoded with a Hill cipher for two-letter blocks, we start with the encoding formulas

$$C_1 = (aP_1 + bP_2) \ \mathrm{mod}\ 26,$$
$$C_2 = (cP_1 + dP_2) \ \mathrm{mod}\ 26,$$

and we solve for P_1 and P_2 using the same ideas we would use to solve for two unknowns in two equations of this form in regular arithmetic. In order to eliminate the variable P_2, we begin by multiplying the first equation by d on both sides and the second equation by b to get

$$dC_1 \equiv adP_1 + bdP_2 \ (\mathrm{mod}\ 26),$$
$$bC_2 \equiv bcP_1 + bdP_2 \ (\mathrm{mod}\ 26).$$

Now subtracting the second equation from the first, it follows that

$$dC_1 - bC_2 \equiv (ad - bc)P_1 \ (\mathrm{mod}\ 26).$$

Recall that we required that $\gcd(ad - bc, 26) = 1$, and we can now see why we needed to make this restriction. To solve for P_1 in the last equation, we need to multiply both sides by $\overline{ad - bc}$ mod 26. Doing this, we get

$$\left(\overline{ad - bc}\right)(dC_1 - bC_2) \equiv \left(\overline{ad - bc}\right)(ad - bc)P_1 \,(\mathrm{mod}\ 26),$$

and therefore

$$P_1 = \left(\left(\overline{ad - bc}\right)(dC_1 - bC_2)\right)\mathrm{mod}\ 26,$$

or

$$P_1 = \left(\left(\overline{ad - bc}\right)dC_1 - \left(\overline{ad - bc}\right)bC_2\right)\mathrm{mod}\ 26.$$

Letting $A = \left(\left(\overline{ad - bc}\right)d\right)\mathrm{mod}\ 26$ and $B = -\left(\left(\overline{ad - bc}\right)b\right)\mathrm{mod}\ 26$, we have

$$P_1 = (AC_1 + BC_2)\,\mathrm{mod}\ 26.$$

We leave it as an exercise to show that in a similar way we can find that

$$P_2 = (CC_1 + DC_2)\,\mathrm{mod}\ 26,$$

where $C = -\left(\left(\overline{ad - bc}\right)c\right)\mathrm{mod}\ 26$ and $D = \left(\left(\overline{ad - bc}\right)a\right)\mathrm{mod}\ 26$. The formulas for encoding and decoding using the Hill cipher for two-letter blocks are summarized next.

The Hill Cipher for Two-Letter Blocks

If a, b, c, and d are integers between 0 and 25 and $\gcd(ad - bc, 26) = 1$, then the Hill cipher for two-letter blocks is given by the enciphering formulas

$$C_1 = (aP_1 + bP_2)\,\mathrm{mod}\ 26,$$
$$C_2 = (cP_1 + dP_2)\,\mathrm{mod}\ 26.$$

It can be deciphered using the formulas

$$P_1 = (AC_1 + BC_2)\,\mathrm{mod}\ 26,$$
$$P_2 = (CC_1 + DC_2)\,\mathrm{mod}\ 26,$$

where

$$A = \left(\left(\overline{ad - bc}\right)d\right)\mathrm{mod}\ 26,$$
$$B = -\left(\left(\overline{ad - bc}\right)b\right)\mathrm{mod}\ 26,$$
$$C = -\left(\left(\overline{ad - bc}\right)c\right)\mathrm{mod}\ 26,$$
$$D = \left(\left(\overline{ad - bc}\right)a\right)\mathrm{mod}\ 26.$$

Notice that the formulas for encoding and decoding with the Hill cipher are of the same form. In the following example, we look at the decoding process for the Hill cipher.

Technology Tip

A spreadsheet or mathematical calculator is a great aide in encoding and decoding affine ciphers and the Hill cipher. It is especially easy with a MOD function or key, but it is also possible to build up your own MOD function with an integer part function or key.

Example 2   *Decoding with the Hill Cipher*

Decipher the following message:

MZ EL BW MQ YM EZ,

which was enciphered using the Hill cipher

$$C_1 = (13P_1 + 2P_2) \bmod 26,$$
$$C_2 = (9P_1 + 1P_2) \bmod 26.$$

Solution: Converting the ciphertext letters into their numerical equivalents, we have

12 25   4 11   1 22   12 16   24 12   4 25.

To find the deciphering formulas, we first compute

$$ad - bc = 13 \cdot 1 - 2 \cdot 9 = -5 = 21 \ (\text{mod } 26),$$

and then we see that

$$\overline{ad - bc} \bmod 26 = \overline{21} \bmod 26 = 5.$$

In our deciphering formula we have

$$A \equiv 5 \cdot 1 \equiv 5 \ (\text{mod } 26),$$
$$B \equiv -5 \cdot 2 \equiv -10 \equiv 16 \ (\text{mod } 26),$$
$$C \equiv -5 \cdot 9 \equiv -45 \equiv 7 \ (\text{mod } 26),$$
$$D \equiv 5 \cdot 13 \equiv 65 \equiv 13 \ (\text{mod } 26),$$

and therefore

$$P_1 = (5C_1 + 16C_2) \bmod 26,$$
$$P_2 = (7C_1 + 13C_2) \bmod 26.$$

Translating the first block, 12 25, we get

$$P_1 \equiv 5 \cdot 12 + 16 \cdot 25 \equiv 460 \equiv 18 \ (\text{mod } 26),$$
$$P_2 \equiv 7 \cdot 12 + 13 \cdot 25 \equiv 409 \equiv 19 \ (\text{mod } 26),$$

and therefore the first block translates into the block 18 19. Similarly, we translate each of the other two-number blocks and the entire message is translated to

18 19   14 15   19 7   4 6   0 12   4 15.

Now changing the numbers to their numerical equivalents, we get

ST OP TH EG AM EP.

The last letter, P, appears to be a dummy letter, so the original message was

STOP THE GAME.

■

The Hill cipher for two-letter blocks is easy to use and harder to break than the affine codes we looked at in the last section. However, it is far from unbreakable. There are only $26^2 = 676$ different two-letter blocks, and the frequencies of these two-letter blocks in the English language are well known. For instance, according to some studies the most common two-letter block in the English language is TH, and the second most common is HE. This kind of information may provide enough clues to enable a third party to decode a private message. The Hill cipher is much more secure if it is used with larger blocks. The idea we saw with two-letter blocks generalizes easily to larger blocks. For instance, with the Hill cipher for three-letter blocks, each block of three plaintext letters $P_1P_2P_3$ is encoded as a block of three ciphertext letters $C_1C_2C_3$, which are given by the three equations

$$C_1 = (aP_1 + bP_2 + cP_3) \bmod 26,$$
$$C_2 = (dP_1 + eP_2 + fP_3) \bmod 26,$$
$$C_3 = (gP_1 + hP_2 + iP_3) \bmod 26,$$

where a through i are integers between 0 and 25 that satisfy a condition that will make decoding possible. The Hill cipher for blocks of other sizes is similar. In general, the larger the block size, the harder the code is to break. For instance, if we use eight-letter blocks, then because there are $26^8 \approx 200$ billion different eight-letter blocks, trying to decode a message based on the frequency of eight-letter blocks is not really practical. Therefore, the Hill cipher seems to be a good encryption method if used with large blocks. However, if a small portion of the original text is also known, the entire cipher is easily cracked. Because of this major drawback, the Hill cipher is rarely used.

The RSA Public Key System

With the encryption methods we have looked at so far, if the enciphering method is known, then the deciphering method is easy to compute, so the enciphering method must remain private. In the age of electronic communications, when a person may receive private communications from various sources, it would be more convenient if the person could publicly state the encryption method all correspondents should use when sending messages to him or her. This way, all of the person's messages would arrive encoded in the same way. However, you might wonder how an encryption method could be made public without giving away the decoding method at the same time. Remarkably, such encoding systems are possible and are called public key systems.

In 1975, Ron Rivest, Adi Shamir, and Len Adelman invented a public key system, known as the **RSA public key system**. With the RSA system, each person who would like to receive encoded messages chooses a pair of very large prime numbers p and q and a positive integer r such that $\gcd(r,(p-1)(q-1)) = 1$. The person then calculates the product $n = pq$ and publicly announces that his or her RSA **public key address** is (r,n). To send a message to this individual, the sender first converts all of the letters into their numerical equivalents, placing a zero before any one-digit number. For instance, the letter A would be translated as 00 and the letter B would become 01. The sender then breaks the message into blocks of some agreed upon number of digits (not letters). Any size blocks will do as long as they are of size at most one less than the number of digits in n. Then each plaintext block P is encoded as a ciphertext number C by using the formula

$$C = P^r \bmod n.$$

Note that C can be any number between 0 and $n - 1$, so it does not necessarily have the same number of digits as P. The magic of this system, as we will see shortly, is that although anyone who knows the address (r, n) can send messages in this way, only a person who knows how to factor the number n as $n = pq$ will be able to decode the message. This might not seem difficult, but if p and q are fairly large primes, factoring n may take much more than a lifetime to carry out, even with the help of today's computer technology. Surprisingly, it is relatively easy to find very large primes, so it is easy to come up with public key addresses. Before looking at how we go about decoding a message that has been encoded with this method, let's look at an example of encoding.

Example 3   *Encoding with the RSA Public Key System*

Encode the message

THERE IS A DELAY

using three-digit blocks and the RSA public key system. Assume the message is being sent to a person with public key address (3, 2497).

Solution: We note first that the person encoding the message does not need to know how to factor 2497, although in this case it is fairly easy to see that $2497 = 227 \cdot 11$ is the product of the prime numbers 227 and 11. Also, notice that

$$\gcd(3, (227 - 1)(11 - 1)) = \gcd(3, 2260) = 1,$$

so the address (3, 2497) is of the correct form.

To encode the message, we first translate all of the letters into two-digit numbers and get

19 07 04 17 04   08 18   00   03 04 11 00 24.

Regrouping the digits into three-digit blocks, we get

190   704   170   408   180   003   041   100   240.

We added a dummy digit 0 to fill out the last block. Using the encryption formula

$$C = P^3 \bmod 2497$$

to encode the first block $P = 190$, we find that

$$C \equiv 190^3 \equiv 6{,}859{,}000 \equiv 2238 \pmod{2497},$$

so the block 190 is encoded as 2238. The second block, $P = 704$, is encoded as 0363 because

$$C \equiv 704^3 \equiv 348{,}913{,}664 \equiv 363 \pmod{2497}.$$

Notice that we added the digit 0 to the front of 363 to make it a four-digit block. We want all of the encoded blocks to be four digits long (equal to the number of digits in the modulus 2497) so that the message is more uniform and can be sent without including spaces if desired. To ensure this, we may sometimes have to attach zeros to the fronts of the encoded blocks. Continuing in this way, we encode the remaining seven blocks and the entire message is encoded as

2238   0363   1401   1409   1505   0027   1502   1200   0608.

This is the message that is sent to the person with public key address (3, 2497). Notice that we do not convert the encoded message into letters with this encryption system. In fact, we really do not have the option of easily converting to letters at this stage because the two-digit blocks are not all between 0 and 25.

◼

To decode a message that has been encoded using the public key address (r,n), with $n = pq$, the recipient uses the simple formula

$$P = C^k \bmod n,$$

where

$$k = \bar{r} \bmod ((p - 1)(q - 1)),$$

the inverse of r in arithmetic mod $((p - 1)(q - 1))$. The proof that this deciphering method works is based on a famous theorem from number theory known as Euler's Theorem. Finding the value of k can be accomplished through a simple algorithm as long as the value $(p - 1)(q - 1)$ is known. However, recall that the only information available to the public is the address (r,n), and finding the value $(p - 1)(q - 1)$ requires knowing the factorization $n = pq$. Because this factorization is only known by the recipient and factoring n is very difficult if p and q are very large, the encoded message can only be deciphered by the recipient. In practice, the recipient may only know the value k, having purchased the public key address and the decoding key k from a company that knows the factorization $n = pq$. In the next example, we see how this decoding procedure works.

Example 4   *Decoding with the RSA Public Key System*

The following message was encoded using three-digit blocks and the public key address (1373, 1817):

0684   1098   0973   0803   0531   1313.

The recipient of the message knows that $1817 = 79 \cdot 23$ and

$$\overline{1373} \bmod ((79 - 1)(23 - 1)) = \overline{1373} \bmod 1716 = 5.$$

Decode the message.

Solution: We use the decoding formula

$$P = C^5 \bmod 1817.$$

Note that the plaintext blocks were three digits long, so when using the decoding formula we have to write the resulting P in the form of a three-digit block, possibly by adding zeros to the front of P. For the first block of digits, 0684, we need to evaluate $P = 684^5 \bmod 1817$. Because 684^5 is a fairly large number that a calculator may not be able to compute exactly, we break down the computation into smaller powers of 684 and get

$$P \equiv 684^5 \equiv 684^2 \cdot 684^2 \cdot 684 \equiv (467,856)(467,856)(684)$$
$$\equiv (887)(887)(684) \equiv 538,149,996 \equiv 21 \ (\bmod\ 1817),$$

so the first three-digit block in plaintext is 021. For the second block of digits, 1098, we have

$$P \equiv 1098^5 \equiv 1098^2 \cdot 1098^2 \cdot 1098$$
$$\equiv (1{,}205{,}604)(1{,}205{,}604)(1098)$$
$$\equiv (933)(933)(1098) \equiv 955{,}796{,}922 \equiv 412 \ (\mathrm{mod}\ 1817),$$

so the second three-digit block in plaintext is 412. Similarly, the remaining blocks are translated and we get the six three-digit plaintext blocks

$$021 \qquad 412 \qquad 040 \qquad 704 \qquad 170 \qquad 400.$$

Breaking them into the two-digit blocks representing the plaintext letters, we have

$$02 \quad 14 \quad 12 \quad 04 \quad 07 \quad 04 \quad 17 \quad 04 \quad 00,$$

and translating these numbers to their letter equivalents we have

C O M E H E R E A.

The letter A appears to have come from dummy digits, so we can see that the original plaintext message was

COME HERE.

The formulas for encoding and decoding using the RSA public key system are summarized next.

The RSA Public Key System

If p and q are prime numbers, $n = pq$, and r is a positive integer with $\gcd(r,(p-1)(q-1)) = 1$, then a message to be sent to the RSA public key address (r,n) is encoded as follows: Each plaintext block P of digits is encoded as a ciphertext block C by using the formula

$$C = P^r \bmod n.$$

The recipient of the message can decode each ciphertext block C using the formula

$$P = C^k \bmod n,$$

where

$$k = \bar{r} \bmod ((p-1)(q-1)).$$

The blocks of plaintext digits must all be of the same designated length, which must be less than the number of digits of n. The blocks of ciphertext digits must be of length equal to the number of digits of n.

As we noted earlier, the security of the RSA public key system is based on the fact that, although it is fairly easy to find large primes p and q, it is difficult to factor the number $n = pq$. For example, it only takes a few minutes of com-

BRANCHING OUT 8.6

CRACKING RSA AND DES: INDIVIDUAL PRIVACY AND NATIONAL SECURITY

The RSA public key system is in widespread use today. It is licensed by over 150 companies and is part of Microsoft, Apple, Sun, and Novell operating systems. It is estimated that there have been 80 million installations of the RSA encryption engine. Although the RSA public key system can be used by itself, it is more common to couple RSA with another encryption system such as one called DES (for Data Encryption Standard). Developed at IBM, DES was established as an official standard by the United States in 1977. With DES, messages may be encoded and decoded over 100 times as fast as RSA, which takes perhaps 10 to 20 seconds per printed page, a significant difference for long messages or large databases such as financial accounts. However, with DES, the same key is used for both encryption and decryption, so keys must be transmitted securely between the parties. This is where RSA comes in. Because RSA is a public key cipher, the sender can use it to transmit the DES key to the recipient. The message itself is encoded with DES using that key. After using the RSA system to decode the key, the recipient is able to decode the DES-encrypted message.

How secure is this system? The only known way to break RSA is to factor the modulus. (There have been theoretical reports on using differences in the time the computer requires to process different pieces of a message or using a specific computer's computational or transmission errors to break the system; these had not been implemented in any way when this text was written.) Derek Atkins of MIT, Michael Graff of Ohio State University, Arjen Lenstra of MIT, and Paul Leyland of Oxford University coordinated a team of volunteers using (mostly) personal computers to crack a 129-digit key over eight months in 1993 and 1994. It was estimated that factoring a 150-digit number can now be done for under $1 million in less than eight months. Due to advances in hardware and software, the cost and time required rapidly decrease. In 1997, RSA Laboratories recommended 232-digit keys for personal use, 309-digit keys

puter time to find primes p and q with 100 digits each. However, in this case, the product $n = pq$ will have about 200 digits, and factoring a number of this size that has no small factors might take as much as a billion years of computer time. As computers become faster and better algorithms are developed, the time required to factor is expected to decrease significantly, so larger values of p and q may become necessary. However, it is never expected to become especially easy to factor such numbers, so the RSA public key system will probably continue to be a secure enciphering method.

for corporate use, and 617-digit keys for the most important uses, expecting these likely to be secure until at least the year 2004. Even the largest of these keys takes only a few seconds to generate, with a similar time to encrypt and decrypt the DES key. Thus, the RSA part of the system is fast and apparently as secure as necessary.

What about a direct attack on the DES-encrypted message itself? In January 1997, Ian Goldberg, a graduate student at the University of California at Berkeley, was able to crack a system based on a 13-digit key in under four hours on a network of 250 workstations, winning a $1000 prize for his efforts. A month later, a Swiss team was able to crack the next level, a system based on a 15-digit key, in less than two weeks. Current government regulations limit exports of DES encryption technology to keys that are at most 17-digits, the next level. The same methods would take perhaps a dozen years to crack a 17-digit key, so systems based on such keys seem secure for now unless long-term privacy is needed.

The State Department and the National Security Agency regulate the export of cryptographic software and hardware by controlling the lengths of the keys. In the RSA cryptosystem, the limit is a key of roughly 150 digits. As we just saw, this is below the level recommended for even personal security, so the regulations are a subject of heated debate. One argument against such regulations is that both RSA and DES are widely available abroad in foreign products, hence the main effect is to harm U.S. companies who develop and sell such technology. This argument has proved unpersuasive to lawmakers and so the restrictions remain.

Another advantage of the RSA public key system is that if a third party obtains a message in both its plaintext and ciphertext versions, he or she would still not know how the decoding is done. Therefore, this third party would be unable to decipher other messages encoded the same way. This would not be the case with the affine cipher or Hill cipher. With these ciphers, having access to even a short message in both its plaintext and ciphertext versions would usually give a third party who knew the general method of encoding enough information to find the exact encoding and decoding formulas.

A CLOSER LOOK 8.3

RAISING A NUMBER TO A HIGH POWER IN MODULAR ARITHMETIC

When encoding and decoding in the RSA cryptosystem, we must raise a number to an exponent in modular arithmetic. The exponents in our examples were small, but in actual practice these exponents may be quite large. To simplify the calculations, successively squaring is helpful. To illustrate this idea, let's calculate 35^{27} mod 57. First we successively square 35 modulo 57.

$$35^2 \equiv 1225 \equiv 28 \ (\text{mod } 57),$$
$$35^4 \equiv 35^2 \cdot 35^2 \equiv 28 \cdot 28 \equiv 784 \equiv 43 \ (\text{mod } 57),$$
$$35^8 \equiv 35^4 \cdot 35^4 \equiv 43 \cdot 43 \equiv 1849 \equiv 25 \ (\text{mod } 57),$$
$$35^{16} \equiv 35^8 \cdot 35^8 \equiv 25 \cdot 25 \equiv 625 \equiv 55 \ (\text{mod } 57).$$

Now using the fact that $27 = 16 + 8 + 2 + 1$, we have

$$35^{27} \equiv 35^{16+8+2+1} \equiv 35^{16} \cdot 35^8 \cdot 35^2 \cdot 35$$
$$\equiv 55 \cdot 25 \cdot 28 \cdot 35 \equiv 1,347,500 \equiv 20 \ (\text{mod } 57).$$

Because any positive integer can be written as a sum of powers of two (this is the base 2 expansion), we can apply this technique in general. Performing the calculation in this way keeps down the total number of multiplications to be performed. For instance, if we had computed 35^{27} by successively multiplying by 35, we would have had to perform 26 multiplications. Using the successive squaring method, we performed only seven multiplications. For larger exponents the savings in multiplication are even more notable. For instance, an exponent of one trillion requires only 51 multiplications using the successive squaring method.

*E*XERCISES FOR SECTION 8.7

In Exercises 1–4, encode the given message using the given Hill cipher for two-letter blocks. Use the letter X as a dummy letter in the last block if necessary.

1. A STORM IS COMING; $C_1 = (4P_1 + 3P_2) \bmod 26$,
$$C_2 = (15P_1 + 5P_2) \bmod 26$$

2. GOOD MORNING; $C_1 = (11P_1 + 4P_2) \bmod 26$,
$$C_2 = (3P_1 + 9P_2) \bmod 26$$

3. GO FOR IT; $C_1 = (5P_1 + 1P_2) \bmod 26$,
$$C_2 = (6P_1 + 13P_2) \bmod 26$$

4. PLEASE CALL; $C_1 = (21P_1 + 8P_2) \bmod 26$,
$$C_2 = (1P_1 + 3P_2) \bmod 26$$

In Exercises 5–8, decode the given message, which was encoded using the given Hill cipher for two-letter blocks with some dummy letter in the last block when needed.

5. DF ZH TZ PJ UI; $C_1 = (5P_1 + 2P_2) \bmod 26$,
$$C_2 = (3P_1 + 7P_2) \bmod 26$$

6. VW ED BC QJ; $C_1 = (1P_1 + 17P_2) \bmod 26$,
$$C_2 = (4P_1 + 9P_2) \bmod 26$$

7. UB NC SP SB VJ; $C_1 = (6P_1 + 7P_2) \bmod 26$,
$$C_2 = (19P_1 + 3P_2) \bmod 26$$

8. AR JB DS GD BI AJ; $C_1 = (16P_1 + 5P_2) \bmod 26$,
$$C_2 = (5P_1 + 8P_2) \bmod 26$$

In Exercises 9–12, encode the given message using three-digit blocks and the RSA public key system where the message is being sent to the person with the given public key address. Use the digit 0 for the dummy digits in the last block if necessary.

9. TELL ME WHY; (3, 3071)

10. SEND MONEY; (3, 4189)

11. I FOUND IT; (5, 7081)

12. GET BACK; (7, 5459)

In Exercises 13–16, decode the given message, which was encoded using the RSA public key system with three-digit blocks and sent to the given address. The digit 0 was used for the dummy digits in the last block if needed.

13. 1472   0012   1099   0944   0038   1098;
Address: (1083, 1711),
where $1711 = 29 \cdot 59$ and $\overline{1083} \bmod ((29 - 1)(59 - 1)) = \overline{1083} \bmod 1624 = 3$

14. 0746   0703   1009   0035   0657   0629   1009;
Address: (1027, 1633),
where $1633 = 23 \cdot 71$ and $\overline{1027} \bmod ((23 - 1)(71 - 1)) = \overline{1027} \bmod 1540 = 3$

15. 3542   1701   1494   2691;
Address: (2423, 4387),
where $4387 = 41 \cdot 107$ and $\overline{2423} \bmod ((41 - 1)(107 - 1)) = \overline{2423} \bmod 4240 = 7$

16. 0998   5408   2751   1719;
Address: (1613, 8249),
where $8249 = 73 \cdot 113$ and $\overline{1613} \bmod ((73 - 1)(113 - 1)) = \overline{1613} \bmod 8064 = 5$

17. Suppose that the formulas
$$C_1 = (5P_1 + 3P_2) \bmod 26,$$
$$C_2 = (2P_1 + 9P_2) \bmod 26$$
are used for encoding.
   (a) Show that the block of letters SH is encoded as HV.
   (b) Show that the block of letters CZ is also encoded as HV.
   (c) Why will the formulas not work as a cipher?

18. A Closer Look 8.3 examines an efficient way to calculate large powers in modular arithmetic. A commonly used exponent for the RSA public key system is the prime $65,537 = 2^{16} + 1$. Explain why an exponent of this form would yield an easier computation than would another exponent of comparable size.

19. How could the RSA public key system be adapted for sending messages that everyone could read but that only one person could send?

20. How could the RSA public key system be adapted so a sender could add a "signature" to a message that allows the receiver to verify that the message came from the actual sender and is not a message from someone else claiming to be the sender?

WRITING EXERCISES

1. Write a letter to the head of your college supporting the use of check digits in student numbers.

2. Discuss some of the practical issues faced by those who create league schedules.

3. Explain important applications of public key ciphers that cannot be accomplished with ciphers that only a recipient who knows the encoding method can decode.

4. Some encryption methods, such as the Hill cipher, may be broken if small sections of plaintext and the corresponding ciphertext are known. Describe how someone attempting to decode a message might gain such knowledge about the encoded message.

5. The U.S. government is considering legislation whereby the government must be given the decoding key for all public key addresses in use. Comment on the positive and negative aspects of such a law.

PROJECTS

1. The Euclidean algorithm, described in the text, is a method for finding the greatest common divisor of two numbers.
   (a) Use the Euclidean algorithm to find $\gcd(1816046, 373678)$.
   (b) Write a spreadsheet or calculator program for computing gcd's using the Euclidean algorithm.

2. Develop your own check digit scheme for a serial number made up of nine digits, the last digit being a check digit. Determine which single-digit errors and errors consisting of the transposition of adjacent digits your scheme detects.

3. For each of the following check digit methods, determine which transposition errors go undetected:
   (a) mod 7 check digit system (the one used on airline tickets),
   (b) banking system check digit method,
   (c) UPC check digit method.

4. Write a spreadsheet or computer program for carrying out the round-robin tournament method described in the text.

5. In some sports such as track, swimming, and golf, there are tri-meets where three teams face each other at the same time.
   (a) With 15 teams, it is possible to schedule 7 rounds of tri-meets so that each team plays in every round and faces each of the other 14 teams exactly once. Create such a schedule.
   (b) Find some basic conditions on the number of teams that must be satisfied in order to have any chance of creating such a schedule.

6. Letter frequency is useful in decoding. Two-letter block frequencies also prove useful, although there are $26 \times 26 = 676$ possible two-letter blocks. Find the two-letter block frequencies on a page of English text out of a book. What patterns appear frequently? Is this what you would expect? How could this information be used to decode a message encoded with the Hill cipher for two-letter blocks?

7. As you would expect, many companies that provide encryption software advertise on the Internet and elsewhere. These advertisements usually describe the system they use, the level of security they provide, the hardware they require, and the cost of the system. Investigate and report on some of these encryption systems.

8. The following passage was encoded with an affine cipher.

GMLAB	VHLMQ	MHZYH	LQHAN	MQMBD	MPCVM	JHHLW
HWBBS	MJWZM	EZMWH	MVMIY	WBHLW	HHLMO	WZMMJ
VAGMV	NOHLM	CZEZM	WHAZG	CHLEM	ZHWCJ	YJWBC
MJWNB	MZCUL	HQHLW	HWSAJ	UHLMQ	MWZMB	CDMBC
NMZHO	WJVHL	MRYZQ	YCHAD	LWRRC	JMQQH	LASWQ
TMDDM	ZQAJ					

(a) Use the frequency distribution of letters in English given in Table 8.11 to decipher the passage.

(b) Find the key to the cipher.

9. Suppose you have intercepted the following message:

K X Z H I A M H G R K L B K G K C K F I O G W P.

Your spy network has discovered that the message was encoded by the Hill cipher with two-letter blocks and that the sender usually begins messages with "Dear." Find the key for decoding the cipher and decode the entire message.

10. In the Hill cipher for three-letter blocks, each block of three plaintext letters $P_1P_2P_3$ is encoded as a block of three ciphertext letters $C_1C_2C_3$, which are given by the three equations

$$C_1 = (aP_1 + bP_2 + cP_3) \bmod 26,$$
$$C_2 = (dP_1 + eP_2 + fP_3) \bmod 26,$$
$$C_3 = (gP_1 + hP_2 + iP_3) \bmod 26,$$

where a through i are integers between 0 and 25 that satisfy a condition that will make decoding possible. Find the decoding formula for this cipher and determine the necessary conditions on the integers a through i.

KEY TERMS

CHAPTER 8 REVIEW EXERCISES

1. Find the prime factorization of 4312.

2. Determine whether or not the given number is prime.
   (a) 437   **(b)** 1553

3. Find the quotient and remainder when the division is performed according to the division algorithm.
   (a) 95 divided by 14   **(b)** −115 divided by 7

4. Evaluate the given greatest common divisor.
   (a) $\gcd(90, 126)$ **(b)** $\gcd(-46, 920)$

In Exercises 5–8, perform the given calculation. In each case, the answer should be a number between 0 and $n - 1$, where n is the modulus.

5. $(7 + 15 + 20) \bmod 9$

6. $(110 - 57) \bmod 25$

7. $(15^{300} + 13^{299}) \bmod 7$

8. $(753 \cdot 124 \cdot 43{,}921 \cdot 100{,}003) \bmod 10$

9. Use casting out nines to determine whether or not 42,098,793,390,115,866,045 is divisible by 9.

10. Use casting out elevens to determine whether or not 9,570,018,702,007 is divisible by 11.

11. Use the test for divisibility by 4 to determine whether or not 47,877,905,008,564,228 is divisible by 4.

12. Use the divisibility tests to determine whether or not the number 153,192,798,948 is divisible by each of the following numbers.
    (a) 2   **(b)** 3   **(c)** 4   **(d)** 5   **(e)** 6
    (f) 7   **(g)** 8   **(h)** 9   **(i)** 10   **(j)** 11

13. Eleven identical bags of potato chips were purchased by check. The checkbook entry reads $20.#9, where the # is an illegible digit. Assuming there was no sales tax, how much did each bag of chips cost?

14. Consider the following calculation:

    $$8{,}109{,}516{,}349 \cdot 4{,}510{,}902{,}130 = 36{,}581{,}234{,}571{,}973{,}023{,}370.$$

    (a) Use casting out nines to determine whether the calculation *could* be correct.

    (b) Use casting out elevens to determine whether the calculation *could* be correct.

15. Suppose an airline ticket has an 11-digit serial number with main part 2780143911 followed by a check digit given by the main part mod 7. What is the check digit?

16. The bank identification number of the Guaranty Federal Bank is 314970664. Suppose it is accidentally typed in as 319470664. Will the error be detected by the check digit?

17. Suppose the serial number for a U.S. Postal Service money order is 40983276587. If the number is accidentally recorded as 40983296587, will the check digit detect the error?

18. The main part of the ISBN number of *How Stella Got Her Groove Back* by Terry McMillan is 0-451-19200. Find the check digit.

19. Suppose you know that the UPC number of a certain product is 07#628071504, where # is a digit that is illegible. What must the digit # be?

20. For a tournament with 22 teams, use our round-robin scheduling method to find the team that will be assigned to play Team 3 in round 18.

21. For a tournament with 19 teams, use our round-robin scheduling method to find the team that will be assigned to play Team 11 in round 6.

22. Decipher the following message, which was enciphered with the Caesar cipher:

VKHNQ RZV.

23. Encode the message

BRING SOME MONEY

by breaking it into five-letter blocks and using the affine cipher $C = (7P + 20) \bmod 26$.

24. Decipher the message

BYPBE EBQAS JW,

which was encoded using the affine cipher $C = (19P + 3) \bmod 26$.

25. Encode the message

TURN LEFT

using the following Hill cipher for two-letter blocks:

$$C_1 = (9P_1 + 4P_2) \bmod 26,$$
$$C_2 = (5P_1 + 23P_2) \bmod 26.$$

26. Decipher the message

OR GJ NW TQ,

which was encoded using the following Hill cipher for two-letter blocks with some dummy letter in the last block:

$$C_1 = (8P_1 + 13P_2) \bmod 26,$$
$$C_2 = (3P_1 + 2P_2) \bmod 26.$$

27. Encode the message

BEGIN

using three-digit blocks and the RSA public key system where the message is being sent to the person with the public key address $(3, 4189)$. Use the digit 0 for the dummy digits in the last block.

28. Decipher the message

2804   2170   0394   1206,

which was encoded using the RSA public key system with three-digit blocks and sent to the address $(689, 3569)$, where $3569 = 43 \cdot 83$ and

$$\overline{689} \bmod ((43 - 1)(83 - 1)) = \overline{689} \bmod 3444 = 5.$$

Suggested Readings

Beutelspacher, Albrecht. *Cryptology*. Washington, D.C.: The Mathematical Association of America, 1994. An introduction to cryptology, including the RSA public key system.

Clausing, Jeri. "White House Yields a Bit on Encryption." *New York Times*, 8 July 1998, C1, 5. A brief report on the national security and business sides of the debate over the export of cryptographic technology.

Collins, David Jarrett, and Nancy Nasuti Whipple. *Using Bar Code, Why It's Taking Over*, 2nd ed. Duxbury, Mass.: Data Capture Institute, 1994. An overview of bar code technology.

Crandall, Richard E. "The Challenge of Large Numbers." *Scientific American* 276:2 (February 1997), 74–78. Large numbers in history and how computers calculate with large numbers, emphasizing number theory and cryptology.

Dewdney, A. K. "On Making and Breaking Codes: Part I." *Scientific American* 259:4 (October 1988), 144–147; "On Making and Breaking Codes: Part II." *Scientific American* 259:5 (November 1988), 142–145. Letter frequencies and the German Enigma machine in Part I, the Data Encryption Standard and the RSA cryptosystem in Part II.

Gallian, Joseph. "Assigning Driver's License Numbers." *Mathematics Magazine* 64:1 (February 1991), 13–22. Methods used by states to assign driver's license numbers, including several check digit schemes.

Gallian, Joseph. "The Mathematics of Identification Numbers." *College Mathematics Journal* 22:3 (May 1991), 194–202. A more technical look at some check digit schemes. Includes a description of Verhoeff's check digit method.

Gallian, Joseph, and Steven Winters. "Modular Arithmetic in the Marketplace." *American Mathematical Monthly* 95:6 (June–July 1988), 548–551. A more technical look at some check digit schemes.

Harmon, Craig K., and Russ Adams. *Reading Between the Lines*, 2nd ed. Peterborough, N.H.: North American Technology, 1985. An overview of bar code technology.

Kahn, David. *The Codebreakers*. New York: Macmillan, 1967. A comprehensive history of cryptology.

Ore, Oystein. *Invitation to Number Theory*. Washington, D.C.: Mathematical Association of America, 1967. An elementary introduction to number theory.

Singh, Simon, and Kenneth A. Ribet. "Fermat's Last Stand." *Scientific American* 277:5 (November 1997), 68–73. The history of Fermat's Last Theorem and the story of Andrew Wiles solving the problem.

Although game playing is as old as mankind, the comprehensive, formal study of the mathematics of games—known as game theory—is barely 50 years old. Consciously or not, you use the ideas underlying game theory nearly every day. Your decision on when to eat dinner in a cafete-

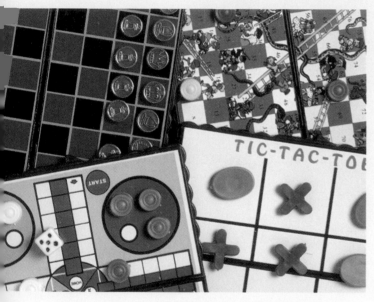

ria or restaurant might involve a bit of game theory, as you balance your ideal time to eat with the best times to avoid having to wait in line or for a table. Rush hour traffic on roads is often variable, particularly on highway routes that suffer from frequent accidents or breakdowns. Your choice of routes driving to school or work might be improved by using game theory, with your decision based on the consequences of being a few minutes late or perhaps a half hour late. Your decision on how late to stay up studying for an exam, where you may trade mental sharpness for knowl-

edge as you stay up later, may be considered game theory. One of the goals of mathematician John von Neumann and economist Oskar Morgenstern's foundational work in game theory in the first half of the twentieth century was to analyze decisions in business and politics. We do not limit game theory to the study of recreational games such as Monopoly or cards, but instead view it in the broader context of the study of decisions or choices. Some simple games can be analyzed and understood completely through game theory. At other times, the "game" is too complex to analyze completely, such as deciding how to campaign for political office, or even the game of chess for that matter. However, even with a complex game, we can often use ideas from game theory to shed light on the game even if we cannot fully "solve" it.

Game theory can be used to analyze serious matters such as nuclear deterrence.

CHAPTER 9

GAME THEORY WITH AN INTRODUCTION TO LINEAR PROGRAMMING

9.1   ALTERNATE MOVE GAMES

We begin by looking at games played by two players who move one at a time. Such games are called **alternate move games**. This framework also includes one-player games of chance such as solitaire card games, where we consider chance or nature to be the second player.

Tic-tac-toe is one of the first games learned by most children. Two players, denoted X and O, take turns choosing squares in a 3 × 3 board or array. We will assume that X goes first. The first player to get three marks in a row, horizontally, vertically, or diagonally, wins the game. If no player succeeds in getting three in a row by the time all nine squares have been chosen, the game is a draw, or tie. An example of a complete game of tic-tac-toe is shown in the sequence of moves in Figure 9.1. We see that player X won this game.

Figure 9.1
A game of tic-tac-toe.

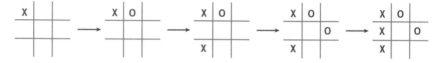

Game Trees

Figure 9.2
The numbering of the tic-tac-toe board.

To analyze tic-tac-toe, it will be helpful to introduce some notation for the different possible moves. If we number the squares 1 through 9 as in Figure 9.2, then we can indicate the move in which a player puts an X in square 2 by X2. Similarly, the move of placing O in square 7 is denoted by O7.

We now consider a game in progress, which has begun X1, O5, X3, O2, X8, O4, as illustrated in Figure 9.3.

Player X now has three choices: square 6, 7, or 9. Player O will have two possible responses, and then Player X is left with one choice for the last move, if the game has not already ended. To put these possible choices into a useful format, we use a **tree diagram**. The first, or main, branches of the tree correspond to the possible first moves. Each branch coming out of one of the first branches represents a possible response to that move. Continuing in this way, we get a tree whose outer branches are the last moves. The tree diagram for the rest of our game of tic-tac-toe is given in Figure 9.4.

Figure 9.3
A game in progress.

Figure 9.4
A tree diagram for the end of a game of tic-tac-toe.

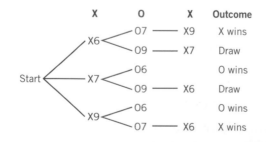

Let's trace through this diagram. Player X has three initial choices: 6, 7, or 9. If he or she chooses 6, then O may choose 7 or 9, and X is left with the other. If O chooses 7, then X's 9 wins the game. If O chooses 9, the game ends in a draw. The situation is similar when X chooses 7 or 9, except that in this case, if O chooses 6, then O wins.

Note that the two players in a game of tic-tac-toe have at most nine squares to choose from at any point in the game. Furthermore, the game must end within nine moves because no squares are left to choose at that stage. Any such game, with only finitely many choices for each turn and always ending within a specific number of turns, can be analyzed by a tree diagram. For convenience, we will always draw our trees "growing" from left to right. By starting at the different finishes of the game and working backwards, we can often determine the lines of play most likely to be chosen by two skillful players. We collect next the terms we use in discussing alternate move games.

> A **move** is a single choice made by one of the players.
>
> A **game tree** is a tree diagram showing all possible sequences of moves for a game.
>
> A **partial game tree** is a tree diagram showing only some of the possible sequences of moves for a game.
>
> A **strategy** is a player's planned choices of moves throughout a game. In the case of an alternate move game it includes responses to all of an opponent's possible moves.
>
> An **optimal strategy** is a strategy that produces the best possible result against the most skillful opponent.

Example 1   *Tic-Tac-Toe*

For a tic-tac-toe game that begins X1, O5, X3, O2, X8, O4, use the game tree to determine the players' optimal strategies and who wins under best play by both players.

Solution: Note that, as described, the game we want to analyze really begins after Player O's third move. We begin with the game tree corresponding to this game, which we have already constructed in Figure 9.5.

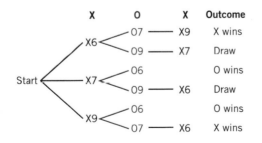

Figure 9.5
A tree diagram for the end of a game of tic-tac-toe.

Because Player X has no choice in the last move, we begin backtracking through the tree diagram with Player O's last move. Observe that O can win in the second and third main branches, those where X chooses 7 or 9 on the first move. However, had X chosen 6 on the first move, O would respond by taking 9, leading to a draw. Eliminating the branches of the game tree that Player O would avoid leaves the partial game tree shown in Figure 9.6.

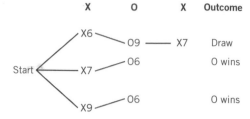

Figure 9.6
Partial game tree after eliminating O's inferior strategies.

So Player X should avoid the two lower branches. However, by first choosing 6, Player X ensures at least a draw. Player O should then choose 9 and the result is a draw. Thus, X's optimal strategy is to choose 6 on his or her first turn. Player O's optimal strategy is to choose 6 if available and 9 if not. Under best play by both players, the result is a draw.

Imagine trying to construct a game tree to analyze tic-tac-toe in its entirety. There are 9 first choices for X, then 8 choices for O, then 7 for X, then 6 for O,

BRANCHING OUT 9.1

COMPUTER CHESS PROGRAMS

When Deep Blue, the IBM chess program and computer, defeated the world chess champion Garry Kasparov with 2 wins, 1 loss, and 3 draws in a match in May 1997, the world took notice. In fact, one of television talk show host David Letterman's Top Ten lists soon after the match speculated on the "wild" ways Deep Blue might have celebrated its victory.

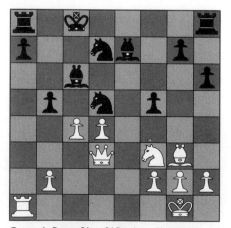

Game 6, Deep Blue (White) vs. Kasparov, when Kasparov resigns.

A chess player may unilaterally declare a draw by moving so as to repeat the positions of all pieces on the board for a third time. Thus, assuming that one player or the other will take advantage of this rule, there are only finitely many possible games of chess. Were the number of possible games small, few would bother to play. (How many adults play tic-tac-toe, except to humor a child?) However, because chess has a finite—be it unimaginably enormous—game tree, both players have optimal strategies. Most chess experts tend to believe that these optimal strategies lead to a draw. Yet, even though the White player tends to have an advantage in games between humans due to having the first move, it is possible that Black could force a win with best play. No one knows.

The existence of optimal strategies is of much practical importance. Chess games between experts will typically last anywhere from 20 moves to nearly 100 moves. Games at this level rarely go to completion; instead a player will resign a position that is a sure loss rather than drag out the match to its bitter end. In games that appear evenly matched, with neither player progressing toward victory, players of-

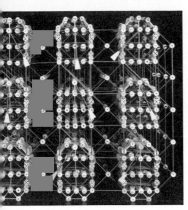

A Tinker Toy
computer that never
loses at tic-tac-toe.

then 5 for X—$9 \cdot 8 \cdot 7 \cdot 6 \cdot 5 = 15{,}120$ different paths so far. The shortest possible games will last five moves. Others will continue on for as many as nine moves. Clearly, analysis by hand is impractical. On the other hand, such a problem is well within reach of computers. Remember that tic-tac-toe is a very simple game. Most games complex enough to hold our interest have game trees that are beyond the reach of even the fastest computers. For instance, in chess there are 400 possible branches representing the first moves by White and Black. The time required for a fast computer to construct a tree representing the first 80 moves of a chess game, 40 by each player, estimated by Claude Shannon to have roughly 10^{120} branches, would make the age of the solar system seem like no time at all.

Although tic-tac-toe is easily analyzed by computer, to write out the thousands of different game trees by hand is a daunting task. However, most of us can play a perfect game of tic-tac-toe because we recognize a few common features to either seek or avoid. We will now consider approaches that simplify the analysis of a game tree. First, we reduce the size of the tree by combining game situations that are strategically the same into one. For instance, the starting moves X1, X3, X7, and X9 are all essentially the same by the symmetry of the game board because in each case Player X is marking a corner of the board on the first

ten agree to a draw. Because analysis of all branches of the tree for up to the first 80 moves of a chess game, 40 by each player, would require more time than the universe has existed (for those who believe in the Big Bang) and because the number of such games exceeds all estimates for the number of atoms in the universe, it is fruitless to attempt a full analysis of chess. Yet large-scale analysis of a game tree is exactly the strength of computer chess programs! Most chess programs examine the game tree to a certain number of moves, this "depth" being dependent on the time allowed to search. Deep Blue was able to search on the order of 16 moves, within its time limit of an average of 3 minutes per move. Therefore, the faster the computer, the more positions it can analyze within its time constraints. Much of the improvement in computer chess programs is due to the increased speeds of computer hardware. Because the computer's search through the game tree stops before the game ends, the second major factor in determining a program's ability is its evaluation of which player is likeliest to win a position arising in the game tree. This evaluation, usually a numerical score, is based on which pieces remain on the board and on their positions relative to each other. Evaluations vary somewhat from program to program. The Deep Blue human support team included Joel Benjamin, one of the world's top players, who helped in developing its evaluation methods. Some programs drop or "prune" unpromising branches from consideration in order to be able to examine more moves in the promising branches. Computer chess programs may differ in many aspects, but all depend on a comprehensive search of the game tree. The good chess programs for personal computers can consistently defeat over 99% of all human players.

The Deep Blue Team: Front, left to right: Joel Benjamin, C. J. Tan; Back, left to right: Jerry Brody, Murray Campbell, Feng-hsiung Hsu, A. Joseph Hoane, Jr.

move. Therefore, one tree branch suffices to represent all four starting choices. When we combine moves that are strategically the same due to the symmetry of the game, we call the game tree a **compressed game tree**. We will call a tree diagram that includes only some of the possible sequences of moves in a compressed game tree a **partial compressed game tree**.

Example 2   *Compressed Game Tree for Tic-Tac-Toe*

Construct a compressed game tree representing the first two moves of a tic-tac-toe game.

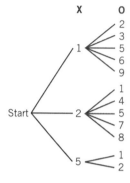

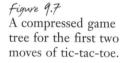

Figure 9.7
A compressed game tree for the first two moves of tic-tac-toe.

Solution: Player X has three strategically different first moves: the center, a corner, or the middle of a side. We let squares 1, 2, and 5 represent these three choices, which form the first branches of the compressed game tree in Figure 9.7. To a first move of 5 by Player X, Player O can respond with either a corner or a side, which we represent by the choice of squares 1 and 2, respectively. To a first move of 1, Player O can respond with the middle of one of two adjacent sides, with one of the two nearer corners, with the center, with the middle of one of the two far sides, or with the far corner. We represent these five strategically distinct cases by 2, 3, 5, 6, and 9, respectively. To a first move of 2, Player O can respond with 1 (or equivalently 3), 4 (or equivalently 6), 5, 7 (or equivalently 9), or 8. Thus, one possibility for the compressed game tree appears in Figure 9.7.

Our exploitation of symmetry has compressed the game tree for the first two moves from one with 72 final branches to a more manageable 12.

Example 3   *Tic-Tac-Toe*

Show that in a game of tic-tac-toe that begins X5, O2, X8, Player O can achieve at least a draw with optimal play

(a) through a verbal argument,
(b) through a partial game tree.

Solution:
(a) We will show that if Player O's first response is O1, then Player O can ensure at least a draw with suitable responses to any subsequent moves Player X might make. The first board of Figure 9.8 shows the position as Player O is about to move. Suppose Player O first responds with O1 as shown in the second board of Figure 9.8. Unless Player X chooses square 3, Player O will take it and win the game. Thus, we may assume X chooses X3, as shown in the third board of Figure 9.8. Player O must now choose O7 or lose. Therefore, we assume Player O chooses O7, as shown in the fourth board of Figure 9.8. Then X is forced to respond with X4, as shown in the fifth board of Figure 9.8. Finally, we have O choose O6 leaving X9 for X and a draw, as shown in the final two boards of Figure 9.8.

Figure 9.8
Sequence of moves.

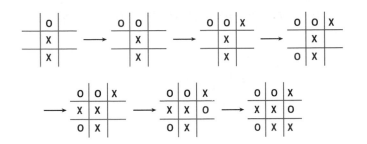

(b) Another way to demonstrate that Player O has a strategy that will at least achieve a draw is to construct a partial game tree showing only the moves Player O would choose if playing the particular strategy and all of Player X's possible responses. We need to show that each line of play leads to a win or draw for Player O. The partial game tree for this situation is shown in Figure 9.9.

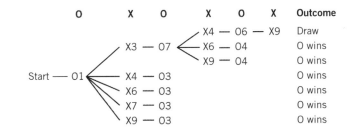

Figure 9.9
Partial game tree for Example 3.

Before leaving tic-tac-toe, we conclude with one more example, where the argument is brief and easy to follow, although not so easy to come up with.

Example 4   *Choosing the Center Square in Tic-Tac-Toe*

Show that by taking the center square with the first move, Player X can achieve at least a draw with optimal play.

1	2	3
4	5	6
7	8	9

Figure 9.10
Paired squares on a tic-tac-toe board.

Solution: The key idea is that if a player has the center square, the player can only lose if his or her opponent has three in a row along one of the four sides. Player X should mentally pair off 1 and 2, 3 and 6, 8 and 9, and 4 and 7, as shown in Figure 9.10. After Player O chooses a square, Player X should choose a square to get three in a row, if possible, but otherwise select the square that is paired with the square O chose. It is easy to see that O cannot get three in a row, so that X must achieve at least a draw.

Be warned that we have shown only that Player X can achieve at least a draw, but we have not addressed the question of whether X has a strategy that actually guarantees a win. (In fact, the best X can do if O plays well is a draw, as you probably know from experience.)

Games can often model real-life situations, as we see in the following simple example.

Example 5   *A Business Decision*

The semiretired president of a company leaves most decisions to two of the company's vice presidents, Allen and Brown. When they agree on a course of action, it becomes the company policy. Should they disagree, the president steps in to choose one of their recommendations. Allen is always first to offer her recommendation, followed by Brown. The president then chooses one alternative if necessary. He never opts for a third possibility. The current issue is where to locate a new manufacturing plant. The options are Des Moines, St. Louis, and Omaha. Allen prefers St. Louis, with Des Moines as her second choice. Brown prefers Des Moines, with Omaha as his second choice. Both have discussed their preferences with the other, and both know the president well enough to know that he would rank the alternatives Omaha first, then St. Louis, and finally Des Moines. What recommendations should Allen and Brown make in order to achieve the best possible outcome from their perspectives?

Solution: To construct a game tree modeling this situation, shown in Figure 9.11, we begin by constructing the branches of the tree corresponding to Allen's choice followed by Brown's choice, where D stands for Des Moines, S for St. Louis, and O for Omaha.

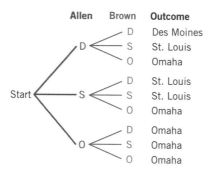

Figure 9.11
Game tree for Example 5.

To determine the outcome of each path in the game tree, we first note that in the three cases in which Allen and Brown make the same recommendation, that recommendation will be the final outcome. In the cases in which Allen and Brown disagree, the outcome will be determined by the president's preference between the two recommendations. To help visualize the preference rankings of Allen, Brown, and the president, we record them in Table 9.1.

Table 9.1	**Preference Rankings**		
	Allen	*Brown*	*President*
First Choice	St. Louis	Des Moines	Omaha
Second Choice	Des Moines	Omaha	St. Louis
Third Choice	Omaha	St. Louis	Des Moines

When Allen recommends Des Moines and Brown recommends St. Louis, the president will choose St. Louis, so St. Louis will be the outcome in this case. The outcomes for the five remaining cases are determined similarly.

To determine the best strategy for each vice president, let's first use the game tree to examine Brown's reaction to Allen's possible recommendations. To Allen's recommendation of Des Moines, Brown should also recommend Des Moines. If Allen recommends St. Louis, Brown should recommend Omaha. Finally, if Allen recommends Omaha, Brown's recommendation is irrelevant, so Brown may as well pick Omaha to curry favor with the boss. (Actually, the description of the game assumes that the vice presidents' only goal is the city to be selected. To bring in anything else is to change the rules of the game!) In order to move on to Allen's options, we construct the partial game tree in Figure 9.12 based on the previous analysis.

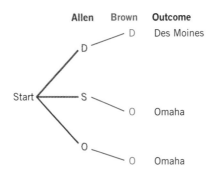

Figure 9.12
Partial game tree for Example 5.

We see that Allen has no hope of St. Louis being selected, so she must settle for Des Moines by recommending Des Moines to the president.

If we were to work through a similar analysis should Brown be the first to speak, we would find that Omaha would end up being the chosen site. If the president were also playing the game, perhaps he would select Brown to speak first. But again, that's another game.

∎

Nim

For our next examples we turn to a variation of the game of Nim, in which two players alternately remove objects, which we will call pebbles, from one or more piles according to some prescribed rules. The last player to remove a pebble either wins or loses, depending on the particular variation of the game. It is among the most frequently analyzed of all games. We denote the players by A and B, with A going first.

Example 6 *Nim with Four Pebbles*

Consider a game of Nim where two players alternate removing one or two pebbles from a single pile. The game starts with a pile of four pebbles, and the last player to take a pebble wins.

(a) Construct the game tree for the game.

(b) Determine who wins the game under best play by both players.

(c) Draw a partial game tree to show the winning strategy.

Solution:

(a) A legal move reduces the pile by one or two. Because it makes it easier to keep track of the game, we will depict a move by the number of pebbles remaining in the pile instead of by how many pebbles a given player takes. This notation has the advantage of describing the exact state of the pile at any instant. Figure 9.13 shows the game tree using this notation.

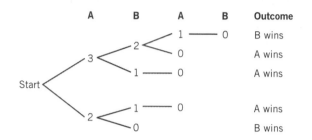

Figure 9.13
Game tree for Nim.

(b) Looking at the top branch, where A takes one pebble to reduce the pile to three pebbles, we see that A can win no matter how B responds. Thus, in beginning by taking one pebble on the first move, A wins the game under best play by both players.

(c) To show the winning strategy for Player A, we need only show the move A would choose for every situation, but we must show all of Player B's responses. This partial game tree is given in Figure 9.14. Every path of this partial game tree results in a win for Player A.

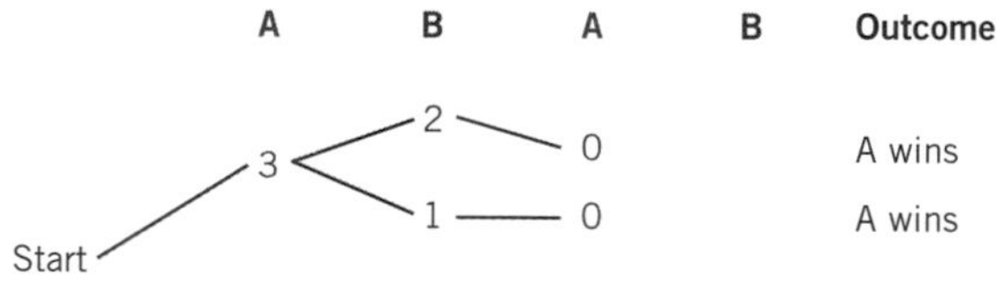

Figure 9.14
Partial game tree showing the winning strategy of Player A.

A	B	A	B	Outcome
Start — 3 < 2 — 0				A wins
1 — 0				A wins

The notation used in Example 6 makes it much easier to consider a game of Nim with more than one pile, because, for instance, we can write three pebbles in pile 1 and five pebbles in pile 2 very concisely as (3, 5). We will look at games of this type in the exercises.

Constructing a game tree for Nim as in Example 6 is feasible only when the size of the pile or piles is very small. In general, we must reason more abstractly, as in the following example.

Example 7   *Nim with n Pebbles*

Consider a game of Nim where two players alternate removing one or two pebbles from a single pile. The game starts with a pile of n pebbles, and the last player to take a pebble wins.

(a) Determine those n for which Player A wins the game under best play by both players.

(b) Who should win a game with a pile of one million pebbles? What should this player do on the first move?

Solution:

(a) It is clear that A, the first player, wins when *n* is 1 or 2. We saw in Example 6 that, by leaving three pebbles for Player B, Player A was able to guarantee a win. Therefore, the case *n* = 3 is a win for the second player to move, here B, under best play. We call a pile with three pebbles a **safe point**, meaning that the player who puts the game in that state can force a win with good play. For *n* > 3, a player cannot take the last pebble in one move, so it is reasonable to view the objective of the game as being the player to leave three pebbles in the pile. In essence, we are reserving three pebbles and playing Nim with the rest of the pile. Thus, games starting with four or five pebbles are a win for Player A under best play, because he or she can reduce the pile to three pebbles. Similarly, six pebbles leads to a win for Player B, seven or eight for Player A, nine for Player B, and so on. We see that Player B should win if *n* is a multiple of 3 and Player A should win if not.

(b) Player A should win because one million is not a multiple of 3. The winning initial move is to take one pebble because 999,999 is a multiple of 3, hence a safe point.

Playing Against Chance

Many games have some randomness to them. For instance, board games often use dice, cards, or a spinner. It is often convenient to think of this random element as represented by a player, one who is not necessarily logical or attempting to win. Because we are restricting our attention in this chapter to two-player games, we will consider only what people might normally call solitaire. For such games, we will call the logical player the Player and the random element, or player, Chance.

Let's look at a game we call Random Chase. The game is played on a collection of line segments that may intersect at their endpoints. We call such a collection a *graph*, one of the line segments an *edge*, and one of the endpoints a *vertex*. The Player and Chance each start with one piece on different, specified vertices. They alternate moving their pieces along adjoining edges to different vertices, the Player however he or she wants, Chance at random, with each adjoining edge equally likely to be chosen. Chance wins if it occupies the same vertex as the Player within some given number of moves. Otherwise, the Player wins. We will number the vertices and denote the move from vertex *a* to vertex *b* by *b*.

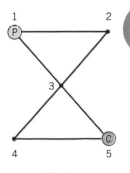

Figure 9.15
Random Chase board.

Example 8   *Random Chase*

In a Random Chase game played on the graph shown in Figure 9.15, the Player moves first and wins if the two pieces do not occupy the same vertex within two moves by each player.

(a) Construct the game tree for this game.

(b) Determine a strategy that will give the Player the best chance to win the game.

Solution:

(a) The game tree for the Random Chase game played on the given graph is shown in Figure 9.16.

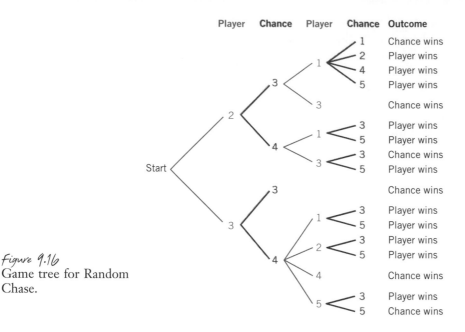

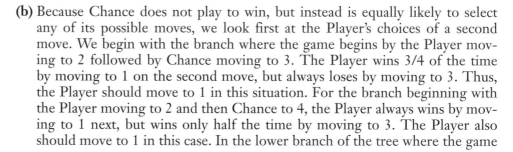

figure 9.16
Game tree for Random
Chase.

(b) Because Chance does not play to win, but instead is equally likely to select any of its possible moves, we look first at the Player's choices of a second move. We begin with the branch where the game begins by the Player moving to 2 followed by Chance moving to 3. The Player wins 3/4 of the time by moving to 1 on the second move, but always loses by moving to 3. Thus, the Player should move to 1 in this situation. For the branch beginning with the Player moving to 2 and then Chance to 4, the Player always wins by moving to 1 next, but wins only half the time by moving to 3. The Player also should move to 1 in this case. In the lower branch of the tree where the game

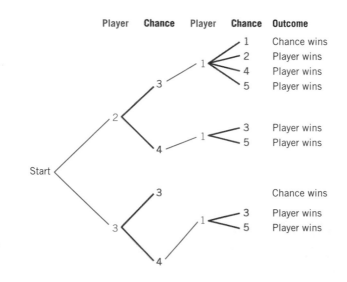

figure 9.17
Partial game tree for
Random Chase.

begins by the Player moving to 3 followed by Chance moving to 4, we see that by moving to either 1 or 2 next, the Player can be assured of a win. To designate a definite strategy, we will call on the Player to move to 1 in this situation; however, a strategy calling for moving to 2 would be equally good. We now record our analysis thus far by drawing the partial game tree showing only the best second moves for the Player in Figure 9.17.

Looking first at the lower branch where the Player moves to vertex 3, we see that the Player wins half of the time, exactly when Chance moves to vertex 4. On the other hand, by first moving to vertex 2, the Player wins the half of the time that Chance moves to vertex 4 and also the $(3/4) \cdot (1/2) = 3/8$ of the time Chance moves to vertex 3 and then to some vertex other than 1. Thus, the Player wins a total of $1/2 + 3/8 = 7/8$ of the time by moving first to vertex 2. The Player's best strategy is to move to vertex 2, and then back to vertex 1.

∎

In the last example, the Player's best strategy turned out to be unique because the Player did not choose 3 for the first move. Furthermore, it had the same second move no matter what Chance's first move turned out to be. You should not expect either of these features to hold for alternate move games in general.

EXERCISES FOR SECTION 9.1

1. Consider the given game tree. (The possible moves have been labeled a through q for convenient reference.)

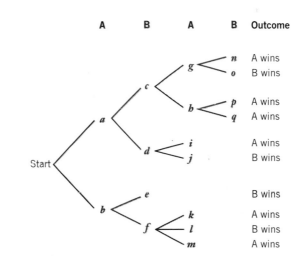

A	B	A	B	Outcome
			n	A wins
		g	o	B wins
	c		p	A wins
		b	q	A wins
a		d	i	A wins
			j	B wins
		e		B wins
b			k	A wins
	f		l	B wins
			m	A wins

(a) Analyze it to determine who wins the game under best play by both players.

(b) Draw a partial game tree to illustrate the winning strategy.

2. Consider the given game tree. (The possible moves have been labeled *a* through *z* for convenient reference.)

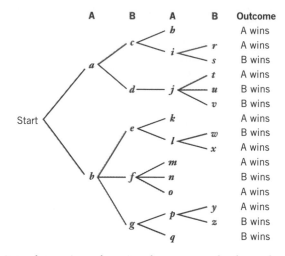

(a) Analyze it to determine who wins the game under best play by both players.

(b) Draw a partial game tree to illustrate the winning strategy.

3. (a) Construct the game tree for the remainder of a tic-tac-toe game that begins X3, O8, X9, O6, X7, O5.

(b) What is the outcome of the game under optimal play?

4. (a) Construct the game tree for the remainder of a tic-tac-toe game that begins X5, O8, X6, O4, X7, O3.

(b) What is the outcome of the game under optimal play?

5. Construct a compressed game tree for the next two moves of a game of tic-tac-toe that begins X5.

6. Construct a compressed game tree for the next three moves of a game of tic-tac-toe that begins X3, O5, X7, O1.

7. Show that Player O can achieve at least a draw with optimal play in a game of tic-tac-toe that begins X1, O5, X6.

8. In a game of tic-tac-toe that begins X5, O8, show that Player X can win with optimal play.

9. Consider a game of Nim in which two players alternate removing one or two pebbles from a single pile that starts with four pebbles, and the player taking the last pebble loses.

(a) Construct the game tree for the game.

(b) Determine who wins the game under best play by both players.

10. Consider a game of Nim in which two players alternate removing two, three, or four pebbles from a single pile that starts with eight pebbles, and the player able to take the last pebble wins. If the pile gets down to one pebble, the game is a draw.

(a) Construct the game tree for the game.

(b) What is the outcome of the game under best play by both players?

11. Consider a game of Nim in which two players alternate removing one or two pebbles from either one of two piles, each of which starts with two pebbles. The player taking the last pebble wins.

(a) Construct the compressed game tree for the game.

(b) Determine who wins the game under best play by both players.

12. Consider a game of Nim in which two players alternate removing one or two pebbles from either one of two piles, each of which starts with two pebbles. The player taking the last pebble loses.

(a) Construct the compressed game tree for the game.

(b) Determine who wins the game under best play by both players.

13. Consider a game of Nim in which two players alternate removing one, two, three, or four pebbles from a single pile that starts with n pebbles and the last player to take a pebble wins.

(a) Determine those n for which the player going first wins the game under best play by both players.

(b) Who should win a game with a pile of one million pebbles? What is the winning strategy?

14. Consider a game of Nim in which two players alternate removing anywhere from 1 through 10 pebbles from a single pile that starts with n pebbles and the last player to take a pebble wins.

(a) Determine those n for which the player going first wins the game under best play by both players.

(b) Who should win a game with a pile of one million pebbles? What is the winning strategy?

15. Determine the safe points in a game of Nim in which two players alternate removing one or more pebbles from one of two piles that have m and n pebbles to start with, and the player taking the last pebble loses.

16. Consider a version of Nim in which two players alternate removing two, three, or four pebbles from a single pile that has n pebbles to start with, and the player taking the last pebble wins. If a single pebble is left in the pile, the game is considered a draw. Assuming best play on the part of both players, determine those n yielding a win for the player going first, those n yielding a win for the player going second, and those n leading to a draw.

The two main factions of a labor union are each to nominate one candidate for the new president from among Lewis, Kennedy, Miller, and Church. If both factions nominate the same candidate, that candidate will win without opposition. Otherwise the general membership votes for one of the two nominees. The "Old Guard" favors Lewis, then Kennedy, then Miller, and then Church. The "Reformers" favor Church, then Lewis, then Kennedy, and then Miller. After extensive polling, both factions know that the general membership prefers Kennedy, then Miller, then Lewis, and then Church. In the scenarios described in Exercises 17 and 18, one or the other group makes its nomination first.

(a) Construct a game tree of the groups' possible nominees leading to the final election results.

(b) Describe the course of events under best play by both factions.

17. The Old Guard decides on and then announces its nomination, followed by the selection of the Reformers' nominee.

18. The Reformers decide on and then announce their nomination, followed by the selection of the Old Guard's nominee.

The game we will analyze in Exercises 19–22 comes in many variations. On a one-row board of squares, two players, Left and Right, alternate moving their pieces in one of the following two ways: Left may move one of its pieces one square to the left into an empty square, or it may jump one of its pieces into an empty square two squares to the left over an opponent's piece occupying the adjacent square on its left (without removing the opponent's piece). Right moves in the same way, only to the right. As soon as one of Left's pieces occupies the far left square, the game is over and Left wins. Similarly, Right wins immediately upon occupying the far right square. A player who is unable to move loses. Left always moves first, and we record a move by writing the number of the destination square. (For example, in Exercise 21 Left's first move may be either 3 or 5.) For the given Left-Right game, who wins the game under optimal play by both sides? Construct a partial game tree to demonstrate that player's strategy.

19.

R		R		L		L
1	2	3	4	5	6	7

20.

R		R	L			L
1	2	3	4	5	6	7

21.

R	R		L		R	L	L
1	2	3	4	5	6	7	8

22.

R		R	L	R		L	L
1	2	3	4	5	6	7	8

In the game of Crosscram or Domineering, players H and V sit at the bottom of a checkerboard (usually rectangular) and alternate placing 2×1 dominoes. Player H moves first, laying a domino horizontally to cover two squares. Player V places his or her tiles vertically (in a single column) on two empty squares. The players then alternate placing dominoes; the last player able to place a domino wins. In Exercises 23–26, label moves by the two squares on which the domino is placed. Exercises 27 and 28 vary the size of the tiles.

23. For Crosscram played on the 2×4 board shown,

1	2	3	4
5	6	7	8

    (a) construct a compressed game tree;
    (b) determine which player should win under best play by both sides, using a partial compressed game tree to show that player's strategy.

24. For Crosscram played on the 3×3 board shown,

1	2	3
4	5	6
7	8	9

    (a) construct a compressed game tree;
    (b) determine which player should win under best play by both sides, using a partial compressed game tree to show that player's strategy.

25. For Crosscram played on the unusually shaped board shown,

1	2	3	
4	5	6	7
8	9	10	

    (a) construct a compressed game tree;
    (b) determine which player should win under best play by both sides, using a partial compressed game tree to show that player's strategy.

26. For Crosscram played on the 2 × 5 board shown,

1	2	3	4	5
6	7	8	9	10

    (a) construct a compressed game tree;

    (b) determine which player should win under best play by both sides, using a partial compressed game tree to show that player's strategy.

27. On an $n \times n$ game board, both players have $k \times 1$ tiles, where $n/2 < k \le n$. For every positive whole number n, which player should win under best play by both players? Describe that player's strategy in detail and explain why it works.

28. On an $n \times n$ game board, Player H has $n \times 1$ tiles whereas Player V has $k \times 1$ tiles, where k is some whole number from 1 to n. For every n, determine those k for which Player H should win under best play by both players. Describe the winning player's strategy in detail and explain why it works.

Football teams often replace certain players in special situations. One instance is near the goal line, where the team with the ball may have its regular offense and a goal line offense. Similarly, the team without the ball may also have a regular defense and a goal line defense. The defense generally has an opportunity to replace its players after the offense has. Therefore, we will assume the offensive coach chooses the regular or goal line offense, and then the defensive coach selects its regular or goal line defense. Suppose that both coaches know the offense's success rate (i.e., the percentage of the time it scores a touchdown) for the four different combinations of offense and defense, given as a table of success rates (the higher the number, the better for the offense, the worse for the defense) in Exercises 29 and 30.

(a) Construct a game tree for the coaches' choices.

(b) Find optimal strategies for both coaches.

(c) Does it help the defense to have the last move?

29.

	Regular Defense	*Goal Line Defense*
Regular Offense	40%	50%
Goal Line Offense	70%	60%

30.

	Regular Defense	*Goal Line Defense*
Regular Offense	40%	50%
Goal Line Offense	60%	30%

In a simplified version of baseball strategy (representing, say, the last inning of a tie game with runners on base), suppose the batter succeeds by getting a hit and the pitcher succeeds if the batter fails to get a hit. Thus, we measure the result of a particular batter facing a particular pitcher by the fraction of the time the batter gets a hit, which we give in the form of a table. Our version of baseball strategy has the following rules: The first decision is made by the manager of the team at bat. Batter A is at bat but may be replaced by batter B if the manager desires. Pitcher X is pitching for the opponents but may be replaced by either pitcher Y or Z if their manager wishes. Finally, if batter A has not yet been replaced, the batters' manager again has the option to substitute batter B. For the data given in Exercises 31 and 32,

(a) construct a game tree for the managers' choices;

(b) find optimal strategies for both managers.

31.

	Pitcher X	*Pitcher Y*	*Pitcher Z*
Batter A	.320	.240	.280
Batter B	.250	.260	.230

32.

	Pitcher X	Pitcher Y	Pitcher Z
Batter A	.200	.290	.270
Batter B	.350	.230	.240

A three-person zoning commission must decide whether a tract of undeveloped residential land will be rezoned as either commercial, industrial, or left as residential. Fong, the chair of the commission, favors leaving the zoning as residential but prefers commercial rezoning to industrial rezoning. Commissioner Green favors rezoning the land and prefers commercial to industrial rezoning. Commissioner Herrera would most prefer the land to be rezoned as industrial but views rezoning to commercial usage as the worst alternative. Green has introduced a motion to rezone the land to commercial, whereas Herrera has introduced a bill to rezone the land to industrial. All are well aware of the other commissioners' preferences. The rules of order permit Fong to determine which bill comes to a vote first. The other two commissioners will then vote on the bill, with Fong breaking a tie. The winning land use in this vote (either the proposed rezoning or residential) then faces the alternative proposed by the second bill in a second vote. The winner of this vote becomes the commission's decision. All three commissioners will vote their true preferences on the second vote, so that the game is Fong's choice of which bill is voted on first and each member's decision how he or she will vote on this first vote. In the scenarios in Exercises 33 and 34, one of the commissioners' behavior is given, so that the zoning decision becomes a two-player game.

(a) Construct a game tree based on the two pertinent commissioners' choices.

(b) Describe the course of events if these two commissioners follow their optimal strategies.

33. Commissioner Herrera has stated that, on the first vote, he will vote for whichever of the two choices he prefers. Thus, the "game" is between Fong and Green.

34. Commissioner Green has stated that, on the first vote, she will vote for whichever of the two choices she prefers. Thus, the "game" is between Fong and Herrera.

35. In the game between a Player and Chance with the given tree diagram, Chance chooses branches at random, with any possible choice equally likely. (The possible moves have been labeled *a* through *z* for convenient reference.) Find the Player's best strategy and the probability that the Player wins the game by following this strategy.

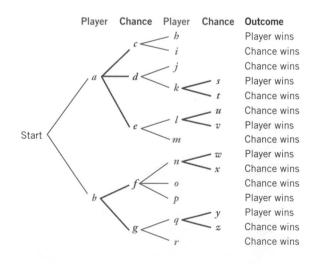

36. In the game between a Player and Chance with the given tree diagram, Chance chooses branches at random, with any possible choice equally likely. (The possible moves have been labeled *a* through *p* for convenient reference.) Find the Player's best strategy and the probability that the Player wins the game by following this strategy.

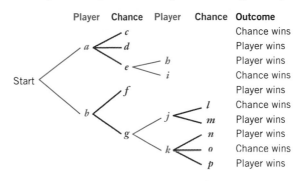

37. In a Random Chase game played on the graph shown, the Player moves first and wins if the two pieces do not occupy the same vertex within two moves by the Player and two moves by Chance.

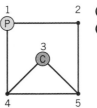

    (a) Construct the game tree for this game.

    (b) Determine a strategy that will give the Player the best chance to win the game and compute this probability.

38. In a Random Chase game played on the graph shown, the Player moves first and wins if the two pieces do not occupy the same vertex within two moves by the Player and two moves by Chance.

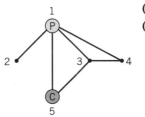

    (a) Construct the game tree for this game.

    (b) Determine a strategy that will give the Player the best chance to win the game and compute this probability.

A simplification of blackjack or baccarat that we call Pass or Spin is played by a single Player using a spinner that has two or more equally likely numbers on it. Chance enters the game via the spinner and as an opponent playing by fixed rules. The object of the game is to have spins adding up to a number that is at most 4 (in our version) and is larger than the total of the opponent's spins. In case of a tie, the first player to reach that total wins. The sequence of play is the following:

• The Player spins.
• (With help from the Player) Chance spins.
• The Player spins or passes.
• Chance spins if in a losing position (meaning that it would lose if no one spins again) and passes otherwise.
• The Player spins or passes.
• Chance spins if in a losing position and passes otherwise, and the game ends.

Because the first player whose spins add up exactly to 4 wins by passing from that point on, the game ends at that point, and that player wins. Similarly, the game ends once either player's total exceeds 4, and that player loses. For each situation described in Exercises 39 and 40,

(a) construct the game tree;

(b) determine the Player's best strategy;

(c) determine the fraction of the time the Player will win using his or her best strategy.

39. The spinner has the numbers 1 and 2. The Player and Chance both get 2's on their first spins.

40. The spinner has the numbers 1, 2, and 3. The Player's first spin is a 3 and Chance's first spin is a 1.

9.2 SIMULTANEOUS MOVE GAMES AND DOMINANCE

James Dean (1931–1955), who ironically died in a car crash.

In the 1955 movie *Rebel Without a Cause*, Jim Stark and Buzz Gunderson, played by James Dean and Corey Allen, face off side by side in their hotrods high on a hill. As the cars raced toward a cliff, each character was faced with the choice of showing his courage by maintaining his course for the longest time, or either swerving to safety or bailing out of his car. Buzz ends up driving over the cliff and dying. These characters in *Rebel Without a Cause* had to carry out their choice of action simultaneously, ignorant of the other's intentions. The outcome of their decisions depended not just on a single character's decision but on the pair of decisions the two made. This situation is an example of a **simultaneous move game**. The uncertainty makes analysis of simultaneous move games more difficult than the analysis of the alternate move games we studied in the last section. We consider games of this type for the remainder of the chapter. We begin with the assumption that players have no communication of any kind with their opponents. In the final section of this chapter, we change this assumption and look at games that have an opportunity for cooperation, negotiation, or threats.

Let's begin with an economic game. Two supermarket chains, Groceries Galore and Friendly Foods, plan to expand by adding one store to one of two neighborhoods that have no major chain. The first neighborhood has 8000 people who would go to a local supermarket chain, and the second neighborhood has 5000 people. If only one store locates in a given neighborhood, that store gains all of the potential customers. If the stores locate in the same neighborhood, then 70% of the customers will go to Groceries Galore, the better-known chain. Each chain is fully aware of all these details and must choose the neighborhood for its store without knowing the choice of its competitor. Each chain has two choices of strategy: to locate in the first neighborhood or the second. The value of this "game" to each chain is the number of customers gained by that chain for the chosen pair of strategies.

We will set up the results of the game in a 2 × 2 array called a payoff matrix, with one row for each of Groceries Galore's two options and one column for each of Friendly Foods' two options, as shown in Table 9.2.

Table 9.2	Skeleton for the Payoff Matrix for the Grocery Stores		
		Friendly Foods	
		First Neighborhood	*Second Neighborhood*
Groceries Galore	*First Neighborhood* *Second Neighborhood*		

When Groceries Galore locates in the first neighborhood and Friendly Foods locates in the second, Groceries Galore gains 8000 customers and Friendly Foods gains 5000. These represent the values to the stores when they choose this pair of strategies. We denote it by (8000, 5000) and enter it in the first row, second column of the table. Similarly, the payoffs for the second row, first column are 5000 for Groceries Galore and 8000 for Friendly Foods, and the entry is (5000, 8000). The other two payoff pairs require computation. When both chains opt to locate in the first neighborhood, the payoff to Groceries Galore is 70% of 8000, or $(8000) \cdot (0.7) = 5600$, and the payoff to Friendly Foods is $8000 - 5600 = 2400$. Similarly, if both stores locate in the second neighborhood, the payoffs to Groceries Galore and Friendly Foods are $(5000)(0.7) = 3500$ and $5000 - 3500 = 1500$, respectively. Table 9.3 shows the payoff matrix once it has been filled in.

Table 9.3	Payoff Matrix for the Grocery Stores		
		Friendly Foods	
		First Neighborhood	*Second Neighborhood*
Groceries Galore	*First Neighborhood* *Second Neighborhood*	(5600, 2400) (5000, 8000)	(8000, 5000) (3500, 1500)

We have just found the payoff matrix for this particular simultaneous move game. In general, simultaneous move games between two players can be described by payoff matrices, which are defined as follows:

The **payoff matrix** of a simultaneous move game is an array whose rows correspond to the strategies of one player, who is sometimes called the Row Player, and whose columns correspond to the strategies of the other player, who is sometimes called the Column Player. Each entry of the array is the result, or **payoff**, obtained when the Row Player chooses the strategy corresponding to the particular row and the Column Player chooses the strategy corresponding to the particular column. This entry is often written as an ordered pair whose first coordinate is a number expressing the payoff from the Row Player's perspective and whose second coordinate represents the payoff from the Column Player's perspective. We adopt the convention that bigger values are better.

In looking at the payoff matrix in Table 9.3, we see that, although neither chain knows what the other intends to do, Groceries Galore should elect to locate in the first neighborhood. If Friendly Foods selects the first neighborhood, Groceries Galore gets 5600 customers by locating in the first neighborhood and only 5000 by locating in the second. Similarly, if Friendly Foods locates in the second neighborhood, Groceries Galore gets 8000 customers by locating in the first neighborhood but only 3500 locating in the second. In other words, locating in the second neighborhood is worse for Groceries Galore regardless of Friendly Foods' strategy. An intelligent and thoughtful player will always eliminate an inferior strategy. The following definitions describe this situation.

A strategy of a player is said to be **dominated** by a second strategy if the second strategy always results in at least as good an outcome for the player, no matter what strategy the player's opponent chooses, and results in a better outcome against at least one of the opponent's strategies. We call the inferior strategy a **dominated strategy**. A strategy that dominates all others is called a **dominant strategy**. It must be unique.

Because there is always a better option than a dominated strategy, it would be poor play to ever select a dominated strategy. Best play demands that players eliminate dominated strategies, which effectively reduces the size of the payoff matrix. After dominated strategies for both players have been eliminated, the smaller matrix may have some new dominated strategies that should now be eliminated. After a series of reductions, we arrive at a matrix with no dominated strategies. This is called the reduced payoff matrix of the game.

The **reduced payoff matrix** of a game is the submatrix of a game where dominated strategies have been eliminated in one or more stages. It gives the relevant portion of the original payoff matrix under the assumption of best play by both players.

Note that dominated strategies in the original payoff matrix should never be played, regardless of the skill of one's opponent. On the other hand, strategies that are dominated in intermediate matrices in the reduction should not be played if one assumes best play on the part of two players who know they are facing a highly skilled opponent, but *may* be good choices against unskilled opponents.

Example 1   *Reducing a Payoff Matrix*

Find the reduced payoff matrix for the game played by Groceries Galore and Friendly Foods. What does the reduced payoff matrix imply about the best strategies for the two chains?

Solution: As we have seen, Groceries Galore does better by locating in the first neighborhood than the second no matter which option Friendly Foods chooses. Thus, for Groceries Galore, the strategy of locating in the second neighborhood is dominated by the strategy of locating in the first neighborhood. On the other hand, neither of Friendly Foods' strategies dominates the other because choosing the first neighborhood is better if Groceries Galore chooses the second neighborhood, but choosing the second neighborhood is better if Groceries Galore chooses the first neighborhood. Eliminating the second neighborhood from the choices for Groceries Galore leaves the smaller matrix of Table 9.4.

Table 9.4	**A Partially Reduced Payoff Matrix for the Grocery Stores**	
		Friendly Foods
	First Neighborhood	*Second Neighborhood*
Groceries Galore *First Neighborhood*	(5600, 2400)	(8000, 5000)

Given that both supermarkets know the payoff matrix, Friendly Foods should anticipate what has happened. Based on the first reduction and preferring 5000 customers to 2400, it will eliminate the option of locating in the first neighborhood, which leaves the reduced payoff matrix in Table 9.5.

Table 9.5	**Reduced Payoff Matrix for the Grocery Stores**
	Friendly Foods
	Second Neighborhood
Groceries Galore *First Neighborhood*	(8000, 5000)

Because we have reduced this particular game to single strategies for both players, best play by both dictates that Groceries Galore should locate in the first neighborhood and Friendly Foods in the second. As we will see, there are many games that cannot be reduced to a single pair of strategies. ∎

Now imagine that Groceries Galore and Friendly Foods have both publicly announced their plans (which of course they based on the reduced payoff matrix!) but that each has a chance to second-guess its decision. Would either have incentive to do so? By reversing its original plans and deciding to locate in the second neighborhood, Groceries Galore would drop from 8000 to 3500 customers. Similarly, Friendly Foods would drop from 5000 to 2400 customers. Neither chain would change strategies unilaterally. By changing **unilaterally**, we mean changing strategy while the opposing player stays with his or her current strategy. A pair of strategies of this type is called an equilibrium point.

An **equilibrium point** of a game is a pair of strategies such that neither player has any incentive to unilaterally change strategies. For this reason it is **stable**, and once it is reached, it will generally persist through repeated plays of the game. When a game is not at an equilibrium point, at least one player has incentive to unilaterally change strategies, so that such a point is called **unstable**. A game may have a unique equilibrium point, more than one equilibrium point, or no equilibrium points.

If an entry in a matrix game is an equilibrium point, then the Row Player cannot move to another entry in the column chosen by the Column Player and achieve a better outcome (although there may be an outcome that is equally good). Thus, the value to the Row Player is at least as large as any other value to the Row Player in the column. Similarly, the Column Player cannot move to another entry in the row chosen by the Row Player and achieve a better outcome, so that the value to the Column Player is at least as large as any in the row. *Remember that the Row Player chooses the row and the Column Player chooses the column.*

Whenever elimination of dominated strategies leads to single strategies for both players, as was the case for Groceries Galore and Friendly Foods, the pair of strategies form an equilibrium point. However, as we will see, equilibrium points may also occur in other ways.

Our next two examples are probably the best known of all matrix games. The first is the Prisoner's Dilemma, the second, the game of chicken.

Example 2   *The Prisoner's Dilemma*

Roberts and Carson are arrested on a burglary charge. Each faces a 4-year prison term. However, police are convinced that the two are perpetrators of a string of burglaries. Unfortunately, they lack sufficient evidence for conviction on these other charges. They have kept Roberts and Carson separated since their arrest. The district attorney offers each of them the following deal: If one of them signs a confession to the string of burglaries implicating the other, the one who confesses will be let off with a 3-year sentence instead of the 4 years he is currently facing. The other will receive a 20-year sentence. If both confess, each will receive a 15-year sentence. Assume Roberts and Carson are only interested in minimizing their jail time and that those described are their only options (no matter whom they get as their lawyers).

(a) Construct a payoff matrix for the Prisoner's Dilemma.

(b) Find all equilibrium points of the game.

(c) Find the reduced payoff matrix for the game.

Solution:

(a) Because time in jail is bad, it is reasonable to set the value of a 4-year sentence to be -4, the payoff of a 3-year sentence to be -3, and so forth. The payoff matrix is given in Table 9.6.

Table 9.6	**Payoff Matrix for the Prisoner's Dilemma**		
		Carson	
		Remain Silent	*Confess*
Roberts	*Remain Silent* *Confess*	$(-4, -4)$ $(-3, -20)$	$(-20, -3)$ $(-15, -15)$

(b) One systematic way to search for potential equilibrium points is to begin by scanning each column for the biggest payoff to the row player, Roberts. Roberts would want to unilaterally switch strategy from any point in a column with a smaller payoff to Roberts than that of some other point in the column. Thus, our only candidate for an equilibrium point in the first column is the strategy pair (confess, remain silent) with payoff -3 to Roberts. However, the column player, Carson, can improve from the payoff -20 in strategy pair (confess, remain silent) to a payoff of -15 by unilaterally switching strategy from remain silent to confess. Therefore, the first column contains no equilibrium points. Similarly, scanning the second column we see that the strategy pair (confess, confess) gives a larger payoff of -15 to Roberts than the -20 payoff corresponding to the strategy pair (remain silent, confess). If Roberts chooses to confess, then by looking across the entries in the second row we see that Carson can obtain a payoff of -20 in the strategy pair (confess, remain silent) and a payoff of -15 in the strategy pair (confess, confess). Thus, we see that in this situation Carson has no incentive to unilaterally switch strategies from confess to remain silent. Therefore, the pair of strategies (confess, confess) is an equilibrium point.

(c) For Roberts, the strategy of remaining silent is dominated by that of confessing because the payoff -4 is less than the payoff -3 and the payoff -20 is less than the payoff -15. Similarly, for Carson the strategy of remaining silent is dominated by that of confessing. Therefore, as shown in Table 9.7, the game again reduces to a single strategy for each criminal.

Table 9.7	**Reduced Payoff Matrix for the Prisoner's Dilemma**	
		Carson
		Confess
Roberts	*Confess*	$(-15, -15)$

We point out that it is only the way in which the criminals would rank the outcomes in order of preference, rather than the actual length of sentence, that determines the optimal strategies in the Prisoner's Dilemma game considered in Example 2. If each criminal ranks the outcomes from 1 through 4, from worst to best, we could rewrite the payoff matrix as shown in Table 9.8.

Table 9.8	**Payoff Matrix of Ranks for the Prisoner's Dilemma**	
		Carson	
		Remain Silent	*Confess*
Roberts	*Remain Silent*	(3, 3)	(1, 4)
	Confess	(4, 1)	(2, 2)

Written this way, the analysis remains exactly the same. Any game with these rankings (up to ordering the strategies) may be called the Prisoner's Dilemma.

An arms race between two countries is a form of the Prisoner's Dilemma, where both countries have incentives to unilaterally and perhaps secretly build up their arms supply, even though they end up worse off than if neither had done so. Companies that do not install pollution controls, consumers who do not re-

BRANCHING OUT 9.2

REPEATED GAMES

Suppose the Prisoner's Dilemma is to be played more than once between the same two players, with rewards or penalties being cumulative. Repeated games of some version of the Prisoner's Dilemma appear so often that they have gained the name of the *Iterated Prisoner's Dilemma*. Let's begin with playing the game just twice. As we did in alternate move games, it pays to trace back the strategies beginning with the players' last moves. In the last round, Roberts and Carson are faced with the original Prisoner's Dilemma. Thus, both players should confess on the second, and final, round. Now let's move back to the first round. They should recognize that both players' optimal strategy is to confess in the second round no matter what transpires in the first round. Given that their choices will not affect play of the second round, it is in both players' best interests to confess in the first round. The Prisoner's Dilemma has simply repeated itself. Now suppose Roberts and Carson are going to play exactly 10 rounds and that both know this. Then, reasoning as previously, both will confess in the last round, leading back to both confessing in the ninth round, and so forth. This reasoning is completely general. When two players play a simultaneous move game without communication some fixed number of times that both players know in advance, their opti-

mal strategies for the repeated game are to repeatedly play their optimal strategies for a single play.

One test of any theory is how it plays out in practice. In experiments consisting of a single play of the Prisoner's Dilemma, subjects generally choose the dominant strategy of confessing. As the number of plays increases, players are more apt to venture off this path and risk remaining silent for a round or two in the hope that the other player follows suit. These players are treating the game as if the number of repetitions is indefinite or as if the game is to be repeated infinitely often. If there is no definite last game, we can no longer apply the argument of reasoning backwards from the last game. In fact, cooperating becomes a realistic option and strategies in which both players remain silent are among the many equilibria.

There are several examples of the Iterated Prisoner's Dilemma in real life. In the days of bloody trench warfare in World War I, where enemy soldiers were sometimes only a few hundred yards apart, there were sometimes uneasy, unofficial truces, in which soldiers walked around fairly openly, well within range of enemy fire. Why did the soldiers not take advantage of these easy targets? At least in part because they knew that the enemy would retaliate, making both sides worse off as far as the individual

cycle, citizens who decline to vote, and fishermen who do not limit their catch are all playing out a multiplayer version of the Prisoner's Dilemma, where it is in no one individual's interest to self-sacrifice, yet each member of the group may be better off if all sacrifice than if none do.

Many people view the conclusion that the criminals' optimal strategy is to confess as paradoxical because both would do much better by remaining silent. These people point out that there is often "honor among thieves" or that criminals who "squeal" may also face retribution. However, that is not the game at hand! The payoff matrix is assumed to represent the total payoff of the game to the players. So if the criminals agreed before the crime that anyone "ratting" would be killed, this becomes part of the payoff matrix and alters the game. *Any and all* factors other than length of sentence have to be built into the payoff matrix. They would affect the outcome of the game only to the extent they changed the order in which the criminals ranked the outcomes. For the given payoff matrix, it is only when they can communicate, so that they have some hope of coordinating their actions, that remaining silent may be a reasonable strategy.

soldiers were concerned. Another example was reported by G. S. Wilkinson in a 1984 *Nature* article. Vampire bats are known to share blood with roostmates whose hunt was unsuccessful on a given night. Because the bats cannot live much longer than 72 hours without food, if we view a bat's choice of whether or not to share as a game lasting several nights and played over many rounds, the interaction between a pair of bats may be viewed as a form of the Iterated Prisoner's Dilemma. The article goes on to report evidence of retaliatory behavior against those bats that fail to share.

When theoretical analysis needs verification or proves overly difficult, scientists often turn to simulation, which in game theory simply means playing the game many times under controlled conditions. In 1981, Robert Axelrod held an Iterated Prisoner's Dilemma computer tournament, inviting experts in game theory to submit a strategy. What is a strategy? The player must choose what to do in the first round. In the second round, the player must choose a play for each of the two choices the opponent could have made in the first round. In the third round, the player must choose a play for each way the first two rounds could have been played. Clearly, possible strategies may be enormously complicated. Sixty-two contestants entered programs, consisting of computer programs written in Basic or Fortran and ranging from 5

to 152 internal statements, for the main tournament. It was designed as a round-robin event, so that every program played every other program as well as playing itself and a random program that flipped a coin to determine its play in each round, in which each iterated game consisted of 200 rounds. The highest total combined score was declared the winner. The winner was the five-statement program, Tit for Tat, written by Anatol Rapoport! It chose to remain silent on the first round and then to mimic its opponent's last move on succeeding rounds. Thus, it attempted to cooperate, instantly punished those who did not, but was quick to forgive and attempt cooperation again. We note that Tit for Tat can never do better than its opponent in any given round. The goal, however, is not to beat one's opponent; it would be fine if both do well.

Axelrod's book, *The Evolution of Cooperation,* is excellent, mostly nontechnical reading. He describes results from several computer tournaments he has conducted playing the Iterated Prisoner's Dilemma. Many strategies are described in detail, with summaries of how they do versus each other. He even describes an ecological model, with the more successful strategies proliferating and the less successful strategies dying out. Some strategies were initially successful by "preying" on the weaker strategies but eventually dying out themselves as they were forced to go up against the stronger strategies more and more often.

Hopefully the examples we have seen so far have not led you to conclude that all payoff matrices reduce to single strategies. If anything, such results are unusual. The next example, the game of chicken, is more typical.

Example 3   *Chicken*

Chicken is played with cars by two players, whom we call the Row Fool and the Column Fool. They drive at high speeds toward each other and either swerve out of the way at the last second or not. Both players most prefer that their opponent swerves out of the way, while they do not. The second best alternative is for both players to swerve, so that neither loses face completely. The worst alternative is for neither player to swerve. Assume that both players only swerve at the last possible second, so that neither player can determine what his or her opponent will do before making his or her own decision.

(a) Construct a payoff matrix for chicken.

(b) Find all equilibrium points of the game.

(c) Find the reduced payoff matrix for the game.

Solution:

(a) Because it is difficult to assign quantitative values to the different outcomes, we fall back on ranking them in order of preference from worst to best and assign the respective payoffs the numbers 1 through 4. The worst option for a player is for neither to swerve, so this outcome will have payoff 1 for the player. The second worst outcome for a player is to swerve while the opponent does not, so this outcome will have a payoff 2 for the player. The third worst outcome, with payoff 3, for a player is that both players swerve. The best outcome for a player is for the player to not swerve and the opponent to swerve, so this outcome will have the top payoff 4 for the player. The resulting payoff matrix of ranks is given in Table 9.9.

Table 9.9	**Payoff Matrix for Chicken**		
		Column Fool	
		Swerve	*Do Not Swerve*
Row Fool	*Swerve*	(3, 3)	(2, 4)
	Do Not Swerve	(4, 2)	(1, 1)

(b) The game of chicken has two equilibrium points, the pair of strategies (do not swerve, swerve) and (swerve, do not swerve). These equilibrium points have corresponding payoffs (4, 2) and (2, 4). Notice that neither player can improve his or her payoff by changing unilaterally from either of these equilibrium points.

(c) In this example, there are no dominated strategies. Therefore, no reduction of the payoff matrix is possible, so the original payoff matrix is also the reduced payoff matrix.

If communication were possible between the two players in the game of chicken, the situation might be different. Observe that if the Row Fool plans not to swerve and could absolutely convince the Column Fool of this, then the Row Fool would obtain his or her most preferred outcome, because the Column Fool would react by choosing to swerve. Of course, if both decide not to swerve and mistakenly believe they have convinced the other of their intentions, the result is disaster. Confrontations such as the Cuban missile crisis may be interpreted as games of chicken. We will return to the game of chicken when we discuss negotiation in Section 9.6.

In each of the games we have seen so far, either the payoff matrix can be reduced to a single pair of strategies through the elimination of dominated strategies or the payoff matrix cannot be reduced at all. There are also games that can be reduced somewhat, but not to a single pair of strategies, as we see in the next example.

President Kennedy meeting with his cabinet and advisors during the Cuban missile crisis in October 1962.

Example 4 *Dodging a Bully*

A school bully has a favorite victim. They live near to each other, so they normally take the same route home. Each day after school, the Bully seeks to confront the Victim. Fortunately, the Victim may sometimes avoid confrontation. He may take the normal route walking home, he may take a longer roundabout route walking home, or he may avoid possible confrontation altogether by getting his mother to drive him home or by hiring a bigger kid to escort him home. Because the Bully must stay after school for detention, he must guess which option the Victim has chosen that day. He can find and catch up to the Victim only if he chooses the same walking route as the Victim. The Bully's main goal is to confront the Victim. The Victim wants to avoid a confrontation above all, but preferably without the embarrassment of needing his mother's protection or of hiring a bigger kid as a bodyguard, which is slightly worse than getting a ride with mom. All else being equal, both the Victim and the Bully would prefer to take the normal route home rather than the roundabout route.

Calvin and Hobbes by Bill Watterson

(a) Construct a payoff matrix for this situation.
(b) Find all equilibrium points of the game.
(c) Find the reduced payoff matrix for the game.

Solution:

(a) We again make use of rankings rather than trying to assign specific values to different outcomes. We avoid making any assumptions about preferences that are not explicitly given, such as that the Bully might prefer the Victim's mother to pick him up if he cannot confront him. Listing the preferences of the Victim from worst to best, along with the rank, we obtain confrontation on the long route (1), confrontation on the short route (2), hiring a bodyguard (3), ride home from mother (4), taking the long route without confrontation (5), and taking the short route without confrontation (6). The Bully's rankings would be taking the long route without confrontation (1), taking the short route without confrontation (2), confrontation on the long route (3), and confrontation on the short route (4). Our payoff matrix is given in Table 9.10.

Table 9.10 Payoff Matrix for the Bully Game

		Bully	
		Short Route	**Long Route**
	Short Route	(2, 4)	(6, 1)
Victim	**Long Route**	(5, 2)	(1, 3)
	Ride from Mom	(4, 2)	(4, 1)
	Hire a Bodyguard	(3, 2)	(3, 1)

(b) There are no equilibrium points. The Victim would always elect to walk home along the route not taken by the Bully, whereas the Bully would always elect to walk home along the same route as the Victim. Thus, one or the other would always want to alter his strategy.

(c) The Victim's strategy of hiring a bodyguard is dominated by the strategy of getting a ride home with mom. No further reduction is possible, and the reduced payoff matrix is given in Table 9.11. The Victim should rule out hiring a bodyguard, but all other options for both Victim and Bully remain feasible.

Table 9.11 Reduced Payoff Matrix for the Bully Game

		Bully	
		Short Route	**Long Route**
	Short Route	(2, 4)	(6, 1)
Victim	**Long Route**	(5, 2)	(1, 3)
	Ride from Mom	(4, 2)	(4, 1)

Simultaneous Move Games with a Random Component

Sometimes a game will have an additional, random feature such as the flip of a coin or the roll of a die. We will then take the payoffs resulting from the chosen pair of strategies to be the average payoff, or expected payoff, obtained. We will use the ideas of average or expected payoff here on an informal basis with a meaning illustrated in the examples. You can find a more extensive treatment in Section 4.7 and in Section 9.4. We look at a game with a random feature in the next example.

Example 5 · *Bidding at an Auction*

In a sealed bid auction, people submit written bids for items, with the highest bid winning the auction. In our case two antique dealers are competing for a chair either could sell for $400. They can bid $100, $200, or $300 for the chair. (We have limited the dollar amount of the bids they can make for simplicity. In a true auction, the bids would not be so limited.) The highest bid will win, and they will flip a coin in the case of a tie. Assume that each player's objective is to obtain the greatest possible profit on the chair and that they are not permitted to communicate with each other about their possible bids. (Collusion between independent dealers is illegal under the Sherman Antitrust Act. In 1997, the U.S. Justice Department subpoenaed the records of several prominent art dealers, as well as a few large auction houses, as part of an investigation of possible antitrust violations.)

An auction at Christie's in London.

(a) Construct a payoff matrix for the auction.

(b) Find all equilibrium points of the game.

(c) Find the reduced payoff matrix for the game.

Solution:

(a) If the first dealer bids $300 and the second dealer bids either $100 or $200, the first dealer wins the auction and obtains a $100 profit, based on the $400 price the dealer could sell the chair for. The corresponding matrix entries will both be (100, 0). On the other hand, if both bid $300, they will flip a coin to see who wins the auction. Each will obtain the $100 profit half of the time, so that their average profit will be $50. The corresponding matrix entry is (50, 50). Computing the rest of the matrix similarly, we get the payoff matrix in Table 9.12.

Table 9.12	Payoff Matrix for the Auction		

		Second Dealer		
		$100	*$200*	*$300*
	$100	(150, 150)	(0, 200)	(0, 100)
First Dealer	*$200*	(200, 0)	(100, 100)	(0, 100)
	$300	(100, 0)	(100, 0)	(50, 50)

(b) One equilibrium point is when both dealers bid $200. The entry in this case is (100, 100). To see that this pair of strategies is an equilibrium point, notice that the first dealer cannot achieve a payoff better than 100 by moving to another entry in the same column of the payoff matrix, and similarly the second dealer cannot improve her payoff by moving to another entry in the same row. The strategies of both dealers bidding $300 with entry (50, 50) is a second equilibrium point.

(c) For both dealers the strategy of bidding $100 is dominated by the strategy of bidding $200. Eliminating the $100 strategies leaves each player with two options, shown in Table 9.13.

Table 9.13	Partially Reduced Payoff Matrix for the Auction	
	Second Dealer	
	$200	**$300**
First Dealer $200	(100, 100)	(0, 100)
$300	(100, 0)	(50, 50)

After this first reduction, for both dealers bidding $200 is dominated by the strategy of bidding $300. Therefore, the auction reduces to a single strategy for each player, as shown in Table 9.14.

Table 9.14	Reduced Payoff Matrix for the Auction
	Second Dealer
	$300
First Dealer $300	(50, 50)

It is interesting to note that both dealers would be better off at the equilibrium point ($200, $200) where they both bid $200 rather than at the equilibrium point ($300, $300) where both bid $300, because the entry (100, 100) is better for both players than the entry (50, 50). However, because the dealers cannot communicate, the optimal strategy for each of them is to bid $300.

It is natural to know your own preferences fairly well. However, it may not be so easy to determine your opponent's preferences in practice. In the exercises, you will be able to examine the effect of one or both players having erroneous views of their opponent's preferences.

When eliminating dominated strategies reduces the game to unique strategies for each player, selecting such strategies will be the result of the game under best play by both players. The situation is much more complicated when the reduced payoff matrix still leaves several possible strategies. However, eliminated strategies should never be selected in a game between two skilled players. Over the next two sections, we consider one approach to such games in a special case, that of zero-sum games.

EXERCISES FOR SECTION 9.2

In Exercises 1–8, for the given payoff matrix, find
(a) all equilibrium points of the game,
(b) the reduced payoff matrix for the game.

1.

		Column Player	
		C1	C2
Row Player	R1	(2, 2)	(4, 1)
	R2	(3, 4)	(1, 3)

2.

		Column Player	
		C1	C2
Row Player	R1	(2, 2)	(3, 4)
	R2	(4, 3)	(1, 1)

3.

		Column Player			
		C1	C2	C3	C4
	R1	(6, 4)	(7, 1)	(8, 6)	(7, 5)
Row Player	R2	(1, 2)	(9, 5)	(4, 7)	(5, 9)
	R3	(8, 8)	(6, 2)	(3, 3)	(5, 7)

4.

		Column Player			
		C1	C2	C3	C4
Row Player	R1	(1, 3)	(3, 2)	(5, 1)	(2, 8)
	R2	(8, 4)	(7, 6)	(4, 5)	(6, 7)

5.

		Column Player		
		C1	C2	C3
	R1	(13, 0)	(5, 6)	(1, 8)
Row Player	R2	(10, 23)	(4, 9)	(0, 9)
	R3	(5, 12)	(3, 7)	(10, 4)

6.

		Column Player			
		C1	C2	C3	C4
	R1	(3, 7)	(−1, 7)	(−3, 6)	(4, 6)
	R2	(−2, 10)	(6, 8)	(7, 10)	(3, −1)
Row Player	R3	(−1, 7)	(5, 8)	(5, 7)	(5, 10)
	R4	(5, 4)	(6, 9)	(4, 3)	(5, 3)

7.

		Column Player			
		C1	C2	C3	C4
	R1	(9, 0)	(5, 4)	(9, 1)	(8, 3)
	R2	(−2, 8)	(3, 5)	(7, 3)	(0, 6)
Row Player	R3	(1, −4)	(6, 6)	(5, 5)	(7, −3)
	R4	(−1, 7)	(5, 8)	(8, 3)	(0, 7)

8.

		Column Player				
		C1	C2	C3	C4	C5
	R1	(0, 10)	(7, 4)	(6, 6)	(7, 3)	(2, 8)
Row Player	R2	(3, 8)	(8, −1)	(8, 2)	(2, 5)	(2, 7)
	R3	(4, 5)	(6, 3)	(6, 7)	(5, 6)	(5, 4)

In Exercises 9 and 10, consider the following: Two pigs, one dominant and one submissive, are put into a cage. One corner of the cage has a lever that either of the pigs can push to release 12 units of food at the opposite corner of the cage. Pressing the lever costs the pig the equivalent of one unit of food in expended energy. If only the dominant pig presses the lever, the submissive pig is able to eat eight units of the food before the dominant pig is able to force the pig aside and eat the last four units. If only the submissive pig presses the lever, it will get little or no food as described in the exercise. If neither pig presses the lever, both go hungry. If both should somehow manage to press the lever together, each expends one unit of energy and the submissive pig is able to eat four units of food and the dominant pig, eight units.

(a) Construct the payoff matrix for the game between the two pigs.

(b) Can you determine the pigs' optimal strategies, assuming best play on the part of the other pig?

9. When the submissive pig pushes the lever, it does not get any food. (This game is sometimes called The Rational Pigs. It has been simulated with real pigs by animal behaviorists. In most cases, the pigs followed their optimal strategies.)

10. When the submissive pig pushes the lever, it manages to get 2 units of food, the dominant pig, 10. (Note that the difference between the answers to Exercises 9 and 10 shows that experiments must be carefully controlled.)

Exercises 11–14 concern cartels or oligopolies. They show that optimal behavior may depend on the market share of the parties. Two rock hounds (amateur geologists) have stumbled onto the remains of a meteor, each having found one large chunk. They uneasily agree to divide their own chunk into a target number of pieces (which will vary in the exercises) that they can sell at a profit of $25 for each piece. However, they can also manage to divide their own chunk into as many as three additional pieces above the target levels. Unfortunately, this extra supply of meteor pieces will depress the price by $2 for each additional chunk either of them sells. For instance, if one exceeds the target level by one piece and the other by two pieces, the profit will be $19 per piece. They do not have any particular loyalty to each other, so their goal is to maximize their own total profit.

(a) Construct the payoff matrix for this cartel of two people.

(b) Find all equilibrium points of this matrix.

(c) Find the reduced payoff matrix for the cartel.

(d) Is it necessarily in both rock hounds' self-interest to cheat on their uneasy agreement?

11. Both rock hounds have a target level of 6 pieces.

12. Both rock hounds have a target level of 5 pieces.

13. One rock hound has a target level of 10 pieces, the other a target level of 9 pieces.

14. One rock hound has a target level of 20 pieces, the other a target level of 7 pieces.

Children are notoriously hard to get to bed on time. In our simplified game between a child and parent, each must choose between two types of behavior. The parent can be either mild mannered or harsh in telling the child that it is bedtime. The child can be either cooperative or uncooperative when asked to go to bed. Assume that the parent most prefers taking a mild-mannered approach when the child intends to cooperate and least prefers taking a harsh approach when the child intends to cooperate. On the other hand, the child most prefers to be uncooperative with the parent taking a mild-mannered approach and least prefers for the parent to take a harsh approach when the child intends to cooperate. Which of the other two outcomes falls second and which third depends on further details as described in Exercises 15–18. Experience has taught both what the other prefers. For each situation,

(a) construct the payoff matrix based on the given rankings;

(b) find all equilibrium points of the game;

(c) find the reduced payoff matrix for the game.

15. The parent prefers to be harsh instead of mild mannered when the child intends to be uncooperative. The child prefers to be cooperative with the parent taking a mild-mannered approach to being uncooperative with the parent taking a harsh approach.

16. The parent prefers to be harsh instead of mild mannered when the child intends to be uncooperative. The child prefers to be uncooperative with the parent taking a harsh approach to being cooperative with the parent taking a mild-mannered approach.

17. The parent prefers to be mild mannered instead of harsh when the child intends to be uncooperative. The child prefers to be cooperative with the parent taking a mild-mannered approach to being uncooperative with the parent taking a harsh approach.

18. The parent prefers to be mild mannered instead of harsh when the child intends to be uncooperative. The child prefers to be uncooperative with the parent taking a harsh approach to being cooperative with the parent taking a mild-mannered approach.

An automobile buying game has two players, a Buyer and a Seller. The Buyer can (i) buy the car (with at most mild attempts at negotiation), (ii) try to negotiate a lower price and go elsewhere if rebuffed, or (iii) try to negotiate a lower price but buy the car anyway if rebuffed. The Seller can (i) refuse to negotiate (if asked), or (ii) agree to a lower price (if asked). The Seller cannot discern if the Buyer will go elsewhere if rebuffed, and once it happens, it is too late. There are four distinct outcomes for each party. The Buyer's most preferred outcome is to successfully negotiate a lower price. The Buyer prefers not trying to negotiate to buying the car after a failed negotiation. The Buyer ranks walking away from an unsuccessful negotiation either above or below all cases where he or she buys the car at the original price. This will vary in Exercises 19–24. The Seller's most preferred outcome is to sell the car with no attempt by the Buyer to negotiate a lower price, followed by refusing to negotiate with a buyer who will purchase the car anyway. Whether agreeing to a lower price or having the buyer go elsewhere comes next will also vary by exercise. Assume that both the Buyer and the Seller are able to successfully assess the other parties' preferences, except as noted in Exercises 23 and 24.

19. The Buyer prefers buying the car at the original price after unsuccessful negotiation to going elsewhere to buy the car. The Seller prefers to sell the car at the lower price rather than not to sell it to the Buyer.
    (a) Construct the payoff matrix based on the given rankings.
    (b) Find all equilibrium points of the game.
    (c) Find the reduced payoff matrix for the game.

20. The Buyer prefers buying the car at the original price after unsuccessful negotiation to going elsewhere to buy the car. The Seller prefers not to sell the car to the Buyer rather than to sell it at the lower price.
    (a) Construct the payoff matrix based on the given rankings.
    (b) Find all equilibrium points of the game.
    (c) Find the reduced payoff matrix for the game.

21. The Buyer prefers going elsewhere to buying the car at the original price. The Seller prefers to sell the car at the lower price rather than not to sell it to the Buyer.
    (a) Construct the payoff matrix based on the given rankings.
    (b) Find all equilibrium points of the game.
    (c) Find the reduced payoff matrix for the game.

22. The Buyer prefers going elsewhere to buying the car at the original price. The Seller prefers not to sell the car to the Buyer rather than to sell it at the lower price.

(a) Construct the payoff matrix based on the given rankings.

(b) Find all equilibrium points of the game.

(c) Find the reduced payoff matrix for the game.

23. The rankings of the Buyer and Seller are as in Exercise 19. However, the Seller mistakenly believes that the Buyer prefers going elsewhere to buying the car at the original price, although the Buyer does not know of this mistake. Analyze how you expect this game to end.

24. The rankings of the Buyer and Seller are as in Exercise 21. However, the Seller mistakenly believes that the Buyer prefers buying the car at the original price after unsuccessful negotiation to going elsewhere to buy the car, although the Buyer does not know of this mistake. Analyze how you expect this game to end.

In Exercises 25 and 26, consider the following: Among the paintings to be simultaneously auctioned in a sealed bid auction are a landscape worth $3000 and a portrait worth $4000. Only two dealers are interested in these paintings. Bids must be multiples of $1000 and ties will be decided by the flip of a coin. Each dealer wishes to maximize her potential profit and will only bid less than the value of a painting. Assume each knows how much the other has to spend.

(a) Construct the payoff matrix for the auction.

(b) Find all equilibrium points.

(c) Find the reduced payoff matrix for the auction.

(d) How should the art dealers bid, assuming best play on the part of the competing dealer?

25. The first dealer has at most $4000 to bid on the two paintings, and the second has at most $5000. Assume they will bid at least $1000 for each painting.

26. The first dealer has at most $4000 to bid on the two paintings, and the second has at most $3000. Assume they will bid at least $1000 for each painting unless the second dealer decides to bid her entire $3000 for the second painting.

27. Construct an example of a game with each player having at least two strategies that has exactly five equilibrium points.

9.3 ZERO-SUM GAMES

When eliminating dominated strategies reduces a game to a single strategy for both players, this pair of strategies is optimal under the assumption of best play by both players and provides a "solution" to the game. However, in games that do not reduce as far, such as the game of chicken in Section 9.2, we seem to be stuck. It is clear that if our strategy is predictable, our opponent has a huge advantage in the game, but beyond hiding our intentions we are at a loss as to how to proceed. We will see one way to arrive at a strategy in a special case referred to as a zero-sum game. In this section we introduce the concept and notation of a zero-sum game. In the next two sections we will look at "mixed" strategies and how they apply to find optimal strategies for zero-sum games.

In many two-player games, for each pair of strategies one player's loss is exactly equal to the other player's gain. You may have played the children's game

Rock-Paper-Scissors. Two players simultaneously select one of rock, paper, or scissors by making specified hand motions. Rock crushes scissors (and wins), scissors cut paper (and wins), and paper covers rock (and wins). If the players' selections are the same, the game is a draw. Otherwise the winner is determined by the rules described. Thus, counting a win as 1, a loss as -1, and a draw as 0, the payoff matrix for Rock-Paper-Scissors is given in Table 9.15.

Table 9.15   Standard Payoff Matrix for Rock-Paper-Scissors

		Column Player		
		Rock	Paper	Scissors
	Rock	$(0, 0)$	$(-1, 1)$	$(1, -1)$
Row Player	Paper	$(1, -1)$	$(0, 0)$	$(-1, 1)$
	Scissors	$(-1, 1)$	$(1, -1)$	$(0, 0)$

Notice that the sum of the pair of payoffs for each matrix entry is 0. A game with this property is called a zero-sum game.

A **zero-sum game** is a matrix game where the pair of payoffs for each matrix entry sums to 0.

With zero-sum games, we can write the payoff matrix in a more abbreviated form. Because for each matrix entry the payoff for the Column Player is just the opposite of the Row Player's payoff, we do not have to write it down explicitly. Instead, we can abbreviate the entries by showing only the payoff of the outcome for the Row Player. For instance, in place of the entry $(1, -1)$ for Rock-Paper-Scissors, we write only 1, and in place of the entry $(-1, 1)$ we write -1. Writing all of the entries of the payoff matrix in Table 9.15 this way, we get the abbreviated form of the payoff matrix shown in Table 9.16.

Table 9.16   Zero-Sum Payoff Matrix for Rock-Paper-Scissors

		Column Player		
		Rock	Paper	Scissors
	Rock	0	-1	1
Row Player	Paper	1	0	-1
	Scissors	-1	1	0

We analyze zero-sum games in the same way as the matrix games we have already seen. Only the way in which we write the payoff matrix is different. For instance, we now recognize an equilibrium point by noting that it must be greater than or equal to anything else in its column and less than or equal to anything else in its row. Note that if the Column Player has a dominated strategy, that strategy will have values greater than or equal to those of some other column.

Example 1   *Rock-Paper-Scissors*

For the Rock-Paper-Scissors game,

(a) find all equilibrium points;

(b) find the reduced payoff matrix.

Solution:

(a) There are no equilibrium points because for each pair of strategies one of the players can benefit by unilaterally switching strategies. Given the Column Player's strategy, the Row Player will always choose the row with value 1. On the other hand, given the Row Player's strategy, the Column Player will choose the column with value −1 (from the Row Player's point of view). Thus, one player or the other will always want to change strategies.

(b) There are no dominated strategies, so the original matrix is reduced.

Our next zero-sum game involves military strategies.

Example 2   *Battle of the Bismarck Sea*

In February 1943, in the middle of World War II, Japanese Admiral Imamura was ordered to deliver reinforcements and supplies via the Bismarck Sea to Japanese forces battling in New Guinea. He had the choice of a northern route or a southern route. General George Churchill Kenney, the U.S. commander, also had two choices: he could send his planes to intercept the Japanese along the northern route or along the southern route. If he anticipated correctly, he could bomb the Japanese convoy for two days along the northern route or three days along the southern route (the difference due to weather conditions). If not, he could recall his planes, send them to the other route, and lose only one day of bombing. The payoffs to the players are the number of days of bombing: positive for Kenney, negative for Imamura.

(a) Construct a payoff matrix for the Battle of the Bismarck Sea.

(b) Find all equilibrium points of the game.

(c) Find the reduced payoff matrix for the game.

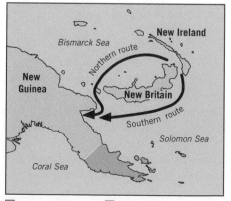

☐ Japanese held areas   ☐ Allied held areas

Solution:

(a) We set the game matrix up as a zero-sum game. With General Kenney as the Row Player, the payoffs are the number of days of bombing the convoy. For Admiral Imamura, the payoffs are the opposite of this. The payoff matrix appears in Table 9.17.

Table 9.17	**Battle of the Bismarck Sea**		
		Admiral Imamura	
		Travel North	*Travel South*
General Kenney	*Search North*	2	2
	Search South	1	3

Source: O.G. Haywood, Jr., "Military decision and game theory." *Journal of the Operations Research Society of America*, 2:1 (November 1954).[1]

(b) The only equilibrium point is when the Japanese convoy takes the northern route and the U.S. forces search the northern route.

(c) General Kenney has no dominated strategies at the start. On the other hand, for Admiral Imamura, who wishes to minimize the number of days of bombing, taking the southern route is dominated by taking the northern route. Upon discerning this, General Kenney should elect to search north, resulting in the reduced payoff matrix in Table 9.18.

Table 9.18	**Reduced Payoff Matrix for the Battle of the Bismarck Sea**
	Admiral Imamura
	Travel North
General Kenney *Search North*	2

You may be curious about how the Battle of the Bismarck Sea turned out. Both commanders selected the northern route. Even though both commanders chose their optimal strategies, the result was heavy losses for the Japanese, which would have been worse with another day of bombing. The military doctrine of decision for U.S. forces is that commanders should select a course of action based on what the enemy is able to do rather than trying to anticipate what the enemy might do. This corresponds exactly to determining a strategy by reducing the game through dominated strategies.

An equilibrium point in a zero-sum game is sometimes called a **saddle point**. Its payoff is the largest for the given column and the smallest for the given row. Thinking of the vertical direction or column as going across a horse's back and the horizontal direction or row as going along a horse's back, with the payoffs or outcomes representing height, then the idea of an equilibrium point is given by a horse's saddle, as illustrated in Figure 9.18.

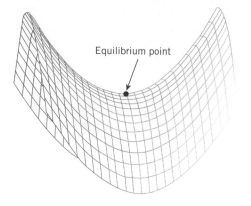

Equilibrium point

Figure 9.18
Equilibrium point
viewed as a saddle
point.

As is true for all matrix games, a zero-sum game may have a unique equilibrium point, more than one equilibrium point, or no equilibrium points. However, in a zero-sum game all of the equilibrium points must have the same payoff (to the Row Player). We leave the verification of this fact as an exercise.

A kind of matrix game that seems more general than a zero-sum game is a **constant-sum game**. In a constant-sum game the pair of payoffs for each matrix entry sums to the same payoff c. However, any constant-sum game whose entry pairs sum to c can be converted to a strategically equivalent zero-sum game by subtracting c from each of the Column Player's payoffs.

Exercises for Section 9.3

In Exercises 1–6, for the given payoff matrix of a zero-sum game, find
(a) all equilibrium points of the game,
(b) the reduced payoff matrix for the game.

1.

		C1	*C2*	*C3*	*C4*
			Column Player		
	R1	5	2	2	4
Row Player	*R2*	6	2	4	3
	R3	−3	1	7	5
	R4	7	0	−1	0

2.

		C1	*C2*	*C3*	*C4*
			Column Player		
	R1	3	−2	−3	5
Row Player	*R2*	1	−4	−2	6
	R3	2	0	1	−1

3.

		C1	C2	C3
			Column Player	
		C1	C2	C3
	R1	−3	12	12
	R2	29	17	25
Row Player	R3	−8	−14	3
	R4	18	−25	−9
	R5	4	0	1

4.

		Column Player			
		C1	C2	C3	C4
	R1	3	−4	−10	−8
	R2	−6	−1	−2	−5
Row Player	R3	0	−13	1	−12
	R4	2	−2	0	−4

5.

		Column Player			
		C1	C2	C3	C4
	R1	3.4	4.0	2.5	2.1
	R2	1.7	1.9	8.0	7.1
Row Player	R3	1.3	5.7	4.7	4.3
	R4	0	3.1	9.1	8.6

6.

		Column Player			
		C1	C2	C3	C4
	R1	7.2	1.0	0	5.2
	R2	6.7	5.3	5.3	5.9
Row Player	R3	3.8	1.2	3.5	2.4
	R4	4.6	3.3	0	4.1
	R5	0	4.7	2.9	7.3

Exercises 7 and 8 involve the following game, which is a miniaturized version of poker. Two players play poker with a deck consisting of four aces and four 2's. To start the game, each player puts $2 into the pot and is dealt one card at random from the deck. The first player can elect to compare cards immediately or to bet by adding a bet of a given size to the pot. If the first player bets this given amount, the second player must decide whether to "fold," surrendering the pot to the first player without any comparison of cards, or to "call" by putting this same amount into the pot. Should the second player call, they then compare cards. In the comparison, a player holding an ace beats a player holding a 2 and wins the entire pot. If both players hold aces or both hold 2's, they divide the pot equally. The payoff matrices, based on the average or expected net winnings of the chosen pair of strategies, for two particular bet sizes are given.

(a) Find all equilibrium points of the game.

(b) Find the reduced payoff matrix for the game.

7. The size of the bet is \$1.

		Second Player			
		Call with Ace or 2	*Call with Ace Only*	*Call with 2 Only*	*Fold with Ace or 2*
First Player	*Bet with Ace or 2*	0	$\frac{1}{7}$	$\frac{13}{7}$	2
	Bet with Ace Only	$\frac{2}{7}$	0	$\frac{5}{7}$	$\frac{3}{7}$
	Bet with 2 Only	$-\frac{2}{7}$	$\frac{1}{7}$	$\frac{8}{7}$	$\frac{11}{7}$
	Do Not Bet	0	0	0	0

8. The size of the bet is \$2.

		Second Player			
		Call with Ace or 2	*Call with Ace Only*	*Call with 2 Only*	*Fold with Ace or 2*
First Player	*Bet with Ace or 2*	0	$-\frac{1}{7}$	$\frac{15}{7}$	2
	Bet with Ace Only	$\frac{4}{7}$	0	1	$\frac{3}{7}$
	Bet with 2 Only	$-\frac{4}{7}$	$-\frac{1}{7}$	$\frac{8}{7}$	$\frac{11}{7}$
	Do Not Bet	0	0	0	0

The games in Exercises 9–10 are somewhat similar to the game Random Chase described in Section 9.1. The game is played on the given graph. We call the line segments edges and the endpoints vertices. A move consists of moving a piece from the vertex it is on along an edge connecting it to another vertex, which we denote by the destination vertex. Each player secretly writes down one move. They then reveal their moves and simultaneously make their moves. The Chaser pays the Runner \$1 for each edge the Chaser would have to travel in order to reach the Runner (\$0 if they end up on the same vertex). For instance, in Exercise 10, if the Chaser ends up on 7 and the Runner ends up on 2, the Chaser pays the Runner \$2.

(a) Set up the payoff matrix as a zero-sum game with the Runner as the row player.

(b) Find the reduced matrix and use it to analyze the moves the players should consider.

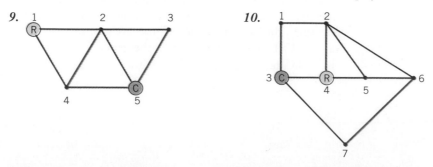

9.   *10.*

Exercises 11–14 involve the following game: In a Battle for Two Cities, Players A and B are fighting to gain control of cities C and D. Each has one army already in City C vying for its control and one army in reserve. Neither has control of, or armies in, city D. The players must simultaneously and secretly allocate their reserve armies to one of the two cities, or continue to hold them in reserve for some future battle. If a player ends up with more armies in a city than his or her opponent, that player gains control of the city and any opponent's army in the city is wiped out. When they have the same number of armies in a city, the outcome will depend on the different scenarios described in the exercises. Players gain one point for each of their opponents' armies wiped out and a certain number of points, depending on the scenario, for each city they control. Their objective is to maximize the difference between their points and their opponent's points.

(a) Let C, D, and H denote each player's options of deploying the reserve army to City C, City D, or holding it in reserve, respectively. Construct the payoff matrix with Player A as the row player and payoffs equal to Player A's points minus Player B's points.

(b) Find all equilibrium points.

(c) Find the reduced payoff matrix for the game.

11. Each city is worth 1 point. When both players occupy a city with the same number of armies, both players' armies are wiped out and neither player gains control of the city.

12. Each city is worth 1 point. When both players occupy a city with the same number of armies, Player B's armies are wiped out and Player A gains control of the city.

13. City C is worth 1 point and City D is worth 2 points. When both players occupy a city with the same number of armies, both players' armies are wiped out and neither player gains control of the city.

14. City C is worth 2 points and City D is worth 1 point. When both players occupy a city with the same number of armies, both players' armies are wiped out and Player A gains control of the city.

Exercises 15 and 16 concern the following scenario: Aaron and Alex engage in a pistol duel to the death. They start 60 paces apart, steadily and slowly walking toward each other. Each has one shot. We simplify matters by assuming a shot either kills the rival or misses completely. Also to simplify matters, assume that Aaron may plan to fire at 60 paces, 40 paces, or 20 paces, whereas Alex may fire at 50 paces, 30 paces, or 10 paces. The probability that each kills his rival will be given in the exercises. Consider a "noisy" duel, meaning that a duelist knows when his rival misses his one shot. Thus, instead of firing as planned, he is then able to continue walking up to his opponent and kill him with absolute certainty. View the duel as a zero-sum game with value 1 for killing one's opponent and -1 for being killed.

(a) Find the payoff matrix for the game with Aaron as the row player.

(b) Eliminate dominated strategies.

(c) Find both players' optimal strategies.

(d) If the duelists select their optimal strategies, find the probability that Alex is the one to be killed.

(In Exercise 15, if Aaron elects to shoot at 40 paces and Alex at 10, Aaron shoots first and kills Alex with probability 0.7 and thus is killed with probability $1 - 0.7 = 0.3$. Aaron's payoff is then $0.7 - 0.3 = 0.4$.)

15.

Aaron			*Alex*	
Strategy	*Probability of Killing Alex*		*Strategy*	*Probability of Killing Aaron*
60 paces	0.6		50 paces	0.5
40 paces	0.7		30 paces	0.65
20 paces	0.8		10 paces	0.85

16.

Aaron			Alex	
Strategy	Probability of Killing Alex		Strategy	Probability of Killing Aaron
60 paces	0.2		50 paces	0.25
40 paces	0.4		30 paces	0.35
20 paces	0.7		10 paces	0.75

17. Explain why all equilibrium points in a zero-sum game have the same payoff (to the Row Player).

9.4 SIMULTANEOUS MOVE GAMES AND MIXED STRATEGIES

When eliminating dominated strategies reduces the payoff matrix to a unique pair of strategies for the players, the result gives the optimal strategy for both players, on the assumption that both players would select their best strategies. Otherwise, we are left without clear strategies for the players and unable to predict what strategies they might use.

To resolve this impasse, imagine that you could estimate the probability that your opponent would play each strategy. Your estimate could be based on knowing how your opponent's mind works or based on experience from previous plays of the game or of similar games. In fact, your opponent can adopt such a list of probabilities as a strategy and use a spinner with sectors whose sizes correspond to the probabilities to select his or her strategy. Such a strategy is a mixture of the original strategies, motivating the following definition:

A **mixed strategy** for a player with strategies numbered from 1 to n is a set of nonnegative numbers $p_1, p_2, \ldots, p_n$, summing to 1. The player using such a strategy selects the ith strategy with probability p_i. The original strategies are called **pure strategies**. A pure strategy may be thought of as a special sort of mixed strategy where one of the probabilities is 1 and the rest are 0.

When a player is using a mixed strategy, the particular strategy played is determined by some randomizing device, such as a spinner, where the probability that the ith strategy is chosen is equal to the probability p_i. A mixed strategy is itself a strategy. However, there are infinitely many mixed strategies whenever a player has at least two pure strategies. It seems at first that we have made analysis of the game infinitely harder and that finding optimal strategies is completely hopeless. Do not despair, for we will learn techniques that make the analysis manageable.

If the payoff matrix has entries that are numerical values, such as years in prison in the Prisoner's Dilemma, and not merely rankings of preferences, then you can compute the average payoff, formally called the **expected payoff**, of playing any of your strategies (pure or mixed) against a mixed strategy. It turns out that optimal mixed strategies exist, and we will find them using algebra and analytic geometry for zero-sum games in this section and the next. We first need to know how to compute averages in general.

> The **expected value** (or "average") of a set of values $x_1, x_2, \ldots, x_n$ that occur with probabilities $p_1, p_2, \ldots, p_n$ (summing to 1) is
>
> $$p_1x_1 + p_2x_2 + \cdots + p_nx_n.$$

Playing Against Known Mixed Strategies

First we apply the definition of expected value to find the expected payoff to a player who chooses a pure strategy to use against an opponent whose play is given by a mixed strategy.

> ### Computing the Expected Payoff to a Player Who Uses a Pure Strategy Against an Opponent's Mixed Strategy
>
> Compute the expected value using the payoffs to the player along the row or column corresponding to the pure strategy as values and using the mixed strategy of the opponent as the probabilities.

It is often reasonable for a player to choose the strategy that will yield the highest expected payoff. However, for some games such as chicken, it seems nearly impossible to assign numerical values to the different outcomes. For other games, the highest expected payoff does not always reflect the most preferred outcome. For instance, most of us would rather have a sure million dollars than a 1-in-500 chance at a billion dollars, even though the expected value of the latter is $2 million. In spite of these limitations, we will restrict our attention to games with numerical payoff matrices and assume that players want to maximize their expected payoffs.

We return to Example 1 from Section 9.2.

Example 1   *One Player Using a Mixed Strategy*

Two supermarket chains, Groceries Galore and Friendly Foods, plan to expand by adding one store to one of two neighborhoods that have no major chain. The payoff matrix giving the number of customers each would gain depending on the neighborhoods they choose is given in Table 9.19.

Table 9.19 | **Payoff Matrix for the Grocery Stores**

		Friendly Foods	
		First Neighborhood	*Second Neighborhood*
Groceries Galore	*First Neighborhood*	(5600, 2400)	(8000, 5000)
	Second Neighborhood	(5000, 8000)	(3500, 1500)

(a) Where should Groceries Galore locate their new store if they believe the probability that Friendly Foods will locate in the first neighborhood is 3/5?

(b) Where should Friendly Foods locate their new store if they believe the probability that Groceries Galore will locate in the first neighborhood is 3/5?

(c) Where should Friendly Foods locate their new store if they believe the probability that Groceries Galore will locate in the first neighborhood is 9/10?

Solution:

(a) For Groceries Galore the strategy of locating in the second neighborhood is dominated by the strategy of locating in the first neighborhood. Thus, they should locate in the first neighborhood no matter what mixed strategy Friendly Foods uses.

(b) Even though we found that Friendly Foods should locate in the second neighborhood when we computed the reduced payoff matrix, that conclusion was based on the assumption of optimal play by both players. Here we are not assuming that Groceries Galore will definitely locate in the first neighborhood, so we must carry out the computation of the expected payoff. Because the probability that Groceries Galore will locate in the first neighborhood is assumed to be 3/5, the probability that they will locate in the second neighborhood must be $1 - 3/5 = 2/5$. If Friendly Foods locates in the first neighborhood, their expected number of new customers will be $(3/5) \cdot 2400 + (2/5) \cdot 8000 = 4640$. If they locate in the second neighborhood, their expected number of new customers will be $(3/5) \cdot 5000 + (2/5) \cdot 1500 = 3600$. Therefore, Friendly Foods should locate in the first neighborhood.

(c) Assuming the probability that Groceries Galore will locate in the first neighborhood is 9/10, the probability that they will locate in the second neighborhood is $1 - 9/10 = 1/10$. If Friendly Foods locates in the first neighborhood, their expected number of new customers will be $(9/10) \cdot 2400 + (1/10) \cdot 8000 = 2960$. If they locate in the second neighborhood, their expected number of new customers will be $(9/10) \cdot 5000 + (1/10) \cdot 1500 = 4650$. This time, Friendly Foods should locate in the second neighborhood.

■

When mixed strategies are used by both players, we can compute the expected payoff of the game to each player.

> ### *Computing the Expected Payoff to a Player When Mixed Strategies Are Used by Both Players*
>
> **1.** Compute the expected payoff to one player for each of that player's pure strategies based on the mixed strategy of the other player.
> **2.** Compute the expected value using the expected payoffs for the player's pure strategies computed in step 1 as values and the mixed strategy of the player as the probabilities. This will give the overall expected payoff to the player.

Thus, the values $x_1, x_2, \ldots, x_n$ in the formula for the expected value are payoffs from one row or column of the payoff matrix in the first step and are expected payoffs themselves in the second step.

We return again to the supermarket chain situation we considered in Example 1 of Section 9.2 and Example 1 of this section.

Example 2 *Both Players Using a Mixed Strategy*

How many customers will each supermarket gain on average if Groceries Galore locates in the first neighborhood with probability 4/5 and Friendly Foods locates in the first neighborhood with probability 1/4? The payoff matrix giving the number of customers each supermarket chain would gain depending on the neighborhoods they choose is given in Table 9.20.

Table 9.20	Payoff Matrix for the Grocery Stores		
		Friendly Foods	
		First Neighborhood	**Second Neighborhood**
Groceries Galore	**First Neighborhood**	(5600, 2400)	(8000, 5000)
	Second Neighborhood	(5000, 8000)	(3500, 1500)

Solution: By locating in the first neighborhood, the expected number of customers Groceries Galore will gain is $(1/4) \cdot 5600 + (3/4) \cdot 8000 = 7400$. Locating in the second neighborhood, the expected number of customers Groceries Galore will gain is $(1/4) \cdot 5000 + (3/4) \cdot 3500 = 3875$. By playing the mixed strategy with probabilities 4/5 and 1/5, the expected number of customers Groceries Galore will gain is $(4/5) \cdot 7400 + (1/5) \cdot 3875 = 6695$.

Similarly, the expected number of customers Friendly Foods will gain is $(4/5) \cdot 2400 + (1/5) \cdot 8000 = 3520$ when it locates in the first neighborhood and $(4/5) \cdot 5000 + (1/5) \cdot 1500 = 4300$ customers when it locates in the second neighborhood. By playing the mixed strategy with probabilities 1/4 and 3/4, the expected number of customers Friendly Foods will gain is $(1/4) \cdot 3520 + (3/4) \cdot 4300 = 4105$.

∎

Technology Tip

Most calculators and statistical software have built-in routines to calculate expected value. You may have to look up *mean* or *weighted mean* in the manual to find them.

For a zero-sum game, the expected payoffs to the two players when both use mixed strategies must be opposites of each other.

Example 3 *Mixed Strategies in Rock-Paper-Scissors*

The payoff matrix for Rock-Paper-Scissors is given in Table 9.21, where 1 denotes a win, −1 a loss, and 0 a draw.

Table 9.21 **Payoff Matrix for Rock-Paper-Scissors**

		Column Player		
		Rock	*Paper*	*Scissors*
	Rock	0	−1	1
Row Player	*Paper*	1	0	−1
	Scissors	−1	1	0

(a) What pure strategy should the Row Player select if the Column Player chooses rock with probability 0.4, paper with probability 0.25, and scissors with probability 0.35?

(b) What pure strategy should the Column Player select if the Row Player chooses rock with probability 0.2, paper with probability 0.3, and scissors with probability 0.5?

(c) Suppose the Row Player chooses rock with probability 0.2, paper with probability 0.3, and scissors with probability 0.5, and the Column Player chooses rock with probability 0.4, paper with probability 0.25, and scissors with probability 0.35. For each of the players, find the expected payoff of the game.

Solution:

(a) If the Row Player chooses to play rock, the expected payoff will be $0.4 \cdot 0 + 0.25 \cdot (−1) + 0.35 \cdot 1 = 0.1$. Should the Row Player select paper, the expected payoff will be $0.4 \cdot 1 + 0.25 \cdot 0 + 0.35 \cdot (−1) = 0.05$. Finally, if the Row Player opts for scissors, the expected payoff will be $0.4 \cdot (−1) + 0.25 \cdot 1 + 0.35 \cdot 0 = −0.15$. Because the largest expected payoff is when the Row Player selects rock, that should be his or her selection.

(b) For a zero-sum game, the payoffs to the Column Player are the opposites of the payoffs in the payoff matrix. Thus, if the Column Player chooses to play rock, the expected payoff to the Column Player is $0.2 \cdot 0 + 0.3 \cdot (−1) + 0.5 \cdot 1 = 0.2$. If the Column Player selects paper or scissors, the expected payoffs are $0.2 \cdot 1 + 0.3 \cdot 0 + 0.5 \cdot (−1) = −0.3$ and $0.2 \cdot (−1) + 0.3 \cdot 1 + 0.5 \cdot 0 = 0.1$, respectively. Because the largest expected payoff is when the Column Player selects rock, that should also be his or her selection.

(c) We average the expected payoffs for the Row Player's pure strategies obtained in (a) based on the probabilities given by the Row Player's mixed strategy. We obtain an expected payoff of $0.2 \cdot 0.1 + 0.3 \cdot 0.05 + 0.5 \cdot (−0.15) = −0.04$ when both players play the given mixed strategies. Because Rock-Paper-Scissors is a zero-sum game, the expected payoff for the Column Player is the opposite of −0.04, or 0.04. We quickly compute the Column Player's expected payoff directly as a check, again obtaining an expected payoff of $0.4 \cdot 0.2 + 0.25 \cdot (−0.3) + 0.35 \cdot 0.1 = 0.04$.

◼

Optimal Mixed Strategies for Zero-Sum Games

When we can eliminate dominated strategies to arrive at single (pure) strategies for both players, we can easily predict the outcome or "solution" of the game under best play by both players. More generally, one could consider an equilibrium point a "solution" in a dynamic sense, meaning that once such a point is reached, neither player would have any incentive to unilaterally change strategies. However, we have seen examples with more than one equilibrium point, so that a study of equilibrium points will not yield a unique optimal pure strategy. Worse yet, even when we restrict to zero-sum games, equilibrium points may not exist. By broadening the players' choices of strategies to include mixed strategies, we might hope to find mixed strategies for the two players that lead to stability, or equilibrium. We therefore define equilibrium points in essentially the same manner as for pure strategies.

> An **equilibrium point** of a game where both players may use mixed strategies is a pair of mixed strategies such that neither player has any incentive to unilaterally change to another mixed strategy.

Do equilibrium points in mixed strategies always exist? We take a short detour to pose several forms of a related question. We will then see that the celebrated Minimax Theorem answers these questions for zero-sum games.

Suppose that our opponent was able to know our strategy, which could be mixed, through experience gained by playing the game many times, through cunning reasoning, or in some other way. What strategy should we adopt? Suppose that we had to announce our strategy, which could be mixed, before our opponent decided on a strategy. What strategy should we adopt? Suppose that no matter what mixed strategy we adopt, we assume that our opponent will find the best strategy against it. What strategy should we adopt? Reading through these questions carefully, you will see that they are really just rephrasing the same question. Our choice of mixed strategy against the backdrop of these questions will be a conservative approach.

Knowing our mixed strategy, our opponent will choose a strategy that maximizes his or her expected payoff or value. We remark that our opponent may always choose a pure strategy to obtain this maximum. For a zero-sum game, one player's loss is the other's gain. The Row Player is trying to maximize the value in the payoff matrix and the Column Player is trying to minimize it. The famous **Minimax Theorem** was proved in 1928 by John von Neumann, one of the founders of the mathematical field of game theory.

John von Neumann
(1903 – 1957).

The Minimax Theorem

For every zero-sum game, there is a number v (for value) and particular mixed strategies for the two players such that (i) the expected payoff to the Row Player will be at least v if the Row Player plays his or her particular mixed strategy, no matter what mixed strategy the Column Player chooses, and (ii) the expected payoff to the Row Player will be at most v if the Column Player plays his or her particular mixed strategy, no matter what mixed strategy the Row Player chooses. This number v is called the **value of the game** and represents the expected advantage to the Row Player (actually a disadvantage if v is negative).

The statement of the Minimax Theorem implies that the expected payoff to the Row Player is v if both players choose these mixed strategies. (The payoff to the Column Player must then be $-v$.) The theorem also says that neither player can do better by unilaterally choosing another strategy—that is, that these mixed strategies form an equilibrium point. Furthermore, because the Row Player can ensure an expected payoff of at least v and the Column Player can ensure a payoff of at most v to the Row Player, this expected payoff to the Row Player must be the result under optimal play by both players.

In 1951, John Nash proved the existence of equilibrium points in mixed strategies for games that are not necessarily zero-sum. Because the players' objectives need not be diametrically opposed to each other, we can no longer say that such equilibrium points comprise optimal strategies for the players. The 1994 Nobel Prize in economics was shared by John Nash, John C. Harsanyi, and Reinhard Selten.

The Minimax Theorem guarantees the existence of optimal mixed strategies for zero-sum games but does not suggest how to find them. Fortunately, there are relatively efficient algorithms for finding equilibrium points using what is known as linear programming. These methods are too involved to undertake in full generality in this chapter. However, in the remainder of this section, we will use a fairly simple algebraic approach to find optimal strategies whenever we can eliminate dominated strategies to reduce the game to only two strategies for each player. As a bonus, this algorithm makes it clear why the Minimax Theorem holds in this special case. In the next section, we will introduce some concepts of linear programming for zero-sum games in which we can eliminate dominated strategies to reduce the game to only two strategies for just one of the players.

Another, more empirical, method of obtaining optimal mixed strategies in zero-sum games was proved by Julia Robinson in 1951. Suppose each player's strategy at every round after the first is chosen to be the best pure strategy against the mixture of strategies exhibited by the opponent in all earlier rounds. Then Robinson's theorem states that, as the number of rounds increases, these mixtures will tend to the optimal mixed strategies referred to in the Minimax Theorem.

We illustrate the reasoning of the Minimax Theorem with a zero-sum game that has the payoff matrix in Table 9.22.

Julia Robinson (1919 – 1985).

Table 9.22	A Representative Zero-Sum Game		
		Column Player	
		C1	*C2*
Row Player	*R1*	3	1
	R2	−2	4

Let q represent the probability that the Column Player selects strategy C1, so that the probability that the Column Player selects strategy C2 is $1 - q$. The expected payoff to the Row Player playing strategy R1 is $3q + 1(1 - q) = 2q + 1$. Similarly, the expected payoff to the Row Player playing R2 is $-2q + 4(1 - q) = -6q + 4$. If the Row Player employs the mixed strategy that chooses strategy R1 with probability p, the expected payoff to the Row Player when both players play their mixed strategies is

$$p(2q + 1) + (1 - p)(-6q + 4) = 8pq - 3p - 6q + 4.$$

This last expression may be written in the form

$$\frac{7}{4} + 8\left(p - \frac{3}{4}\right)\left(q - \frac{3}{8}\right).$$

(We will not show the algebraic procedure for doing this because we will be using general formulas after this example. However, you can check our claim by expanding out the final result.)

By setting $p = 3/4$, we see that the Row Player can guarantee an expected payoff of $7/4$. If the Row Player chooses any other value of p, the Column Player can choose q so that the expression $(p - 3/4)(q - 3/8)$ is negative, lowering the expected payoff to the Row Player. Thus, the Row Player's optimal mixed strategy is when $p = 3/4$. Similarly, the Column Player's optimal strategy is when $q = 3/8$. In this case the value of the game to the Row Player is $7/4$ ($-7/4$ to the Column Player), and the game is to the Row Player's advantage. You might also observe that if the Row Player sets $p = 3/4$, the expected payoff to the Row Player is $7/4$, no matter what strategy, pure or mixed, the Column Player chooses. Similarly, if the Column Player sets $q = 3/8$, the expected payoff to the Row Player is $7/4$, whatever strategy the Row Player chooses. The technique we used in this special case will work for any zero-sum game that when completely reduced has two pure strategies for each player as long as the matrix entries are not the same. The results are as follows:

Optimal Mixed Strategies and the Value of a Zero-Sum Game

Suppose that when completely reduced the payoff matrix of a game is of the form shown in Table 9.23, where the matrix entries a, b, c, and d are not all the same.

Table 9.23	The General Payoff Matrix for 2 × 2 Zero-Sum Games		

		Column Player	
		C1	**C2**
Row Player	**R1**	a	b
	R2	c	d

1. The Row Player's probability of playing strategy R1 in the optimal mixed strategy is given by

$$p = \frac{d - c}{a - b - c + d}.$$

The Row Player's probability of playing strategy R2 is then $1 - p$.

2. The Column Player's probability of playing strategy C1 in the optimal mixed strategy is given by

$$q = \frac{d - b}{a - b - c + d}.$$

The Column Player's probability of playing strategy C2 is then $1 - q$.

3. The value of the game to the Row Player is

$$v = \frac{ad - bc}{a - b - c + d}.$$

The expected value of the game to the Column Player is the opposite of this number.

Notice that the formulas for p, q, and v do not make sense if $a - b - c + d = 0$ because we would have to divide by zero. Fortunately, the only way to have $a - b - c + d = 0$ for a reduced payoff matrix is if all of the entries, a, b, c, and d, are equal. In this case, because all of the payoffs are the same, any mix of strategies by the players will result in this same payoff, a pretty dull game.

Observe that the denominators for p, q, and v in these formulas are the same, so this number need only be computed once. The expected value v of the game to the Row Player is the outcome of his or her optimal mixed strategy against either of the Column Player's pure strategies. If one player plays the optimal strategy and the other any mixture of the two strategies, the expected payoff will be v to the Row Player and $-v$ to the Column Player. However, should one player include a strategy that should have been eliminated through domination, the other player may do better. Furthermore, if a player chooses a different mixed strategy figuring that it cannot hurt, the opponent may now have incentive to choose a different strategy as well.

Example 4 *Tennis*

Martina Hingis serves at the French Open.

Suppose we model a simplified game of tennis as follows: The server must decide whether to serve to the receiver's forehand or the backhand. The receiver must anticipate whether the serve will come to the forehand or backhand. Suppose that the server wins 70% of serves to the receiver's forehand when the receiver guesses incorrectly, but only 40% when the receiver guesses correctly. Also suppose that the server wins 80% of serves to the receiver's backhand when the receiver guesses incorrectly, but only 60% when the receiver guesses correctly.

(a) Set up the payoff matrix for this game based on the given payoffs.

(b) Find the players' optimal mixed strategies assuming that both players know these percentages.

(c) Find the value of the game.

Solution:

(a) The game is a constant-sum game where the percentage of points won by the two players sums to 100%. We write the payoff matrix, shown in Table 9.24, in the form of a zero-sum game, letting the entries be the percentage of points ultimately won by the server.

Table 9.24 **Payoff Matrix for Tennis**			
		Receiver	
		Guess Forehand	*Guess Backhand*
Server	*Serve to Forehand*	40	70
	Serve to Backhand	80	60

(b) After noting that there are no dominated strategies for either player, using our formula for p we find that

$$p = \frac{60 - 80}{40 - 70 - 80 + 60} = \frac{-20}{-50} = \frac{2}{5}.$$

Therefore, the first player should serve to the receiver's forehand with probability 2/5 and to the backhand with probability 3/5. Similarly,

$$q = \frac{60 - 70}{-50} = \frac{-10}{-50} = \frac{1}{5}.$$

Therefore, the second player should anticipate forehand with probability 1/5 and backhand with probability 4/5. These probabilities seem to reflect that the receiver's backhand is somewhat weaker than his or her forehand, although it could conceivably reflect the server's abilities.

(c) The value of the game to the server is

$$v = \frac{40 \cdot 60 - 70 \cdot 80}{-50} = \frac{-3200}{-50} = 64.$$

Thus, under optimal play by both players, the server should win 64% of the points.

∎

You might note that if the server served exclusively to the forehand or backhand, the receiver would always guess correctly and the server would win only 40% and 60% of the points, respectively. On the other hand, if the receiver were to always guess forehand or backhand, the server would serve to the other side and win 80% and 70% of the points, respectively. This confirms our expectation that both players gain by being unpredictable, in other words, by playing a mixed strategy. If the receiver's backhand improves, it is possible that the server's best strategy is to serve more often to the receiver's backhand, a somewhat paradoxical result! We will explore such possibilities in the exercises.

Example 5   *Political Campaign Strategies*

The campaign staffs of two competing political candidates are planning the style of campaign. The time frame for the campaign is very short, so that both staffs must commit to a strategy for logistical and scheduling reasons before knowing their opponent's strategy. The status quo is for the candidates to give formal speeches to large gatherings, but they could also opt for more personal, informal gatherings, or for taped radio and television messages. The first candidate is more personable and able to improvise, whereas the second is more thoughtful and able to come up with detailed goals given time to prepare. We will let the payoff of a pair of strategies to the first candidate be the percentage of voters that will shift from the second candidate to the first candidate due solely to these media strategies, and we get the payoff matrix in Table 9.25.

Table 9.25 **Payoff Matrix for the Election Game**

		Second Candidate		
		Formal Speeches	*Informal Settings*	*Radio and TV*
	Formal Speeches	0	3	−1
First Candidate	*Informal Settings*	2	3	−2
	Radio and TV	1	2	4

Find the candidates' optimal strategies and determine the value of the game.

Solution: For the second candidate, the strategy of informal settings is dominated by appearing in formal speeches. Eliminating this strategy, we see that the first candidate should not elect to campaign through formal speeches, because this strategy will now be dominated by radio and TV messages. The reduced payoff matrix is shown in Table 9.26.

Table 9.26 Reduced Payoff Matrix for the Election Game

		Second Candidate	
		Formal Speeches	*Radio and TV*
First Candidate	*Informal Settings*	2	−2
	Radio and TV	1	4

We now compute the optimal mixed strategies. The first candidate should select informal settings with probability

$$p = \frac{4 - 1}{2 - (-2) - 1 + 4} = \frac{3}{7},$$

and radio and TV with probability 4/7. The second candidate should select formal speeches with probability

$$q = \frac{4 - (-2)}{7} = \frac{6}{7},$$

and radio and TV with probability 1/7. We find that the value of the game for the first candidate is

$$v = \frac{2 \cdot 4 - (-2) \cdot 1}{7} = \frac{10}{7}.$$

Thus, the first candidate has an expected gain of 10/7 of a percent under optimal play.

■

The overriding game in Example 5 is to win the election. Thus, if the candidates have accurate information about their standings in the polls, their strategies may be unrelated to the game of Example 5. For instance, if the first candidate has a lead that is less than 1%, the first candidate's best strategy would be to always choose radio and TV messages so as to ensure victory.

EXERCISES FOR SECTION 9.4

1. Consider the following payoff matrix:

		Column Player	
		C1	*C2*
Row Player	*R1*	(5, 1)	(3, 2)
	R2	(3, 4)	(4, 3)

(a) What pure strategy should the Row Player choose if the Column Player selects strategy C1 with probability 0.4 and C2 with probability 0.6?

(b) What pure strategy should the Column Player choose if the Row Player selects strategy R1 with probability 0.7 and R2 with probability 0.3?

(c) Suppose the Row Player chooses strategy R1 with probability 0.7 and R2 with probability 0.3 and the Column Player chooses strategy C1 with probability 0.4 and C2 with probability 0.6. For each player, find the expected payoff of the game.

2. Consider the following payoff matrix:

		Column Player	
		C1	**C2**
Row Player	**R1**	(2, 2)	(3, 7)
	R2	(5, 3)	(1, 0)

(a) What pure strategy should the Row Player choose if the Column Player selects strategy C1 with probability 0.75 and C2 with probability 0.25?

(b) What pure strategy should the Column Player choose if the Row Player selects strategy R1 with probability 0.5 and R2 with probability 0.5?

(c) Suppose the Row Player chooses strategy R1 with probability 0.5 and R2 with probability 0.5 and the Column Player chooses strategy C1 with probability 0.75 and C2 with probability 0.25. For each player, find the expected payoff of the game.

3. Consider the following payoff matrix:

		Column Player	
		C1	**C2**
Row Player	**R1**	(8, 6)	(−1, −3)
	R2	(4, −2)	(5, 7)

(a) What pure strategy should the Row Player choose if the Column Player selects strategy C1 with probability 1/3 and C2 with probability 2/3?

(b) What pure strategy should the Column Player choose if the Row Player selects strategy R1 with probability 4/5 and R2 with probability 1/5?

(c) Suppose the Row Player chooses strategy R1 with probability 4/5 and R2 with probability 1/5 and the Column Player chooses strategy C1 with probability 1/3 and C2 with probability 2/3. For each player, find the expected payoff of the game.

4. Consider the following payoff matrix:

		Column Player	
		C1	**C2**
Row Player	**R1**	(2, 2)	(−3, −4)
	R2	(−4, −3)	(1, 1)

(a) What pure strategy should the Row Player choose if the Column Player selects strategy C1 with probability 1/4 and C2 with probability 3/4?

(b) What pure strategy should the Column Player choose if the Row Player selects strategy R1 with probability 1/6 and R2 with probability 5/6?

(c) Suppose the Row Player chooses strategy R1 with probability 1/6 and R2 with probability 5/6 and the Column Player chooses strategy C1 with probability 1/4 and C2 with probability 3/4. For each player, find the expected payoff of the game.

5. Consider the following payoff matrix:

		Column Player		
		C1	**C2**	**C3**
	R1	$(9, 0)$	$(5, 4)$	$(9, 1)$
Row Player	**R2**	$(-2, 8)$	$(3, 5)$	$(7, 3)$
	R3	$(1, -4)$	$(6, 6)$	$(5, 5)$
	R4	$(-1, 7)$	$(5, 8)$	$(8, 3)$

(a) What pure strategy should the Row Player choose if the Column Player selects strategy C1 with probability 0.4, C2 with probability 0.5, and C3 with probability 0.1?

(b) What pure strategy should the Column Player choose if the Row Player selects strategy R1 with probability 0.1, R2 with probability 0.2, R3 with probability 0.4, and R4 with probability 0.3?

(c) Suppose the Row Player chooses strategy R1 with probability 0.1, R2 with probability 0.2, R3 with probability 0.4, and R4 with probability 0.3, and suppose the Column Player chooses strategy C1 with probability 0.4, C2 with probability 0.5, and C3 with probability 0.1. For each player, find the expected payoff of the game.

6. Consider the following payoff matrix:

		Column Player			
		C1	**C2**	**C3**	**C4**
	R1	$(9, 0)$	$(5, 4)$	$(9, 1)$	$(8, 3)$
Row Player	**R2**	$(-2, 8)$	$(3, 5)$	$(7, 3)$	$(0, 6)$
	R3	$(1, -4)$	$(6, 6)$	$(5, 5)$	$(7, -3)$

(a) What pure strategy should the Row Player choose if the Column Player selects strategy C1 with probability 0.3, C2 with probability 0.3, C3 with probability 0.1, and C4 with probability 0.3?

(b) What pure strategy should the Column Player choose if the Row Player selects strategy R1 with probability 0.6, R2 with probability 0.1, and R3 with probability 0.3?

(c) Suppose the Row Player chooses strategy R1 with probability 0.6, R2 with probability 0.1, and R3 with probability 0.3, and suppose the Column Player chooses strategy C1 with probability 0.3, C2 with probability 0.3, C3 with probability 0.1, and C4 with probability 0.3. For each player, find the expected payoff of the game.

7. Consider the following payoff matrix for a zero-sum game:

		Column Player	
		C1	**C2**
Row Player	**R1**	2	5
	R2	7	3

(a) What pure strategy should the Row Player choose if the Column Player selects strategy C1 with probability 2/7 and C2 with probability 5/7?

(b) What pure strategy should the Column Player choose if the Row Player selects strategy R1 with probability 7/10 and R2 with probability 3/10?

(c) Suppose the Row Player chooses strategy R1 with probability 7/10 and R2 with probability 3/10 and the Column Player chooses strategy C1 with probability 2/7 and C2 with probability 5/7. For each player, find the expected payoff of the game.

8. Consider the following payoff matrix for a zero-sum game:

		Column Player	
		C1	*C2*
Row Player	*R1*	1	2
	R2	8	4

(a) What pure strategy should the Row Player choose if the Column Player selects strategy C1 with probability 2/5 and C2 with probability 3/5?

(b) What pure strategy should the Column Player choose if the Row Player selects strategy R1 with probability 1/4 and R2 with probability 3/4?

(c) Suppose the Row Player chooses strategy R1 with probability 1/4 and R2 with probability 3/4 and the Column Player chooses strategy C1 with probability 2/5 and C2 with probability 3/5. For each player, find the expected payoff of the game.

9. Consider the following payoff matrix for a zero-sum game:

		Column Player			
		C1	*C2*	*C3*	*C4*
	R1	1	3	3	1
Row Player	*R2*	2	−1	0	2
	R3	3	2	3	−1

(a) What pure strategy should the Row Player choose if the Column Player selects strategy C1 with probability 0.2, C2 with probability 0.4, C3 with probability 0.1, and C4 with probability 0.3?

(b) What pure strategy should the Column Player choose if the Row Player selects strategy R1 with probability 0.2, R2 with probability 0.3, and R3 with probability 0.5?

(c) Suppose the Row Player chooses strategy R1 with probability 0.2, R2 with probability 0.3, and R3 with probability 0.5, and suppose the Column Player chooses strategy C1 with probability 0.2, C2 with probability 0.4, C3 with probability 0.1, and C4 with probability 0.3. For each player, find the expected payoff of the game.

10. Consider the following payoff matrix for a zero-sum game:

		Column Player				
		C1	*C2*	*C3*	*C4*	*C5*
Row Player	*R1*	−1	2	1	2	1
	R2	2	−5	3	−3	3

(a) What pure strategy should the Row Player choose if the Column Player selects strategy C1 with probability 0.2, C2 with probability 0.1, C3 with probability 0.1, C4 with probability 0.5, and C5 with probability 0.1?

(b) What pure strategy should the Column Player choose if the Row Player selects strategy R1 with probability 0.25 and R2 with probability 0.75?

(c) Suppose the Row Player chooses strategy R1 with probability 0.25 and R2 with probability 0.75, and suppose the Column Player chooses strategy C1 with probability 0.2, C2 with probability 0.1, C3 with probability 0.1, C4 with probability 0.5, and C5 with probability 0.1. For each player, find the expected payoff of the game.

In Exercises 11–16, for the given payoff matrix, find

(a) the optimal strategies, which may end up pure or mixed,

(b) the value of the game.

11.

		Column Player	
		C1	C2
Row Player	R1	2	5
	R2	7	3

12.

		Column Player	
		C1	C2
Row Player	R1	1	2
	R2	8	4

13.

		Column Player	
		C1	C2
Row Player	R1	−2	5
	R2	7	−3

14.

		Column Player	
		C1	C2
Row Player	R1	1	−2
	R2	−8	4

15.

		Column Player	
		C1	C2
Row Player	R1	0.75	0.1
	R2	0.4	0.4

16.

		Column Player	
		C1	C2
Row Player	R1	0.3	0.25
	R2	0.2	0.55

In Exercises 17–20, the given payoff matrix represents a zero-sum game.

(a) Eliminate dominated strategies to find the reduced payoff matrix.

(b) Use this reduced payoff matrix to determine the players' optimal mixed strategies.

(c) Determine the value of the game.

17.

		Column Player		
		C1	C2	C3
	R1	−2	−3	2
Row Player	R2	1	1	−1
	R3	0	2	3

18.

		Column Player		
		C1	*C2*	*C3*
	R1	1	−1	0
Row Player	*R2*	−4	4	−5
	R3	3	−2	−1

19.

		Column Player			
		C1	*C2*	*C3*	*C4*
	R1	1	3	3	1
	R2	2	−1	0	2
Row Player	*R3*	3	2	3	−1
	R4	4	0	0	4

20.

		Column Player				
		C1	*C2*	*C3*	*C4*	*C5*
	R1	−1	2	1	2	1
Row Player	*R2*	2	−5	3	−3	3
	R3	−2	1	0	2	−2

Exercises 21 and 22 concern the following scenario: The plaintiff and defendant in an imminent civil trial must choose their legal representation. Each must elect to have no lawyer or to retain a lawyer, not knowing the other's decision. The payoff matrix for these possibilities (in dollars), taking into account the results of litigation and the lawyers' fees, is given here.

		Defendant	
		No Lawyer	**Hire a Lawyer**
Plaintiff	**No Lawyer**	(3000, −3000)	(0, −800)
	Hire a Lawyer	(2400, −3200)	(2000, −3600)

21. (a) If the probability that the plaintiff retains a lawyer is 2/3, what strategy should the defendant choose?

(b) If the probability that the defendant retains a lawyer is 1/2, what strategy should the plaintiff choose?

(c) If the probability that the plaintiff retains a lawyer is 2/3 and the probability that the defendant retains a lawyer is 1/2, what is the expected payoff of the trial to each party?

22. (a) If the probability that the plaintiff does not retain a lawyer is 1/10, what strategy should the defendant choose?

(b) If the probability that the defendant does not retain a lawyer is 1/3, what strategy should the plaintiff choose?

(c) If the probability that the plaintiff does not retain a lawyer is 1/10 and the probability the defendant does not retain a lawyer is 1/3, what is the expected payoff of the trial to each party?

Exercises 23 and 24 return to the sealed bid auction between two antique dealers in Example 5 of Section 9.2. It had the following payoff matrix:

		Second Dealer		
		$100	*$200*	*$300*
First Dealer	*$100*	(150, 150)	(0, 200)	(0, 100)
	$200	(200, 0)	(100, 100)	(0, 100)
	$300	(100, 0)	(100, 0)	(50, 50)

23. (a) What pure strategy should the first dealer select if the second dealer bids $100 with probability 0.7, $200 with probability 0, and $300 with probability 0.3?

   (b) What pure strategy should the second dealer select if the first dealer bids $100 with probability 0.1, $200 with probability 0.2, and $300 with probability 0.7?

   (c) Suppose the first dealer bids $100 with probability 0.1, $200 with probability 0.2, and $300 with probability 0.7 and the second dealer bids $100 with probability 0.7, $200 with probability 0, and $300 with probability 0.3. For each dealer, find the expected profit of the game.

24. (a) What pure strategy should the first dealer select if the second dealer bids $100 with probability 0.8, $200 with probability 0.15, and $300 with probability 0.05?

   (b) What pure strategy should the second dealer select if the first dealer bids $100 with probability 0.7, $200 with probability 0.2, and $300 with probability 0.1?

   (c) Suppose the first dealer bids $100 with probability 0.7, $200 with probability 0.2, and $300 with probability 0.1 and the second dealer bids $100 with probability 0.8, $200 with probability 0.15, and $300 with probability 0.05. For each dealer, find the expected profit of the game.

For Exercises 25 and 26, the zero-sum payoff matrix for Rock-Paper-Scissors is given here, where 1 denotes a win, -1 a loss, and 0 a draw for the Row Player.

		Column Player		
		Rock	*Paper*	*Scissors*
Row Player	*Rock*	0	-1	1
	Paper	1	0	-1
	Scissors	-1	1	0

25. (a) What pure strategy should the Row Player select if the Column Player chooses rock with probability 0.7, paper with probability 0.2, and scissors with probability 0.1?

   (b) What pure strategy should the Column Player select if the Row Player chooses rock with probability 0.2, paper with probability 0.4, and scissors with probability 0.4?

   (c) Suppose the Row Player chooses rock with probability 0.2, paper with probability 0.4, and scissors with probability 0.4 and the Column Player chooses rock with probability 0.7, paper with probability 0.2, and scissors with probability 0.1. For each player, find the expected payoff of the game.

26. (a) What pure strategy should the Row Player select if the Column Player chooses rock with probability 0.4, paper with probability 0.35, and scissors with probability 0.25?

   (b) What pure strategy should the Column Player select if the Row Player chooses rock with probability 0.35, paper with probability 0.3, and scissors with probability 0.35?

(c) Suppose the Row Player chooses rock with probability 0.35, paper with probability 0.3, and scissors with probability 0.35 and the Column Player chooses rock with probability 0.4, paper with probability 0.35, and scissors with probability 0.25. For each player, find the expected payoff of the game.

For Exercises 27 and 28, consider the following: Expected value is one of the major aspects of probability theory. When a person is faced with choosing a strategy in an attempt to maximize expected value, it is possible to phrase the decision in the context of a game against chance. Suppose you plan to invest $1000 for a 4-year period. Your choices are a stock market mutual fund or a 6% CD. Your opponent, Chance, may "choose" a bull market, a mediocre market, or a bear market for the period. You estimate the value of your investment in the payoff matrix shown here.

			Chance	
		Bull	*Mediocre*	*Bear*
You	*Stocks*	$1700	$1200	$900
	CD	$1200	$1200	$1200

27. What should your strategy be if Chance chooses the mixed strategy of a bull market with probability 0.6, a mediocre market with probability 0.3, and a bear market with probability 0.1?

28. What should your strategy be if Chance chooses the mixed strategy of a bull market with probability 0.2, a mediocre market with probability 0.4, and a bear market with probability 0.4?

In one of many versions of the game of Two-Finger Morra, two players simultaneously put out one or two fingers (with neither player able to react to the other's choice). If the total number of fingers is even, the first player wins. If the total number of fingers is odd, the second player wins. Exercises 29 and 30 look at two possibilities for the amount paid to the winner by the loser.

(a) Set up the payoff matrix for this game based on the given payoffs.

(b) What should the first player do if the second player chooses one finger with probability 3/5?

(c) If the first player chooses one finger with probability 2/3 and the second player chooses one finger with probability 3/5, what is the expected payoff of the game to the players?

(d) Find the players' optimal mixed strategies.

(e) Find the value of the game.

29. The loser pays the winner $1.

30. The loser pays the winner an amount in dollars equal to the total number of fingers held out by both players.

In Exercises 31–34, consider the simplified game of tennis discussed in Example 4. We examine what happens when a player's ability changes so that new probabilities hold. We still assume that both players know all percentages.

(a) Set up the payoff matrix for this game based on the given payoffs.

(b) Find the players' optimal strategies, which may be pure or mixed.

(c) Find the value of the game.

31. Suppose the receiver's backhand has improved, so that the server now wins only 72% of serves to the receiver's backhand when the receiver guesses incorrectly and only 54% when the receiver guesses correctly. The server still wins 70% of serves to the receiver's forehand when the receiver guesses incorrectly and 40% when the receiver guesses correctly.

32. Suppose the server's accuracy has improved so that the server wins more often when the receiver guesses incorrectly. Assume that the server wins 75% of serves to the forehand and 85% of serves to the backhand when the receiver guesses incorrectly. The server still wins 40% of serves to the forehand and 60% of serves to the backhand when the receiver guesses correctly.

33. Suppose the receiver has an elbow injury that hampers his or her backhand, so that the server wins 90% of serves to the receiver's backhand when the receiver guesses incorrectly and 75% when the receiver guesses correctly. The server still wins 70% of serves to the receiver's forehand when the receiver guesses incorrectly and 40% when the receiver guesses correctly.

34. Suppose the server's overall game has gone up so that he or she wins 10% more often than before. In other words, the server wins 77% of serves to the receiver's forehand when the receiver guesses incorrectly and 44% when the receiver guesses correctly, and 88% of serves to the receiver's backhand when the receiver guesses incorrectly and 66% when the receiver guesses correctly.

For Exercises 35–38, consider the following: The Battle of the Avranches Gap in France occurred in August 1944. Our description is taken indirectly from the autobiography of American general, Omar Bradley, as reported in "Military decision and game theory" by O. G. Haywood, Jr. The U.S. forces had four divisions in reserve, which they could use to reinforce the gap, to move eastward to attack, or to hold in reserve for a day. The exposed German forces, commanded by General von Kluge, could either attack the gap or withdraw to the east. The likely outcomes and preference rankings are given in the following payoff matrix, which has been augmented with a description of the outcomes.

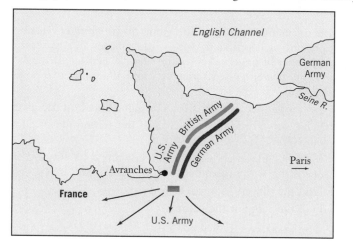

Army positions near Avranches.

		von Kluge	
		Attack Gap	*Withdraw*
	Reinforce Gap	Gap holds (2, 5)	Weak pressure on Germans (3, 4)
Bradley	*Move Eastward*	Gap broken through (1, 6)	Strong pressure on Germans (5, 2)
	Hold in Reserve	Gap holds, possibility of encircling Germans (6, 1)	Moderate pressure on Germans (4, 3)

Source: O.G. Haywood, Jr., "Military decision and game theory." *Journal of the Operations Research Society of America*, 2:1 (November 1954).[2]

The payoff matrix does not reduce completely, implying that mixed strategies are called for. The information needed to set this up as a zero-sum game with different values for the rankings is provided in Exercises 35–38. (In 1944, Bradley decided to hold the unit in reserve and von Kluge initially decided to withdraw. However, Hitler overruled von Kluge, ordering the German forces to attack. The results were disastrous for the Germans and for von Kluge, who committed suicide shortly after the battle.)

(a) For the given values for the rankings, set up the zero-sum game based on these values.

(b) Reduce the payoff matrix by eliminating dominated strategies.

(c) Find the optimal mixed strategies for the players.

(d) Find the value of the game and compare this to the result of Bradley's decision and von Kluge's initial decision.

35.

Bradley's Ranking	1	2	3	4	5	6
Bradley's Value	1	2	3	4	5	6

36.

Bradley's Ranking	1	2	3	4	5	6
Bradley's Value	0	0.4	0.7	1	2	3

37.

Bradley's Ranking	1	2	3	4	5	6
Bradley's Value	-5	0	1	3	7	10

38.

Bradley's Ranking	1	2	3	4	5	6
Bradley's Value	-10	0	3	5	12	25

A tax cheat plans to attempt to defraud the government by altering exactly one of two lines on her tax form. The cheat will be audited by an Internal Revenue Service (IRS) agent who will check only one of the two lines and will catch any fraud on this line. Assume that both are aware of the procedure we have described and that both know the dollar amounts involved, which are given in Exercises 39–44. Make the (unrealistic) assumption that each wants only to maximize expected revenue.

(a) Set up the payoff matrix for the game, letting the tax cheat be the row player and viewing an undetected theft as a payment from the IRS to the cheat and any additional fines when detected as a payment from the cheat to the IRS.

(b) Find both parties' optimal mixed strategies.

(c) Find the expected gain, or loss, by the cheat (i.e., find the value of the game).

39. The cheater's potential gain on the first line is $1000. On the second line, it is $4000. The penalties if caught are $2000 (and no fraudulent gain).

40. The cheater's potential gain on the first line is $1000. On the second line, it is $4000. The penalties if caught are $5000 (and no fraudulent gain).

41. The cheater's potential gain on the first line is $1000. On the second line, it is $4000. The penalties if caught are twice the potential gain (and no fraudulent gain).

42. The cheater's potential gain on the first line is $1000. On the second line, it is $8000. The penalties if caught are twice the potential gain (and no fraudulent gain).

43. The cheater's potential gain on the first line is $1000. On the second line, it is $4000. The penalties if caught are five times the potential gain (and no fraudulent gain).

44. The cheater's potential gain on the first line is $1000. On the second line, it is $8000. The penalties if caught are five times the potential gain (and no fraudulent gain).

A bank robber has just driven out of town down Main Street. A short way out of town, Main Street ends and the robber must turn either north or south. The sheriff is chasing the robber but cannot see the robber's car when it makes this turn. Thus, the sheriff must guess north or south. If the sheriff chases the robber in the wrong direction, the robber is sure to escape. Even if the sheriff correctly guesses which direction the robber has gone, there is a good chance the robber will get away. In Exercises 45 and 46, compute the optimal mixed strategies and the probability that the robber escapes for the given probabilities.

45. If both turn north, the robber escapes with probability 2/3. If both turn south, the robber escapes with probability 3/4.

46. If both turn north, the robber escapes with probability 3/4. If both turn south, the robber escapes with probability 3/4.

In top-level soccer, when a penalty kick is taken, the goalie must anticipate the direction of the kick in order to have a chance of stopping it. In our simplified version, the kicker must choose whether to kick left or right, whereas the goalie must decide whether to move left or right a split second before the ball is kicked. Neither player can afford to wait to react to the other's decision. In Exercises 47–50, compute the players' optimal mixed strategies and the probability that the kicker scores when the optimal strategies are used.

47. The kicker scores 90% of the time when the goalie guesses incorrectly and 50% of the time when the goalie guesses correctly.

48. When the goalie guesses incorrectly, the kicker scores 95% of the time on kicks to the left and 85% of the time on kicks to the right. When the goalie guesses correctly, these percentages drop to 55% and 45%, respectively.

France vs. Brazil in the finals of the 1998 World Cup.

49. When the goalie guesses incorrectly, the kicker scores 95% of the time on kicks to the left and 85% of the time on kicks to the right. When the goalie guesses correctly, the kicker only scores 50% of the time.

50. The kicker scores 90% of the time when the goalie guesses incorrectly, but only 55% and 45% of the time when the goalie correctly anticipates a kick to the right and left, respectively.

9.5 MIXED STRATEGIES AND LINEAR PROGRAMMING

In the last section we found optimal mixed strategies by algebraic methods for zero-sum games that reduce to a 2 × 2 matrix. In this section we examine a more geometric approach, a rudimentary form of linear programming, that lets us find optimal mixed strategies whenever at least one of the players has only two strategies after the payoff matrix has been reduced.

Let's begin by returning to the tennis example from Section 9.4, whose payoff matrix we repeat in Table 9.27.

Table 9.27 **Payoff Matrix for Tennis**			
		Receiver	
		Guess Forehand	**Guess Backhand**
Server	**Serve to Forehand**	40	70
	Serve to Backhand	80	60

Assume that the server uses a mixed strategy of serving to the forehand with probability p and serving to the backhand with probability $1 - p$. If the receiver guesses forehand, the percentage of points the server wins is $p \cdot 40 + (1 - p) \cdot 80 = 80 - 40p$. The graph of this winning percentage as a function of p is a line. If you have a graphing calculator, you might use it here. In case you do not, we will see an easy way to graph the line by hand. In either case, remember that we are only interested in values of p between 0 and 1. When $p = 0$ (no serves to the forehand) and the receiver guesses forehand, the server wins 80% of the time. We plot the point $(0, 80)$. Similarly, when $p = 1$ (all serves to the forehand) and the receiver guesses forehand, the server wins only 40% of the time. We plot the point $(1, 40)$. We would also get the values 80 and 40 by substituting $p = 0$ and $p = 1$ into $80 - 40p$. We get the relevant part of the line by connecting these two points with a straight-line segment as in Figure 9.19. We call this line segment a **strategy line**.

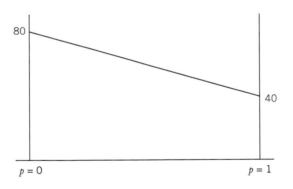

Figure 9.19
Strategy line for the receiver's guess of forehand.

We repeat the process against the receiver's strategy of guessing backhand, arriving at Figure 9.20.

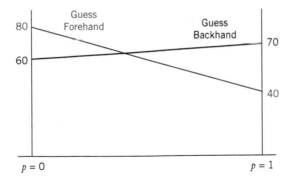

Figure 9.20
The two strategy lines for the receiver in tennis.

The premise that leads to the Minimax Theorem is that the receiver knows the server's mixed strategy, and vice versa. Knowing p, the receiver would guess so as to minimize the server's winning percentage. Geometrically, this corresponds to choosing the lower of the two strategy lines against a particular p. Thus, we may represent the receiver's optimal strategy as a function of p by the highlighted line segments in Figure 9.21.

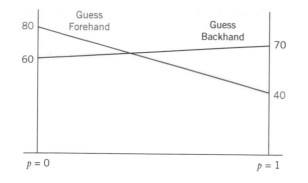

Figure 9.21
The receiver's optimal strategy as a function of p.

The server should therefore choose p to give the highest point on these highlighted segments, the intersection of the two line segments in this case. The two line segments will intersect for the value p giving equal payoffs for the two strategies. Setting the payoffs equal gives the equation

$$p \cdot 40 + (1 - p) \cdot 80 = p \cdot 70 + (1 - p) \cdot 60.$$

Simplifying yields

$$80 - 40p = 60 + 10p,$$
$$(-40 - 10)p = 60 - 80,$$
$$-50p = -20,$$
$$p = \frac{-20}{-50} = \frac{2}{5}.$$

We have found the server's optimal mixed strategy by a completely different method than that of Section 9.4. Applying the argument from the point of view of the server knowing the receiver's probability q of guessing forehand, we rediscover that the receiver's optimal mixed strategy is to guess forehand with probability $q = 1/5$.

Technology Tip

You can plot strategy lines and either find or approximate points of intersection with a graphing calculator or mathematical graphics software.

Although this geometric approach may have increased our understanding of mixed strategies and the Minimax Theorem, at this point you could reasonably ask why we go to this effort when we already have easy formulas for computing the optimal mixed strategies. Our payback comes when exactly one of the players has more than two strategies in the reduced payoff matrix. Set up the payoff matrix with this player as the Column Player. It is easy to put all of the Column Player's strategies into the mix by drawing a strategy line for each strategy. The Column Player's optimal strategy against the Row Player's mixed strategy of playing the first of two strategies with probability p is that strategy corresponding to the lowest line for the particular value of p. The different lowest lines, which may now consist of more than two line segments, trace out a path representing the

Column Player's optimal strategies. The Row Player should choose p to achieve the highest value on this path. Generally, the lines for two of the Column Player's strategy lines will intersect at this point, although on rare occasions three or more of the lines may intersect at this point. Assuming that only two strategy lines intersect at this highest point, the Column Player should choose a mixed strategy involving only these two strategies. Other strategy lines pass above this point, so if the Row Player plays his or her optimal mixed strategy, the Column Player would do worse by choosing one of the other strategies. Now use the formulas from Section 9.4 to compute the mixed strategies for the 2×2 game involving only these two pairs of strategies. We summarize the step-by-step procedure next.

Solving Reduced $2 \times n$ Zero-Sum Games

1. Double check that all dominated strategies have been eliminated.
2. Set up the game so that the Row Player has only two strategies.
3. Plot the strategy lines for each of the Column Player's strategies.
4. Highlight the path consisting of the smallest payoffs for each value of p. This corresponds to the optimal strategy for the Column Player as a function of p.
5. Choose the highest point on this path. This gives the Row Player's optimal value of p graphically.
6. Choose the Column Player's two strategies that intersect at this point. If there are more than two, choose the "outside" two, meaning the two that enclose the others. Eliminate all of the Column Player's other strategies from consideration. (Although these other strategies are not chosen, they are not dominated.)
7. Use the formulas for 2×2 games to find the optimal mixed strategies for the players and the value of the game.

Example 1 *Tennis*

Suppose that in our tennis example the receiver also has the strategy of not guessing. With this strategy, the server wins 45% of serves to the forehand and 65% of serves to the backhand.

(a) Construct the payoff matrix for this expanded game.

(b) Find the optimal mixed strategies for the server and the receiver.

(c) What percentage of points should the server win under best play by both players?

Solution:

(a) We need to add a third column to the original payoff matrix, as in Table 9.28.

Table 9.28 | **Tennis Payoff Matrix with Three Strategies for the Receiver**

		Receiver		
		Guess Forehand	*Guess Backhand*	*No Guess*
Server	Serve to Forehand	40	70	45
	Serve to Backhand	80	60	65

(b) After checking that there are no dominated strategies for either player, we plot the receiver's three strategy lines and highlight the lowest path, as in Figure 9.22.

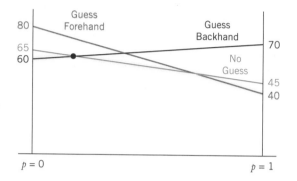

Figure 9.22
Receiver's three strategy lines in tennis.

We have put a dot at the highest point on this path, from which we see that the receiver should mix the strategies of guess backhand and no guess, dropping the guess forehand strategy as in Table 9.29.

Table 9.29	**Tennis Payoff Matrix After Dropping the Unused Strategy**	
		Receiver	
		Guess Backhand	*No Guess*
Server	*Serve to Forehand*	70	45
	Serve to Backhand	60	65

From our Section 9.4 formulas, we find that the optimal mixed strategies are given by

$$p = \frac{65 - 60}{70 - 45 - 60 + 65} = \frac{5}{30} = \frac{1}{6},$$

$$q = \frac{65 - 45}{30} = \frac{2}{3}.$$

Note that this means that the server should serve to the forehand with probability 1/6, to the backhand with probability 5/6, and that the receiver should guess forehand with probability 0, guess backhand with probability 2/3, and not guess with probability 1/3.

(c) Following their optimal strategies, the percentage of points won by the server is

$$v = \frac{70 \cdot 65 - 45 \cdot 60}{30} = \frac{1850}{30} \approx 61.67.$$

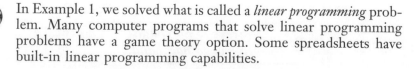

Technology Tip

In Example 1, we solved what is called a *linear programming* problem. Many computer programs that solve linear programming problems have a game theory option. Some spreadsheets have built-in linear programming capabilities.

The following example illustrates the unusual situation in which three of the Column Player's strategies intersect at the Row Player's optimal mixed strategy.

Example 2   *A Money Game*

One of two players has a $1 bill, a $5 bill, and a $10 bill. Out of sight of the second player, the first player puts one of the three bills in a blue envelope and the other two bills in a white envelope. The second player chooses one of the two envelopes and gets to keep the money in that envelope. The first player gets the money in the remaining envelope.

(a) Construct a payoff matrix representing this game as a zero-sum game.

(b) Find the players' optimal mixed strategies.

(c) Find the value of the game.

Solution:

(a) Because the player stuffing the envelopes has three strategies and the other player has only two, we let the stuffer be the column player, and call the other player the chooser. There is $16 to be divided, so the game is a constant-sum game. We may write it in the form of a zero-sum game by letting the payoffs be the amount obtained by the chooser. (View the stuffer as having $16 and having to pay the chooser some of it.) The payoff matrix is then given in Table 9.30.

Table 9.30	**Payoff Matrix for the Money-in-Envelopes Game**			
			Stuffer	
		$1	*$5*	*$10*
Chooser	**Blue Envelope**	$1	$5	$10
	White Envelope	$15	$11	$6

(b) First we check that no strategies are dominated. Next, we graph the stuffer's three strategy lines in Figure 9.23.

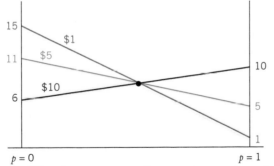

Figure 9.23
Stuffer's strategy lines for the money-in-envelopes game.

The three strategy lines appear to meet in a single point. Assuming this is the case, the stuffer should mix the outside two of these strategies—that is, stuffing $1 or $10. We use the 2 × 2 formulas to find the optimal mixed strategies:

$$p = \frac{6 - 15}{1 - 10 - 15 + 6} = \frac{-9}{-18} = \frac{1}{2},$$

$$q = \frac{6 - 10}{-18} = \frac{-4}{-18} = \frac{2}{9}.$$

(c) The value of the game is

$$v = \frac{1 \cdot 6 - 10 \cdot 15}{-18} = \frac{-144}{-18} = 8.$$

We can now check that all three strategies intersect at a single point. The strategy line for stuffing \$5 in the blue envelope is plotted from $5p + 11(1 - p)$. Setting $p = 1/2$, we get $5 \cdot (1/2) + 11 \cdot (1 - 1/2) = 8$, the value of the game when optimal strategies are chosen. This is what we expect. ∎

A CLOSER LOOK 9.1

WHAT IS LINEAR PROGRAMMING?

Our geometric approach to $2 \times n$ zero-sum games provides one of the most elementary examples of linear programming. In 1980, computer scientist Laszlo Lovasz, then at Hungary's University of Szeged, said that, "If one would take statistics about which mathematical problem is using up most of the computer time in the world, then (not including database handling problems like sorting and searching) the answer would probably be linear programming." The statement may or may not be true today, but the number of applications of linear programming has increased dramatically in the intervening years. The variety of applications of linear programming to real-life problems is enormous. It ranges from matching a collection of employees with varying abilities to a collection of jobs, to scheduling production, to formulating a diet with certain nutritional requirements in a cost-effective manner, to transporting a product from many distribution centers to many retail outlets, to using a network of pipelines at maximal capacity.

In linear programming problems, some quantity is to be optimized subject to certain restrictions. The "linear" in linear programming means that the geometric setting of the problem involves lines or planes, or higher dimensional generalizations, rather than curves or curved surfaces. Algebraically, this means that variables x, y, and z might appear in the form $2x - 3y + 11z$, but you would not see anything like x^2, yz, 2^y, or $\sin z$. In game theory, we wish to choose the probabilities for our mixed strategies in order to maximize our payoff, with restrictions provided by our opponent's strategy lines. In other cases, we may wish to minimize cost or time or to maximize flow in a network, subject to restrictions such as the minimum daily requirements of vitamins and minerals or the capacities of the different pipes in a network.

Whereas we had only one variable—the probability p of choosing the first of our two strategies—and perhaps only three or four restrictions coming from our opponent's strategies, many real-life applications have hundreds or even thousands of variables and restrictions. The picture becomes a nightmare. (Imagine hundreds of strategy lines intersecting in thousands of points.) Using the theory of matrices, in 1947 George B. Dantzig invented a method for going from one point of intersection to another in a way that rapidly reaches the optimal point of intersection. This method is called the *simplex method*. Implemented on today's computers, it solves many typical problems with a few thousand variables and constraints within several minutes. An alternate method was proposed by Narendra K. Karmarkar of AT&T Bell Laboratories in 1984. Karmarkar's discovery has prompted improvements in the simplex method, so that it remains unclear which is the better method overall or if perhaps some other approach is better still. One could think of the simplex method as traveling along roads and looking down the different roads at every intersection to decide how to proceed. With Karmarkar's method, we cut through fields and forests, periodically gauging our progress and changing direction before we return to a road or intersection.

EXERCISES FOR SECTION 9.5

In Exercises 1–4, find the optimal strategies for the two players in the (reduced) zero-sum game with the given payoff matrix.

1.

		Column Player		
		C1	C2	C3
Row Player	R1	8	2	3
	R2	2	7	5

2.

		Column Player		
		C1	C2	C3
Row Player	R1	1	5	4
	R2	8	3	6

3.

		Column Player			
		C1	C2	C3	C4
Row Player	R1	3.0	−8.9	1.3	7.9
	R2	1.0	6.1	5.0	−1.1

4.

		Column Player		
		C1	C2	C3
Row Player	R1	77	−14	−4
	R2	−34	96	−24

In Exercises 5–8, reduce the games by eliminating dominated strategies. All will reduce so that one player will have only two strategies. If necessary, switch the positions of Player A and Player B and transform the payoff matrix so that the row player has only two strategies. Find both players' optimal strategies.

5.

		Player B			
		B1	B2	B3	B4
	A1	−3	0	−2	−1
Player A	A2	1	−4	1	−4
	A3	4	0	5	−3
	A4	−5	6	−5	4

6.

		Player B					
		B1	B2	B3	B4	B5	B6
	A1	−1	0	−4.4	4.6	1.6	1
Player A	A2	3	0	2.2	−4.8	−0.4	−2
	A3	1.8	−1.2	0	−5	−2.8	−3.4

7.

		B1	B2	B3	B4	B5
				Player B		
Player A	A1	7	−19	25	−7	48
	A2	−39	20	−41	−28	−50
	A3	−36	44	−20	−21	−47
	A4	−5	−32	20	−10	23

8.

		B1	B2	B3	B4	B5
				Player B		
Player A	A1	0	−21	1	−10	−2
	A2	−4	2	−3	4	−7
	A3	−11	4	−2	5	−14
	A4	−2	−8	2	0	−3

Suppose a game of Morra is played, where the first player puts out one or two fingers and the second player simultaneously puts out one, two, or three fingers (with neither player able to react to the other's choice). The loser pays the winner a dollar for each finger the two players have put out. The winner varies in Exercises 9 and 10.

(a) Set up the payoff matrix for this game.

(b) Find the players' optimal mixed strategies.

(c) Find the value of the game.

9. If the total number of fingers is even, the first player wins. If the total number of fingers is odd, the second player wins.

10. If the total number of fingers is odd, the first player wins. If the total number of fingers is even, the second player wins.

A baseball pitcher throws four pitches: fastball, change-up, curve, and slider. Assume that the hitter's two strategies are to look for a fastball or not anticipate any particular pitch. A reasonable measure of the payoff for their confrontation is the expected number of runs the batter creates. (Don't worry about how we might measure this.) Given the payoff matrices for the players' strategies in Exercises 11–14, compute the optimal mixed strategies for the batter and pitcher and the value of the game.

11.

		Fastball	Change-Up	Curve	Slider
			Pitcher		
Batter	Look for Fastball	0.48	0.29	0.31	0.35
	Do Not Anticipate	0.33	0.44	0.41	0.38

12.

		Fastball	Change-Up	Curve	Slider
			Pitcher		
Batter	Look for Fastball	0.53	0.34	0.36	0.43
	Do Not Anticipate	0.37	0.49	0.46	0.44

13.

		Fastball	Change-Up	Curve	Slider
			Pitcher		
Batter	Look for Fastball	0.50	0.38	0.44	0.41
	Do Not Anticipate	0.40	0.48	0.45	0.46

14.

		Pitcher			
		Fastball	*Change-Up*	*Curve*	*Slider*
Batter	*Look for Fastball*	0.61	0.46	0.33	0.43
	Do Not Anticipate	0.35	0.55	0.48	0.47

In Exercises 15–18, consider the following: An invading army is headed for one of a country's two border cities, either Northridge or Southport. The defending country has three units available to allocate to the defense of the cities and is able to send each of these units to one of the two cities. Geography and communication is such that the invader must decide which city to attack and the defender must decide how many units to allocate to each city before knowing the enemy's plans. If the invader attacks three defending units, it suffers a sound defeat. If it attacks two units, the result is a draw, worth no points to either side. If it attacks one unit, it captures the city and gets the value of the city. If it attacks a defenseless city, it gets the value of the city plus bonus points for minimizing its casualties and destruction of the city. Based on the values in each exercise,

(a) set up the payoff matrix for the game viewing it as a zero-sum game with the attacker as the row player;

(b) find the combatants' optimal mixed strategies;

(c) use the value of the game to determine whether the attacker should invade the defending country at all.

15. Sound Defeat = −10   Bonus = 2
      Northridge = 8   Southport = 3

16. Sound Defeat = −2   Bonus = 2
      Northridge = 10   Southport = 5

17. Sound Defeat = −1   Bonus = 3
      Northridge = 15   Southport = 7

18. Sound Defeat = −1   Bonus = 3
      Northridge = 6   Southport = 4

19. Draw a graph of the strategy lines to illustrate a 2×3 game where the Column Player has a dominated strategy.

20. Draw a graph of the strategy lines to illustrate a 2×3 game where the Row Player has a dominated strategy.

21. Show that if three of the Column Player's strategy lines intersect at the point representing the Row Player's optimal mixed strategy, then there is a mixed strategy involving the two "outside" lines that yields exactly the same payoffs as the "middle" pure strategy. Explain why this means the Column Player need never play the "middle" strategy.

9.6 SIMULTANEOUS MOVE GAMES WITH COMMUNICATION

In our treatment of simultaneous move games, we have consistently stressed that players could not communicate in any way, so that there was no room for any sort of cooperation or bargaining. In this section we consider three ways players can communicate with each other: "psychologically," in one direction only, and through two-way negotiation. The U.S. Defense Strategic Command authored a study entitled "Essentials of Post-Cold War Deterrence" in 1995. It asserted that "Because of the value that comes from the ambiguity of what the U.S. may do to an adversary if the acts we seek to deter are carried out, it hurts to portray ourselves as too fully rational and cool-headed." They write that conveying the impression that parts of the government "appear to be potentially 'out of control' can be beneficial to creating and reinforcing fears and doubts within the minds of an adversary," and go on to state "That the U.S. may become irrational and vindictive should be a part of the national persona we project to all adversaries." We will only begin to touch on the complexities introduced by communication.

For two-person, zero-sum games, the use of mixed strategies and the Minimax Theorem provide a fully satisfactory solution. Introducing communication adds nothing in this case because one player's loss corresponds exactly to the other player's gain, and vice versa. Thus, it cannot be to both players' benefit to depart from their optimal mixed strategies, so there is no room to negotiate. This is emphatically not the case when we depart from the zero-sum case.

Coordination Games and Focal Points

Suppose we ask you and another person to name a country and promise that if you both write the same country, we will pay each of you $100. No communication of any form is allowed. Probably each of you could name any one of 50 or more countries, which makes it look as if you have only a slight chance of winning. However, our $200 is probably as good as gone. What country did you think of? The United States, or whatever country you happen to be in? This particular game is an example of a **coordination game**, in which players have interests that largely coincide. Prominent strategies, such as naming the country you are in, are called **focal points**, or **Schelling points**, after Thomas Schelling, an economist who drew attention to them with compelling examples such as the following "quiz." Imagine that you are one of a small group of people taking the quiz and that you have been assigned a partner, but you do not know whom.

Thomas Schelling's Original Quiz

1. Name "heads" or "tails." If you and your partner name the same, you both win a prize.

2. Circle one of the numbers listed in the line below. You win if you all succeed in circling the same number.

7   100   13   261   99   555

3. Put a check mark in one of the sixteen squares. You win if you all succeed in checking the same square.

4. You are to meet somebody in New York City. You have not been instructed where to meet; you have no prior understanding with the person on where to meet; and you cannot communicate with each other. You

are simply told that you will have to guess where to meet and that he is being told the same thing and that you will just have to try to make your guesses coincide.

5. You were told the date but not the hour of the meeting in No. 4; the two of you must guess the exact minute of the day for meeting. At what time will you appear at the meeting place that you elected in No. 4?

6. Write some positive number. If you all write the same number, you win.

7. Name an amount of money. If you all name the same amount, you can have as much as you named.

8. You are to divide $100 into two piles, labeled A and B. Your partner is to divide another $100 into two piles labeled A and B. If you allot the same amounts to A and B, respectively, that your partner does, each of you gets $100; if your amounts differ from his, neither of you gets anything.

9. On the first ballot, candidates polled as follows.

Smith	19	Robinson	29
Jones	28	White	9
Brown	15		

The second ballot is about to be taken. You have no interest in the outcome, except that you will be rewarded if someone gets a majority on the second ballot and you vote for the one who does. Similarly, all voters are interested only in voting with the majority, and everybody knows that this is everybody's interest. For whom do you vote on the second ballot?

Source: Reprinted by permission of the publisher from THE STRATEGY OF CONFLICT by Thomas C. Schelling, Cambridge, Mass.: Harvard University Press, Copyright © 1960, 1980 by the President and Fellows of Harvard College.

Go back and try the quiz, if you have not already. The "correct" answers are whatever respondents say they are. Schelling's original sample yielded the following results.

Results of Schelling's Quiz

1. 36 of 42 chose heads.

2. 7 was chosen slightly more often than 100, followed by 13. Only 4 of 41 chose one of the last three numbers.

3. 24 of 41 chose the upper left corner; only 3 were off this main diagonal.

4. A majority said Grand Central Station.

5. Almost all said noon.

6. 40% said 1, the rest varied.

7. 12 of 41 said $1,000,000; only 3 did not list a power of 10.

8. 36 of 41 split the total $50–$50.

9. 20 of 22 said Robinson. (Another 18 were given a variant of the question with Jones and Robinson tied. Of these, 16 named Jones.)

Schelling's questions successfully showed how groups could coordinate action given a prominent option. Although there was no communication among players, there was prior "communication" of some nature due to shared experiences and common cultures. In countries where the written language reads from right to left, the upper right corner was a common answer to question 3. Coordination becomes much more difficult without a clear prominent choice. The Statue of Liberty or the Empire State Building also come to mind as potential meeting places in New York. What about meeting in Washington, D.C.? Do you choose the White House, the Washington Monument, the Capital, the Lincoln Memorial, some other place? Suppose you were required to allocate three large diamonds of approximately the same value between yourself and another player, with you and the other player getting this money only if you both made the same allocation. Probably a two-to-one split comes to mind, but which do you take? Here, even with direct two-way communication, it might not be so easy to agree. In fact, communication back and forth may make agreement even harder. However, one-way communication would almost certainly result in two diamonds for the communicator and one for the other person.

One-Way Communication: Threats

Turning from psychological communication based on shared experiences, we now look at more direct communication, beginning with one-way communication. For communication to have any potential benefit, a player must seek an outcome better than what he or she can guarantee unilaterally. This guaranteed outcome is called the security level of the player.

> The security level for one strategy of a player is the worst possible outcome the player could obtain by playing that strategy. A player's **security level** for a game is the best (largest) possible security level for any strategy the player could choose. It represents the largest payoff the player can unilaterally guarantee himself or herself.

$\mathbf{E}$xample 1   *Security Levels in the Prisoner's Dilemma*

Find the security levels of the players in the Prisoner's Dilemma game given by the payoff matrix in Table 9.31.

Table 9.31	Payoff Matrix for the Prisoner's Dilemma	
		Carson	
		Remain Silent	**Confess**
Roberts	**Remain Silent**	$(-4, -4)$	$(-20, -3)$
	Confess	$(-3, -20)$	$(-15, -15)$

Solution: By remaining silent, the worst payoff Roberts could receive would be -20 (20 years in jail). By confessing, the worst Roberts could do is -15. His security level is -15. Similarly, Carson's security level is -15.

Recall that the game of chicken described in Example 3 of Section 9.2 had the payoff matrix in Table 9.32, where 4 was the most preferred outcome and 1 the least preferred.

Table 9.32	Payoff Matrix for Chicken		
		Driver 2	
		Swerve	Do Not Swerve
Driver 1	Swerve	(3, 3)	(2, 4)
	Do Not Swerve	(4, 2)	(1, 1)

Is there a way for Driver 1 to obtain his or her best outcome? Perhaps. Suppose Driver 1 states that he or she will positively, absolutely not swerve. If Driver 2 believes this announcement, then it effectively reduces the game to the one in Table 9.33.

Table 9.33	Payoff Matrix for Modified Chicken		
		Driver 2	
		Swerve	Do Not Swerve
Driver 1	Do Not Swerve	(4, 2)	(1, 1)

In this game, Driver 2 has little choice; he or she should swerve. Driver 1's statement, which we will call a threat, has clear consequences in the sense that if Driver 2 believes the threat and plays logically, it is in Driver 1's interest to carry it out.

> A **threat** is a statement made before the play of the game that a player intends to select a specific strategy regardless of what the other player plans to select.

Threats need not be especially threatening or harmful to a player's opponent. However, they are an aggressive action that changes the nature of the game. In the face of a threat by Driver 1 that Driver 2 believes, Driver 2 responds in his or her best interest to Driver 1's choice of strategy. The threat is credible if Driver 1 has no reason to unilaterally change strategies on the basis of this response. In other words, for a threat to be **credible**, the result of the opponent's best response must be an equilibrium point. When the opponent has more than one best response, the threat is credible if at least one such response yields an equilibrium point. On the other hand, if a player's opponent chooses his or her best response to a threat that is not credible, the player would rather not go through with the threat.

It appears that Driver 1 gains an advantage simply by committing to a strategy. However, this version of chicken is completely symmetric. Thus, it does not seem right that Driver 1 should have an advantage. Of course, things are not so simple. First of all, maybe Driver 2 could beat Driver 1 to the punch by being the first to commit to not swerving. Or maybe Driver 2 could retort, "Well, I'm

not swerving either," placing the burden right back on Driver 1. Even without such options, Driver 2 must evaluate the credibility of Driver 1's statement. Perhaps Driver 1's rankings place both players' not swerving way behind the other three outcomes, which he or she views as nearly equivalent. Maybe Driver 1 has made such claims before and then swerved in the end. Then it seems likely that Driver 1 will actually swerve and is making such statements in the hope of obtaining the second best outcome, both players swerving, instead of the third best outcome of Driver 1 swerving and Driver 2 not swerving. In this scenario, the statement may not be very credible. On the other hand, if Driver 1 has never swerved in previous games of chicken, the announcement should be taken very seriously and Driver 2 should probably swerve. Finally, the threat could be made very credible by locking the steering wheel so it would not turn.

Making a credible threat was the premise behind the 1964 movie *Dr. Strangelove*. There the Soviet Union had a Doomsday Device that would automatically and unstoppably render Earth uninhabitable in reaction to a U.S. nuclear attack. When a mad U.S. general launches such an attack, the predictable cataclysm occurs.

In making a threat or promise there is always the potential that you will not be believed. If you ultimately back off from the threat or promise, it may ruin your credibility in future games. That is one of the reasons a large company such as Harley-Davidson would pursue a trademark infringement case against small-fry businesses, such as the owner of a pub named Harley's, who have negligible financial impact on the company. Although the legal costs in the particular case far outweigh any potential loss due to the infringement or recovery of damages, the financial harm to the company from numerous infringements could be substantial. By establishing a reputation for pursuing offenders, a company may save itself future trouble. A similar argument is made by countries that refuse to negotiate with terrorists. An especially common way to establish credibility is to agree to a contract, usually between the two players, that establishes penalties sufficient to prevent any change of heart. Of course, the contract itself must be enforceable.

Let's consider a specific version of a game known as the Battle of the Sexes.

Example 2   *Battle of the Sexes*

A woman and a man must each decide whether to go to an opera, to a movie, or bowling on a Friday night. The woman most of all prefers to go out with the man and prefers opera, then the movie, and then bowling. The man also prefers to go out with the woman, but he cannot stand opera. He prefers bowling to a movie if they go together but prefers a movie to bowling if he is by himself. A reasonable payoff matrix, based on their rankings, is shown in Table 9.34, where a bigger number is better, as usual.

Table 9.34	**Payoff Matrix for the Battle of the Sexes**		
		Man	
	Opera	*Movie*	*Bowling*
Woman *Opera*	(6, 2)	(3, 4)	(3, 3)
Movie	(2, 1)	(5, 5)	(2, 3)
Bowling	(1, 1)	(1, 4)	(4, 6)

(a) Find the woman's security level.

(b) Find the man's security level.

(c) If the woman threatens to go to the opera (no matter what) and the man believes her, how should he react? Is this a credible threat?

(d) If the woman threatens to go to the movie and the man believes her, how should he react? Is this a credible threat?

(e) If the woman threatens to go bowling and the man believes her, how should he react? Is this a credible threat?

Solution:

(a) The woman's security level is 3, achieved by going to the opera. Going to the movie could result in a payoff as low as 2; going bowling, as low as 1.

(b) The man's security level is 4, achieved by going to the movie.

(c) The man's response to the woman's going to the opera would be to go to the movie. Because the woman would then rather back down and go to the movie also, her threat is not credible.

(d) The man's response to the woman's going to a movie would be to join her. This is an equilibrium point, so her threat is credible.

(e) The man's response to the woman's going bowling would be to join her. This is also an equilibrium point, so her threat is credible. This is not much of a "threat" because the man obtains his most preferred outcome.

It is clear that the Battle of the Sexes is a type of coordination game to a large extent, in which the rewards are greatest if the woman and man can agree on where to go. Unfortunately, there is no obvious focal point to help in reaching an agreement. The woman could threaten to go to the movie no matter what. The man could threaten to go bowling no matter what. If neither gives in, neither ends up particularly happy. This game is similar to chicken in this respect.

Two-Way Negotiations

We have started to shift our attention to two-way negotiation, where there is the give and take between the players of true negotiating or bargaining. We will assume that agreements are binding and enforceable, although we will take note of when a player has incentive to cheat. The security levels are the payoffs that the players can ensure unilaterally; thus, they form a baseline for the negotiations. Neither player should accept a deal with a smaller payoff than his or her security level. We will call the outcomes that meet this standard the negotiation set.

> The **negotiation set** of a game consists of the pairs of strategies for which the payoff to each player is at least his or her security level.

There may be pairs of strategies in the negotiation set that are not efficient, in the sense that there may be another point in the negotiation set that both players prefer. An efficient outcome is said to be *Pareto optimal* in the language of economics. The precise definition of optimal in our context follows:

> The **optimal negotiation set** consists of those pairs of strategies for which there is not another point in the negotiation set that makes one player better off without making the other player worse off.

For example, if the negotiation set has *payoffs* $(4, 8)$, $(5, 5)$, $(7, 6)$, $(8, 4)$, and $(9, 4)$, then the optimal negotiation set is those strategy pairs with payoffs $(4, 8)$, $(7, 6)$, and $(9, 4)$. The reason is that both players prefer $(7, 6)$ to $(5, 5)$, and the first player prefers $(9, 4)$ to $(8, 4)$ while the second player is indifferent. Therefore, neither $(5, 5)$ nor $(8, 4)$ are payoffs for points in the optimal negotiation set.

If we view the negotiation set as the possible intermediate stages in negotiation, the optimal negotiation set represents the possible final outcomes for these negotiations. Once a point in the optimal negotiation set is reached, one or both players would be hurt by any change in the potential agreement, so would resist this change. In theory, parties should restrict negotiations to those pairs of strategies in the optimal negotiation set, but this ignores the possibility of an impasse in the discussions.

We will not attempt to determine which point in the optimal negotiation set is the likeliest final resolution of negotiations. This depends to a very large extent on the relative negotiation skills of the two players. Often the end result of negotiations will be some focal point of the game, such as a 50-50 split of money. However, focal points may vary according to time and circumstance. "In sight, in mind," the opposite of the well-known expression "out of sight, out of mind," seems appropriate terminology. Economists Judith Mehta, Chris Starmer, and Robert Sugden conducted an experiment published in 1990 in which two subjects were ultimately to divide £10 (British pounds), with neither receiving anything if no agreement was reached within a specified time. The subjects were allowed to discuss the division face to face. For subjects in other experiments, the focal point of a £5-£5 split would be an overwhelming choice. However, the three economists led up to the negotiation in a suggestive way. Each of the two players was dealt four cards from an eight-card deck consisting of four aces and four 2's. They were told that all four aces combined were worth £10, with three or fewer aces worth nothing, and that they must agree how to divide the £10 and turn in all four aces in order to collect anything. The £5-£5 split was still the most common division, but many with one ace ended up with £2.50 and those with more aces tended to end up with a larger share on average.

On the other hand, if a point is in the negotiation set, but not the optimal negotiation set, then a different pair of strategies would help one player and not hurt the other, so that further negotiation is clearly advantageous. One or both players would pursue a more desirable outcome and the other player would have no reason to oppose it, other than to try to negotiate a different outcome, more desirable from his or her own perspective, and no less desirable than the original point in the negotiation set for the other player. We have the following definitions:

> Points in the negotiation set are **stable** if they are equilibrium points and **unstable** if they are not equilibrium points.

By the definition of equilibrium points, neither player has an incentive to unilaterally change strategies, or *cheat*, from a stable point in the negotiation set. On the other hand, at least one player has incentive to cheat, meaning to change strategies, from an unstable point in the negotiation set. Optimal and stable are independent concepts, neither implying the other, as we will see in the examples that follow.

Example 3   *Simplified Battle of the Sexes*

In a simpler version of the Battle of the Sexes, both the woman and the man prefer to go out together above all. Their choices are a piano concert or dancing. Both prefer dancing if they go together, but the concert if they go separately.

(a) Construct the payoff matrix for this version of the Battle of the Sexes.

(b) Find the security levels of the woman and the man.

(c) Find the negotiation set.

(d) Find the optimal negotiation set.

(e) Find the stable points in the negotiation set.

Solution:

(a) Based on the rankings of outcomes from worst (lowest) to best (highest), we find the payoff matrix given in Table 9.35.

Table 9.35 2 × 2 **Battle of the Sexes**

		Man	
		Concert	*Dancing*
Woman	*Concert*	(3, 3)	(2, 1)
	Dancing	(1, 2)	(4, 4)

(b) The security level for each is 2, achieved by going to the concert regardless of what the other chooses.

(c) To be in the negotiation set, each player must receive a payoff of at least 2. The two such points are (concert, concert) and (dancing, dancing).

(d) Only the point (dancing, dancing) is optimal, because both prefer to go dancing together over going to the concert together.

(e) Both points in the negotiation set are stable because both are equilibrium points. Thus, this example shows that stable points need not be optimal. ∎

Example 4   *Expanded Game of Chicken*

Suppose that in an expanded game of chicken, players could swerve early, swerve late, or not swerve. Assume that if both swerve, they miss each other, but if one swerves late and the other does not swerve, then they scratch up their cars a bit. The payoff matrix might look like the one given in Table 9.36.

Table 9.36	Payoff Matrix for an Expanded Game of Chicken		
		Driver 2	
	Swerve Early	Swerve Late	Do Not Swerve
Driver 1 Swerve Early	(5, 5)	(3, 8)	(2, 9)
Swerve Late	(8, 3)	(6, 6)	(4, 7)
Do Not Swerve	(9, 2)	(7, 4)	(1, 1)

(a) Find the negotiation set.
(b) Find the optimal negotiation set.
(c) Find the stable points in the negotiation set.

Solution:
(a) First, we must find the drivers' security levels, which are the same by the symmetry of the game. Swerving late guarantees a payoff rank of at least 4, the best they can guarantee. Therefore, the security level for both drivers is 4. There are four pairs of strategies in which both drivers do at least this well: (swerve early, swerve early), (swerve late, swerve late), (swerve late, do not swerve), and (do not swerve, swerve late). These four pairs of strategies form the negotiation set.
(b) Both drivers prefer (swerve late, swerve late) to (swerve early, swerve early), so the latter cannot be optimal. The other three pairs of strategies in the negotiation set are all optimal.
(c) Each driver would want to switch to not swerving (or even swerving late) if both had decided to swerve early. Thus, (swerve early, swerve early) is not stable. Similarly, if both decide to swerve late, each has incentive to change strategies and not swerve at all. The two pairs (swerve late, do not swerve) and (do not swerve, swerve late) are equilibrium points—that is, they are stable. The pair (swerve late, swerve late) is not stable but is optimal. Thus, this example shows that optimal points need not be stable.

If the entries of the payoff matrix of a game represent numerical values, not just rankings of preferences, a player could elect to play a mixed strategy instead of one pure strategy. In this case, we could generalize the term security level to be the highest security level for any mixed strategy, not just any pure strategy. The resulting security level cannot be lower than that based only on pure strategies. Comparing outcomes may be like comparing apples and oranges, so it may be very difficult to impose a numerical scale to outcomes. This problem is known as the issue of *interpersonal comparisons of utility*, an issue that has troubled economists for as long as there have been economists, and will almost certainly continue to do so. A common solution is to put a monetary value on every outcome. This usually works fairly well in theory, although it may be difficult to determine appropriate monetary values in practice. Successfully assigning monetary values to outcomes brings an added benefit. We can now broaden the negotiations to allow for *side payments*, or monetary exchanges between the players. For instance, an outcome worth $500 to one player and $100 to another player could be negotiated to a $300-$300 split, a $400-$200 split, a $600-$0 split, or even a −$200-$800 split. This generalization is complicated by the technical issue of computing security levels based on mixed strategies. On the other hand, describing the optimal negotiation set becomes simpler. We find pairs of strategies yielding the largest combined payoff, and then imagine negotiating the division of this total subject to each player receiving at least his or her security level. As previously, the final agreement will depend on factors such as the relative bargaining abilities of the players.

EXERCISES FOR SECTION 9.6

1. When two people are trying to name the same date of the year, what are some likely focal points?

2. When two people are trying to name an ice cream flavor, what are some likely focal points?

3. When two people are trying to name a pre-1900 president of the United States, what are some likely focal points?

4. When two people are trying to name the same letter of the alphabet, what are some likely focal points?

5. When two people are trying to name the same pizza topping, what are some likely focal points?

6. You have called a friend on the phone when the line is suddenly disconnected. You and the other person must each decide whether to call the other back or to wait for the other person to call. If you both attempt to call, you will get a busy signal. Write a possible payoff matrix for this game. Let the payoffs be how you and your friend would rank the outcomes, and assume that if the second call gets through, neither of you cares who made it. What do you think is the focal point of this game and why?

In Exercises 7–10, answer the following questions for the given payoff matrix.

(a) For each of the Row Player's possible threats, how should a Column Player who believes the threat react? Which threats are credible?

(b) For each of the Column Player's possible threats, how should a Row Player who believes the threat react? Which threats are credible?

7.

		Column Player		
		C1	*C2*	*C3*
	R1	(−2, 9)	(4, 8)	(7, 10)
Row Player	*R2*	(−1, 7)	(5, 8)	(5, 7)
	R3	(5, 4)	(6, 9)	(4, 3)

8.

		Column Player		
		C1	*C2*	*C3*
	R1	(6, −6)	(7, 3)	(2, 8)
Row Player	*R2*	(8, 2)	(2, −5)	(2, −7)
	R3	(6, 7)	(5, 6)	(5, 4)

9.

		Column Player			
		C1	C2	C3	C4
Row Player	R1	(6, 7)	(7, 1)	(8, 6)	(7, 5)
	R2	(1, 2)	(9, 5)	(4, 7)	(5, 9)
	R3	(8, 8)	(6, 2)	(3, 3)	(5, 7)

10.

		Column Player			
		C1	C2	C3	C4
Row Player	R1	(5, 2)	(2, 6)	(2, 7)	(4, 5)
	R2	(6, 0)	(2, 7)	(4, 3)	(3, 4)
	R3	(−3, 9)	(1, 6)	(7, 1)	(5, 3)
	R4	(7, −1)	(0, 8)	(−1, 8)	(0, 6)

In Exercises 11–14, answer the following for the given payoff matrix.

(a) Find the security levels.

(b) Find the negotiation set.

(c) Find the optimal negotiation set.

(d) Find all stable points in the negotiation set.

11.

		Column Player		
		C1	C2	C3
Row Player	R1	(−2, 9)	(4, 8)	(7, 10)
	R2	(−1, 7)	(5, 8)	(5, 7)
	R3	(5, 4)	(6, 9)	(4, 3)

12.

		Column Player		
		C1	C2	C3
Row Player	R1	(6, −6)	(7, 3)	(2, 8)
	R2	(8, 2)	(2, −5)	(2, −7)
	R3	(6, 7)	(5, 6)	(5, 4)

13.

		Column Player			
		C1	C2	C3	C4
Row Player	R1	(5, 2)	(2, 6)	(2, 7)	(4, 5)
	R2	(6, 0)	(2, 7)	(4, 3)	(3, 4)
	R3	(−3, 9)	(1, 6)	(7, 1)	(5, 3)
	R4	(7, −1)	(0, 8)	(−1, 8)	(0, 6)

14.

		Column Player			
		C1	C2	C3	C4
Row Player	R1	(6, 7)	(7, 1)	(8, 6)	(7, 5)
	R2	(1, 2)	(9, 5)	(4, 7)	(5, 9)
	R3	(8, 8)	(6, 2)	(3, 3)	(5, 7)

Three towns—Riverside, Mountain View, and Sunnyvale—have no local newspaper. A local newspaper in these cities would have circulations of 40,000, 30,000, and 20,000, respectively. Two competing newspaper chains, the *Times* and the *Post*, are considering starting local papers in some of the three towns. Use the additional information provided in Exercises 15–18 to answer the following:

(a) Construct the payoff matrix based on the circulation the papers would obtain.

(b) Find the reduced payoff matrix to see what might happen with no negotiation.

(c) Find each player's credible threats.

(d) Find the security levels.

(e) Find the negotiation set.

(f) Find the optimal negotiation set.

(g) Find all stable points in the negotiation set.

15. Each chain will start a paper in one town. If they choose the same town, the *Times* will get 75% of the readers, the *Post*, 25%.

16. Each chain will start a paper in two towns. If they choose the same town, the *Times* will get 60% of the readers, the *Post*, 40%.

17. The *Times* will start papers in two towns, the *Post* in one. If they choose the same town, each paper will get 50% of the readers.

18. The *Times* will start papers in two towns, the *Post* in one. If they choose the same town, the *Times* will get 70% of the readers, the *Post*, 30%.

The plaintiff and defendant in a civil trial are preparing to go to trial and must choose the form of their legal representation. They cannot avoid a trial at this point. Each may elect to have no lawyer, to hire a mediocre lawyer costing $2000, or to hire a good lawyer costing $4000. They must pay their own lawyers no matter who wins the case. Both are very close-lipped about what their strategies will be. Assume that if one of the litigants pays more for a lawyer, that person will win the case. For each given situation in Exercises 19–22, answer the following:

(a) Construct the payoff matrix for the case, taking into account both any judgment the defendant must pay the plaintiff if the plaintiff wins and any legal costs either incurs.

(b) Find the reduced payoff matrix to see what might happen with no negotiation.

(c) Find the security levels of the litigants.

(d) Find the negotiation set.

(e) Find the optimal negotiation set.

(f) Find all stable points in the negotiation set.

19. If the plaintiff wins, the defendant must pay him or her $10,000. The plaintiff has a slightly stronger case, so will win when they pay the same amount for legal representation.

20. If the plaintiff wins, the defendant must pay him or her $4000. The defendant has a slightly stronger case, so will win when they pay the same amount for legal representation.

21. If the plaintiff wins, the defendant must pay him or her $4000. When they pay the same amount for legal representation, each has a 50% probability of winning the case.

22. If the plaintiff wins, the defendant must pay him or her $10,000. When they pay the same amount for legal representation, each has a 50% probability of winning the case.

Two countries each claim ownership of an island. Each has the choice of no defense buildup, a moderate defense buildup, and a large defense buildup. Each most prefers a successful invasion of the independent island, next prefers no successful invasion by either country, and least prefers to have the other country successfully invade it. Within these three broad types of outcomes, they would prefer to spend as little money as possible on their defense buildup. The difference in military capabilities is described in Exercises 23–26. For each given situation, answer the following:

(a) Construct the payoff matrix for the military buildup strategies based on the countries' rankings of the outcomes.

(b) Reduce the matrix by eliminating dominated strategies to see how the game might be played with no negotiation.

(c) Find the security levels of the countries.

(d) Find the negotiation set.

(e) Find the optimal negotiation set.

(f) Find all stable points in the negotiation set.

23. Each country will successfully invade the island if and only if it has a greater defense buildup.

24. Each country will successfully invade the island if and only if it has a large defense buildup while the other has no buildup.

25. The first country will successfully invade the island if and only if it has a greater defense buildup, but the second country cannot successfully invade the island under any circumstances.

26. The first country will successfully invade the island if and only if it has a greater defense buildup, whereas the second will successfully invade the island if and only if it has a large defense buildup while the first has no buildup.

In Exercises 27–32, consider the following: In politics, logrolling, or the trading of votes, is a fairly common practice. In its simplest form, two politicians each support one or more bills that the other opposes. However, the support of both politicians is needed in order for any of the bills to pass. We assume that when both support one of these bills, then it will pass. If both politicians decide how to vote on the bills without communication, then their dominant strategies are to vote exactly for those bills they favor, so that nothing is passed. However, if they prefer some particular combination of "good" bills and "bad" bills, then by each agreeing to vote for some of the other's bills, the combination that is passed may be better than no bills at all from both of their points of view. Let A, B, C, and D denote the bills. Denote by ABD the outcome where A, B, and D pass, while C fails. Other combinations of bills passing and failing will be denoted similarly. For each given situation, answer the following:

(a) Construct the payoff matrix of rankings (bigger is better and 1 is the worst). Circle the outcome at the equilibrium point that arises by eliminating dominated strategies. (The payoffs at this point are the politicians' security levels.)

(b) Find all possible outcomes the politicians could agree to that both feel would be better than no bills at all. (These, along with no bills, form the outcomes of the negotiation set.)

In Exercises 27 and 28 there are two bills, A and B. Politician 1 favors A and opposes B, whereas Politician 2 favors B and opposes A. Both politicians have four strategies: vote for A and B, vote for A only, vote for B only, vote for neither.

27. Both favor passing both, AB, to passing nothing.

28. Only Politician 1 favors passing both to passing nothing.

In Exercises 29 and 30 there are three bills, A, B, and C. Politician 1 favors A and B and opposes C, whereas Politician 2 favors C and opposes A and B. Assume that both politicians are committed to vote for the bills they favor and must only decide how to vote on those they oppose. Thus, Politician 1 has two strategies and Politician 2 has four strategies.

29. Politician 1 strongly supports A but thinks C is quite bad, ranking the outcomes from best to worst as AB, ABC, A, B, AC, BC, nothing, C. Politician 2 desperately wants C, ranking the outcomes from best to worst as C, AC, BC, ABC, nothing, A, B, AB.

30. Politician 1 thinks C is not so bad, ranking the outcomes from best to worst as AB, ABC, A, B, AC, BC, nothing, C. Politician 2 hates A, ranking the outcomes from best to worst as C, BC, nothing, B, AC, ABC, A, AB.

In Exercises 31 and 32 there are four bills, A, B, C, and D. Politician 1 favors A and B and opposes C and D, whereas Politician 2 favors C and D and opposes A and B. Assume that both politicians are committed to vote for the bills they favor and must only decide how to vote on those they oppose. Thus, both politicians have four strategies.

31. Politician 1 ranks the outcomes from best to worst as AB, ABC, A, AC, ABD, ABCD, AD, B, ACD, BC, nothing, C, BD, BCD, D, CD. Politician 2 ranks the outcomes from best to worst as CD, BCD, D, C, BD, BC, ACD, nothing, ABCD, B, AD, AC, ABD, ABC, A, AB.

32. Politician 1 ranks the outcomes from best to worst as AB, ABC, A, AC, B, BC, ABD, nothing, ABCD, C, AD, ACD, BD, BCD, D, CD. Politician 2 ranks the outcomes from best to worst as CD, ACD, C, D, BCD, AC, AD, ABCD, nothing, BC, BD, A, ABC, ABD, B, AB.

33. Explain why the payoffs at an equilibrium point must be at least the players' security levels.

WRITING EXERCISES

1. Discuss a variety of reasons people may not choose what seems to be their optimal strategy in a real-life game.

2. Describe some real-life instances of the "Tragedy of the Commons," where individuals agreeing to one course of action yields the greatest common good, yet each individual has incentive to choose a different course of action, often resulting in a less desirable outcome for all. This is exemplified for two players by the Prisoner's Dilemma and for more players by unrestricted use of public ranch land, where it may always be to each rancher's individual advantage to have his or her sheep or cattle graze on public lands, but if all ranchers act in their own self-interest, the result is overgrazing and nearly permanent loss of the land for grazing.

3. Describe decisions you have made in your life for which a mixed strategy would be appropriate.

4. Martin Shubik, one of the world's experts in game theory, stated in 1964 that "The writing down of zero-sum payoff matrices does not make an experiment a two-person zero-sum game." Suppose that you, the experimenter, set up some sort of two-person game with monetary payoffs, where one subject's gain is paid by the other subject. Comment on how Dr. Shubik's observation might pertain to your game.

5. In 1960, physicist William Newcomb posed what has become known as Newcomb's paradox: In front of you are two boxes whose contents you cannot determine. Box 1 contains either $1,000,000 or nothing. Box 2 contains $1000. You may either take both boxes and their contents or only Box 1 and its contents. The contents of Box 1 were decided yesterday by a special entity (perhaps God, perhaps a psychic) who is able to predict the future with an extraordinarily high degree of accuracy, though not necessarily perfectly. If the entity predicts that you will take only Box 1, it puts $1,000,000 in it. If it predicts that you will take both boxes, it leaves Box 1 empty. What do you do? Present an argument in support of taking only Box 1. Present an argument in support of taking both boxes.

PROJECTS

Projects 1–5 involve experiments with participants playing games. You may run these with only hypothetical payments or, in order to recruit participants and increase realism, actually pay off the game's payoffs or award prizes for top performances.

1. Get an even number of people to play the following (or some similar) symmetric version of the Prisoner's Dilemma. Make sure that the players do not communicate with each other.

		Column Player	
		C1	*C2*
Row Player	*R1*	($0.50, $0.50)	($0, $1)
	R2	($1, $0)	($0.25, $0.25)

After all subjects have chosen a pure strategy, determine the outcomes for all possible pairings of opponents. Analyze your results.

2. Get an even number of people to play the following (or some similar) symmetric version of chicken. Make sure that the players do not communicate with each other.

		Column Player	
		C1	*C2*
Row Player	*R1*	($0.25, $0.25)	($0.10, $1)
	R2	($1, $0.10)	($0, $0)

After all subjects have chosen a pure strategy, determine the outcomes for all possible pairings of opponents. Analyze your results.

3. The following game simulates the "Tragedy of the Commons," described in Writing Exercise 2. Get 11 people to play the following 11-player game with two strategies for each player. Make sure that the players do not communicate with each other. They may choose Strategy 1 in which you pay them $1 minus 10¢ for each opponent who chooses Strategy 2, or Strategy 2 in which they are paid twice what they would receive under Strategy 1. Analyze optimal strategies for the game and your experimental results.

4. Get four players for the following multiplayer game. Make sure that the players do not communicate with each other. Have each player select a multiple of 25¢. Pay them the amount they chose so long as the total amount chosen by the four does not exceed $5.50, and pay nothing if the total exceeds $5.50. What are the range of amounts from which it makes sense for a player to choose? Describe the equilibrium points for the game. Analyze your experimental results. If several students in the class try this experiment, it may be interesting to vary the number of players, the cap on the total amount, and the increments players may choose.

5. Cities Arlington, Bridgeport, and Centerville need to construct new wastewater treatment facilities. Three individual facilities would cost $300 million for Arlington, $200 million for Bridgeport, and $100 million for Centerville. By building joint facilities, the cities can save some money provided they can agree on how to split the cost. A facility built jointly by Arlington and Bridgeport costs $450 million, one for Arlington and Centerville costs $350 million, and one for Bridgeport and Centerville costs $250 million. A facility serving all three cities would cost $530 million.

(a) Get several groups of three to play the roles of the three cities. Give them a fixed time to negotiate a cost-sharing agreement for a joint plant shared by two or three cities. Barring an agreement on how to split the costs, the cities must build three individual plants.

(b) Describe the outcomes in the negotiation set and optimal negotiation set.

(c) Analyze your results from part (a) in relation to what you computed in part (b).

6. One form of a bilateral monopoly is comprised of a wholesaler and a retailer, each without direct competition in their respective domains. The game is played by the wholesaler announcing a price and the retailer deciding how much to buy at that price. Their decisions depend on the wholesaler's costs of producing a given amount and the price at which the retailer could sell that amount, as given in the following tables:

Wholesaler's Production	Wholesaler's Total Cost (dollars)	Retailer's Unit Price (dollars)	Quantity Sold
3,200	146,000	80	10,000
4,000	162,000	100	6,500
6,500	219,000	120	4,000
10,000	300,000	140	3,200

Both want to maximize their profit. Suppose that the wholesaler can set a unit price of $50, $60, $70, or $80 and that the retailer can respond by setting a retail price of $80, $100, $120, or $140 and ordering just enough to sell them all.

(a) What are the optimal strategies for the wholesaler and retailer?

(b) How might the ability to negotiate (collude) alter the outcome?

7. Consider the duel between Aaron and Alex described in Exercise 16 of Section 9.3. Instead of a noisy duel, assume that the duel is "silent," meaning that a duelist is unaware that his rival has missed with his one shot. Thus, even if his rival has fired, he will fire as originally planned (if still alive). View the duel as a zero-sum game with value 1 for killing one's opponent, -1 for being killed, and 0 if both live.

(a) Find the payoff matrix for the game with Aaron as the row player.

(b) Eliminate dominated strategies.

(c) Find both players' optimal strategies.

(d) The value of the game is no longer enough to calculate probabilities that the players die. Why not?

(e) Calculate the probabilities that Aaron dies, that Alex dies, and that both live, assuming that both players play their optimal strategies.

(Calculating probabilities is harder here than in Exercise 16. If Aaron elects to shoot at 40 paces and Alex at 10, Aaron shoots first and kills Alex with probability 0.4. Therefore, Aaron misses with probability 0.6 and Alex kills Aaron 0.75 or 75% of this fraction of the time, yielding a probability of $(0.6) \cdot (0.75) = 0.45$ that Alex kills Aaron. The payoff is then $0.4 - 0.45 = -0.05$. Similar calculations are also required to answer part (e). More details on the techniques from probability that we have used are given in Section 4.4.)

Aaron	
Strategy	Probability of Killing Alex
60 paces	0.2
40 paces	0.4
20 paces	0.6

Alex	
Strategy	Probability of Killing Aaron
50 paces	0.25
30 paces	0.3
10 paces	0.75

8. How could you generalize the idea of the payoff matrix for a two-player game to analogous games with three players? How would you recognize dominated strategies under your formulation? What is an equilibrium point? Create examples that show games that reduce to single strategies for each player and games that do not reduce and have zero, one, and more than one equilibrium points.

9. Games sometimes arise in which the players know their own payoffs, but one or both players do not know the payoff for the opponent. Consider the payoff matrix shown here.

		Column Player	
		C1	**C2**
Row Player	**R1**	(14, ?)	(−7, ?)
	R2	(5, ?)	(4, ?)
	R3	(−1, ?)	(6, ?)
	R4	(4, ?)	(2, ?)
	R5	(−3, ?)	(9, ?)

(a) Are there any strategies the Row Player should not select?

(b) Laplace's method is to choose the row with the highest average—in other words, to treat each column as equally likely. What strategy would the Row Player select on this basis?

(c) The optimistic case of Hurwicz's method is to assume that each choice of strategy will result in the best possible payoff. What strategy would the Row Player select on this basis?

(d) Wald's method, also known as the minimax method, is to treat the game as a zero-sum game. It may also be thought of as expecting the worst and is the pessimistic case of Hurwicz's method. (In full generality, Hurwicz's method allows the player to select a "degree of optimism.") What mixed strategy would the Row Player select on this basis?

(e) Savage's method, also known as the minimax regret method, is to compute the "regret" matrix by subtracting the largest value in a column from every entry in the column and applying the Wald method to this new matrix. What mixed strategy would the Row Player select on this basis?

(f) Suppose the Column Player selects strategy C1 with probability q. Determine the payoff (in terms of q) obtained by a Row Player using each of the four methods. Plot these lines for $0 \leq q \leq 1$ and comment on any conclusions you can draw in this particular example.

(g) In analyzing these methods, one can check whether they satisfy certain desirable properties. For instance, one is that the strategy selected should not change if the rows and columns are rearranged. All four methods just described satisfy this property. Try to think of some other properties a "good" method *should* satisfy, and see if you can determine which of the four methods satisfy these properties.

KEY TERMS

alternate move game, 542
compressed game tree, 546
constant-sum game, 580
coordination game, 614

credible (threat), 617
dominant strategy, 562
dominated, 562
dominated strategy, 562

equilibrium point, 564, 589
expected payoff, 584
expected value, 585
focal point, 614

CHAPTER 9 REVIEW EXERCISES

1. A couple's four pets are being divided up in a divorce settlement. The wife is allowed the first choice, the husband the second choice, the wife the third choice, and the husband the final choice. The wife's favorite is Fluffy, followed by Fido, Spot, and Rover. The husband's favorite is Rover, followed by Fido, Spot, and Fluffy. Of course, each knows the other's preferences. Each has getting his or her favorite pets as a goal and has no interest in preventing the spouse from getting his or her favorite pets. Assume that neither wife nor husband would voluntarily choose her or his last choice. Construct a game tree for this settlement and analyze it to find the optimal strategy for the wife and husband.

2. Labor is in the middle of a strike against management that is proving costly to both sides. In our simple version, there are three coins to be divided between the two parties. Both sides want to get the greatest possible number of coins for themselves. Management begins with a proposal to divide the coins (no fractions allowed). Labor may either accept the division or propose a division of two coins, with the other coin representing a loss of business during the strike. This time, management may either accept this division or propose a "division" of a single coin, which labor may either accept or refuse. In either case, the game ends. Assume neither management nor labor would forward a proposal that all remaining coins go to the other side. Assume further that both sides are benevolent in the sense that if they are getting some given amount, they would rather the other party get more rather than less. Construct a game tree and analyze it to determine the outcome of the game under best play by both labor and management.

Exercises 3–6 involve the game of Dawson's Kayles, in which a continuous line of squares is laid out. Each of two players takes turns placing a 1×2 domino on two adjacent squares. The last player able to place a domino wins.

3. For a line of six squares, draw a compressed game tree and analyze it to determine which player has a winning strategy.

1	2	3	4	5	6

4. For a line of 10 squares, draw a partial compressed game tree that shows the winning strategy for whichever player, either the first or second to move, has one.

1	2	3	4	5	6	7	8	9	10

5. For a game with a line of n squares, explain why the first player has a winning strategy if the second player has a winning strategy for either a game with $n - 2$ squares or a game with $n - 3$ squares.

6. Describe a winning strategy for the first player whenever the line has an even number of squares.

7. Consider a game of Nim where Chance and a Player alternate removing one, two, or three pebbles from a single pile. The game starts with a pile of five pebbles, and the last player to take a pebble wins. In this game the first to move is Chance, who plays randomly with each allowable move equally likely.

    (a) Construct the game tree for the game, labeling a move by the number of pebbles that remain in the pile.

    (b) Determine the strategy that gives the Player the best probability of winning and compute this probability.

In Exercises 8 and 9, the game is somewhat similar to the game Random Chase described in Section 9.1. The game is played on the given graph. We call the line segments edges and the endpoints vertices. The Chaser starts on the lowest numbered vertex and the Runner on the highest numbered vertex. A move consists of moving a piece from the vertex it is on along an edge connecting it to another vertex. The Chaser can only move along edges to higher numbers; the Runner can only move along edges to lower numbers. Each player secretly writes down two consecutive moves. They then reveal their moves and simultaneously make their first and then their second moves. The Chaser wins by occupying the same vertex at the end of a move or if they switch vertices ("meeting on the connecting edge") within the two moves. If this happens on the first move, the Runner pays the Chaser $2. If it happens on the second, $1. Otherwise, the Chaser pays the Runner $2. We will number the vertices and denote the move from vertex a to vertex b by b. Thus, we write the two moves from vertex 1 to vertex 2 and then to vertex 3 as 2, 3.

(a) Set up the payoff matrix as a zero-sum game with the Chaser as the row player.

(b) Find the optimal strategies for the players.

(c) Find the value of the game to the Chaser.

Exercises 10 and 11 involve a slight variant of the television game show *Jeopardy!* in which there are only two players in Final Jeopardy! Each player begins the last round with a certain number of points. The player with the most points after this round wins $1000, which the players split should they tie. Each writes down the number of points he or she wishes to wager on the final question; it must be an integer between 0 and the number of points each has, inclusive. The number of points they wager is added to their scores if they get the question right and subtracted from their scores if they get it wrong. For these exercises, assume that the probability that both players get the question right is 0.35, that both get it wrong is 0.3, that only the leader gets it right is 0.2, and that only the leader gets it wrong is 0.15. The game is a constant-sum game, so may be viewed as a zero-sum game with payoffs equal to the Row Player's expected winnings. Based on this viewpoint, the payoff matrix has been constructed in each exercise. (If you have some background in probability, such as Chapter 4 in this text, you might want to verify the payoffs.) Compute the optimal strategies for the players.

10. The Row Player has 3 points and the Column Player has 5 points. The payoff matrix is as follows:

		Column Player					
		0	**1**	**2**	**3**	**4**	**5**
	0	0	0	225	450	450	450
Row Player	**1**	0	75	150	300	450	450
	2	250	150	150	150	300	450
	3	500	325	150	150	150	300

11. The Row Player has 4 points and the Column Player has 6 points. The payoff matrix is as follows:

		Column Player						
		0	**1**	**2**	**3**	**4**	**5**	**6**
	0	0	0	225	450	450	450	450
	1	0	75	150	300	450	450	450
Row Player	**2**	250	150	150	150	300	450	450
	3	500	325	150	150	150	300	450
	4	500	500	325	150	150	150	300

12. Suppose a game of Morra is played in which two players simultaneously put out one or two fingers (with neither player able to react to the other's choice). One of the players, again simultaneously, calls out "even" or "odd." This player wins by correctly guessing whether the total number of fingers is even or odd, and loses with an incorrect guess. The loser pays the winner a dollar for each finger the two players have put out.

(a) Set up the payoff matrix for this game.

(b) Find the players' optimal mixed strategies.

(c) Find the value of the game.

13. We considered a game of chicken in Example 3 of Section 9.2. Suppose the game is slightly altered so that the Row Fool has clearly quicker reflexes than the Column Fool, thus giving the Row Fool the option of swerving once he has determined it is too late for the Column Fool to swerve. Assume that both players are aware of this.

(a) Construct a new payoff matrix for the game that reflects the Row Fool's new option of swerving late. For the purposes of the payoffs, consider a late swerve the same as a normal swerve.

(b) Assuming no communication between players, find the players' optimal strategies, if they exist.

Exercises 14 and 15 treat two versions of the game of Nuts. It simulates consumption of a renewable resource. The game has two players and a bowl with four nuts. The game consists of two rounds. In the first round, each player attempts to grab some number of nuts (zero to four) from the bowl. Assume that the players simultaneously grab one nut at a time, proceeding until both have taken all the nuts they intended to take or until the bowl is empty. At the end of this round, the nuts remaining in the bowl grow by some factor and the players divide them equally.

(a) Set up the payoff matrix for this game.

(b) Assuming there is no communication between players, find the players' optimal strategies, if they exist.

(c) Changing the game so that the players can communicate, find the security levels of the players and the negotiation and optimal negotiation sets. Which points in the negotiation set are stable?

14. The nuts remaining in the bowl double after the first round.

15. The nuts remaining in the bowl quadruple after the first round.

16. Suppose that the prisoners Roberts and Carson in Example 2 of Section 9.2 are able to receive word of the other's confession while still under interrogation. Each then has a third strategy of confessing if and only if the other has confessed.

(a) Add this strategy for each prisoner to create a new payoff matrix.

(b) Assuming no further communication between players, find the players' optimal strategies, if they exist.

(c) Find all equilibrium points in the game.

Now assume that the prisoners are somehow able to communicate back and forth.

(d) Find their security levels, the negotiation set, and the optimal negotiation set.

(e) Which points in the negotiation set are stable?

(f) Which of Roberts' three strategies are credible threats?

17. When two people are trying to choose the same square in the picture shown, what are some likely focal points?

18. Two people share a pile of 12 tokens. If they agree on how to divide the pile, the first player will be paid $1 for each of his or her tokens and the second player will be paid $3 for each of his or her tokens. Name two possible focal points of the negotiation.

SUGGESTED READINGS

Axelrod, Robert. *Conflict of Interest: A Theory of Divergent Goals with Applications to Politics.* Chicago: Markham Publishing Company, 1970. Bargaining, with emphasis on a graphical viewpoint.

Axelrod, Robert. *The Evolution of Cooperation.* New York: Basic Books, 1984. Centers around the Iterated Prisoner's Dilemma.

Beasley, John D. *The Mathematics of Games.* Oxford, England: Oxford University Press, 1989. A wide variety of topics, including bluffing and Nim.

Berlekamp, Elwyn R., John H. Conway, and Richard K. Guy. *Winning Ways for Your Mathematical Plays.* London: Academic Press, 1982. Two-volume treatise including Nim, Crosscram, Dots-and-Boxes.

Brams, Steven J. *Game Theory and Politics.* New York: The Free Press, 1975. Political examples of alternate move games and 2 × 2 games.

Dewdney, A. K. "A Tinkertoy Computer That Plays Tic-Tac-Toe," *Scientific American* 261:4 (October 1989), 120–123. A report on the project.

Dixit, Avinash K., and Barry J. Nalebuff. *Thinking Strategically: The Competitive Edge in Business, Politics, and Everyday Life.* New York: W. W. Norton, 1991. Interesting, and occasionally hilarious, case studies of game theory in action.

Hardin, Garrett. "The Tragedy of the Commons." *Science* 162:3859 (December 13, 1968), 1243–1248. A multiplayer Prisoner's Dilemma for those interested in environmental and conservation issues.

Heap, Shaun P. Hargreaves, and Yanis Varoufakis. *Game Theory: A Critical Introduction.* London: Routledge, 1995. Bargaining and repeated games, evolutionary games, experimental studies. Mainly narrative, with some good examples and a few mathematically technical spots.

Hofstadter, Douglas R. "Computer Tournaments of the Prisoner's Dilemma Suggest How Cooperation Evolves." *Scientific American* 248:5 (May 1983), 16–26. A report on Axelrod's tournament.

Kagel, John H., and Alvin E. Roth, eds. *The Handbook of Experimental Economics.* Princeton, N. J.: Princeton University Press, 1995. A survey of research on how people actually play games, which is not always optimally.

Lucas, William F., and Louis J. Billera. "Modeling Coalitional Values." In *Political and Related Models*, vol. 2, edited by Steven J. Brams, William F. Lucas, and Philip D. Straffin, Jr. New York: Springer-Verlag, 1983, 66–97. Examples of games, especially multiplayer games with negotiation.

McMillan, John. *Games, Strategies, and Managers.* New York: Oxford University Press, 1992. Many real-life examples, with the high-level math deferred to an appendix.

Murnighan, J. Keith. *Bargaining Games: A New Approach to Strategic Thinking in Negotiations.* New York: William Morrow, 1992. Pragmatic approach, with a limited amount of game theory.

Poundstone, William. *Prisoner's Dilemma.* New York: Doubleday, 1992. A historical narrative, with a lot on the brilliant John von Neumann, one of the founders of game theory. Includes the Iterated Prisoner's Dilemma, chicken, and the dollar auction.

Riker, William H. *The Art of Political Manipulation.* New Haven, Conn.: Yale University Press, 1986. Game theory is only implicit, but great reading.

Shubik, Martin, ed. *Game Theory and Related Approaches to Social Behavior.* New York: John Wiley, 1964. A varied collection of articles, many reprinted from other sources.

Straffin, Philip D. *Game Theory and Strategy.* Washington, D.C.: The Mathematical Association of America, 1993. Very readable introductory mathematical treatment, with varied applications. Two-person games, with and without bargaining, multiperson games. Games with incomplete payoff information.

Taylor, Alan D. *Mathematics and Politics: Strategy, Voting, Power and Proof.* New York: Springer-Verlag, 1995. Escalation and conflict games, the Vickrey auction.

Williams, J. D. *The Compleat Strategyst.* New York: McGraw-Hill, 1954. A collection of examples of zero-sum games. Very computational, but lively.

Appendix A

Values for the Standard Normal Distribution

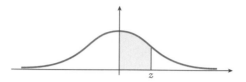

The values in the table below represent the shaded area or, equivalently, $P(0 \le Z \le z)$.

| z | \multicolumn{10}{c}{Second decimal place of z} |
	0.00	0.01	0.02	0.03	0.04	0.05	0.06	0.07	0.08	0.09
0.0	0.0000	0.0040	0.0080	0.0120	0.0160	0.0199	0.0239	0.0279	0.0319	0.0359
0.1	0.0398	0.0438	0.0478	0.0517	0.0557	0.0596	0.0636	0.0675	0.0714	0.0753
0.2	0.0793	0.0832	0.0871	0.0910	0.0948	0.0987	0.1026	0.1064	0.1103	0.1141
0.3	0.1179	0.1217	0.1255	0.1293	0.1331	0.1368	0.1406	0.1443	0.1480	0.1517
0.4	0.1554	0.1591	0.1628	0.1664	0.1700	0.1736	0.1772	0.1808	0.1844	0.1879
0.5	0.1915	0.1950	0.1985	0.2019	0.2054	0.2088	0.2123	0.2157	0.2190	0.2224
0.6	0.2257	0.2291	0.2324	0.2357	0.2389	0.2422	0.2454	0.2486	0.2517	0.2549
0.7	0.2580	0.2611	0.2642	0.2673	0.2704	0.2734	0.2764	0.2794	0.2823	0.2852
0.8	0.2881	0.2910	0.2939	0.2967	0.2995	0.3023	0.3051	0.3078	0.3106	0.3133
0.9	0.3159	0.3186	0.3212	0.3238	0.3264	0.3289	0.3315	0.3340	0.3365	0.3389
1.0	0.3413	0.3438	0.3461	0.3485	0.3508	0.3531	0.3554	0.3577	0.3599	0.3621
1.1	0.3643	0.3665	0.3686	0.3708	0.3729	0.3749	0.3770	0.3790	0.3810	0.3830
1.2	0.3849	0.3869	0.3888	0.3907	0.3925	0.3944	0.3962	0.3980	0.3997	0.4015
1.3	0.4032	0.4049	0.4066	0.4082	0.4099	0.4115	0.4131	0.4147	0.4162	0.4177
1.4	0.4192	0.4207	0.4222	0.4236	0.4251	0.4265	0.4279	0.4292	0.4306	0.4319
1.5	0.4332	0.4345	0.4357	0.4370	0.4382	0.4394	0.4406	0.4418	0.4429	0.4441
1.6	0.4452	0.4463	0.4474	0.4484	0.4495	0.4505	0.4515	0.4525	0.4535	0.4545
1.7	0.4554	0.4564	0.4573	0.4582	0.4591	0.4599	0.4608	0.4616	0.4625	0.4633
1.8	0.4641	0.4649	0.4656	0.4664	0.4671	0.4678	0.4686	0.4693	0.4699	0.4706
1.9	0.4713	0.4719	0.4726	0.4732	0.4738	0.4744	0.4750	0.4756	0.4761	0.4767
2.0	0.4772	0.4778	0.4783	0.4788	0.4793	0.4798	0.4803	0.4808	0.4812	0.4817
2.1	0.4821	0.4826	0.4830	0.4834	0.4838	0.4842	0.4846	0.4850	0.4854	0.4857
2.2	0.4861	0.4864	0.4868	0.4871	0.4875	0.4878	0.4881	0.4884	0.4887	0.4890
2.3	0.4893	0.4896	0.4898	0.4901	0.4904	0.4906	0.4909	0.4911	0.4913	0.4916
2.4	0.4918	0.4920	0.4922	0.4925	0.4927	0.4929	0.4931	0.4932	0.4934	0.4936
2.5	0.4938	0.4940	0.4941	0.4943	0.4945	0.4946	0.4948	0.4949	0.4951	0.4952
2.6	0.4953	0.4955	0.4956	0.4957	0.4959	0.4960	0.4961	0.4962	0.4963	0.4964
2.7	0.4965	0.4966	0.4967	0.4968	0.4969	0.4970	0.4971	0.4972	0.4973	0.4974
2.8	0.4974	0.4975	0.4976	0.4977	0.4977	0.4978	0.4979	0.4979	0.4980	0.4981
2.9	0.4981	0.4982	0.4982	0.4983	0.4984	0.4984	0.4985	0.4985	0.4986	0.4986
3.0	0.4987	0.4987	0.4987	0.4988	0.4988	0.4989	0.4989	0.4989	0.4990	0.4990
3.1	0.4990	0.4991	0.4991	0.4991	0.4992	0.4992	0.4992	0.4992	0.4993	0.4993
3.2	0.4993	0.4993	0.4994	0.4994	0.4994	0.4994	0.4994	0.4995	0.4995	0.4995
3.3	0.4995	0.4995	0.4995	0.4996	0.4996	0.4996	0.4996	0.4996	0.4996	0.4997
3.4	0.4997	0.4997	0.4997	0.4997	0.4997	0.4997	0.4997	0.4997	0.4997	0.4998

Other approximate values are 0.4998 for $z = 3.5$, 0.49997 for $z = 4$, 0.499997 for $z = 4.5$, and 0.4999997 for $z = 5$.

Answers to Selected Exercises and Review Exercises

CHAPTER 1

Section 1.1
1. Chinese **3. (a)** 12 **(b)** 20 **(c)** 23 **5.** 62 **7. (a)** Jackson or Adams **(b)** 14.62% **9.** 50% **11. (a)** carpet **(b)** wood
13. (a) no curfew **(b)** 1 A.M. **15. (a)** 40 hours **(b)** 40 hours **(c)** no **(d)** Yes, they could vote so that 20 hours wins.
17. (a) volleyball **(b)** Yes, they could vote so that softball wins. **19.** volleyball **21.** softball **23.** yes **25.** football

Section 1.2
1. Cardona **3.** Horned Frogs **5.** Monday **7.** Tilson **9. (a)** Brokaw **(b)** tie between Jennings and Rather **(c)** tie
between Jennings and Rather **11. (a)** football **(b)** football **(c)** football **(d)** Yes, they could vote so that baseball wins.
13. (a) Cullors **(b)** Yes, they could vote so that Allen wins. **(c)** no

Section 1.3
1. (a) Sanders **(b)** Bauer **(c)** Sanders **(d)** Sanders **3.** Mexican **5.** none **7.** Hungry Boar **9. (a)** optional uniforms
(b) no uniforms **(c)** no uniforms **(d)** optional uniforms **11. (a)** beagle **(b)** beagle **(c)** beagle **(d)** none **13.** no
15. (a) public development **(b)** private development **(c)** recommitment **(d)** none **17.** (a) and (b) **19.** yes

Section 1.4
1. Freeman **3. (a)** A **(b)** 0.956% **5.** De Castro **7. (a)** eighteen **(b)** eighteen **(c)** eighteen **(d)** eighteen **(e)** eighteen
9. (a) Bailey **(b)** Holmes **(c)** Holmes **(d)** Holmes **(e)** Bailey **(f)** no **11. (a)** white **(b)** ivory **(c)** tie between white and
ivory **(d)** ivory **(e)** ivory **(f)** Yes, they could vote so that white wins. **13. (b)** 9.4% **(c)** 15.3% **(d)** 42.3% **(e)** Reagan
14%, Carter 23%, Anderson 63% **(f)** Reagan 217.2, Carter 206.4, Anderson 176.4; Reagan wins **15. (a)** Wilson
(b) Wilson **(c)** Wilson **(d)** Roosevelt **(e)** Wilson **(f)** Wilson **17. (a)** Taft **(b)** Roosevelt **(c)** Wilson **(d)** Roosevelt
(e) Wilson **(f)** Roosevelt **19. (a)** Baldwin **(b)** Baldwin **(c)** Baldwin **(d)** Baldwin **21. (a)** Musselman **(b)** Watkins
(c) Watkins **(d)** Watkins **23. (a)** Buckley **(b)** Goodell **(c)** Goodell

Chapter 1 Review Exercises
1. 77 **2. (a)** 18 **(b)** 27 **3. (a)** bananas **(b)** bananas **(c)** grapes **(d)** bananas **4. (a)** basketball court **(b)** baseball field
(c) baseball field **(d)** baseball field **5.** jazz trio **6. (a)** *Romeo and Juliet* **(b)** *Romeo and Juliet* **(c)** *The Fantasticks*
(d) none **(e)** no **(f)** Yes, they could vote so that *Death of a Salesman* wins. **7. (a)** tie between Mintz and Zukoff
(b) Zukoff **(c)** Zukoff **(d)** Zukoff **(e)** Mintz **(f)** Yes, they could vote so that Mintz wins. **(g)** Yes, they could vote so that
Zukoff wins. **8.** If every voter prefers A to B, then B has no first-place votes and cannot make the runoff, so cannot
win the election.

9. One example is

	Number of Voters			
	1	**1**	**1**	**1**
A	1	4	3	2
B	2	1	4	3
C	3	2	1	4
D	4	3	2	1

CHAPTER 2

Section 2.1

1. Country	**(a) Hamilton**	**(b) Lowndes**
Denmark	10	10
Finland	10	10
Iceland	0	1
Norway	8	8
Sweden	17	16

3.

Company	(a) Hamilton	(b) Lowndes
Alpha Software	10	10
Beta Technology	1	2
Gamma Computing	9	8
Delta Development	6	6

(c) The shift was from a larger company to a smaller company.

5.

District	(a) Hamilton	(b) Lowndes
1	9	9
2	11	11
3	7	7
4	11	11
5	5	5
6	7	7

7.

County	(a) Hamilton	(b) Lowndes
Cheshire	6	6
Grafton	3	3
Hillsborough	6	6
Rockingham	8	8
Strafford	5	5

(c) The three counties with the largest fractional parts were also the smallest three counties.

9.

Party	(a) Hamilton	(b) Lowndes
Labor	44	44
Likud	32	31
Energy	12	12
Zionist	7	7
Sephardic Jews	7	7
National Religious	6	6
United Torah	4	5
Democratic Front	3	3
Fatherland	3	3
Arab Democratic	2	2

(c) no

Section 2.2

1.

Candidate	(a) Jefferson	(b) Webster
Grover Cleveland	4	4
Benjamin Harrison	4	4
James R. Weaver	1	1
John Bidwell	0	0

3.

School	(a) Jefferson	(b) Webster
Eddyville	0	1
Newport	8	8
Taft	9	8
Toledo	6	6
Waldport	4	4

(c) The shift goes from schools with more students to those with fewer students.

5.

County	(a) Jefferson	(b) Webster	(c) Hamilton	(d) Lowndes
Hawaii	4	5	4	4
Honolulu	33	31	32	31
Kalawao	0	0	0	1
Kauai	2	2	2	2
Maui	3	4	4	4

(e) Jefferson's method (f) Lowndes' method

7.

Dormitory	(a) Jefferson	(b) Webster
Couch	5	5
Galloway	3	3
Hardin	4	4
Martin	4	4
Raney	3	3
Veasey	4	4

9.

Party	(a) Jefferson	(b) Webster
Social Democrats	43	43
Christian Democrats	42	41
Free Democrats	9	10
Greens	7	7

(c) no (d) yes, with Webster's method

11.

Party	(a) Jefferson	(b) Webster	(c) Hamilton	(d) Lowndes
African National Congress	254	253	253	252
National Party	83	82	82	82
Inkatha Freedom Party	42	42	42	43
Freedom Front	8	9	9	9
Democratic Party	7	7	7	7
Pan Africanist Congress	5	5	5	5
African Christian Democratic Party	1	2	2	2

(e) The Jefferson's method apportionment violates the Quota Property. **(f)** yes, with Lowndes' method

13.

Region	(a) Rounded Percentages	(b) Webster (percent)	(c) Jefferson (percent)
North America	17	18	18
Central and South America	9	9	9
Western Europe	10	10	10
Eastern Europe and former U.S.S.R.	11	11	11
Middle East	30	30	30
Africa	11	11	11
Far East and Oceania	11	11	11
Total	99	100	100

15.

Candidate	(a) Jefferson	(b) Webster
Bob Dole	14	14
Pat Buchanan	6	6
Steve Forbes	3	3

17.

County	(a) Jefferson	(b) Webster
Hawaii	4	4
Honolulu	32	31
Kalawao	1	1
Kauai	2	2
Maui	3	4

Section 2.3

1. 6.48074   **3.** 780.49984

5.

Candidate	Hill–Huntington
Woodrow Wilson	12
Theodore Roosevelt	14
William H. Taft	9
Eugene V. Debs	3

7.

County	Hill–Huntington
Fairfield	20
Hartford	21
Litchfield	4
Middlesex	3
New Haven	19
New London	6
Tolland	3
Windham	3

9.

Country	(a) Hill–Huntington	(b) Hamilton	(c) Lowndes	(d) Jefferson	(e) Webster
Algeria	9	9	9	9	9
Egypt	20	20	20	21	20
Libya	2	2	2	1	2
Morocco	9	10	9	10	10
Tunisia	3	3	3	3	3
Western Sahara	1	0	1	0	0

(f) Jefferson's method

11.

Precinct	(a) Hill–Huntington	(b) Hamilton	(c) Lowndes	(d) Jefferson	(e) Webster
1	58	58	58	59	58
2	12	12	12	12	12
3	34	34	34	34	34
4	29	29	29	29	29
5	24	24	24	23	24
6	18	18	18	18	18

13.

Party	Hill–Huntington
Red Party	33
National White Party	31
Progressive Encounter	30
New Space	5

15.

State	Hill–Huntington
Connecticut	6
Delaware	2
Georgia	2
Kentucky	2
Maine	3
Maryland	9
Massachusetts	10
New Hampshire	4
New Jersey	5
New York	9
North Carolina	10
Pennsylvania	12
Rhode Island	2
South Carolina	7
Vermont	2
Virginia	20

17.

Commodity	(a) Rounded Percentages	(b) Hill–Huntington (percent)
Food and agriculture	7.6	7.6
Beverages and tobacco	1.4	1.4
Crude materials	6.0	6.0
Mineral fuels and related products	1.8	1.8
Chemicals and related products	10.6	10.6
Machinery and transport equipment	48.4	48.3
Manufactured goods and other	24.3	24.3
Total	100.1	100.0

19.

County	Adams
Fairfield	20
Hartford	20
Litchfield	4
Middlesex	4
New Haven	19
New London	6
Tolland	3
Windham	3

21. threshold divisor for n seats under Dean's method $= \dfrac{\text{population of the state}}{(n-1)n/(n-0.5)}$

23.

County	Dean
Fairfield	20
Hartford	20
Litchfield	4
Middlesex	4
New Haven	19
New London	6
Tolland	3
Windham	3

25.

County	Condorcet
Fairfield	20
Hartford	20
Litchfield	4
Middlesex	4
New Haven	19
New London	6
Tolland	3
Windham	3

Section 2.4

1.

		Quota Property	House Size Property	Population Property
Quota Methods	Hamilton	✓		
	Lowndes	✓		
Divisor Methods	Jefferson		✓	✓
	Webster		✓	✓
	Hill–Huntington		✓	✓

3. None exists. **5.** State A's natural quota is 14.9603 and it is allocated 16 seats. State B's natural quota is 12.9757 and it is allocated 14 seats. **7.** State D receives 4 seats in a 24-seat house but only 3 seats in a 25-seat house. **9.** State A loses population and gains a seat, whereas state C gains population and loses a seat.

Chapter 2 Review Exercises

1.

Company	(a) Hamilton	(b) Lowndes	(c) Jefferson	(d) Webster	(e) Hill–Huntington
East Michigan	6	6	6	6	6
Great Lakes	9	9	9	9	9
Mid Michigan	4	5	4	4	4
Southeast Michigan	16	15	16	16	16

2.

Party	(a) Hamilton	(b) Lowndes	(c) Jefferson	(d) Webster	(e) Hill–Huntington
Siumut	12	12	12	12	12
Inuit Ataqatigiit	7	7	6	6	7
Atassut	9	9	10	10	9

3.

School	(a) Hamilton	(b) Lowndes	(c) Jefferson	(d) Webster	(e) Hill–Huntington
Canyon Hills	56	55	56	56	56
Magnolia	49	49	49	49	49
Ramona	49	49	49	49	49
Townsend	45	45	45	44	44
Woodcrest	26	27	26	27	27

(f) Lowndes' method **(g)** no

4.

Candidate	(a) Hamilton	(b) Lowndes	(c) Jefferson	(d) Webster	(e) Hill–Huntington
Bob Dole	8	7	8	7	7
Steve Forbes	4	3	4	4	3
Pat Buchanan	3	3	3	3	3
Phil Gramm	2	2	2	2	2
Lamar Alexander	1	1	1	1	1
Alan Keyes	0	1	0	1	1
Richard Lugar	0	1	0	0	1

(f) Hamilton's method and Jefferson's method give the actual apportionment.

5. With 10 seats, one example is

State	Population
A	244
B	251
C	252
D	253

6. If the natural divisor works for Webster's method, then rounding up all natural quotas with fractional parts at least 0.5 and rounding down all natural quotas with fractional parts less than 0.5 yields the correct total number of seats. Because Hamilton's method allocates an extra seat to as many states as necessary to obtain the correct total number of seats, it too will allocate an extra seat exactly to those states with fractional parts at least 0.5. **7.** Any final divisor for Jefferson's method is at most the natural divisor (and must be smaller unless all natural quotas are integers). Therefore, a state's final allocation is at least its initial allocation, which in turn is its natural quota rounded down. **8.** Suppose state A gains population and state B loses population. Because the total population does not change, A's natural quota must increase while B's must decrease. For A to lose a seat, the integer part of its natural quota must remain the same as the population changes. For B to gain a seat, the integer part of its natural quota must remain the same. Therefore, in order for A to lose a seat while B gains a seat, A's relative fractional part must change from being greater than B's to being smaller. However, this is incompatible with the increase in A's natural quota and the decrease in B's.

CHAPTER 3

Section 3.1
1. 2.02319948 **3.** 1.04035789 **5.** 1.84639651 **7.** 6.06388151 **9.** −3.79376085 **11.** −0.31546488 **13.** 0.93578497 **15.** 0.17216130 **17.** 0.01857668 **19.** 0.00921670 **21.** 12.66977089 **23.** 0.01729323 **25.** 3.02246118 **27.** 19.59094589 **29.** 10.75121514

Section 3.2
1. $600 **3.** $727.94 **5.** $1848 **7.** $358.40 **9.** $51,530.24 **11.** $1913.88 **13.** $57,397.96 **15.** 5.38 years **17.** 16.59% **19.** $2237.93 **21.** 80.32% **23.** 1.82 years **25.** 4320 talents **27.** $137.50 **29.** $560 **31.** 1300%

Section 3.3
1. $7895.59 **3.** $9750.44 **5.** $2458.65 **7.** $1151.61 **9.** $4197.54 **11.** $150.56 **13.** 8.83 years **15.** 15.58 years **17.** 10.67% **19.** 6.58% **21.** 4.80% **23.** 4.03% **25.** (a) 5.09% (b) 5.12% (c) 5.13% **27.** (a) $793.44 (b) $802.41 (c) $804.52 (d) $805.55 **29.** Bank of America **31.** 11.31% **33.** $0.936 million or $936,000.66 **35.** $664.04 **37.** 3.59 years **39.** $45,993.34 **41.** $49,452.13 **43.** $83,856.13 **45.** 17.68 years **47.** $1755.44 **49.** (a) $643,000 (b) $2.62172775 \cdot 10^{31}$ **51.** $17,720.09 **53.** (a) $11,882.34 (b) 7.23%

Section 3.4
1. $59,561.68 **3.** $151,898.73 **5.** $687.42 **7.** $222.61 **9.** 1.29 years **11.** 5.53 years **13.** (a) $14,513.57 (b) $9000.00 (c) $5513.57 **15.** 2.62 years **17.** (a) $471,591.05 (b) $243,994.01 **19.** (a) $279.67 (b) $1072.53 **21.** lump sum option: $38,397.48; $185 payments for 5-year option: $38,325.69; $80 payments for 18-year option: $38,406.91 **23.** $48,478.85 **25.** (b) $1725.92

Section 3.5
1. $14,332.92 **3.** $1731.54 **5.** $862.00 **7.** $1383.07 **9.** 3.15 years **11.** 8.42 years **13.** $7923.19 **15.** (a) $95.46 (b) $11,455.20 (c) $2455.20 **17.** $2330.74 **19.** 3.03 years **21.** (a) $932.31 (b) $66,191.24 (c) $56,338.57 **23.** the 0.9% option **25.** (a) $1055.47 (b) 17.69 years (c) $113,458.03 **27.** (a) $726.81 (b) $929.29 (c) $94,379.40

29.

Payment Number	Payment	Interest Paid	Principal Paid	Balance
				8000.00
1	267.42	120.00	147.42	7852.58
2	267.42	117.79	149.63	7702.95
3	267.42	115.54	151.88	7551.07
4	267.42	113.27	154.15	7396.92

31.

Payment Number	Payment	Interest Paid	Principal Paid	Balance
				5429.38
1	927.20	38.01	889.19	4540.19
2	927.20	31.78	895.42	3644.77
3	927.20	25.51	901.69	2743.08
4	927.20	19.20	908.00	1835.08
5	927.20	12.85	914.35	920.73
6	927.18	6.45	920.73	0

33.

Payment Number	Payment	Interest Paid	Principal Paid	Balance
				7300.00
1	2574.57	209.88	2364.69	4935.31
2	2574.57	141.89	2432.68	2502.63
3	2574.58	71.95	2502.63	0

35. (a) $346,604.40 **(b)** $103,651.91 **(c)** $331,419.60 **(d)** yes **37. (a)** $8758.80 **(b)** yes **(c)** $583.92 **(d)** $93,092.48 **(e)** $93,222.81 **(f)** no **39. (a)** yes **(b)** no **41.** $54,772.73 **43.** 6.67 years **45.** $439,474.67 **47. (a)** $237,100.84 **(b)** $5,927,521 **49. (b)** $368.64

Chapter 3 Review Exercises

1. $650.81 **2.** 80% **3. (a)** $666,876.87 **(b)** $92,000.00 **(c)** $574,876.87 **4. (a)** $3311.32 **(b)** $278,150.88 **(c)** $67,050.88 **5.** $4221.59 **6.** 17.35 years **7. (a)** $70,558.73 **(b)** $62,066.33 **8.** $815.75 **9.** 27,000 pounds

10.

Payment Number	Payment	Interest Paid	Principal Paid	Balance
				943.61
1	160.87	6.13	154.74	788.87
2	160.87	5.13	155.74	633.13
3	160.87	4.12	156.75	476.38
4	160.87	3.10	157.77	318.61
5	160.87	2.07	158.80	159.81
6	160.85	1.04	159.81	0

11. (a) $15,485.54 **(b)** $4461.28 **12.** 30,508 francs **13. (a)** $172.83 **(b)** 6.92 years **(c)** $2235.80 **14.** 18.73% **15.** 5.90% **16.** 1.30 years **17.** 5.46 years **18.** 10% down payment option: $10,408.32; no down payment option: $10,666.56 **19.** $4674.94 **20. (a)** 9.40% **(b)** $71,810.39 **(c)** the $128 monthly payment option

CHAPTER 4

Section 4.1

1. (a) 1/6 **(b)** 1/2 **(c)** 1/3 **(d)** 0 **3.** 3/5 **5.** 0.766 **7. (a)** 1/13 **(b)** 3/13 **9.** 12/13 **11.** 5/8 **13. (a)** H1, H2, H3, H4, H5, H6, T1, T2, T3, T4, T5, T6 **(b)** 1/12 **(c)** 1/4 **15.** 1/145 **17.** 3/8 **19.** 1/9 **21.** 31/36 **23.** 1/5 **25. (a)** 0.25 **(b)** 0.607

27.

(1, 1)	(1, 2)	(1, 2)	(1, 3)	(1, 3)	(1, 4)
(3, 1)	(3, 2)	(3, 2)	(3, 3)	(3, 3)	(3, 4)
(4, 1)	(4, 2)	(4, 2)	(4, 3)	(4, 3)	(4, 4)
(5, 1)	(5, 2)	(5, 2)	(5, 3)	(5, 3)	(5, 4)
(6, 1)	(6, 2)	(6, 2)	(6, 3)	(6, 3)	(6, 4)
(8, 1)	(8, 2)	(8, 2)	(8, 3)	(8, 3)	(8, 4)

(a) 1/36 **(b)** 1/18 **(c)** 1/12 **(d)** 1/9 **(e)** 5/36 **(f)** 1/6 **(g)** 5/36 **(h)** 1/9 **(i)** 1/12 **(j)** 1/18 **(k)** 1/36 **29.** 0.8077 **31. (a)** 0.000000451 **(b)** 0.000000378 **(c)** No, for otherwise you would expect the probability in part (b) to be larger than the probability in part (a). **33.** 2/3

Section 4.2

1. 1 to 1 **3.** 2/9 **5.** 3 to 17 **7.** 2/11 **9.** 3 to 1 **11.** 5 to 3 **13.** 1/6 **15.** 34 to 1 **17.** $0.80 **19.** $750 **21. (a)** 10 to 9 **(b)** no **23.** yes

Section 4.3

1. 25% **3. (a)** 7/10 **(b)** 3/10 **5.** 82.3% **7.** 28% **9.** 7/13 **11.** 0.1266 **13.** 13/24 **15.** 1/12

Section 4.4

1. (a) 0.5528 **(b)** 0.3939 **(c)** 0.6805 **(d)** yes **3. (a)** 0.385 **(b)** 0.2892 **(c)** 0.4595 **(d)** 0.4463 **5.** 1/3 **7.** 1/12 **9.** 2/11 **11.** 0.000000000001% **13.** 1/26 **15. (a)** 1/27 **(b)** 8/27 **17.** 1/7140 **19.** 0.002641 **21.** 0.133 **23.** 1/1000 **25.** 0.002767 **27.** 0.178 **29. (a)** 0.000616 **(b)** 0.274 **31. (a)** 1/4 **(b)** 30% **33.** 1.84% **35. (a)** 0.172 **(b)** 0.134 **37.** 0.250 **39.** 0.00132

Section 4.5

1. 100 **3.** 20,358,520 **5. (a)** 792 **(b)** 336 **(c)** 56 **7.** 36,400 **9. (a)** 1287 **(b)** 5148 **11.** 1,746,360 **13.** 32,768 **15.** 490,776 **17. (a)** 60,466,176 **(b)** 1,679,616 **19.** 216 **21.** 2,600,000 **23.** 6000 **25.** 13,624,345 **27.** 120 **29. (a)** 128 **31.** 120 **33.** 2496 **35.** 168,168,000 **37.** 80,089,128

Section 4.6

1. 1/990 **3.** 1/324,632 **5.** 1/3,628,800 **7.** 1/220 **9.** 0.008583 **11. (a)** 1/1024 **(b)** 63/256 **(c)** 1023/1024 **13.** 0.00009793 **15.** 1/24 **17.** 0.9395 **19. (a)** 0.05929 **(b)** 0.09684 **(c)** 0.23715 **21.** 0.003034 **23.** 0.001479 **25.** 0.1447 **27. (a)** 0.048265 **(b)** 0.047797 **29.** 0.004559 **31.** 0.07853 **33.** 25.29% **35.** 88.86% **37.** 1/80,089,128 **39.** 0.001654 **41.** 0.004843

Section 4.7

1. 1/6 **3.** 3.5 **5.** −$1.33 **7.** −$0.275 **9.** $3.75 **11. (a)** −$0.0526 **(b)** 5.26% **13. (a)** expected loss is $0.05 **(b)** expected gain is $0.375 **15. (a)** −$0.281 **(b)** 28.1% **17. (a)** $0.6016 **(b)** $0.2016 **(c)** $0.1016 **19. (a)** for 20 games, $0.095; for 16 games, $0.153; for 12 games, $0.122; for 10 games, $0.049; for 8 games, $0.059; for 5 games, $0.156 **(b)** 5 **(c)** no **21.** 268.49 **23. (a)** the first lawyer

Section 4.8

1. (a)

	T	s
s	Ts	ss
s	Ts	ss

(b) 1/2 **(c)** 1/2 **3.** 1 **5. (a)** 3/4 **(b)** 2/3 **7.** 2/3 **9. (a)** 1/16 **(b)** 15/16 **11. (a)** 1/2 **(b)** 0 **(c)** 1/2 **13. (a)** 1/4 **(b)** 1/4 **(c)** 1/2 **15.** 1/2 **17. (a)** 1/2 **(b)** 1/4 **19.** 1/2 **21. (a)** 1/2 **(b)** 1/2 **(c)** 0 **(d)** 0 **23. (a)** 0 **(b)** 1/2 **(c)** 1/2 **(d)** 0 **25.** 1/4 **27.** 1/3

Chapter 4 Review Exercises

1. 1/12 **2.** 96% **3.** 0.8433 **4.** 1/6001 **5.** 2 to 1 **6.** $5.00 **7.** no **8.** 69% **9.** 0.7354 **10. (a)** 0.4444 **(b)** 0.0870 **(c)** 0.15625 **11.** 0.05882 **12.** 0.476 **13.** 0.218 **14.** 1000 **15.** 21 **16.** 1/362,880 **17.** 0.3054 **18.** 0.9975 **19.** 0.2734 **20.** 34.69% **21.** 0.1301 **22.** 0.03497 **23.** the first choice **24. (a)** −$0.447 **(b)** 44.7% **25. (a)** 1/4 **(b)** 175/256 **26.** 9/16 **27. (a)** 1/4 **(b)** 1/4

CHAPTER 5

Section 5.1

1. (a)

Grade	Frequency	Relative Frequency
A	3	0.2143
B	3	0.2143
C	5	0.3571
D	2	0.1429
F	1	0.0714

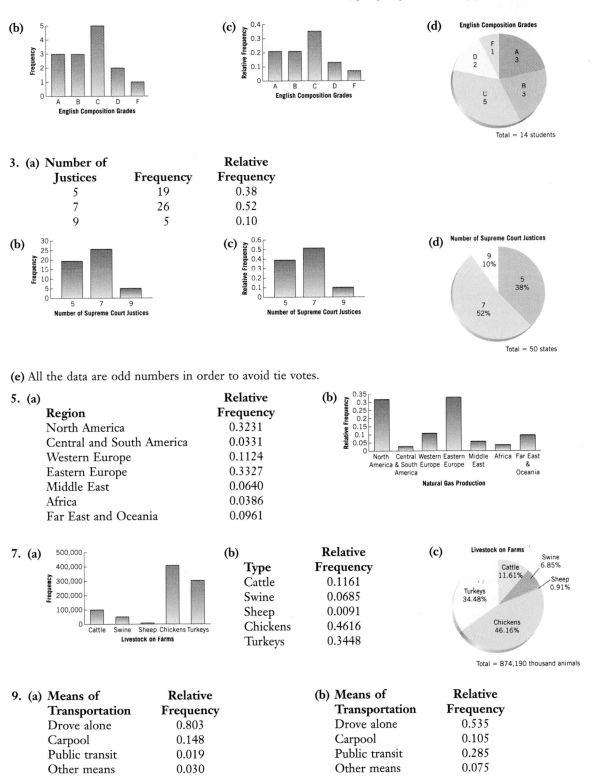

(b)

(c)

(d) English Composition Grades

Total = 14 students

3. (a)

Number of Justices	Frequency	Relative Frequency
5	19	0.38
7	26	0.52
9	5	0.10

(b)

(c)

(d) Number of Supreme Court Justices

Total = 50 states

(e) All the data are odd numbers in order to avoid tie votes.

5. (a)

Region	Relative Frequency
North America	0.3231
Central and South America	0.0331
Western Europe	0.1124
Eastern Europe	0.3327
Middle East	0.0640
Africa	0.0386
Far East and Oceania	0.0961

(b)

Natural Gas Production

7. (a)

Livestock on Farms

(b)

Type	Relative Frequency
Cattle	0.1161
Swine	0.0685
Sheep	0.0091
Chickens	0.4616
Turkeys	0.3448

(c) Livestock on Farms

Total = 874,190 thousand animals

9. (a)

Means of Transportation	Relative Frequency
Drove alone	0.803
Carpool	0.148
Public transit	0.019
Other means	0.030

(b)

Means of Transportation	Relative Frequency
Drove alone	0.535
Carpool	0.105
Public transit	0.285
Other means	0.075

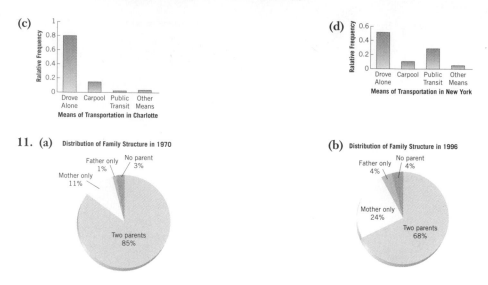

(c) There was a significantly lower percentage of two-parent families in 1996 than in 1970.

13. (a) One possibility is as follows:

Age	Frequency	Relative Frequency
40–47	5	0.1220
48–55	18	0.4390
56–63	13	0.3171
64–71	5	0.1220

(b)

(c) Age of Presidents at Inauguration

Total = 41 presidents

15. (a) One possibility is as follows:

Sales (in dollars)	Frequency	Relative Frequency
15,000–18,999	6	0.1176
19,000–22,999	30	0.5882
23,000–26,999	13	0.2549
27,000–30,999	1	0.0196
31,000–34,999	1	0.0196

(b)

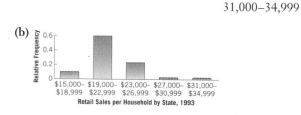

(c) Retail Sales per Household by State, 1993

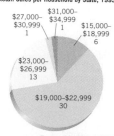

Total = 50 states and D.C.

17. (a)

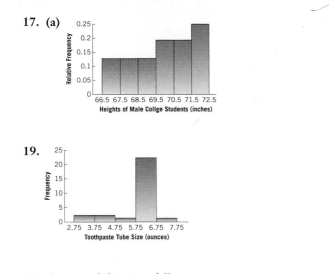

(b)

19.

21. One possibility is as follows:

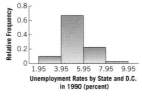

Section 5.2

1. C **3. (a)** 9 wins **(b)** 8 wins **(c)** 8 wins **5. (a)** 60.95°F **(b)** 58.55°F **7. (a)** 4.3% **(b)** 4.53% **9. (a)** $105,250
(b) $126,594 **(c)** The mean is greatly influenced by the one high sales price, so the median seems more representative.
11. (a) 66.8 **(b)** 67.46 **(c)** The birth rate seems to be increasing. **13. (a)** 45.7% **(b)** 52.0% **(c)** 37.9% **(d)** The years
in part (b) are presidential election years, whereas those in part (c) are not. **15. (a)** 10¢ and 11¢ (bimodal) **(b)** 12.5¢
(c) 14.06¢ **17. (a)** 16 years old **(b)** 16 years old **(c)** 16.5 years old **19. (a)** 18 years old **(b)** 17 years old **(c)** 16.9
years old **21. (a)** 4064 lb **(b)** 3200 lb **(c)** Weights decreased significantly. **23. (a)** 31.6 years old **(b)** 31.7 years old
(c) 31.8 years old **(e)** not overly critical

Section 5.3

1. (a) 209 sq. mi **(b)** 388 sq. mi **(c)** 143.8 sq. mi **3. (a)** $2.7 billion **(b)** $20 billion **(c)** $5.8 billion **5. (a)** 224.9 lb
(b) 126 lb **(c)** 35.0 lb **7. (a)** 6'7.67″ **(b)** 12″ **(c)** 3.50″ **(d)** The mean height of the Dream Team is well above the
national mean. **9. (a)** 8.49% **(b)** 1.9% **(c)** 0.61% **11. (a)** 3.10 inches **(b)** 3.07 inches **(c)** range, 5.15 inches; standard
deviation, 1.77 inches **(d)** range, 1.39 inches; standard deviation, 0.47 inch **(e)** The monthly precipitation is more
variable (seasonal) in Seattle than in Pittsburgh. **13. (a)** 89.0 sq. mi **(b)** 133.54 sq. mi **(c)** The low-density states
have the largest areas, whereas the high-density states have the smallest areas. **(d)** 365.5 sq. mi **(e)** 110.05 sq.mi
15. (a) 26.67 **(b)** 12 **(c)** 3.35 **17.** mean, 8.69; standard deviation, 0.12 **19.** mean, 9742; standard deviation, 2791
21. Using 70 for "60 and over," $\mu = 39.8$ hours, $\sigma = 14.2$ hours

Section 5.4

1. 0.4474 **3.** 0.0972 **5.** 0.6520 **7.** 0.0012 **9.** 0.9976 **11.** 0.8400 **13.** 0.8790 **15.** 0.1370 **17.** 0.1038
19. 0.9452 **21.** 0.25 **23.** −1.04 **25.** 1.555 **27.** 2.65 **29.** −13.885 **31.** 91.883 **33.** 61st **35.** 0.39%
37. 15.87% **39.** 89.44% **41.** 109 **43.** 97th **45.** 13.59% **47.** 0.13% **49.** 5'7.4″ **51.** 1st **53.** 66.78%
55. 5'1.2″ **57. (a)** 86 **(b)** 73 **(c)** 50 **59.** 12.17% **61.** 29 years old **63.** 0.59% **65.** 1.02% **67.** 77.34%
69. 41st **71.** 6.56% **73.** 6th **75.** 58.78% **77.** mean, 67.7; standard deviation, 15.1

Section 5.5

1.

n	Confidence Level			
	60%	**70%**	**80%**	**90%**
50	1.19	1.47	1.81	2.33
100	0.84	1.04	1.28	1.65
500	0.38	0.47	0.57	0.74
1000	0.27	0.33	0.40	0.52

3. (a) 0.021 mg/l **(b)** 0.179 to 0.221 mg/l **5.** 657.3 to 663.7 ppm **7.** 2.70 to 2.78 g/cc **9.** 2.17 to 2.27 g/cc **11.** 1.6 to 2.2 months **13.** 3.11 to 4.99 μg/g **15. (a)** 11.7 to 11.9 g/dl **(b)** 10.8 to 11.4 g/dl **(c)** Yes, the intervals do not overlap at all. **(d)** 1.32% **(e)** 22.36% **(f)** They seem close enough to support the assumption. **17.** 84.76% to 98.50% **21.** 1415 or more **23.** 1.63 to 2.13 siblings **25.** 299 to 385 μg/l **27.** 121° to 127°

Section 5.6

1.

n	Confidence Level			
	60%	**70%**	**80%**	**90%**
50	5.94%	7.35%	9.05%	11.63%
100	4.20%	5.20%	6.40%	8.23%
500	1.88%	2.33%	2.86%	3.68%
1000	1.33%	1.64%	2.02%	2.60%

3. 3% **5. (a)** 2.62% **7.** 22.4% to 29.2% **9.** 54.7% to 70.3% **11.** 1037 or more **13.** 784 **15.** 95% **17.** 94% (or 95%) **19.** 153

Chapter 5 Review Exercises

1. (a)

Country	Frequency	Relative Frequency
Italy	3	0.2000
Uruguay	1	0.0667
Germany	3	0.2000
Brazil	4	0.2667
England	1	0.0667
Argentina	2	0.1333
France	1	0.0667

(b)

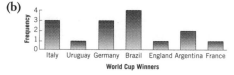

(c)

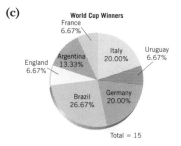

2. (a)

Branch	Relative Frequency
Army	0.3542
Navy	0.3027
Marine Corps	0.1218
Air Force	0.2213

(b)

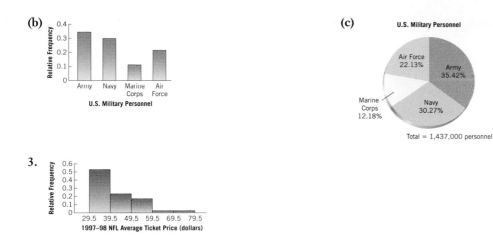

(c)

U.S. Military Personnel

Air Force 22.13%

Army 35.42%

Marine Corps 12.18%

Navy 30.27%

Total = 1,437,000 personnel

3.

(Relative Frequency chart)

1997–98 NFL Average Ticket Price (dollars)

4. (a) 618 **(b)** 572.7 **(c)** 631 **(d)** 236.4   **5. (a)** 22.90 mm **(b)** 21.13 mm **(c)** range, 3.8 mm; standard deviation, 1.03 mm **(d)** range, 2.4 mm; standard deviation, 0.72 mm   **6. (a)** 15 hours **(b)** 14.6 hours **(c)** 1.7 hours
7. (a) $94,199.50 **(b)** $18,851   **8.** 24.44%   **9.** 45.62%   **10.** 43.7 years old   **11.** 25th   **12. (a)** 3.25 to 3.53 mg/kg
(b) 2.94%   **13. (a)** 2.52 to 3.60 ppm **(b)** 29.3 to 42.7 ppm **(c)** Selenium content seems to depend heavily on the region, as evidenced by the gap between confidence intervals.   **14.** 4%   **15.** 73.2% to 80.6%   **16.** 66,307 or more
17. 350 or more

CHAPTER 6

Section 6.1
1. A, B **3.** B, C, D, E **5.** It has an Eulerian circuit. One possibility is as follows:

7. It has four odd vertices, $A, B, C,$ and D, so it has neither an Eulerian circuit nor an Eulerian path.
9. It has an Eulerian circuit. One possibility is as follows:

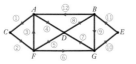

11. It has two odd vertices, A and D, so it does not have an Eulerian circuit. It does have an Eulerian path. One possibility is as follows:

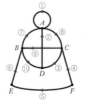

13. It has two odd vertices, A and C, so it does not have an Eulerian circuit. It does have an Eulerian path. One possibility is as follows:

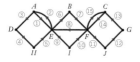

15. It is possible to take such a walk. One such walk is as follows:

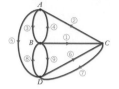

17. The graph has an Eulerian circuit. One possibility is as follows:

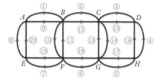

19. The graph has two odd vertices, B and F, so it does not have an Eulerian circuit. It does have an Eulerian path. One possibility is as follows:

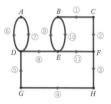

21. (a)

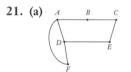

(b) The graph has two odd vertices, A and D, so it is not possible.

23. One possible eulerization is as follows:

25. One possible eulerization is as follows:

27. One possible eulerization is as follows:

29. One possible eulerization is as follows:

31. (a)

(b) One possible eulerization is as follows:

(c) The eulerization shown in (b) gives the shortest tour of the park.

Section 6.2

1. One Hamiltonian circuit is shown here.

3. One Hamiltonian circuit is shown here.

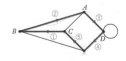

5. There are no Hamiltonian circuits on this graph because any circuit would have to go through the edge between vertices *D* and *E* twice and therefore visit both vertex *D* and vertex *E* twice.

7. (a)

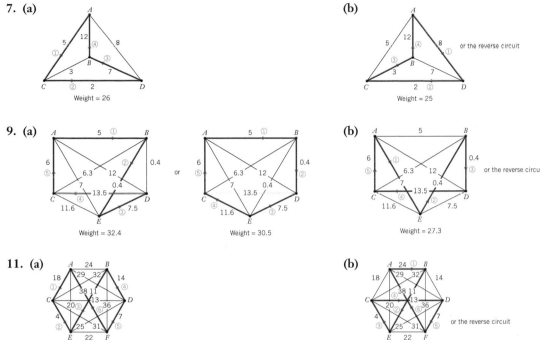

11. (a)

Weight = 92

(b)

Weight = 97

13. (a)

Weight = 86

(b) the same as the circuit in part (a) or the reverse circuit; weight = 86

15. (a)

Weight = 30

(b) or the reverse circuit

Weight = 29

(c) or the reverse circuit

Weight = 27

17. (a)

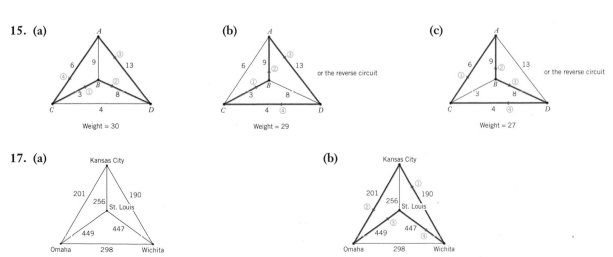

(b)

Length = 1287 miles

(c) the same as the circuit in part (b) or the reverse circuit; length = 1287 miles

(d) It is 94 miles shorter.

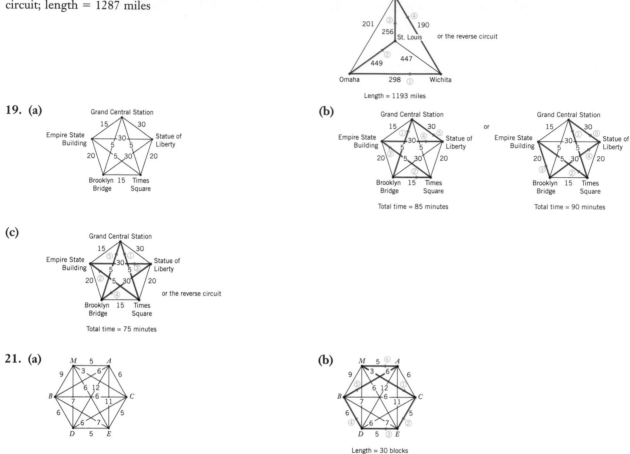

or the reverse circuit

Length = 1193 miles

19. (a)

(b) Total time = 85 minutes   or   Total time = 90 minutes

(c)

or the reverse circuit

Total time = 75 minutes

21. (a)

(b) Length = 30 blocks

(c) the same as the circuit in part (a) or the reverse circuit; length = 30 blocks   **23.** Base, Barnard 3, Shoemaker–Levy 7, Gunn, Swift, Neujmin 2, du Toit–Hartley, Harrington–Wilson, Base; total angular change = 419.9°

25. (a)

(b) Length = $\sqrt{5} + \sqrt{17} + \sqrt{10} + \sqrt{32} \approx 15.18$ cm

(c) 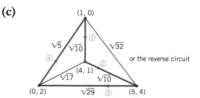 or the reverse circuit

Length = $\sqrt{10} + \sqrt{10} + \sqrt{29} + \sqrt{5} \approx 13.95$ cm

27. One Hamiltonian circuit is as follows:

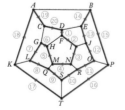

29.

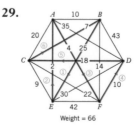

Weight = 66

Section 6.3

1. not a tree because it includes a circuit   **3.** a tree, but not a spanning tree   **5.** a spanning tree

7. Three possibilities are as follows:

9.

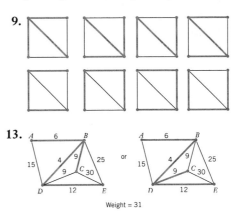

11.

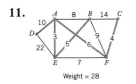

Weight = 28

13.

Weight = 31

15.

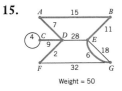

Weight = 50

17.

A 17 B 1 C 7 D

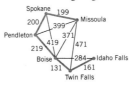

Weight = 47

19.

Length = 29.9 miles

21. The following flights should be retained.

23. The cable should be installed as shown here.

Length = 1150 miles

25. The minimum cost network is as follows:

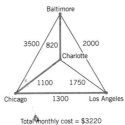

Total monthly cost = $3220

Chapter 6 Review Exercises

1. It has two odd vertices, *A* and *B*, so it does not have an Eulerian circuit. It has an Eulerian path. One possibility is as follows:

2. It has four odd vertices, *B*, *C*, *D*, and *I*, so it has neither an Eulerian circuit nor an Eulerian path.

3. It has six odd vertices, *A*, *B*, *C*, *E*, *F*, and *H*, so it has neither an Eulerian circuit nor an Eulerian path.

4. It has an Eulerian circuit. One possibility is as follows:

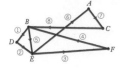

5. One possible eulerization is shown here.

6. One possible eulerization is shown here.

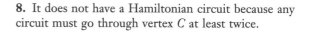

7. One Hamiltonian circuit is shown here.

8. It does not have a Hamiltonian circuit because any circuit must go through vertex *C* at least twice.

9. (a)

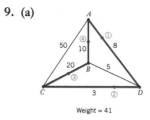

Weight = 41

(b)

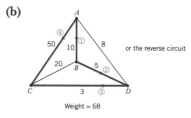

Weight = 68

10. (a)

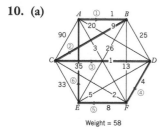

Weight = 58

(b)

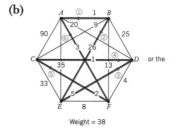

Weight = 38

11.

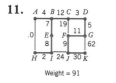

Weight = 91

12.

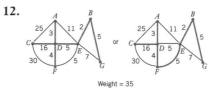

Weight = 35

13. The graph has an Eulerian circuit. One possibility is as follows:

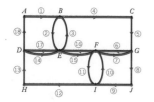

14. We first label the rooms of the house and the area outside the house as shown here.

Letting the rooms and the area outside the house be vertices and drawing an edge between any two vertices that have a door or window connecting them, we get the following graph.

This graph has two odd vertices, *B* and *C*, so it cannot have an Eulerian circuit. Therefore, the burglar must be lying.

15. (a)   **(b)**   **(c)**

Length = 2380 miles   Length = 2545 miles

16. The cheapest way of installing the pipes is shown here.   **17.** for *n* odd

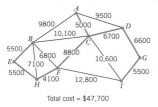

Total cost = $47,700

CHAPTER 7

Section 7.1
1. yes   **3.** no   **5.** no   **7.** no   **9.** concave; not regular; heptagon   **11.** convex; regular; nonagon   **13.** convex; not regular; quadrilateral   **15.** convex; regular; 11-gon   **17.** 1440°   **19.** 2700°   **21.** 1260°   **23.** 135°   **25.** 128.5714°   **27.** 168°   **29.** 21°   **31.** 185°

Section 7.2
1. not edge-to-edge   **3.** Not all vertices are of the same type.   **5.** The tiles are not all regular polygons and not all vertices are of the same type.   **7.** The four different vertex types are shown here.

9. The two different vertex types are shown here.

17. an equilateral triangle and a regular octagon

19.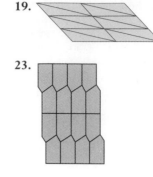

21.

23.

Section 7.3

1. convex   **3.** concave   **5.** $V - E + F = 10 - 17 + 9 = 2$ **7.** $V - E + F = 8 - 14 + 8 = 2$
9. $V - E + F = 4 - 6 + 4 = 2$ **11.** $V - E + F = 6 - 12 + 8 = 2$ **13.** $V - E + F = 12 - 30 + 20 = 2$ **15.** 12
17. All of the faces of the polyhedron are triangles.

19. The polyhedron either has 4 triangular faces and 2 quadrilateral faces or it has 5 triangular faces and 1 pentagonal face.

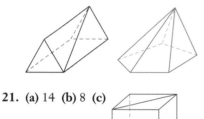

21. **(a)** 14 **(b)** 8 **(c)**

23. $V - E + F = 16 - 24 + 10 = 2$ **25.** $V = 16, E = 32, F = 16$ **27.** $V = 40, E = 84, F = 40$ **31.** **(a)** 8 **(b)** 6
33. 4 equilateral triangle faces and 4 regular hexagon faces   **35.** cube   **37.** dodecahedron   **43.** $V = 6, E = 12$;
$V = 7, E = 13; V = 8, E = 14; V = 9, E = 15; V = 10, E = 16; V = 11, E = 17; V = 12, E = 18$

Chapter 7 Review Exercises

1.

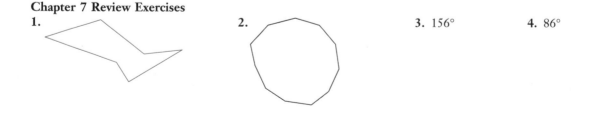

2.

3. 156°

4. 86°

5.

6.

7. The three different vertex types are shown here.

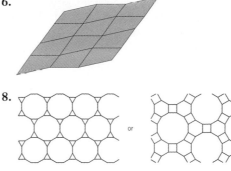

8. or

9. an equilateral triangle and a regular heptagon   **10.** The other polygons about vertex A must be two hexagons. However, this forces hexagons and triangles to alternate about vertex B, a different vertex type.

11. One of the vertices is surrounded by a pentagon, an octagon, and a hexagon. If they were all regular, the sum of the angles would be

$$\left(180° - \frac{360°}{5}\right) + \left(180° - \frac{360°}{8}\right) + \left(180° - \frac{360°}{6}\right) = 108° + 135° + 120° = 363°.$$

Because the angles do not sum to 360°, these polygons cannot all be regular.
12. $V - E + F = 7 - 12 + 7 = 2$ **13.** 25   **14.** 24   **15.** 1 quadrilateral face and 4 triangular faces

16. **(a)** 20 **(b)** 12 **(c)** $V - E + F = 60 - 90 + 32 = 2$

CHAPTER 8

Section 8.1
1. T   **3.** F   **5.** F   **7.** F   **9.** T   **11.** T   **13.** $2^2 \cdot 3 \cdot 11$ **15.** $5^3 \cdot 13$ **17.** $2^4 \cdot 5^2 \cdot 17$ **19.** 2, 3, 5, 7, 11, 13, 17, 19, 23, 29, 31, 37, 41, 43, 47, 53, 59, 61, 67, 71, 73, 79, 83, 89, 97   **21.** prime   **23.** not prime   **25.** prime
29. 3 + 13 or 5 + 11   **31.** 3 + 43 or 5 + 41 or 17 + 29 or 23 + 23   **33.** 19 + 79 or 31 + 67 or 37 + 61   **35.** One possibility is 86 = 7 + 13 + 23 + 43.   **37.** One possibility is 232 = 3 + 11 + 19 + 31 + 37 + 41 + 43 + 47.
39. $q = 9, r = 5$ **41.** $q = 17, r = 0$ **43.** $q = -8, r = 4$ **45.** $q = -14, r = 3$ **47.** $q = -9, r = 3$ **49.** 3   **51.** 24
53. 8   **55.** 1   **57.** 25   **59.** 21   **65.** **(a)** all positive integers **(b)** integers of the form $4k$ or $4k - 1$, where k is a positive integer

Section 8.2
1. 3   **3.** 10   **5.** 4   **7.** 0   **9.** 0   **11.** 2   **13.** 7   **15.** 1   **17.** 7   **19.** 19   **21.** 3   **23.** 8   **25.** 2   **27.** 3   **29.** 6
31. 2   **33.** 5   **35.** 18   **37.** 1   **39.** 43   **41.** 9   **43.** 1   **45.** 9803

Section 8.3
1. divisible by 9   **3.** not divisible by 9   **5.** divisible by 9   **7.** not divisible by 11   **9.** not divisible by 11   **11.** divisible by 11   **13.** not divisible by 4   **15.** divisible by 4   **17.** **(a)** yes **(b)** yes **(c)** yes **(d)** no **(e)** yes **(f)** no **(g)** yes **(h)** no **(i)** no **(j)** no   **19.** **(a)** no **(b)** yes **(c)** no **(d)** yes **(e)** no **(f)** yes **(g)** no **(h)** yes **(i)** no **(j)** no   **25.** 49¢   **29.** It could be correct.   **31.** It could not be correct.

Section 8.4

1. 4 **3.** yes **5.** no **7.** 111000025 **9.** no **11.** 5 **13.** yes **15.** 3 **17.** X **19.** 4179489192 **21. (a)** yes **(b)** replacing a 0 by a 9 or vice versa **23.** W **25.** yes

Section 8.5

1.

	\multicolumn{9}{c}{Round}

	1st	2nd	3rd	4th	5th	6th	7th	8th	9th
Team 1	9	Bye	2	3	4	5	6	7	8
Team 2	8	9	1	Bye	3	4	5	6	7
Team 3	7	8	9	1	2	Bye	4	5	6
Team 4	6	7	8	9	1	2	3	Bye	5
Team 5	Bye	6	7	8	9	1	2	3	4
Team 6	4	5	Bye	7	8	9	1	2	3
Team 7	3	4	5	6	Bye	8	9	1	2
Team 8	2	3	4	5	6	7	Bye	9	1
Team 9	1	2	3	4	5	6	7	8	Bye

3.

	\multicolumn{9}{c}{Round}

	1st	2nd	3rd	4th	5th	6th	7th	8th	9th
Team 1	9	10	2	3	4	5	6	7	8
Team 2	8	9	1	10	3	4	5	6	7
Team 3	7	8	9	1	2	10	4	5	6
Team 4	6	7	8	9	1	2	3	10	5
Team 5	10	6	7	8	9	1	2	3	4
Team 6	4	5	10	7	8	9	1	2	3
Team 7	3	4	5	6	10	8	9	1	2
Team 8	2	3	4	5	6	7	10	9	1
Team 9	1	2	3	4	5	6	7	8	10
Team 10	5	1	6	2	7	3	8	4	9

5.

	\multicolumn{15}{c}{Round}

	1st	2nd	3rd	4th	5th	6th	7th	8th	9th	10th	11th	12th	13th	14th	15th
Team 1	15	Bye	2	3	4	5	6	7	8	9	10	11	12	13	14
Team 2	14	15	1	Bye	3	4	5	6	7	8	9	10	11	12	13
Team 3	13	14	15	1	2	Bye	4	5	6	7	8	9	10	11	12
Team 4	12	13	14	15	1	2	3	Bye	5	6	7	8	9	10	11
Team 5	11	12	13	14	15	1	2	3	4	Bye	6	7	8	9	10
Team 6	10	11	12	13	14	15	1	2	3	4	5	Bye	7	8	9
Team 7	9	10	11	12	13	14	15	1	2	3	4	5	6	Bye	8
Team 8	Bye	9	10	11	12	13	14	15	1	2	3	4	5	6	7
Team 9	7	8	Bye	10	11	12	13	14	15	1	2	3	4	5	6
Team 10	6	7	8	9	Bye	11	12	13	14	15	1	2	3	4	5
Team 11	5	6	7	8	9	10	Bye	12	13	14	15	1	2	3	4
Team 12	4	5	6	7	8	9	10	11	Bye	13	14	15	1	2	3
Team 13	3	4	5	6	7	8	9	10	11	12	Bye	14	15	1	2
Team 14	2	3	4	5	6	7	8	9	10	11	12	13	Bye	15	1
Team 15	1	2	3	4	5	6	7	8	9	10	11	12	13	14	Bye

7.

															Round
	1st	**2nd**	**3rd**	**4th**	**5th**	**6th**	**7th**	**8th**	**9th**	**10th**	**11th**	**12th**	**13th**	**14th**	**15th**
Team 1	15	16	2	3	4	5	6	7	8	9	10	11	12	13	14
Team 2	14	15	1	16	3	4	5	6	7	8	9	10	11	12	13
Team 3	13	14	15	1	2	16	4	5	6	7	8	9	10	11	12
Team 4	12	13	14	15	1	2	3	16	5	6	7	8	9	10	11
Team 5	11	12	13	14	15	1	2	3	4	16	6	7	8	9	10
Team 6	10	11	12	13	14	15	1	2	3	4	5	16	7	8	9
Team 7	9	10	11	12	13	14	15	1	2	3	4	5	6	16	8
Team 8	16	9	10	11	12	13	14	15	1	2	3	4	5	6	7
Team 9	7	8	16	10	11	12	13	14	15	1	2	3	4	5	6
Team 10	6	7	8	9	16	11	12	13	14	15	1	2	3	4	5
Team 11	5	6	7	8	9	10	16	12	13	14	15	1	2	3	4
Team 12	4	5	6	7	8	9	10	11	16	13	14	15	1	2	3
Team 13	3	4	5	6	7	8	9	10	11	12	16	14	15	1	2
Team 14	2	3	4	5	6	7	8	9	10	11	12	13	16	15	1
Team 15	1	2	3	4	5	6	7	8	9	10	11	12	13	14	16
Team 16	8	1	9	2	10	3	11	4	12	5	13	6	14	7	15

9. Team 17   **11.** Team 21   **13.** Round 17   **15.** Round 7

17. The schedule is given in Exercise 3. Home teams: Round 1: 1, 2, 3, 4, 10; Round 2: 6, 7, 8, 9, 10; Round 3: 2, 3, 4, 5, 6; Round 4: 1, 2, 7, 8, 9; Round 5: 3, 4, 5, 6, 10; Round 6: 1, 2, 8, 9, 10; Round 7: 4, 5, 6, 7, 8; Round 8: 1, 2, 3, 4, 9; Round 9: 5, 6, 7, 8, 10

19. The schedule is given in Exercise 5. Home teams: Round 1: 1, 2, 3, 4, 5, 6, 7; Round 2: 9, 10, 11, 12, 13, 14, 15; Round 3: 2, 3, 4, 5, 6, 7, 8; Round 4: 1, 10, 11, 12, 13, 14, 15; Round 5: 3, 4, 5, 6, 7, 8, 9; Round 6: 1, 2, 11, 12, 13, 14, 15; Round 7: 4, 5, 6, 7, 8, 9, 10; Round 8: 1, 2, 3, 12, 13, 14, 15; Round 9: 5, 6, 7, 8, 9, 10, 11; Round 10: 1, 2, 3, 4, 13, 14, 15; Round 11: 6, 7, 8, 9, 10, 11, 12; Round 12: 1, 2, 3, 4, 5, 14, 15; Round 13: 7, 8, 9, 10, 11, 12, 13; Round 14: 1, 2, 3, 4, 5, 6, 15; Round 15: 8, 9, 10, 11, 12, 13, 14

23. The schedule is given in Exercise 3. Home teams: Round 1: 3, 4, 5, 8, 9; Round 2: 3, 4, 5, 9, 10; Round 3: 1, 4, 5, 6, 9; Round 4: 1, 4, 5, 6, 10; Round 5: 1, 2, 5, 6, 7; Round 6: 1, 2, 6, 7, 10; Round 7: 1, 2, 3, 7, 8; Round 8: 2, 3, 7, 8, 10; Round 9: 2, 3, 4, 8, 9

25. The schedule is given in Exercise 5. Home teams: Round 1: 5, 6, 7, 12, 13, 14, 15; Round 2: 5, 6, 7, 8, 13, 14, 15; Round 3: 1, 6, 7, 8, 13, 14, 15; Round 4: 1, 6, 7, 8, 9, 14, 15; Round 5: 1, 2, 7, 8, 9, 14, 15; Round 6: 1, 2, 7, 8, 9, 10, 15; Round 7: 1, 2, 3, 8, 9, 10, 15; Round 8: 1, 2, 3, 8, 9, 10, 11; Round 9: 1, 2, 3, 4, 9, 10, 11; Round 10: 2, 3, 4, 9, 10, 11, 12; Round 11: 2, 3, 4, 5, 10, 11, 12; Round 12: 3, 4, 5, 10, 11, 12, 13; Round 13: 3, 4, 5, 6, 11, 12, 13; Round 14: 4, 5, 6, 11, 12, 13, 14; Round 15: 4, 5, 6, 7, 12, 13, 14

29.

					Round		
	1st	**2nd**	**3rd**	**4th**	**5th**	**6th**	**7th**
Team 1	Bye	3	5	7	2	4	6
Team 2	7	Bye	4	6	1	3	5
Team 3	6	1	Bye	5	7	2	4
Team 4	5	7	2	Bye	6	1	3
Team 5	4	6	1	3	Bye	7	2
Team 6	3	5	7	2	4	Bye	1
Team 7	2	4	6	1	3	5	Bye

The rounds have been scrambled.

Section 8.6

1. DQBWL PHLVR NDB   **3.** YOU GOT IT RIGHT   **5.** EGKAX GKRDA FOE   **7.** KFSFD LEKFL FTSFE
9. PLAY IT AGAIN   **11.** DON'T THINK TWICE

Section 8.7

1. CM OR AD IC YW UM SR   **3.** SK NE PY OX   **5.** VOTE TODAY   **7.** HIT THE ROAD   **9.** 1457
0434   1036   1420   0843   1528   **11.** 3521   4185   3906   4101   0867   2704   **13.** PARDON ME   **15.** BE SURE

Chapter 8 Review Exercises

1. $2^3 \cdot 7^2 \cdot 11$ **2.** (a) not prime (b) prime   **3.** (a) $q = 6, r = 11$ (b) $q = -17, r = 4$ **4.** (a) 18 (b) 46   **5.** 6   **6.** 3
7. 0   **8.** 6   **9.** divisible by 9   **10.** not divisible by 11   **11.** divisible by 4   **12.** (a) yes (b) yes (c) yes (d) no (e) yes
(f) yes (g) no (h) no (i) no (j) no   **13.** $1.89   **14.** (a) It could be correct. (b) It could not be correct.   **15.** 6
16. no   **17.** yes   **18.** 1   **19.** 8   **20.** Team 15   **21.** Team 14   **22.** SHE KNOWS   **23.** BJYHK QOAWA OHWG
24. EXCELLENT JOB   **25.** RJ XU LR RU   **26.** FOR SALE   **27.** 1000   4141   3627   1895   **28.** BE NICE

CHAPTER 9

Section 9.1

1. **(a)** If Player A chooses *a* first, he or she can win by responding to *c* or *d* with *h* or *i*, respectively.

(b)

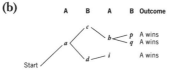

(b) Player O wins.

3. (a)

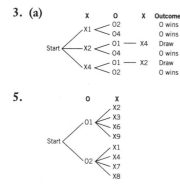

5.

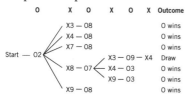

7. One way to show that Player O can achieve at least a draw with optimal play is with a partial game tree, showing only O's optimal responses.

9. **(a)** Denote a move by the number of pebbles left in the pile.   **(b)** Player B

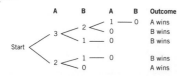

11. (a) Write a move by the number of pebbles left in the two piles, compressing the game tree by putting the larger pile first.

(b) Player B

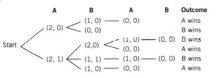

13. (a) n not divisible by 5 **(b)** The second player, by always leaving a multiple of 5 pebbles in the pile at the end of his or her turn. **15.** $(1, 0), (0, 1)$, and (n, n), with $n > 1$

17. (a)

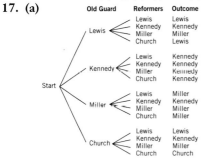

(b) The Old Guard should nominate Lewis, and the Reformers should nominate Lewis or Church in response. In either case, Lewis will be elected.

19. Right wins under best play. One possible partial game tree is

21. Left wins under best play.

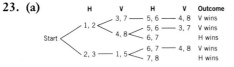

23. (a)

(b) H wins under best play.

25. (a)

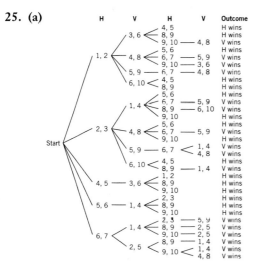

(b) H wins under best play. One possibility is

27. H wins by placing a tile in the middle row if n is odd and one of the middle two rows if n is even.

29. (a)

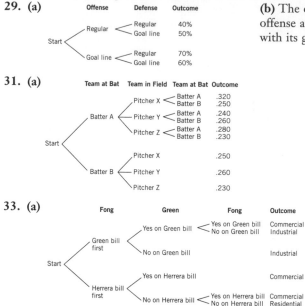

Offense	Defense	Outcome
Regular	Regular	40%
	Goal line	50%
Goal line	Regular	70%
	Goal line	60%

(b) The offense should choose its goal line offense and the defense should respond with its goal line defense.

(c) no

31. (a)

Team at Bat	Team in Field	Team at Bat	Outcome
Batter A	Pitcher X	Batter A	.320
		Batter B	.250
	Pitcher Y	Batter A	.240
		Batter B	.260
	Pitcher Z	Batter A	.280
		Batter B	.230
Batter B	Pitcher X		.250
	Pitcher Y		.260
	Pitcher Z		.230

(b) The team at bat should leave Batter A up and replace the batter with B only if the team in the field replaces the pitcher with Y, which it should do. If the team at bat brings in Batter B right away, the team in the field should switch to Pitcher Z.

33. (a)

Fong	Green	Fong	Outcome
Green bill first	Yes on Green bill	Yes on Green bill	Commercial
		No on Green bill	Industrial
	No on Green bill		Industrial
Herrera bill first	Yes on Herrera bill		Commercial
	No on Herrera bill	Yes on Herrera bill	Commercial
		No on Herrera bill	Residential

(b) Fong can set the vote on either bill first and Green should vote yes. Fong would vote no on the Green bill if it comes up first.

35. The Player should choose b and then p if Chance plays f and q if Chance plays g. The probability that the Player wins the game is then 3/4.

37. (a)

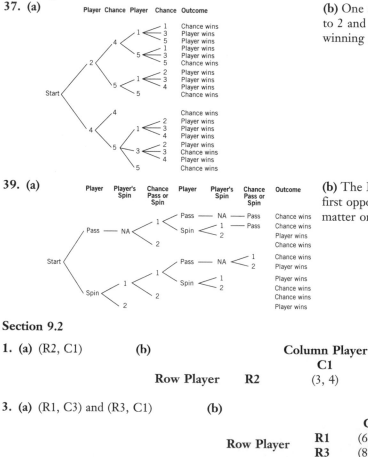

Player	Chance	Player	Chance	Outcome
2	4	1	1	Chance wins
			3	Player wins
			5	Player wins
		5	1	Player wins
			3	Player wins
			5	Chance wins
	5	1	2	Player wins
			3	Player wins
			4	Player wins
		5		Chance wins
4	4			Chance wins
		1	2	Player wins
			3	Player wins
			4	Player wins
	5	3	2	Player wins
			3	Chance wins
			4	Player wins
		5		Chance wins

(b) One such strategy is that the Player should move first to 2 and then back to 1. The Player's probability of winning is 5/6.

39. (a)

Player	Player's Spin	Chance Pass or Spin	Player	Player's Spin	Chance Pass or Spin	Outcome
Pass	NA	1	Pass	NA	Pass	Chance wins
			Spin	1	Pass	Chance wins
				2		Player wins
		2				Chance wins
Spin	1	1	Pass	NA	1	Chance wins
					2	Player wins
			Spin	1		Player wins
				2		Chance wins
	2					Chance wins
						Player wins

(b) The Player should spin at the first opportunity, and it does not matter on the second.

(c) 5/8

Section 9.2

1. (a) (R2, C1)

(b)

		Column Player
		C1
Row Player	**R2**	(3, 4)

3. (a) (R1, C3) and (R3, C1)

(b)

		Column Player		
		C1	**C3**	**C4**
Row Player	**R1**	(6, 4)	(8, 6)	(7, 5)
	R3	(8, 8)	(3, 3)	(5, 7)

5. (a) none   **(b)**

		Column Player		
		C1	**C2**	**C3**
Row Player	**R1**	(13, 0)	(5, 6)	(1, 8)
	R3	(5, 12)	(3, 7)	(10, 4)

7. (a) (R3, C2)   **(b)**

		Column Player
		C2
Row Player	**R3**	(6, 6)

9. (a)

	Submissive Pig	
	Push Lever	**Do Not Push Lever**
Dominant Pig **Push Lever**	(7, 3)	(3, 8)
Do Not Push Lever	(12, −1)	(0, 0)

(b) The dominant pig should press the lever and the submissive pig should not.

11. (a)

		Rock Hound 2			
		6 Pieces	**7 Pieces**	**8 Pieces**	**9 Pieces**
Rock Hound 1	**6 Pieces**	(150, 150)	(138, 161)	(126, 168)	(114, 171)
	7 Pieces	(161, 138)	(147, 147)	(133, 152)	(119, 153)
	8 Pieces	(168, 126)	(152, 133)	(136, 136)	(120, 135)
	9 Pieces	(171, 114)	(153, 119)	(135, 120)	(117, 117)

(b) (8 pieces, 8 pieces)

(c)

		Rock Hound 2
		8 Pieces
Rock Hound 1	**8 Pieces**	(136, 136)

(d) yes

13. (a)

		Rock Hound 2			
		9 Pieces	**10 Pieces**	**11 Pieces**	**12 Pieces**
Rock Hound 1	**10 Pieces**	(250, 225)	(230, 230)	(210, 231)	(190, 228)
	11 Pieces	(253, 207)	(231, 210)	(209, 209)	(187, 204)
	12 Pieces	(252, 189)	(228, 190)	(204, 187)	(180, 180)
	13 Pieces	(247, 171)	(221, 170)	(195, 165)	(169, 156)

(b) (10 pieces, 11 pieces) and (11 pieces, 10 pieces)

(c)

		Rock Hound 2	
		10 Pieces	**11 Pieces**
Rock Hound 1	**10 Pieces**	(230, 230)	(210, 231)
	11 Pieces	(231, 210)	(209, 209)

(d) no

15. (a)

		Child	
		Cooperative	**Uncooperative**
Parent	**Mild Mannered**	(4, 3)	(2, 4)
	Harsh	(1, 1)	(3, 2)

(b) (harsh, uncooperative)   **(c)**

		Child
		Uncooperative
Parent	**Harsh**	(3, 2)

17. (a)

		Child	
		Cooperative	**Uncooperative**
Parent	**Mild Mannered**	(4, 3)	(3, 4)
	Harsh	(1, 1)	(2, 2)

(b) (mild mannered, uncooperative)   **(c)**

		Child
		Uncooperative
Parent	**Mild Mannered**	(3, 4)

19. (a)

		Seller	
		Do Not Negotiate	**Lower Price if Asked**
	Buy	(3, 4)	(3, 4)
Buyer	**Negotiate and Go Elsewhere if Fail**	(1, 1)	(4, 2)
	Negotiate and Buy if Fail	(2, 3)	(4, 2)

(b) (buy, do not negotiate) and (negotiate and go elsewhere if fail, lower price if asked)

(c)

		Seller
		Do Not Negotiate
Buyer	**Buy**	(3, 4)

21. (a)

		Seller	
		Do Not Negotiate	**Lower Price if Asked**
	Buy	(2, 4)	(2, 4)
Buyer	**Negotiate and Go Elsewhere if Fail**	(3, 1)	(4, 2)
	Negotiate and Buy if Fail	(1, 3)	(4, 2)

(b) (negotiate and go elsewhere if fail, lower price if asked)

(c)

		Seller
		Lower Price if Asked
Buyer	**Negotiate and Go Elsewhere if Fail**	(4, 2)

23. The Buyer will buy without negotiation (although the Seller would lower the price).

25. (a) Write the bids as a pair in thousands of dollars, with the bid for the landscape first.

		Second Dealer					
		1 & 1	**2 & 1**	**1 & 2**	**2 & 2**	**1 & 3**	**2 & 3**
	1 & 1	(2.5, 2.5)	(1.5, 2.5)	(1, 3)	(0, 3)	(1, 2)	(0, 2)
	2 & 1	(2.5, 1.5)	(2, 2)	(1, 2)	(0.5, 2.5)	(1, 1)	(0.5, 1.5)
First Dealer	**1 & 2**	(3, 1)	(2, 1)	(2, 2)	(1, 2)	(1, 2)	(0, 2)
	2 & 2	(3, 0)	(2.5, 0.5)	(2, 1)	(1.5, 1.5)	(1, 1)	(0.5, 1.5)
	1 & 3	(2, 1)	(1, 1)	(2, 1)	(1, 1)	(1.5, 1.5)	(0.5, 1.5)

(b) (1 & 2, 1 & 2), (2 & 2, 2 & 2), (2 & 2, 2 & 3), (1 & 3, 1 & 3), (1 & 3, 2 & 3)

(c)

		Second Dealer
		2 & 3
First Dealer	**2 & 2**	(0.5, 1.5)

(d) The first dealer should bid $2000 for each painting, and the second dealer should bid $2000 for the landscape and $3000 for the portrait.

Section 9.3

1. (a) (R1, C2) and (R2, C2) **(b)**

		Column Player		
		C1	**C2**	**C3**
	R2	6	2	4
Row Player	**R3**	−3	1	7
	R4	7	0	−1

3. (a) (R2, C2) **(b)**

		Column Player
		C2
Row Player	**R2**	17

5. (a) none **(b)**

		Column Player	
		C1	**C4**
	R1	3.4	2.1
Row Player	**R2**	1.7	7.1
	R4	0	8.6

7. (a) none **(b)**

		Second Player	
		Call with Ace or 2	**Call with Ace Only**
	Bet with Ace or 2	0	$\frac{1}{7}$
First Player			
	Bet with Ace Only	$\frac{2}{7}$	0

9. (a)

		Chaser		
		2	**3**	**4**
Runner	**2**	0	1	1
	4	1	2	0

(b)

		Chaser	
		2	**4**
Runner	**2**	0	1
	4	1	0

The Runner should choose between 2 and 4, as should the Chaser.

11. (a)

		Player B		
		C	**D**	**H**
	C	0	1	2
Player A	**D**	−1	0	1
	H	−2	−1	0

(b) (C, C)   **(c)**

		Player B
		C
Player A	**C**	0

13. (a)

		Player B		
		C	**D**	**H**
	C	0	0	2
Player A	**D**	0	0	2
	H	−2	−2	0

(b) (C, C), (C, D), (D, C), (D, D)

(c)

		Player B	
		C	**D**
Player A	**C**	0	0
	D	0	0

15. (a)

		Alex		
		50	**30**	**10**
	60	0.2	0.2	0.2
Aaron	**40**	0	0.4	0.4
	20	0	−0.3	0.6

(b)

		Alex
		50
Aaron	**60**	0.2

(c) Aaron should shoot at 60 paces and Alex at 50. **(d)** 0.6

Section 9.4

1. (a) R1 **(b)** C2 **(c)** 3.74 to Row Player, 2.14 to Column Player   **3. (a)** R2 **(b)** C1 **(c)** $2\frac{8}{15}$ to Row Player, $\frac{4}{5}$ to Column Player   **5. (a)** R1 **(b)** C2 **(c)** 3.41 to Row Player, 4.3 to Column Player   **7. (a)** either **(b)** C1 **(c)** $4\frac{1}{7}$ to Row Player, $-4\frac{1}{7}$ to Column Player   **9. (a)** R1 **(b)** C4 **(c)** 1.28 to Row Player, −1.28 to Column Player   **11. (a)** R1: 4/7, R2: 3/7; C1: 2/7, C2: 5/7 **(b)** $4\frac{1}{7}$ **13. (a)** R1: 10/17, R2: 7/17; C1: 8/17, C2: 9/17 **(b)** $1\frac{12}{17}$ **15. (a)** R2 always, C2 always **(b)** 0.4

17. (a)

		Column Player	
		C1	**C3**
Row Player	**R2**	1	−1
	R3	0	3

(b) R1: 0, R2: 3/5, R3: 2/5; C1: 4/5, C2: 0, C3: 1/5   **(c)** 3/5

19. (a)

		Column Player	
		C2	C4
Row Player	R1	3	1
	R4	0	4

(b) R1: 2/3, R2: 0, R3: 0, R4: 1/3; C1: 0, C2: 1/2, C3: 0, C4: 1/2 **(c)** 2

21. (a) hire a lawyer **(b)** hire a lawyer **(c)** $1966.67 to plaintiff, −$2900 to defendant **23. (a)** $200
(b) $300 **(c)** first dealer, $98; second dealer, $30 **25. (a)** paper **(b)** scissors **(c)** 0.02 to the Row Player, −0.02 to the Column Player **27.** stocks

29. (a)

		Second Player	
		One Finger	Two Fingers
First Player	One Finger	1	−1
	Two Fingers	−1	1

(b) one finger **(c)** 1/15 to the first player, −1/15 to the second player **(d)** Each should choose one finger or two fingers with probability 1/2. **(e)** 0

31. (a)

		Receiver	
		Guess Forehand	Guess Backhand
Server	Serve to Forehand	40	70
	Serve to Backhand	72	54

(b) Server: serve to forehand: 3/8, serve to backhand: 5/8; Receiver: guess forehand: 1/3, guess backhand: 2/3 **(c)** 60%

33. (a)

		Receiver	
		Guess Forehand	Guess Backhand
Server	Serve to Forehand	40	70
	Serve to Backhand	90	75

(b) The server should serve to the backhand; the receiver should guess backhand. **(c)** 75%

35. (a)

		von Kluge	
		Attack Gap	Withdraw
Bradley	Reinforce Gap	2	3
	Move Eastward	1	5
	Hold in Reserve	6	4

(b)

		von Kluge	
		Attack Gap	Withdraw
Bradley	Move Eastward	1	5
	Hold in Reserve	6	4

(c) Bradley: reinforce gap: 0, move eastward: 1/3, hold: 2/3; von Kluge: attack: 1/6, withdraw: 5/6 **(d)** $4\frac{1}{3}$, slightly better for the Allies than the commanders' initial decisions

37. (a)

		von Kluge	
		Attack Gap	Withdraw
Bradley	Reinforce Gap	0	1
	Move Eastward	−5	7
	Hold in Reserve	10	3

(b)

		von Kluge	
		Attack Gap	Withdraw
Bradley	Move Eastward	−5	7
	Hold in Reserve	10	3

(c) Bradley: reinforce gap: 0, move eastward: 7/19, hold: 12/19; von Kluge: attack: 4/19, withdraw: 15/19 **(d)** $4\frac{9}{19}$, somewhat better for the Allies than the commanders' initial decisions

39. (a)

		IRS	
		Check First Line	**Check Second Line**
Cheat	**Cheat on First Line**	−2000	1000
	Cheat on Second Line	4000	−2000

(b) Cheat: cheat on first line: 2/3, cheat on second line: 1/3; IRS: check first line: 1/3, check second line: 2/3 **(c)** $0

41. (a)

		IRS	
		Check First Line	**Check Second Line**
Cheat	**Cheat on First Line**	−2000	1000
	Cheat on Second Line	4000	−8000

(b) Cheat: cheat on first line: 4/5, cheat on second line: 1/5; IRS: check first line: 3/5, check second line: 2/5
(c) −$800

43. (a)

		IRS	
		Check First Line	**Check Second Line**
Cheat	**Cheat on First Line**	−5000	1000
	Cheat on Second Line	4000	−20,000

(b) Cheat: cheat on first line: 4/5, cheat on second line: 1/5; IRS: check first line: 7/10, check second line: 3/10
(c) −$3200   **45.** Robber: north: 3/7, south: 4/7; Sheriff: north: 3/7, south: 4/7; probability of escape: 6/7

47.

		Goalie	
		Guess Left	**Guess Right**
Kicker	**Kick Left**	50	90
	Kick Right	90	50

Kicker: kick left: 1/2, kick right: 1/2; Goalie: guess left: 1/2, guess right: 1/2; probability of goal: 70%

49.

		Goalie	
		Guess Left	**Guess Right**
Kicker	**Kick Left**	50	95
	Kick Right	85	50

Kicker: kick left: 7/16, kick right: 9/16; Goalie: guess left: 9/16, guess right: 7/16; probability of goal: 69.6875%

Section 9.5

1. R1: 3/8, R2: 5/8; C1: 1/4, C2: 0, C3: 3/4   **3.** R1: 3/10, R2: 7/10; C1: 0, C2: 3/8, C3: 0, C4: 5/8   **5.** A1: 0, A2: 0, A3: 9/16, A4: 7/16; B1: 7/16, B2: 0, B3: 0, B4: 9/16   **7.** A1: 65/77, A2: 0, A3: 12/77, A4: 0; B1: 0, B2: 2/11, B3: 0, B4: 9/11, B5: 0

9. (a)

		Second Player		
		One Finger	**Two Fingers**	**Three Fingers**
First Player	**One Finger**	2	−3	4
	Two Fingers	−3	4	−5

(b) First Player: one: 7/12, two: 5/12; Second Player: one: 7/12, two: 5/12, three: 0 **(c)** −1/12   **11.** Batter: look for fastball: 5/18, do not anticipate: 13/18; Pitcher: fastball: 1/6, change-up: 0, curve: 0, slider: 5/6; $v = 223/600$ **13.** Batter: look for fastball: 2/5, do not anticipate: 3/5; Pitcher: fastball: 1/2, change-up: 1/2, curve: 0, slider: 0; $v = 11/25$

15. (a)

		Defender			
		0 to N	**1 to N**	**2 to N**	**3 to N**
Attacker	**Attack N**	10	8	0	−10
	Attack S	−10	0	3	5

(b) Attacker: attack N: 3/7, attack S: 4/7; Defender: 0 to N: 3/7, 1 to N: 0, 2 to N: 0, 3 to N: 4/7 **(c)** should not invade

17. (a)

		Defender			
		0 to N	1 to N	2 to N	3 to N
Attacker	Attack N	18	15	0	−1
	Attack S	−1	0	7	10

(b) Attacker: attack N: 7/22, attack S: 15/22; Defender: 0 to N: 0, 1 to N: 7/22, 2 to N: 15/22, 3 to N: 0 **(c)** should invade

Section 9.6

1. January 1, July 4, October 31, December 25 **3.** Washington, Lincoln **5.** pepperoni, sausage **7. (a)** R1: C3, R2: C2, R3: C2; R1, R3 **(b)** C1: R3, C2: R3, C3: R1; C2, C3 **9. (a)** R1: C1, R2: C4, R3: C1; R3 **(b)** C1: R3, C2: R2, C3: R1, C4: R1; C1 **11. (a)** Row Player: 4, Column Player: 8 **(b)** (R1, C2), (R1, C3), (R2, C2), (R3, C2) **(c)** (R1, C3) **(d)** (R1, C3), (R3, C2) **13. (a)** Row Player: 2, Column Player: 6 **(b)** (R1, C2), (R1, C3), (R2, C2) **(c)** (R1, C3), (R2, C2) **(d)** (R2, C2)

In Exercises 15, 17, and 19, we let R, MV, and S stand for Riverside, Mountain View, and Sunnyvale, respectively, and express our payoffs in terms of thousands of people.

15. (a)

		Post		
		R	MV	S
	R	(30, 10)	(40, 30)	(40, 20)
Times	MV	(30, 40)	(22.5, 7.5)	(30, 20)
	S	(20, 40)	(20, 30)	(15, 5)

(b)

		Post
		MV
Times	R	(40, 30)

(c) *Times*: R, MV; *Post*: R, MV **(d)** *Times*: 30; *Post*: 10 **(e)** (R, R), (R, MV), (R, S), (MV, R), (MV, S) **(f)** (R, MV), (MV, R) **(g)** (R, MV), (MV, R)

17. (a)

		Post		
		R	MV	S
	R & MV	(50, 20)	(55, 15)	(70, 20)
Times	R & S	(40, 20)	(60, 30)	(50, 10)
	MV & S	(50, 40)	(35, 15)	(40, 10)

(b)

		Post	
		R	MV
Times	R & MV	(50, 20)	(55, 15)
	R & S	(40, 20)	(60, 30)

(c) *Times*: all; *Post*: all **(d)** *Times*: 50; *Post*: 20 **(e)** (R & MV, R), (R & MV, S), (R & S, MV), (MV & S, R) **(f)** (R & MV, S), (R & S, MV), (MV & S, R) **(g)** (R & MV, R), (R & MV, S), (R & S, MV), (MV & S, R)

In Exercises 19 and 21, we give the payoffs in units of $1000.

19. (a)

		Defendant		
		No Lawyer	Mediocre Lawyer	Good Lawyer
	No Lawyer	(10, −10)	(0, −2)	(0, −4)
Plaintiff	Mediocre Lawyer	(8, −10)	(8, −12)	(−2, −4)
	Good Lawyer	(6, −10)	(6, −12)	(6, −14)

(b)

		Defendant		
		No Lawyer	Mediocre Lawyer	Good Lawyer
	No Lawyer	(10, −10)	(0, −2)	(0, −4)
Plaintiff	Mediocre Lawyer	(8, −10)	(8, −12)	(−2, −4)
	Good Lawyer	(6, −10)	(6, −12)	(6, −14)

(c) Plaintiff: 6; Defendant: −10 **(d)** (no lawyer, no lawyer), (mediocre lawyer, no lawyer), (good lawyer, no lawyer) **(e)** (no lawyer, no lawyer) **(f)** none

21. (a)

		Defendant		
		No Lawyer	Mediocre Lawyer	Good Lawyer
	No Lawyer	(2, −2)	(0, −2)	(0, −4)
Plaintiff	Mediocre Lawyer	(2, −4)	(0, −4)	(−2, −4)
	Good Lawyer	(0, −4)	(0, −6)	(−2, −6)

(b)

		Defendant
		No Lawyer
Plaintiff	No Lawyer	(2, −2)

(c) Plaintiff: 0; Defendant: −4 **(d)** (no lawyer, no lawyer), (no lawyer, mediocre lawyer), (no lawyer, good lawyer), (mediocre lawyer, no lawyer), (mediocre lawyer, mediocre lawyer), (good lawyer, no lawyer) **(e)** (no lawyer, no lawyer) **(f)** (no lawyer, no lawyer), (no lawyer, mediocre lawyer), (mediocre lawyer, no lawyer), (mediocre lawyer, mediocre lawyer)

23. (a)

		Country 2		
		No Buildup	Moderate Buildup	Large Buildup
Country 1	No Buildup	(5, 5)	(2, 7)	(2, 6)
	Moderate Buildup	(7, 2)	(4, 4)	(1, 6)
	Large Buildup	(6, 2)	(6, 1)	(3, 3)

(b)

		Country 2
		Large Buildup
Country 1	Large Buildup	(3, 3)

(c) Country 1: 3; Country 2: 3 **(d)** (no buildup, no buildup), (moderate buildup, moderate buildup), (large buildup, large buildup) **(e)** (no buildup, no buildup) **(f)** (large buildup, large buildup)

25. (a)

		Country 2		
		No Buildup	Moderate Buildup	Large Buildup
Country 1	No Buildup	(3, 5)	(3, 4)	(3, 3)
	Moderate Buildup	(5, 2)	(2, 4)	(2, 3)
	Large Buildup	(4, 2)	(4, 1)	(1, 3)

(b)

		Country 2		
		No Buildup	Moderate Buildup	Large Buildup
Country 1	No Buildup	(3, 5)	(3, 4)	(3, 3)
	Moderate Buildup	(5, 2)	(2, 4)	(2, 3)
	Large Buildup	(4, 2)	(4, 1)	(1, 3)

(c) Country 1: 3; Country 2: 3 **(d)** (no buildup, no buildup), (no buildup, moderate buildup), (no buildup, large buildup) **(e)** (no buildup, no buildup) **(f)** none

27. (a)

		Politician 2			
		For A & B	For A	For B	For Neither
Politician 1	For A & B	(3, 3)	(4, 1)	(1, 4)	(2, 2)
	For A	(4, 1)	(4, 1)	((2, 2))	(2, 2)
	For B	(1, 4)	(2, 2)	(1, 4)	(2, 2)
	For Neither	(2, 2)	(2, 2)	(2, 2)	(2, 2)

(b) AB

29. (a)

		Politician 2			
		For A & B	For A	For B	For None
Politician 1	For C	(7, 5)	(4, 7)	(3, 6)	(1, 8)
	For None	(8, 1)	(6, 3)	(5, 2)	((2, 4))

(b) ABC, AC, BC

31. (a)

		Politician 2			
		For A & B	For A	For B	For None
Politician 1	For C & D	(11, 8)	(8, 10)	(3, 15)	(1, 16)
	For C	(15, 3)	(13, 5)	(7, 11)	(5, 13)
	For D	(12, 4)	(10, 6)	(4, 12)	(2, 14)
	For None	(16, 1)	(14, 2)	(9, 7)	((6, 9))

(b) ACD, BC

Chapter 9 Review Exercises

1.

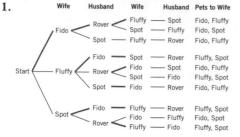

Highlighting the best choices at each juncture and choosing arbitrarily when results are the same, we see that the wife should choose Fido, and the husband should then choose Rover; the wife should then choose Fluffy, leaving Spot for the husband.

2. The game tree, with management shares first and best choices highlighted, is shown here. The likely outcome under best play is for Management to offer a (2, 1) split, which Labor will accept.

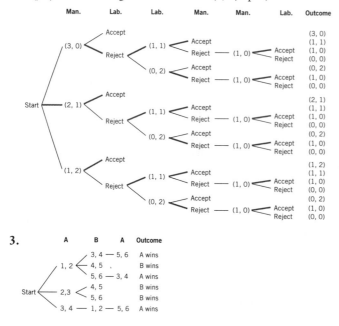

3.

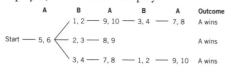

By choosing squares 3 and 4, Player A will win. (Note that Player B's responses of 1, 2 and 5, 6 are strategically the same, so they have been compressed into one branch of the tree.)

4. We take advantage of the symmetry of the game to construct a partial compressed game tree, showing that A, the first player, wins under best play.

	A	B	A	B	A	Outcome
		1, 2 — 9, 10 — 3, 4 — 7, 8				A wins
Start — 5, 6	2, 3 — 8, 9					A wins
		3, 4 — 7, 8 — 1, 2 — 9, 10				A wins

5. By covering squares 1 and 2, the first player becomes the second to move on a board with $n - 2$ squares. By covering squares 2 and 3, neither player can cover square 1 and the game becomes one on $n - 3$ squares with the original first player moving second. Thus, if the second player has a winning strategy in one of the two cases, the first player will win the game with n squares under best play. **6.** Take the middle two squares and then mirror the second player's moves on the squares on the opposite side of the middle two squares.

7. (a)

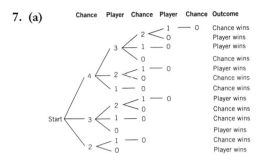

Chance	Player	Chance	Player	Chance	Outcome

(b) From 2 or 3 left after Chance's first move, the Player should win. From 4 left, the Player should take 1 and then win if possible. The Player's probability of winning is 8/9.

8. (a)

		Runner		
		3, 1	**3, 2**	**4, 3**
	2, 3	−2	1	1
Chaser	**3, 4**	2	2	1
	3, 5	2	2	−2

(b) The Chaser should move to vertex 3 and then to vertex 4. The Runner should move to vertex 4 and then to vertex 3. (c) 1

9. (a)

		Runner		
		2, 1	**3, 1**	**3, 2**
	2, 3	2	−2	1
Chaser	**2, 4**	2	−2	−2
	3, 4	−2	2	2

(b) Chaser: 2, 3 and 3, 4 each with probability 1/2 and 2, 4 with probability 0; Runner: 2, 1 and 3, 1 each with probability 1/2 and 3, 2 with probability 0 (c) 0

10. Row Player: 0: 7/16, 1: 0, 2: 0, 3: 9/16; Column Player: 0: 0, 1: 3/16, 2: 13/16, 3: 0, 4: 0, 5: 0

11. Row Player: 0: 7/16, 1: 0, 2: 0, 3: 0, 4: 9/16; Column Player: 0: 3/8, 1: 0, 2: 0, 3: 5/8, 4: 0, 5: 0, 6: 0

12. (a)

		Calling Player			
		One Finger, "Odd"	**One Finger, "Even"**	**Two Fingers, "Odd"**	**Two Fingers, "Even"**
Silent Player	**One Finger**	2	−2	−3	3
	Two Fingers	−3	3	4	−4

(b) Silent Player: one finger: 7/12, two fingers: 5/12; Calling Player: one finger, "odd": 7/12, one finger, "even": 0, two fingers, "odd": 5/12, two fingers, "even": 0 (c) −1/12

13. (a)

		Column Fool	
		Swerve	**Don't Swerve**
	Swerve	(3, 3)	(2, 4)
Row Fool	**Do Not Swerve**	(4, 2)	(1, 1)
	Swerve Late	(4, 2)	(2, 4)

(b) The Row Fool should swerve late; the Column Fool should not swerve.

14. (a) Let the strategy be the number grabbed.

		Player B				
		0	**1**	**2**	**3**	**4**
	0	(4, 4)	(3, 4)	(2, 4)	(1, 4)	(0, 4)
	1	(4, 3)	(3, 3)	(2, 3)	(1, 3)	(1, 3)
Player A	**2**	(4, 2)	(3, 2)	(2, 2)	(2, 2)	(2, 2)
	3	(4, 1)	(3, 1)	(2, 2)	(2, 2)	(2, 2)
	4	(4, 0)	(3, 1)	(2, 2)	(2, 2)	(2, 2)

(b) Both players should try to grab at least two nuts. **(c)** The security levels are 2 for each, with everything in the negotiation set except one player grabbing 3 or 4 and the other 0 or 1. The optimal negotiation set is for each to grab 0. Every point in the negotiation set is stable.

15. (a) Let the strategy be the number grabbed.

		Player B				
		0	**1**	**2**	**3**	**4**
	0	(8, 8)	(6, 7)	(4, 6)	(2, 5)	(0, 4)
	1	(7, 6)	(5, 5)	(3, 4)	(1, 3)	(1, 3)
Player A	**2**	(6, 4)	(4, 3)	(2, 2)	(2, 2)	(2, 2)
	3	(5, 2)	(3, 1)	(2, 2)	(2, 2)	(2, 2)
	4	(4, 0)	(3, 1)	(2, 2)	(2, 2)	(2, 2)

(b) Neither player should grab any nuts. **(c)** The security levels are 2 for each, with everything in the negotiation set except for the strategy pairs (0, 4), (1, 3), (1, 4), (3, 1), (4, 0), and (4, 1). The optimal negotiaton set is for each to grab 0. The stable points are when both try to grab 3 or 4 and when both grab 0.

16. (a)

		Carson		
		Remain Silent	**Confess**	**Confess Only if Roberts Has**
	Remain Silent	(−4, −4)	(−20, −3)	(−4, −4)
Roberts	**Confess**	(−3, −20)	(−15, −15)	(−15, −15)
	Confess Only if Carson Has	(−4, −4)	(−15, −15)	(−4, −4)

(b) Each should confess only if the other has. **(c)** Both confessing and both confessing only if the other has are the two equilibrium points. **(d)** The security levels are −15 for each. The negotiation set is everything except one confessing while the other remains silent. The optimal negotiation set is any combination of remaining silent and confessing only if the other has. **(e)** Both confessing and both confessing only if the other has are stable. **(f)** confess and confess only if Carson has **17.** the left and middle squares **18.** 6 each and a 9-3 split (so each gets $9)

Photo Credits

Cover and chapter openers
Baseball scoreboard: Eunice Harris/Photo Researchers. Subway map: ©MTA. Image courtesy New York City Transit Authority. Reproduced with permission. Mozaic: George Holton/Photo Researchers. Ballot: Andy Sacks/Tony Stone Images, New York. Capitol building: Jack Hamilton/Photo Researchers.

Chapter 1
Page 2 (center): Andy Sacks/Tony Stone Images/New York, Inc. Page 2 (bottom): Peter Turnley/Corbis-Bettmann. Page 8: Paula Bronstein/Tony Stone Images/ New York, Inc. Page 9: D. Halstead/Gamma Liaison. Page 10: Richard Francis/ Tony Stone Images/New York, Inc. Page 25: Courtesy Downtown Athletic Club, NY. Page 26: Al Tielemans/Duomo Photography, Inc. Page 35 (top): Granger Collection. Page 35 (bottom): ©Fred Mulhearn. Reproduced with permission. Page 44: Culver Pictures, Inc. Pages 48–49: Courtesy Van Cliburn Foundation. Page 50: Courtesy Smithsonian Institution. Page 59: Kenneth Arrow.

Chapter 2
Page 70: Vivian Ronay/Gamma Liaison. Pages 74, 75, 84, and 93: Granger Collection. Page 98: Corbis-Bettmann. Page 105: Courtesy Harvard Archives.

Chapter 3
Page 126 (center): Granger Collection. Page 126 (bottom): Laurence Dutton/Tony Stone Images/New York, Inc. Page 132: Jeffrey M. Hamilton/Gamma Liaison. Page 136: Courtesy Colorado State Lottery. Page 138: Courtesy Intel. Page 143: Copyright Bank United. Page 147: Art Wolfe/Tony Stone Images/New York, Inc. Page 148: UPI/Corbis-Bettmann. Page 149: Courtesy National Bank of Hungary. Page 152: Jonathan Kirn/Gamma Liaison. Page 154: Mark Scott/FPG International. Page 158: Carlos Alejandro/The Vanguard Group, Inc. Page 169: Courtesy Chevrolet. Page 172 (left): Courtesy Colorado State Lottery. Page 172 (center): Courtesy Minnesota State Lottery. Page 172 (right): Courtesy New Mexico Lottery.

Chapter 4
Page 184 (top right): Doug Armand/Tony Stone Images/New York, Inc. Page 184 (center): Steven Peters/Tony Stone Images/New York, Inc. Page 184 (bottom): Fred Lyon/Photo Researchers. Page 191: Courtesy Rhonda Hatcher and George Gilbert. Page 198 (top): Jamie Squire/Allsport. Page 198 (center): Armen Kachaturian/Gamma Liaison. Page 199: Doug Penginger/Allsport. Page 200: Courtesy Colorado State Lottery. Page 212: Mitchell Layton/Duomo Photography, Inc. Page 219: Courtesy Masterlock. Page 227: Courtesy New York State Lottery. Page 233 (top): ©United Media. Page 233 (center): Markel/Gamma Liaison. Pages 233 (bottom) and 237: Courtesy Michael Kustermann. Page 241: Photo by AP/Wide World Photos. Text reproduced with permission of *The New York Times.* Page 253: ©Dale Blackwell/Fort Worth Star Telegram. Page 264: Courtesy Michael Kustermann. Page 267: AP/Wide World Photos.

Chapter 5
Page 270 (center): ©VCG/FPG International. Page 270 (bottom): ©Duomo Photography, Inc. Page 279: ©Kenneth Liao, John Wiley & Sons Photo Archive. Page 282: From Doron Witztum, Eliyahu Rips, and Yoav Rosenberg, "Equidistant Letter Sequences in the Book of Genesis," *Statistical Science*, 9:3, August 1994, page 430. Reproduced with permission of Institute of Mathematical Statistics, Hayward, CA. Page 299: Courtesy TU Electric. Page 309: Jed Jacobson/Allsport. Page 324: Culver Pictures, Inc. Page 337: Alexander Lowry/Photo Researchers. Page 345: Sidney Harris. Page 346: Stephen Krasemann/Tony Stone Images/New York, Inc. Page 351: UPI/Corbis-Bettmann. Page 353: Corbis-Bettmann.

Chapter 6
Page 362 (top right): Nick Gunderson/Tony Stone Images/New York, Inc. Page 362 (bottom left): Randy Wells/Tony Stone Images/New York, Inc. Page 364: Granger Collection. Page 380: Science Photo Library/Photo Researchers. Page 395: From *New Scientist*, June 27, 1992. Art courtesy William Cook, Rice University. Reproduced with permission. Page 399: ©MTA. Image courtesy New York City Transit Authority. Reproduced with permission. Page 409: ©Quesada/Burke Studios.

Chapter 7
Page 420 (center): Paul Johnson/Tony Stone Images/New York, Inc. Page 420 (bottom left): Jeremy Burgess/Photo Researchers. Page 420 (bottom right): Sylvain Grandadam/Tony Stone Images/New York, Inc. Page 421: George Holton/Photo Researchers. Page 422: Gian Berto Vanni/Art Resource. Page 428 (left): "Hexagon Flower Garden" quilt made by Tima Garrett, circa 1935. Photo is courtesy of Quilter's Newsletter Magazine. Photography by Mellisa Karlin Mahoney. Reproduced with permission. Page 428 (bottom): Courtesy Rhonda Hatcher and George Gilbert. Page 431: Holt Studios Int'l/Photo Researchers. Page 436: ©M.C. Escher/Cordon Art. Page 437: Granger Collection. Page 441: ©Mario Ruiz/Time Magazine. Page 445 (left): Photo by Violet Anderson, ©Earth Science Department, Royal Ontario Museum. Reproduced with permission. Page 445 (center): Will & Deni McIntyre/Tony Stone Images/New York, Inc. Page 445 (right): Courtesy Societe de Transport de la Commuaute Urbaine de Montreal. Page 446: Courtesy Mony Harden. Page 451 (top): Corbis-Bettmann. Page 451 (center): Charles D. Winters/Photo Researchers. Page 451 (bottom): Manfred Kage/Peter Arnold, Inc. Page 452: Andy Washnik. Page 454: Duomo Photography, Inc. Page 455: Courtesy Dale Taylor/Star-Telegram.

Chapter 8
Page 466: Peter Poulides/Tony Stone Images/New York, Inc. Page 474: Granger Collection. Page 475: Courtesy Denise Applewhite, Princeton University. Page 481: Courtesy Deutsche Bundesbank. Page 495: Courtesy American Airlines. Page 496: Courtesy U.S. Post Office. Page 498: Courtesy Rhonda Hatcher and George Gilbert. Page 499 (top): KELLOGG'S® CRISPIX® is a trademark of Kellogg Company. All rights reserved. Used with permission. Page 499 (bottom): Courtesy Campbell Soup Co. Page 501: Q-Tips® is a registered trademark of Chesebrough-Ponds, Inc. Reproduced courtesy of Chesebrough-Pond's USA Co. Page 503: Courtesy American Express Travelers Cheque, American Express Company. Page 504: Frederic Nebinger/Allsport. Page 518 (top): ©AKG London. Page 518

(bottom left): ©Dorling Kindersley Media Library. Page 518 (bottom right): Courtesy Silvio A. Bedini. Page 519 (bottom left): ©Dorling Kindersley Media Library. Page 519 (bottom right): Courtesy Dallas Semiconductor. Page 530: Courtesy RSA. Page 531: DILBERT reprinted by permission of United Feature Syndicate, Inc.

Chapter 9
Page 540 (center): Paul Seheult/Corbis-Bettmann. Page 540 (bottom): Courtesy U.S. Department of Energy. Page 545 (top): Courtesy The Computer Museum History Center (www.computerhistory.org). Reproduced with permission. Page 545 (bottom): Courtesy International Business Machines Corporation. Page 560: Corbis-Bettmann. Page 569 (top): UPI/Corbis-Bettmann. Page 569 (bottom): Bill Waterson/Universal Press Syndicate. Page 571: Gamma Liaison. Page 589: UPI/ Corbis-Bettmann. Page 590: Courtesy George Bergman, Dept. of Mathematics, U.C. Berkeley. Page 592: Paul J. Sutton/Duomo Photography, Inc. Page 601: Jon Riley/Tony Stone Images/New York, Inc. Page 604: Stu Forster/Allsport. Page 618: Photofest.

Sources

Chapter 1

[1]International Olympic Committee, Lausanne, Switzerland.

[2]*Congressional Record, Proceedings and Debate of the Second Session of the 68th Congress, Volume LXVI-Part 2.* Washington, D.C.: U.S. Government Printing Office, 1925.

[3]Institute of Electrical and Electronics Engineers, Piscataway, NJ.

[4]*Gallup Opinion Index,* Report No. 90 (December 1972).

[5]*Congressional Record, Proceedings and Debate of the 84th Congress Second Session.* Washington, D.C.: U.S. Government Printing Office, 1955.

[6]Samuel Merrill, *Making Multicandidate Elections More Democratic.* Princeton, NJ: Princeton University Press, 1988.

Chapter 2

[1]U.S. Bureau of the Census Web site: http://www.census.gov/ipc/www /idbprint.html.

[2]U.S. Census Office, *Return of the Whole Number of Persons within the Several Districts of the United States* [1st Census]. Philadelphia: Childs and Swaine, 1791.

[3]U.S. Dept. of Commerce, *World Population Profile: 1994.* Washington, D.C.: U.S. Government Printing Office, 1994.

[4]WPL Holdings, Inc. news release, Cedar Rapids, IA, November 11, 1995.

[5]U.S. Census Office, *Return of the Whole Number of Persons within the Several Districts of the United States* [1st Census]. Philadelphia: Childs and Swaine, 1791.

[6]U.S. Census Office, *Return of the Whole Number of Persons within the Several Districts of the United States* [1st Census]. Philadelphia: Childs and Swaine, 1791.

[7]Wilfried P. C. G. Derksen Web site: http://www.agora.stm.it/elections /election/country/at.htm and the Austrian Parliament Web site: http://www .parlinkom.gv.at/.

[8]Wilfried P. C. G. Derksen Web site: http://www.universal.nl/users/derksen /election/home.htm.

[9]U.S. Census Office, *Return of the Whole Number of Persons within the Several Districts of the United States* [1st Census]. Philadelphia: Childs and Swaine, 1791.

[10]U.S. Census Office, *Return of the Whole Number of Persons within the Several Districts of the United States* [1st Census]. Philadelphia: Childs and Swaine, 1791.

[11]*1995 Information Please® Almanac* © 1994 Information Please LLC. All rights reserved.

[12]U.S. Census Office, *Fifth Census.* Washington, D.C.: Duff Green, 1832.

[13]U.S. Department of Agriculture, Statistical Reporting Service, Economic Research Service. As reported in *1995 Information Please® Almanac.* Boston: Houghton Mifflin Company, 1995.

[14]Lincoln County School District Web site: http://taft.k14.ojgse.edu/meta /lcsd/schools.html.

[15]U.S. Bureau of the Census, *Population of Counties by Decennial Census: 1900 to 1990.* Web site: http://www.census.gov/population/cencounts/.

[16]U.S. Bureau of the Census Web site: http://www.census.gov/population /www/estimates/co_97_4.html.

[17]Hendrix College Web site: http://www.hendrix.edu/.

[18]The American Institute for Contemporary German Studies, The Johns Hopkins University. Web site: http://www.jhu.edu/~aicgsdoc/wahlen/rp96.htm.

[19]The American Institute for Contemporary German Studies, The Johns Hopkins University. Web site: http://www.jhu.edu/~aicgsdoc/wahlen/sh96.htm.

[20]Republic of South Africa Web site: http://www.polity.org.za/.

[21]U.S. Census Office, *Return of the Whole Number of Persons within the Several Districts of the United States* [1st Census]. Philadelphia: Childs and Swaine, 1791.

[22]Energy Information Administration, Production—Crude Oil (annual data). Web site: http://www.eia.doe.gov/emeu/international/main.html.

[23]U.S. Bureau of the Census Web site: http://www.census.gov/ftp/pub/epcd /ssel_tabs/view/tab3.txt.

[24]Wilfried P. C. G. Derksen Web site: http://www.agora.stm.it/elections /election/country/tr.htm.

[25]U.S. Bureau of the Census, *Sixteenth Census of the United States.* Washington, D.C.: U.S. Government Printing Office, 1942.

[26]U.S. Bureau of the Census, *Population of Counties by Decennial Census: 1900– 1990.* Web site: http://www.census.gov/population/cencounts/.

[27]U.S. Bureau of the Census, *Population of Counties by Decennial Census: 1900– 1990.* Web site: http://www.census.gov/population/cencounts/.

[28]U.S. Bureau of the Census, *Current Population Reports,* Series P60-200, Washington, D.C.: U.S. Government Printing Office, 1998.

[29]U.S. Dept. of Commerce, *World Population Profile: 1994.* Washington, D.C.: U.S. Government Printing Office, 1994.

[30]Wilfried P. C. G. Derksen Web site: http://www.agora.stm.it/elections /election/country/uy.htm.

[31]Wilfried P. C. G. Derksen Web site: http://www.agora.stm.it/elections /election/country/na.htm.

[32]U.S. Census Office, *Return of the Whole Number of Persons within the Several Districts of the United States* [1st Census]. Philadelphia: Childs and Swaine, 1791.

[33]U.S. Bureau of the Census, FT925/95-A, *U.S. Merchandise Trade: Exports, General Imports, and Imports for Consumption, 1995 Annual,* Standard International Trade Classification Commodity by Country. Washington, D.C.: U.S. Government Printing Office, 1996.

[34]U.S. Bureau of the Census, FT925/95-A, *U.S. Merchandise Trade: Exports, General Imports, and Imports for Consumption, 1995 Annual,* Standard International Trade Classification Commodity by Country. Washington, D.C.: U.S. Government Printing Office, 1996.

[35]Blue Cross Blue Shield Blue Care Network of Michigan Web site: http:// www.bcbsm.com/htdocs/bcbsm/bcn.shtml.

[36]Wilfried P. C. G. Derksen Web site: http://www.agora.stm.it/elections /election/country/gl.htm.

[37]Chino Valley Unified School District Web site: http://www.sbcssk12.ca.us /chino/.

Chapter 4

[1]The tables in *Branching Out 4.3* were reprinted with permission from *The American Statistician.* Copyright 1995 by the American Statistical Association. All rights reserved.

[2]U.S. Bureau of the Census, *Current Population Reports,* Series P20-479. Washington, D.C.: U.S. Government Printing Office, 1993.

Chapter 5

[1]"The kings of cool," Copyright 1998 by Consumers Union of U.S., Inc., Yonkers, NY 10703-1057. Adapted with permission from CONSUMER REPORTS, January 1998.*

[2]U.S. Bureau of the Census, *Current Population Reports, Consumer Income*, Series P60-193. Washington, D.C.: U.S. Government Printing Office, 1996.

[3]H. Becker and F. Piper, *Cipher Systems: The Protection of Communications*. New York: John Wiley & Sons, Inc., 1982.

[4]U.S. Bureau of the Census, *Statistical Abstract of the United States 1997*. Washington, D.C.: U.S. Government Printing Office, 1997.

[5]*1998 Information Please® Almanac* © 1997 Information Please LCC. All rights reserved.

[6]*1998 Information Please® Almanac* © 1997 Information Please LLC. All rights reserved.

[7]U.S. Energy Information Administration, *International Energy Annual 1996*. Washington, D.C.: U.S. Government Printing Office, 1996.

[8]*1998 Information Please® Almanac* © 1997 Information Please LLC. All rights reserved.

[9]U.S. Department of Agriculture, *Agricultural Statistics 1998*. Washington, D.C.: U.S. Government Printing Office, 1998.

[10]Federal Highway Administration, *Journey to Work Trends in the United States and Its Major Metropolitan Areas, 1960–1990*. Springfield, VA: National Technical Information Service, 1993.

[11]U.S. Bureau of the Census, *Current Population Reports*, Series P20-491, and *Statistical Abstract of the United States 1995*. Washington, D.C.: U.S. Government Printing Office, 1996 and 1995, respectively.

[12]U.S. Bureau of the Census, *Current Population Reports*, Series P20-484, and *Statistical Abstract of the United States 1997*. Washington, D.C.: U.S. Government Printing Office, 1996 and 1997, respectively.

[13]U.S. Bureau of the Census, *Statistical Abstract of the United States 1995*. Washington, D.C.: U.S. Government Printing Office, 1995.

[14]*1998 Information Please® Almanac* © 1997 Information Please LLC. All rights reserved.

[15]"Soup's on," Copyright 1993 by Consumers Union of U.S., Inc., Yonkers, NY 10703–1057. Adapted with permission from CONSUMER REPORTS, November 1993.*

[16]Market Statistics and Sales & Marketing Management, *The Survey of Buying Power Data Service*, 1993.

[17]"Toothpastes," Copyright 1992 by Consumers Union of U.S., Inc., Yonkers, NY 10703-1057. Adapted with permission from CONSUMER REPORTS, September 1992.*

[18]U.S. Bureau of the Census, *Statistical Abstract of the United States 1995*. Washington, D.C.: U.S. Government Printing Office, 1995.

[19]Center for the American Woman and Politics, Eagleton Institute of Politics, Rutgers University, New Brunswick, NJ.

[20]Center for the American Woman and Politics, Eagleton Institute of Politics, Rutgers University, New Brunswick, NJ.

[21]U.S. Bureau of the Census, *Statistical Abstract of the United States 1995*. Washington, D.C.: U.S. Government Printing Office, 1995.

[22]"The Best Hotels," Copyright 1994 by Consumers Union of U.S., Inc., Yonkers, NY 10703-1057. Adapted with permission from CONSUMER REPORTS, July 1994.*

[23]U.S. National Oceanic and Atmospheric Administration, *Climatography in the United States*. Washington, D.C.: U.S. Government Printing Office, 1995.

[24]"Frozen Light Entrées," Copyright 1993 by Consumers Union of U.S., Inc., Yonkers, NY 10703-1057. Adapted with permission from CONSUMER REPORTS, January 1993.*

[25]Federal Election Commission Web site: http://www.fed.gov/pages/htmlto5 .htm.

[26]"The wizards of oats," Copyright 1996 by Consumers Union of U.S., Inc., Yonkers, NY 10703-1057. Adapted with permission from CONSUMER REPORTS, October 1996.*

[27]Insurance Institute for Highway Safety, *State Law Facts*. Web site: http:// www.hwysafety.org/facts/pdf/ydriver.pdf.

[28]U.S. Bureau of Justice, *Capital Punishment 1997*. Web site: http://www.ojp .usdoj.gov/bjs/cp.htm.

[29]*1998 Information Please® Almanac* © 1997 Information Please LLC. All rights reserved.

[30]U.S. Environmental Protection Agency, *Automotive Technology and Full Economic Trends Through 1991*, EPA/AA/CTAB/91-02. Washington, D.C.: U.S. Government Printing Office, 1991.

[31]U.S. Bureau of the Census, *Statistical Abstract of the United States 1997*. Washington, D.C.: U.S. Government Printing Office, 1997.

[32]U.S. Federal Bureau of Investigation, *Crime in the United States*, annual. Washington, D.C.: U.S. Government Printing Office, 1996.

[33]U.S. Bureau of the Census, *Statistical Abstract of the United States 1997*. Washington, D.C.: U.S. Government Printing Office, 1997.

[34]U.S. Department of Transportation Web site: http://www.bts.gov/btsprod /nts/chp3/tbl3x28.html.

[35]"Peanut Butter It's Not for Kids Anymore," Copyright 1995 by Consumers Union of U.S., Inc., Yonkers, NY 10703-1057. Adapted with permission from CONSUMER REPORTS, September 1995.*

[36]"Get a grip," Copyright 1998 by Consumers Union of U.S., Inc., Yonkers, NY 10703-1057. Adapted with permission from CONSUMER REPORTS, March 1998.*

[37]Congressional Budget Office Web site: http://aspe.os.dhhs.gov/94gb/apenj .txt.

[38]U.S. National Oceanic and Atmospheric Administration, *Climatography in the United States*. Washington, D.C.: U.S. Government Printing Office, 1995.

[39]U.S. National Oceanic and Atmospheric Administration, *Climatography in the United States*. Washington, D.C.: U.S. Government Printing Office, 1995.

[40]U.S. Bureau of the Census, *Statistical Abstract of the United States 1997*. Washington, D.C.: U.S. Government Printing Office, 1997.

[41]NASA/Goddard Space Flight Center Web site: http://sunearth.gsfc.nasa .gov/eclipse/SEcat/SE2001-2100.html.

[42]National Center for Health Statistics, *Monthly Vital Statistics Report* 45:3 supp. Hyattsville, MD: National Center for Health Statistics, 1996.

[43]U.S. Bureau of the Census, *Statistical Abstract of the United States 1995*. Washington, D.C.: U.S. Government Printing Office, 1995.

[44]U.S. Federal Bureau of Investigation, *Crime in the United States*, annual. Washington, D.C.: U.S. Government Printing Office, 1994.

[45]U.S. Federal Bureau of Investigation, *Crime in the United States*, annual, Washington, D.C.: U.S. Government Printing Office, 1996.

[46]U.S. Bureau of the Census, *Statistical Abstract of the United States 1995*. Washington, D.C.: U.S. Government Printing Office, 1995.

[47]L. H. C. Tippett, *The Methods of Statistics*, 4th ed. New York: John Wiley & Sons, Inc., 1952.

[48]*1998 Information Please® Almanac* © 1997 Information Please LLC. All rights reserved.

[*]Although this material originally appeared in CONSUMER REPORTS, the selective adaptation and resulting conclusions presented are those of the authors and are not sanctioned or endorsed in any way by Consumers Union, the publisher of CONSUMER REPORTS.

Chapter 9

[1]Reprinted by permission, O. G. Haywood, Jr., "Military decision and game theory," *Journal of the Operations Research Society*, Volume 2, Number 1, November 1954. Copyright 1954, The Institute of Management Sciences (currently INFORMS), 901 Elkridge Landing Road, Suite 400, Linthicum, MD 21090-2909 USA.

[2]Reprinted by permission, O. G. Haywood, Jr., "Military decision and game theory," *Journal of the Operations Research Society*, Volume 2, Number 1, November 1954. Copyright 1954, The Institute of Management Sciences (currently INFORMS), 901 Elkridge Landing Road, Suite 400, Linthicum, MD 21090-2909 USA.

Index